Lecture Notes in Bioinformatics 10330

Subseries of Lecture Notes in Computer Science

More information about this series at http://www.springer.com/series/5381

Zhipeng Cai · Ovidiu Daescu
Min Li (Eds.)

Bioinformatics Research and Applications

13th International Symposium, ISBRA 2017
Honolulu, HI, USA, May 29 – June 2, 2017
Proceedings

 Springer

Editors
Zhipeng Cai
Georgia State University
Atlanta, GA
USA

Min Li
Central South University
Changsha
China

Ovidiu Daescu
University of Texas at Dallas
Richardson, TX
USA

ISSN 0302-9743   ISSN 1611-3349   (electronic)
Lecture Notes in Bioinformatics
ISBN 978-3-319-59574-0   ISBN 978-3-319-59575-7   (eBook)
DOI 10.1007/978-3-319-59575-7

Library of Congress Control Number: 2017941549

LNCS Sublibrary: SL8 – Bioinformatics

Printed on acid-free paper

This Springer imprint is published by Springer Nature
The registered company is Springer International Publishing AG
The registered company address is: Gewerbestrasse 11, 6330 Cham, Switzerland

Preface

On behalf of the Program Committee, we would like to welcome you to the proceedings of the 13th edition of the International Symposium on Bioinformatics Research and Applications (ISBRA 2017), held in Honolulu, Hawaii, May 29 to June 2, 2017. The symposium provides a forum for the exchange of ideas and results among researchers, developers, and practitioners working on all aspects of bioinformatics and computational biology and their applications. This year we received 118 submissions in response to the call for extended abstracts. The Program Committee decided to accept 27 of them for full publication in the proceedings and oral presentation at the symposium. We also accepted 24 of them for oral presentation and short abstract publication in the proceedings. Furthermore, we also received 18 submissions in response to the call for short abstracts.

The technical program invited keynote talks by Prof. Michael Q. Zhang from The University of Texas at Dallas and Tsinghua University. Prof. Zhang reviewed the history of computational genome regulation and then introduced some new biochemical (BL-Hi-C), biophysical (super-resolution imaging), and bioinformatics (MICC, 3CPET, FIND) technology developments that may be used for studying 3D genomes and disease markers in the near future.

We would like to thank the Program Committee members and the additional reviewers for volunteering their time to review and discuss symposium papers. We would like to extend special thanks to the steering and general chairs of the symposium for their leadership, and to the finance, publicity, workshops, local organization, and publications chairs for their hard work in making ISBRA 2017 a successful event. Last but not least we would like to thank all authors for presenting their work at the symposium.

April 2017
Zhipeng Cai
Ovidiu Daescu
Min Li

Organization

Steering Committee

Dan Gusfield	University of California Davis, USA
Ion Mandoiu	University of Connecticut, USA
Yi Pan (Chair)	Georgia State University, USA
Marie-France Sagot	INRIA, France
Ying Xu	University of Georgia, USA
Alexander Zelikovsky	Georgia State University, USA

General Chair

Alexander Zelikovsky	Georgia State University, USA

Program Chairs

Zhipeng Cai	Georgia State University, USA
Ovidiu Daescu	The University of Texas at Dallas, USA
Min Li	Central South University, China

Finance Chair

Anu G. Bourgeois	Georgia State University, USA

Publications Chair

Pavel Skums	Georgia State University, USA

Publicity Chairs

Chunyu Ai	University of South Carolina Upstate, USA
Shaoliang Peng	National University of Defense Technology, China
Xiang Wan	Hong Kong Baptist University, China
Gangman Yi	Dongguk University, Korea

Workshop Chairs

Yaohang Li	Old Dominion University, USA
Anu G. Bourgeois	Georgia State University, USA
Wooyoung Kim	University of Washington Bothell, USA

Award Chair

Raj Sunderraman   Georgia State University, USA

Publication Chair

Pavel Skums   Georgia State University, USA

Webmasters

Igor Mandric   Georgia State University, USA
Sergey Knyazev   Georgia State University, USA

Program Committee

Kamal Al Nasr   Tennessee State University, USA
Max Alekseyev   George Washington University, USA
Mukul S. Bansal   University of Connecticut, USA
Robert Beiko   Dalhousie University, Canada
Paola Bonizzoni   Università di Milano-Bicocca, Italy
Zhipeng Cai   Georgia State University
Doina Caragea   Kansas State University, USA
Xing Chen   National Center for Mathematics and Interdisciplinary
                                 Sciences, Chinese Academy of Sciences, China
Ovidiu Daescu   University of Texas at Dallas, USA
Fei Deng   University of California, Davis
Lei Deng   Central South University
Oliver Eulenstein   Iowa State University, USA
Lin Gao   Xidian University, China
Olga Glebova   Georgia State University
Jiong Guo   Shandong University, China
Xuan Guo   Oak Ridge National Laboratory, USA
Jieyue He   Southeast University
Zengyou He   Hong Kong University of Science and Technology,
                                 SAR China
Steffen Heber   NCSU
Xing Hua
Jinling Huang   East Carolina University, USA
Mingon Kang   Kennesaw State University, USA
Wooyoung Kim   University of Washington Bothell, USA
Danny Krizanc   Wesleyan University, USA
Xiujuan Lei   Shaanxi Normal University, China
Jing Li   Case Western Reserve University, USA
Min Li   Central South University
Shuai Cheng Li   City University of Hong Kong, SAR China
Yaohang Li   Old Dominion University, USA
Yingshu Li   Georgia State University, Atlanta, USA

Xiaowen Liu   Indiana University-Purdue University Indianapolis,
                            USA
Ion Mandoiu   University of Connecticut, USA
Fenglou Mao   National Institute of Health, USA
Giri Narasimhan   Florida International University
Chongle Pan   Oak Ridge National Laboratory, USA
Steven Pascal   Old Dominion University, USA
Andrei Paun   University of Bucharest, Romania
Nadia Pisanti   Universita di Pisa, Italy and Erable Team, Italy; Inria,
                            France
Mukul S. Bansal   University of Connecticut, USA
Russell Schwartz   Carnegie Mellon University, USA
Joao Setubal   University of São Paulo, Brazil
Xinghua Shi   University of North Carolina at Charlotte, USA
Yi Shi   Shanghai Jiao Tong University, China
Pavel Skums   Georgia State University, USA
Ileana Streinu   Smith College, Northampton, USA
Chia-Yu Su   Taipei Medical University, Taiwan
Wing-Kin Sung   National University of Singapore, Singapore
Sing-Hoi Sze   Texas A&M University, USA
Weitian Tong   Georgia Southern University, USA
Gabriel Valiente   Technical University of Catalonia, Spain
Xiang Wan   HKBU
Jianxin Wang   Central South University
Li-San Wang   University of Pennsylvania
Lusheng Wang   City University of Hong Kong, SAR China
Peng Wang   Shanghai Advanced Research Institute, Chinese
                            Academy of Science, China
Seth Weinberg   Virginia Commonwealth University, USA
Fangxiang Wu   University of Saskatchewan, Canada
Yufeng Wu   University of Connecticut, USA
Zeng Xiangxiang
Xiuchun Xiao   Fudan University, China
Minzhu Xie   Hunan Normal University, China
Dechang Xu   Harbin Institute of Technology, China
Can Yang   HKBU
Ashraf Yaseen   Texas A&M University, Kingsville, USA
Guoxian Yu   Southwest University
Ning Yu   Georgia State University
Alex Zelikovsky   Georgia State University, USA
Chi Zhang   Indiana University, USA
Fa Zhang   Institute of Computing Technology
Le Zhang   Southwest University
Xue Zhang   Tufts University, USA
Yanqing Zhang   Georgia State University, USA
Leming Zhou   University of Pittsburgh
Quan Zou   Tianjin University, China

Additional Reviewers

Abdelrasoul, Maha
Aldabagh, Hind
Alexeev, Nikita
Antipov, Dmitry
Artyomenko, Alexander
Arunachalam, Harish Babu
Avdeyev, Pavel
Biswas, Abhishek
Chen, Wei
Chu, Chong
Daescu, Kelly
Della Vedova, Gianluca
Diaz Tula, Antonio
Elhefnawy, Wessam
Farhana, Effat
Frith, Martin
Glebova, Olga
He, Jing
Hu, Jialu
Hu, Xiaoming
Hu, Xihao
Icer, Pelin
Ionescu, Vlad
Knyazev, Sergey
Lan, Wei
Li, Jin
Li, Leon
Li, Xin
Liu, Bin

Llabrés, Mercè
Mandric, Igor
Melnyk, Andrii
Moon, Jucheol
Muntean, Radu
Olariu, Ciprian
Patterson, Murray
Pei, Jingwen
Peng, Xiaoqing
Perkins, Patrick
Ren, Xianwen
Rizzi, Raffaella
Sheng, Tao
Shi, Jian-Yu
Sun, Yazhou
Trivette, Andrew
Vyatkina, Kira
Wan, Changlin
Wu, Hao
Wu, Yue
Xiangxiang, Zeng
Yang, Frank
Yuan, Xiguo
Zaccaria, Simone
Zhao, Junfei
Zhao, Qi
Zhu, Shanfeng
Zhu, Zexuan

Abstract of Invited Papers

Copy Number Aberration Based Cancer Type Prediction with Convolutional Neural Networks

Yuchen Yuan[1,2], Yi Shi[2], Xianbin Su[2], Xin Zou[2], Qing Luo[2], Weidong Cai[1], Zeguang Han[2], and David Dagan Feng[1]

[1] School of Information Technologies,
The University of Sydney, Sydney, NSW 2008, Australia
{yuchen.yuan,tom.cai,dagan.feng}@sydney.edu.au
[2] Key Laboratory of Systems Biomedicine,
Shanghai Center for Systems Biomedicine,
Shanghai Jiaotong University, Shanghai 200240, China
{yishi,xbsu,x.zou,simonluo,hanzg}@sjtu.edu.cn

Abstract. Cancer is a category of disease that causes abnormal cell growths and immortality. It usually incarnates into tumor form that potentially invade or metastasize to remote parts of human body [1]. During the past decade, with the developments of DNA sequencing technology, large amounts of sequencing data have become available which provides unprecedented opportunities for advanced association studies between somatic mutations and cancer types/subtypes [2–7], which may contribute to more accurate somatic mutation based cancer typing (SMCT). In existing SMCT methods however, the absence of feature quantification and high-level feature extraction is a major obstacle in improving the classification performance. To address this issue, we propose DeepCNA, an advanced convolutional neural network (CNN) based classifier, which utilizes copy number aberrations (CNAs) [8–10] and HiC data [11] for cancer typing. DeepCNA consists of two steps: firstly, the CNA data is pre-processed by clipping, zero padding and reshaping; secondly, the processed data is fed into a CNN classifier, which extracts high-level features for accurate classification [12].

We conduct experiments on the newly proposed COSMIC CNA dataset, which contains 25 types of cancer. Controlled variable experiments indicate that the 2D CNN with both cell lines of HiC data (hESC and IMR90) contributes to the optimal performance. We then compare DeepCNA with three widely adopted data classifiers, the results of which exhibit the remarkable advantages of DeepCNA, which has achieved significant performance improvements in terms of testing accuracy (78%) against the comparison methods. We have demonstrated the advantages and potentials of the DeepCNA model for somatic point mutation based gene data processing, and suggest that the model can be extended and transferred to other complex genotype-phenotype association studies, which we believe will benefit many related areas [13, 14].

Yuchen Yuan, Yi Shi—These authors contribute equally as co-first authors.

References

1. Feuerstein, M.: Defining cancer survivorship. J. Cancer Survivorship **1**(1), 5–7, (2007)
2. Yang, K., Li, J., Cai, Z., Lin, G.: A model-free and stable gene selection in microarray data analysis. In: IEEE Symposium of BioInformatics BioEngineering (BIBE), Minneapolis, MN, USA, pp. 3–10 (2005)
3. Yang, K., Cai, Z., Li, J., Lin, G.: A stable gene selection in microarray data analysis. BMC Bioinform. **7**(1), 228 (2006)
4. Cai, Z., Goebel, R., Salavatipour, M.R., Lin, G.: Selecting dissimilar genes for multi-class classification, an application in cancer subtyping. BMC Bioinform. **8**(1), 206 (2007)
5. Cai, Z., Xu, L., Shi, Y., Salavatipour, M.R., Goebel, R., Lin, G.: Using gene clustering to identify discriminatory genes with higher classification accuracy. In: IEEE Symposium of BioInformatics BioEngineering (BIBE), Arlington, VA, USA, pp. 235–242 (2006)
6. Cai, Z., Zhang, T., Wan, X.-F.: A computational framework for influenza antigenic cartography. PLoS Comput. Biol. **6**(10), e1000949 (2010)
7. Cai, Z., Ducatez, M.F., Yang, J., Zhang, T., Long, L., Boon, A.C., Webby, R.J., Wan, X.: Identifying antigenicity associated sites in highly pathogenic H5N1 Influenza Virus Hemagglutinin by using sparse learning. J. Mol. Biol. **422**(1), 145–155 (2012)
8. Bakhoum, S.F., Swanton, C.: Chromosomal instability, aneuploidy, and cancer. Frontiers Oncol. **4**, 161 (2014)
9. Burrell, R.A., McGranahan, N., Bartek, J., Swanton, C.: The causes and consequences of genetic heterogeneity in cancer evolution. Nature **501**(7467), 338–345 (2013)
10. Zack, T.I., Schumacher, S.E., Carter, S.L., Cherniack, A.D., Saksena, G., Tabak, B., et al.: Pan-cancer patterns of somatic copy number alteration. Nat. Genet. **45**(10), 1134–1140 (2013)
11. Dixon, J.R., Selvaraj, S., Yue, F., Kim, A., Li, Y., Shen, Y., et al.: Topological domains in mammalian genomes identified by analysis of chromatin interactions. Nature **485**(7398), 376–380 (2012)
12. Yuan, Y., Shi, Y., Li, C., Kim, J., Cai, W., Han, Z., et al.: DeepGene: an advanced cancer type classifier based on deep learning and somatic point mutations. BMC Bioinform. **17**(17), 243 (2016)
13. Longo, D.L.: Tumor heterogeneity and personalized medicine. N. Engl. J. Med. **366**(10), 956–957 (2012)
14. Franken, B., de Groot, M.R., Masthoom, W.J., Vermes, I., van der Palen, J., Tibbe, A.G., et al.: Circulating tumor cells, disease recurrence and survival in newly diagnosed breast cancer. Breast Cancer Res. **14**(5), 1–8 (2012)

Predicting Human Microbe-Disease Associations via Binary Matrix Completion

Jian-Yu Shi[1], Hua Huang[2], Yan-Ning Zhang[3], and Siu-Ming Yiu[4]

[1] School of Life Sciences, Northwestern Polytechnical University, Xi'an, China
jianyushi@nwpu.edu.cn
[2] School of Software and Microelectronics,
Northwestern Polytechnical University, Xi'an, China
1363351294@qq.com
[3] School of Computer Science,
Northwestern Polytechnical University, Xi'an, China
ynzhang@nwpu.edu.cn
[4] Department of Computer Science, the University of Hong Kong,
Pok Fu Lam, Hong Kong
smyiu@cs.hku.hk

With the help of sequencing techniques (e.g. 16S ribosomal RNA sequencing) [1], Human Microbiome Project has revealed that there are diverse communities of microbes in a human intestine, which provides a nutrient-rich and temperature-fixed habitat for microbes. The sequential works have observed that there exists a significant mutual influence between microbes and their host. It is surprising that except for conventional infectious diseases, a wide range of noninfectious diseases is closely associated with microbes, such as cancer, obesity [2], diabetes, kidney stones and systemic inflammatory response syndrome. On the one side, the tremendous amount of microbiome genes and their products can lead a diverse range of biological activities, which serve as a physiological complement in their host body in a wide range, involving metabolic capabilities, pathogens, immune system, and gastrointestinal development [3]. On the other side, the microbes can be greatly influenced by their dynamic habitat in the human body, which undergoes frequent changes caused by diverse environmental variables, such as season, host diet, smoking, hygiene and use of antibiotics. Thus, this mutual association between the host and its microbiota can further modify transcriptomic, proteomic and metabolic profiles of the human host. However, the identification of microbe-noninfectious disease associations (MDAs) requires time-consuming and costly experiments and always bears the limitation of microbe cultivation. Even worse, many bacteria cannot be cultivated at all by current culturing bio-techniques. Fortunately, the number of MDAs found in both experiments and clinic is growing. For example, Ma et al. published the first MDA database, Human Microbe-Disease Association Database (HMDAD) recently, by collecting a large number of MDAs from previously published literature [4]. The growing number of MDAs enables us to perform a systematic analysis, discovery and understanding on the mechanism of microbe-related non-infectious diseases in a new insight. As one of the most important steps to achieve that goal, the discovery or prediction of potential MDAs provides an approach to understand the mechanism of non-infectious disease formation and development and develop novel methods for disease diagnosis and therapy. As the promising complement of experiment-based approaches, computational approaches, especially machine learning-based approaches, are able to predict MDA candidates among a large number of microbe-disease pairs. They cannot only reduce the cost and time of relevant experiments, but also output the candidates, of which even though the involving microbes cannot cultured. Nevertheless, a few of efforts have been made to develop computational models for MDA prediction on a large scale. Very recently, a pioneering work constructing an MDA network based on HMDAD develops an approach KATZHMDA for predicting potential MDAs [5]. KATZHMDA regards the prediction of MDS as link prediction on the constructed MDA network. In this work, we first model MDA prediction as a problem of matrix completion (Fig. 1), then propose a new approach based on Binary Matrix Completion (BMCMDA) to predict potential MDAs. BMCMDA is able to predict new MDAs on a large scale, by only using known microbe-disease association network. Its performance is evaluated by both leave-one-out cross validation (LOOCV) and 5-fold cross validation (5-CV) on HMDAD database, where the whole procedure of 5-CV was repeated 100 times and both the mean and the standard deviation of predicting performance over 100 rounds of 5-CVs were recorded. Finally, in terms of Area Under Receiver-Operating Characteristics, BMCMDA achieves 0.9049 in LOOCV and 0.8954 ± 0.0034 in 5CV, while the state-of-the-art KATZHMDA only achieves 0.8382 and 0.8301 ± 0.0033 respectively. The significantly outperformed prediction achieved by BMCMAD demonstrates its superiority for predicting microbe-disease associations on a large scale.

Acknowledgments. This work was supported by RGC Collaborative Research Fund (CRF) of Hong Kong (C1008-16G), National High Technology Research and Development Program of China (No. 2015AA016008), the Fundamental Research Funds for the Central Universities of China (No. 3102015ZY081), the Program of Peak Experience of NWPU (2016) and partially supported by the National Natural Science Foundation of China (No. 61473232, 91430111).

References

1. Huttenhower, C., Gevers, D., Knight, R., Abubucker, S., Badger, J.H., Chinwalla, A., Creasy, H.H., Earl, A.M., Fitzgerald, M., Fulton, R.S.: Structure, function and diversity of the healthy human microbiome. Nature **486**, 207–214 (2012)
2. Zhang, H., Dibaise, J.K., Zuccolo, A., Kudrna, D., Braidotti, M., Yu, Y., Parameswaran, P., Crowell, M.D., Wing, R.A., Rittmann, B.E.: Human gut microbiota in obesity and after gastric bypass. Proc. Nat. Acad. Sci. U.S.A. **106**, 2365–2370 (2009)
3. Ventura, M., O'Flaherty, S., Claesson, M.J., Turroni, F., Klaenhammer, T.R., Van, S.D., O'Toole, P.W.: Genome-scale analyses of health-promoting bacteria: probiogenomics. Nat. Rev. Microbiol. **7**, 61–72 (2009)
4. Ma, W., Zhang, L., Zeng, P., Huang, C., Li, J., Geng, B., Yang, J., Kong, W., Zhou, X., Cui, Q.: An analysis of human microbe–disease associations. Briefings Bioinform. **18**, 85–97 (2017)
5. Chen, X., Huang, Y.A., You, Z.H., Yan, G.Y., Wang, X.S.: A novel approach based on KATZ measure to predict associations of human microbiota with non-infectious diseases. Bioinformatics **33**, 733–739 (2017)

Characterization of Kinase Gene Expression and Splicing Profile in Prostate Cancer with RNA-Seq Data

Huijuan Feng[1], Tingting Li[2], and Xuegong Zhang[1,3]

[1] MOE Key Laboratory of Bioinformatics,
Bioinformatics Division/Center for Synthetic and Systems Biology,
TNLIST and Department of Automation, Tsinghua University,
Beijing 100084, China
fhj11@mails.tsinghua.edu.cn
[2] Department of Biomedical Informatics, Institute of Systems Biomedicine,
School of Basic Medical Sciences,
Peking University Health Science Center, Beijing 100191, China
litt@hsc.pku.edu.cn
[3] School of Life Sciences, Tsinghua University, Beijing 100084, China
zhangxg@tsinghua.edu.cn

Abstract. Alternative splicing is a ubiquitous post-transcriptional process in most eukaryotic genes. Aberrant splicing isoforms and abnormal isoform ratios can contribute to cancer development. Kinase genes are key regulators of many cellular processes. Multiple kinases are found to be oncogenic. RNA-Seq provides a powerful technology for genome-wide study of alternative splicing. But this potential has not been fully demonstrated on cancers yet. We characterized the transcriptome profile of prostate cancer using RNA-Seq data on both differential expression and differential splicing, with an emphasis on kinase genes and their splicing variations. We identified distinct gene groups from differential expression and splicing analysis, which suggested that alternative splicing adds another level to gene regulation in cancer. Enriched GO terms of differentially expressed and spliced kinase genes were found to play different roles in regulation of cellular metabolism. Function analysis showed that differentially spliced exons of these genes are significantly enriched in protein kinase domains. Among them, we found that gene CDK5 has isoform switching between prostate cancer and benign tissues, which may affect cancer development by changing androgen receptor (AR) phosphorylation. The observation was validated in another RNA-Seq dataset of prostate cancer cell lines. Our work brings new understanding to the role of alternatively spliced kinases in prostate cancer and demonstrates the use of RNA-Seq data in studying alternative splicing in cancer.

Keywords: Prostate cancer · Alternative splicing · Kinase · CDK5 · Isoform switching

Identifying Conserved Protein Complexes Across Multiple Species via Network Alignment

Bo Song[1], Jianliang Gao[1,2], Xiaohua Hu[1], Yu Sheng[2], and Jianxin Wang[2]

[1] College of Computing and Informatics, Drexel University, Philadelphia, USA
[2] School of Information Science and Engineering,
Central South University, Changsha, China
gaojianliang@csu.edu.cn

A protein complex is a bimolecular that contains a number of proteins interacting with each other to perform different cellular functions [1]. The identification of protein complexes in a protein-protein interaction (PPI) network [2] can, therefore, lead to a better understanding of the roles of such a network in different cellular systems. The protein complex identification problem has received a lot of attentions, and a considerable number of techniques have been proposed to address such problem. By representing a PPI network as a graph [3], whose vertices represent proteins and edges as interactions between proteins, these algorithms are able to identify clusters in single PPI network based on different graph properties [4]. For example, an uncertain graph model based method is proposed to detect protein complex from a PPI network [5]. However, they focused on finding protein complexes in a single PPI network, and finding conserved protein complexes from multiple PPI networks still remain challenging.

In this paper, we identify the problem of finding conserved protein complexes via aligning multiple PPI networks. In this way, the knowledge of protein complexes in well-studied species can be extended to that of poor-studied species. Then, we propose an efficient method to find conserved protein complexes from multiple PPI networks. By taking the feature of subnetwork connectivity into consideration, the proposed method improves the coverage significantly without compromising of the consistency in the aligned results.

Given the multiple PPI networks $(G_1, G_2, \ldots, G_\xi)$ and target protein complex M_0 from the target PPI network G_t, the alignment process mainly includes:

(1) Generate initial candidate pools. Only those proteins that have links with given protein complex can be selected as candidate proteins since links represent the biological similarity between proteins across PPI networks. For each aligned network G_i, $1 \leq i \leq \xi$, we construct a pool for a given protein complex M_0, where $M_0 \in G_t$. Every vertex $v \in G_i$ is put into the pool of G_i if it has link with any vertex in M_0. Then, the initial subnetworks M are selected randomly from the pools.

(2) Optimal determination by simulated annealing. Simulated annealing process adopts iteration method for global optimal solution. In each loop, a protein from the candidate pool is chosen randomly to be determined as aligned protein in the

corresponding PPI network. There are two kinds of proteins that are possible to be moved out from the current alignment solution. The first kind is the protein whose score is the lowest in the current solution. The other kind is the protein whose corresponding vertex in the current subnetwork is not connected with other vertices, i.e., its degree is zero. If the new candidate solution achieves higher score, it will take place the previous solution. If not, it still has chance to replace the prior solution with a probability of $(rand(0,1) < e^{\frac{\Delta\Phi}{T_i}})$, where $\Delta\Phi$ is the amount of change score, T_i is the temperature of simulated annealing. Finally, the algorithm returns the best solution as the alignment of protein complexes $M = \{M_1, M_2, \ldots, M_\xi\}$. Overall, we utilize both the biological similarity between proteins and the topological structure to assign scores on subnetworks for simulated annealing process. Formally, given a protein complex of target network $M_0 \subseteq G_t$, its match result $\{M_1, M_2, \ldots, M_\xi\}$ in aligned networks, where $M_k \subseteq G_k$, is assigned a real-valued score Φ:

$$\Phi = \sum_{k \in \{1,\ldots,\xi\}} \sum_{v_j \in V_{M_k}} \left(\alpha * \delta_{bio}(v_j) + (1-\alpha) * \delta_{topo}(v_j) \right) \qquad (1)$$

where ξ is the number of PPI networks, V_{M_k} is the set of proteins in M_k, α is a coefficient to trade off biological and topological scores, δ_{bio} and δ_{topo} are the biological and topological scores respectively. The biological score of a protein consists of: (1) the number of links with the subnetwork M_0, (2) the number of links with the subnetwork M_h, and (3) the number of threads among these three subnetworks which contain the current protein. The topological score of a vertex consists of (1) the degree of current vertex; (2) the size of the maximal component that includes the current vertex. As the same with biological score, we adopt a transform techniques by multiplying a coefficient.

References

1. Hu, A.L., Chan, K.C.: Utilizing both topological and attribute information for protein complex identification in ppi networks. IEEE/ACM Trans. Comput. Biol. Bioinform. **10**(3), 780–792 (2013)
2. Li, M., Chen, X., Ni, P., Wang, J., Pan, Y.: Identifying essential proteins by purifying protein interaction networks. In: International Symposium on Bioinformatics Research and Applications (ISBRA), pp. 106–116 (2016)
3. Song, B., Gao, J., Ke, W., Hu, X. Achieving high k-coverage and k-consistency in global alignment of multiple PPI networks. In: IEEE International Conference on Bioinformatics and Biomedicine (BIBM), pp. 303–307 (2016)
4. Malod-Dognin, N., Przulj, N. L-GRAAL: Lagrangian graphlet-based network aligner. Bioinformatics **31**(13), 2182–2189 (2015)
5. Zhao, B., Wang, J., Li, M., Wu, F.X., Pan, Y.: Detecting protein complexes based on uncertain graph model. IEEE/ACM Trans. Comput. Biol. Bioinform. **11**(3), 486–497 (2014)

Constructing an Integrative MicroRNA eQTL Network on Ovarian Cancer: A Label Propagation Approach Utilizing Multiple Networks

Benika Hall, Andrew Quitadamo, and Xinghua Shi

University of North Carolina at Charlotte, Charlotte 28213, USA
{bjohn157,aquitada,x.shi}@uncc.edu

Abstract. Expression quantitative trait loci (eQTL) network construction has been an important task in understanding functional relationships in genomics. In this paper, we construct an integrative microRNA eQTL network based on a label propagation framework using TCGA ovarian cancer data. Label propagation is a robust semi-supervised learning algorithm capable of handling multiple heterogeneous networks reflecting different types of genetic interactions. Elucidation of the interactions involved in multiple networks provide more insight in the dynamics of cancer progression.

Keywords: microRNAs eQTLs · Regulatory networks · Protein protein interaction networks · Network expansion · Label propagation · Ovarian cancer

1 Introduction

Ovarian cancer is the fifth most deadliest cancer among cancer deaths and is responsible for over five percent of cancer deaths in women [1]. MicroRNAs (miRNAs) are small non-coding RNAs that are approximately 22 nucleotides in length and contribute the progression of ovarian cancer through various functional roles such as cell differentiation, apoptosis and tumoriogenesis. Here, we propose a robust semi-supervised learning approach to model the complex relationships between miRNAs, eQTLs and their regulated genes. Expression quantitative trait loci (eQTLs) are genomic regions that can influence gene expression locally or in a distant manner. Thus, we conduct miRNA eQTL analysis to assess the effect of miRNAs on gene expression [2–5].

2 Methods

We downloaded miRNA and gene expression data from TCGA [6], InWeb network [7], a gene regulatory network from RegNetwork database [8] consisting of experimentally verified targets. We conducted eQTL analysis between miRNAs and gene expression and discovering correlations between miRNAs as well as correlations between genes Lastly, we use our eQTL genes as seed nodes and expand our network with two additional networks, the Inweb and RegNetwork using a label propagation framework.

3   Results

We generated a multi-layered eQTL network including miRNA eQTLs, miRNA correlations, gene correlations, Protein-protein interactions and a gene regulatory network. This integrative network allowed us to capture many facets of gene regulation in ovarian cancer. In the integrated network we have 174 miRNAs and 2,180 genes. These miRNAs and genes are connected through 803 regulatory edges, 1313 protein-protein edges, 9 correlated miRNAs, 18 correlated gene edges and a total of 855 miRNA eQTL edges.

4   Conclusion

We created an integrated miRNA eQTL network utilizing multiple networks. Our integrated network included a miRNA eQTL network, a protein-protein interaction network (InWeb), a gene regulatory network(RegNetwork), and correlation networks on miRNAs and genes respectively. A single miRNA or target usually does not impact the phenotypic outcome individually. To exploit the large scope of regulation, we applied a network based learning approach to integrate multiple networks containing multiple regulatory elements in ovarian cancer.

References

1. Siegel, R.L., Miller, K.D., Jemal, A.: Cancer statistics, 2017. CA Cancer J. Clin. **67**(1), 7–30 (2017)
2. Hall, B., Quitadamo, A., Shi, X., Identifying microrna and gene expression networks using graph communities. Tsinghua Sci. Technol. **21**(2), 176–195 (2016)
3. Quitadamo, A., Tian, L., Hall, B., Shi, X.: An integrated network of microrna and gene expression in ovarian cancer. BMC Bioinform. **16**(5), 1 (2015)
4. Huan, et al.: Genome-wide identification of microrna expression quantitative trait loci. Nat. Commun. **6** (2015)
5. Gamazon, et al.: Genetic architecture of microrna expression: implications for the transcriptome and complex traits. Am. J. Hum. Genet. **90**(6), 1046–1063 (2012)
6. Cancer Genome Atlas Research Network: Integrated genomic analyses of ovarian carcinoma. Nature **474**, 609–615 (2011)

7. Lage, K., Karlberg, E.O., Størling, Z.M., Olason, P.I., Pedersen, A.G., Rigina, O., Hinsby, A.M., Tümer, Z., Pociot, F., Tommerup, N., et al.: A human phenome-interactome network of protein complexes implicated in genetic disorders. Nat. Biotechnol. **25**(3), 309–316 (2007)
8. Liu, Z.-P., Wu, C., Miao, H., Wu, H.: Regnetwork: an integrated database of transcriptional and post-transcriptional regulatory networks in human and mouse. Database 2015:bav095 (2015)

Clustering scRNA-Seq Data Using TF-IDF

Marmar Moussa and Ion Măndoiu

Computer Science and Engineering Department,
University of Connecticut, Storrs, CT, USA
{marmar.moussa,ion}@engr.uconn.edu

Abstract. Single cell RNA sequencing (scRNA-Seq) is critical for understanding cellular heterogeneity and identification of novel cell types. We present
novel computational approaches for clustering scRNA-seq data based on the
TF-IDF transformation.

Introduction

In this abstract, we propose several computational approaches for clustering
scRNA-Seq data based on the Term Frequency - Inverse Document Frequency
(TF-IDF) transformation that has been successfully used in the field of text analysis.
Empirical evaluation on simulated cell mixtures with different levels of complexity
suggests that the TF-IDF methods consistently outperform existing scRNA-Seq
clustering methods.

Methods

We compared eight scRNA-Seq methods, including three existing methods and five
proposed methods based on the TF-IDF transformation. All methods take as input the
raw *Unique Molecular Identifier (UMI)* counts generated using 10X Genomics' Cell-
Ranger pipeline [4]. *Existing scRNA-Seq clustering methods* are: the recommended
workflow for the Seurat package [3], the Expectation-Maximization (EM) algorithm
implemented in the mclust package [2], and a K-means clustering approach similar to
that implemented in the CellRanger pipeline distributed by 10X Genomics [1]. Two
types of *TF-IDF based methods* were explored. In first type of methods, TF-IDF scores
were used to select a subset of the most informative genes that were then clustered with
EM and spherical K-means. In the second type all genes were used for clustering, but
the expression data was first binarized using a TF-IDF based cutoff. The binary
expression level signatures were clustered using: hierarchical clustering with Jaccard
distance, and hierarchical clustering with cosine distance with or without an additional
cluster aggregation step.

Experimental Setup and Results

To assess accuracy we used mixtures of real scRNA-Seq profiles generated from FACS
sorted cells [4]. We selected five cell types: CD8+ cytotoxic T cells (abbreviated as C),

CD4+/CD45RO+ memory T cells (M), CD4+/CD25+ regulatory T cells (R), CD4+ helper T cells (H), and CD19+ B cells (B). We generated mixtures comprised of 5,000 cells sampled from all five cell types in equal proportions. Box-plots of classification accuracy achieved by the eight compared methods are shown in Fig. 1. TF-IDF based hierarchical clustering with cosine distance and cluster aggregation performs better than all other methods, with a mean accuracy of 0.7418, followed by the TF-IDF based spherical K-means, with a mean accuracy of 0.7125.

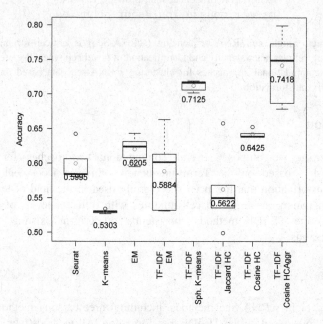

Fig. 1. Accuracy for the B:R:H:M:C datasets with 1:1:1:1:1 ratio.

Acknowledgements. This work was partially supported by NSF Award 1564936 and a UConn Academic Vision Program Grant.

References

1. Cell Ranger R Kit Tutorial. http://s3-us-west-2.amazonaws.com/10x.files/code/cellrangerrkit-PBMC-vignette-knitr-1.1.0.pdf
2. Fraley, C., Raftery, A., Murphy, T., Scrucca, L.: mclust version 4 for R: normal mixture modeling for model-based clustering, classification, and density estimation. University of Washington, Seattle (2012)
3. Satija, R., Farrell, J.A., Gennert, D., Schier, A.F., Regev, A.: Spatial reconstruction of single-cell gene expression data. Nat. Biotechnol. **33**(5), 495–502 (2015)
4. Zheng, G.X.Y., et al.: Massively parallel digital transcriptional profiling of single cells. Nat. Commun. **8**, 14049 (2017)

CircMarker: A Fast and Accurate Algorithm for Circular RNA Detection

Xin Li, Chong Chu, Jingwen Pei, Ion Măndoiu, and Yufeng Wu

Computer Science and Engineering Department,
University of Connecticut, Storrs, CT, USA
{xin.li,chong.chu,jingwen.pei,ion.mandoiu,
yufeng.wug}@uconn.edu

Circular RNA (or circRNA) is a type of RNA which forms a covalently closed continuous loop. It is now believed that circRNA plays important biological roles in some diseases. Within the past several years, several experimental methods, such as RNase R, have been developed to enrich circRNA while degrading linear RNA. Some useful software tools for circRNA detection have been developed as well. However, these tools may miss many circRNA. Also, existing tools are slow for large data because those tools often depend on reads mapping.

In this paper, we present a new computational approach, named CircMarker, based on k-mers rather than reads mapping for circular RNA detection as shown in Fig. 1. The algorithm has two parts, including reference genome proprocessing and annotations (part 1) and circular RNA detection (part 2).

In part 1, CircMarker creates a table for storing the k-mers within the reference genome that are near the exon boundaries as specified by the annotations. The k-mer table is designed to be space-efficient. We only record five types of information for each k-mer, including chromosome index, gene index, transcript index, exon index and part tag. The "part tag" specifies whether a k-mer comes from the head part or the tail part of the exon.

Part 2 is divided into five steps. (1) Sequence reads processing: examine k-mers contained in a read and search for a match in the k-mer table. (2) Filtering by hit number: short exons should be fully covered by the reads more than one time. Otherwise, the reads should be within both boundaries of the hit exons. (3) Filtering by part tags: we collect part tags from start to end, and condense the tags which belong to the same exons based on the number of hits. (4) Calling circRNA: both self-circular case (single exon) and regular-circular case (multiple exon) are considered. In the regular-circular case, we consider if the exon index increases/decreases monotonically and identify the circular joint junction at the position of the first deceasing/increasing position. (5) Refining circular RNA candidates (optional): only the candidates with support number smaller than a predefined threshold will be viewed as correct one.

We use both simulated and real data for evaluation. We compared CircMarker with three other tools, including CIRI [1], Find circ [3], and CIRCexplorer [4] in terms of the number of called circular RNA, accuracy, consensus-based sensitivity, bias and running time. The results are shown in Fig. 1.

– **Simulated Data.** The simulated data is generated by the simulation script released by CIRI. The reference genome is chromosome 1 in human genome (GRCh37). The annotation file is version 18 (Ensembl 73). Two different cases are simulated, including 10X circRNA & 100X linear RNA, and 50X for both circular and linear RNA.

– **Real data: RNase R treated reads with public database.** We choose CircBase [2] as the standard circRNA database of homo sapiens. The reference genome and annotation file come from homo sapiens GRCm37 version 75. The RNA-Seq reads are from SRR901967.

– **Real Data: RNase R treated/untreated Reads.** The reference genome and annotation file are from *Mus Musculus GRCm38 Release79*. RNase R treated/untreated reads are from SRR2219951 and SRR2185851 respectively.

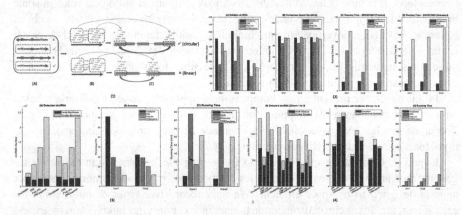

Fig. 1. High Level Approach and Results: (1) High level approach: a fast check for finding circRNA relevant reads, scanning k-mer sequentially from the beginning to the end for each read, and calling circRNA using various criteria and filters. (2) Results of real data based on RNase R treated/untreated reads. (3) Results of simulated data. (4) Results of real data based on RNase R treated reads with public database.

The results show that CircMarker runs much faster and can find more circular RNA than other tools. In addition, CircMarker has higher consensus-based sensitivity and high accuracy/reliable ratio compared with others. Moreover, the circRNAs called by CircMarker often contain most circRNAs called by other tools in the real data we tested. This implies that CircMarker has low bias. CircMarker can be downloaded at: https://github.com/lxwgcool/CircMarker.

References

1. Gao, Y., Wang, J., Zhao, F.: Ciri: an efficient and unbiased algorithm for de novo circular rna identification. Genome Biol. **16**(1), 4 (2015)

2. Glažar, P., Papavasileiou, P., Rajewsky, N.: circbase: a database for circular rnas. RNA **20**(11), 1666–1670 (2014)
3. Memczak, S., Jens, M., Elefsinioti, A., et al.: Circular rnas are a large class of animal rnas with regulatory potency. Nature **495**(7441), 333–338 (2013)
4. Zhang, X.O., Wang, H.B., Zhang, Y., Lu, X., Chen, L.L., Yang, L.: Complementary sequence-mediated exon circularization. Cell **159**(1), 134–147 (2014)

Multiple Model Species Selection
for Transcriptomic and Functional Analysis

Kuan-Hung Li[1], Cin-Han Yang[1,2], Tun-Wen Pai[1], Chi-Hua Hu[3],
Han-Jia Lin[3], Wen-Der Wang[4], and Yet-Ren Chen[5]

[1] Department of Computer Science and Engineering,
National Taiwan Ocean University, Keelung, Taiwan
twp@mail.ntou.edu.tw
[2] Center of Excellence for the Oceans,
National Taiwan Ocean University, Keelung, Taiwan
[3] Department of Bioscience and Biotechnology,
National Taiwan Ocean University, Keelung, Taiwan
[4] Department of Bioagricultural Science, National Chiayi University,
Chiayi City, Taiwan
[5] Agricultural Biotechnology Research Center, Academia Sinica, Taipei, Taiwan

Abstract. Transcriptomic sequencing (RNA-seq) related applications allow for rapid explorations due to their high-throughput and relatively fast experimental capabilities, providing unprecedented progress in gene functional annotation, gene regulation analysis, and environmental factor verification. However, with increasing amounts of sequenced reads and reference model species, the selection of appropriate reference species for gene annotation has become a new challenge. In this study, we proposed a combinatorial approach for finding the most effective reference model species through taxonomic associations and ultra-conserved orthologous (UCO) gene comparisons among species. An online system of multiple species selection (MSS) for RNA-seq differential expression analysis was developed and evaluated. In the designed system, a set of 291 reference model eukaryotic species with comprehensive genomic annotations were selected from the RefSeq, KEGG, and UniProt databases. Using the proposed MSS pipeline, gene ontology and biological pathway enrichment analysis can be efficiently and effectively achieved, especially in the case of transcriptomic analysis of non-model organisms. Regarding the experimental results of selecting appropriate reference model species by analyzing taxonomic relationships and comparing UCOs, accurate evolutionary distances are calculated using sequence alignment and applied to compensate for indistinguishable characteristics of the taxonomic tree. Here, we performed RNA-seq experiments in four non-model species, and the results confirmed that evolutionary distances between species could be ascertained using UCO gene sets. We also performed enrichment analysis of the identified differentially expressed genes using Gene Ontology (GO) and KEGG biological pathway approaches. For example, though GO analysis of *Corbicula fluminea* under hypoxic conditions, we identified additional significant GO terms, including the Notch signaling pathway, cytoskeletal protein binding, and hydrolase activity. These additionally identified GO terms have been found to be associated with hypoxia in previous reported studies. For KEGG biological pathway analysis, additional significant biological pathways could be also identified, such as the CAM pathway, by increasing the number of appropriate reference species. Therefore, pertinent selection of multiple reference species for transcriptomic analysis can reduce required computational hours and unnecessary searches against the non-redundant gene dataset. In addition, selecting multiple appropriate species as reference model species helps to reduce missing crucial annotation information, allowing for more comprehensive results than those obtained with a single model reference species.

Keywords: RNA-seq · Reference model species · Differential expression analysis · Ultra-conserved orthologous genes · Gene ontology · Biological pathway

RNA Transcriptome Mapping
with GraphMap

Krešimir Križanović[1], Ivan Sović[2], Ivan Krpelnik[1], and Mile Šikić[1,3]

[1] Faculty of Electrical Engineering and Computing,
University of Zagreb, Zagreb, Croatia
mile.sikic@fer.hr
[2] Ruđer Bošković Institute, Zagreb, Croatia
[3] Bioinformatics Institute, A*STAR, Singapore, Singapore

The advent of Next Generation Sequencing (NGS) methods has popularized sequencing in various fields of research such as medicine, pharmacy, food technology and agriculture. Aside from DNA sequencing, NGS also enabled RNA sequencing using sequencing-by-synthesis approach. While 3rd generation sequencing technologies are rapidly taking over their share of DNA sequencing market, due to the fact that read length is less important for RNA data analysis, RNA sequencing is still predominately done using NGS. However, it seems likely that at least some aspects of RNA analysis would benefit from increased read length.

Of the currently available RNA-seq aligners BBMap [1] claims to support both PacBio and ONT data, while PacBio GitHub pages offer instructions for working with STAR [2] and GMap [3]. Several available DNA aligners, such as BWA-MEM [4] have been proven to work well with PacBio and ONT data, but they do not offer support for mapping RNA reads to a transcriptome.

In this paper we present an updated version of GraphMap [5] that uses given annotations to generate a transcriptome, and then maps RNA reads to the generated transcriptome using a DNA mapping algorithm. Afterwards, the mapping results are translated back into the genome coordinates. Since initial alignments are calculated for the transcriptome, there is no need to consider spliced alignments and alternative gene splicing. In this way, we can leverage the mapping quality of a proven DNA aligner designed for long and erroneous reads without the need for additional computation to determine exon junctions.

We have compared the new version of GraphMap to three RNA aligners claiming support for 3rd generation sequencing data: BBMap, GMap and STAR. All aligners were tested on three synthetic datasets simulated using a PacBio DNA simulator PBSIM [6]. Since PBSIM is a DNA simulator, to simulate RNA reads it was applied to a transcriptome generated from gene annotations. PBSIM model for CLR reads was used for simulations, and parameters were set for PacBio ROI (Reads of Insert). Alignment results were evaluated by comparing them to MAF files containing information on read origins generated by PBSIM as a part of simulation.

This work has been supported in part by Croatian Science Foundation under the project UIP-11-2013-7353 "Algorithms for Genome Sequence Analysis".

The results displayed in Table 1 show that GraphMap outperforms other aligners by all criteria successfully aligning a read to all exons from its origin (**hit all**) for over 80% of reads and successfully aligning a read to at least one exon of its origin (**hit one**) for over 90% of the reads. It surpasses the results of other aligners by 5–10% on all datasets.

Table 1. Aligner evaluation results. The table shows the percentage of reads for which alignment overlaps all exons from read origin (**hit all**) and the percentage of reads for which alignment overlaps at least one exon from read origin (**hit one**).

Aligner	STAR		BBMap		GMap		Graphmap	
Dataset	Hit all	Hit one	Hit all	Hit one	Hit all	Hit one	Hit all	Hit one
1	46.7%	47.1%	87.0%	88.1%	84.7%	85.7%	93.5%	94.1%
2	32.1%	35.2%	54.4%	78.4%	73.0%	85.4%	82.0%	94.1%
3	33.1%	35.7%	26.8%	61.2%	70.0%	83.8%	85.7%	94.5%

The research presented in this paper demonstrates that the idea to use an appropriate DNA aligner and gene annotations to map RNA reads to a transcriptome and then to transform the mapping results back to genome coordinates is very feasible. Updated GraphMap clearly outperforms other tested splice aware aligners on all datasets. The results suggest that by implementing splice aware mapping logic into a DNA mapper which works well with third generation sequencing data could also work well for de novo RNA spliced mapping.

Keywords: RNA · Transcriptome · Gene annotations · RNA alignment

References

1. Bushnell, B., Egan, R., Copeland, A., Foster, B., Clum, A., Sun, H., et al: BBMap: A Fast, Accurate, Splice-Aware Aligner (2014)
2. Dobin, A., Davis, C.A., Schlesinger, F., Drenkow, J., Zaleski, C., Jha, S., et al.: STAR: ultrafast universal RNA-seq aligner. Bioinformatics **29**, 15–21 (2013). doi:10.1093/bioinformatics/bts635
3. Wu, T.D., Watanabe, C.K.: GMAP: a genomic mapping and alignment program for mRNA and EST sequences. Bioinformatics **21**, 1859–1875 (2005). doi:10.1093/bioinformatics/bti310
4. Li, H.: Aligning sequence reads, clone sequences and assembly contigs with BWA-MEM (2013)
5. Sović, I., Šikić, M., Wilm, A., Fenlon, S.N., Chen, S., Nagarajan, N.: Fast and sensitive mapping of nanopore sequencing reads with GraphMap. Nat Commun. **7**, 11307 (2016). doi:10.1038/ncomms11307
6. Ono, Y., Asai, K., Hamada, M.: PBSIM: PacBio reads simulator–toward accurate genome assembly. Bioinformatics **29**, 119–21 (2013). doi:10.1093/bioinformatics/bts649

A Graph-Based Approach for Proteoform Identification and Quantification Using Homogeneous Multiplexed Top-Down Tandem Mass Spectra

Kaiyuan Zhu[1] and Xiaowen Liu[2,3]

[1] School of Informatics and Computing, Indiana University Bloomington,
Bloomington, USA
[2] School of Informatics and Computing,
Indiana University-Purdue University Indianapolis, Indianapolis, USA
xwliu@iupui.edu
[3] Center for Computational Biology and Bioinformatics,
Indiana University School of Medicine, Indianapolis, USA

Although protein separation techniques have been significantly advanced, it is still a challenging problem to separate proteoforms with similar weights and similar chemical properties, especially those with the same amino acid sequence, but different post-translational modification (PTM) patterns, in top-down mass spectrometry [1]. Tandem mass spectrometry analysis of two or more proteoforms that are not separated by protein separation methods and have similar molecular masses results in a *multiplexed tandem mass (MTM) spectrum*, which is a superimposing of the tandem mass spectra of the proteoforms [4]. There are two types of MTM spectra: *heterogeneous* multiplexed tandem mass (HetMTM) spectra are generated from proteoforms of two or more different proteins; *homogeneous* multiplexed tandem mass (HomMTM) spectra from proteoforms of the same protein with different PTM patterns.

We focus on the study of the identification and quantification of modified proteoforms using HomMTM spectra, in which purified proteins are often analyzed and the target protein is often known. Let P be a unmodified target protein sequence and S a HomMTM spectrum generated from k modified proteoforms of P. Denote Q as the set of modified proteoforms of P that match the precursor mass of S. The *HomMTM spectral identification problem* is to find k proteoforms in Q and their relative abundances such that the peaks (their m/z values and intensities) in spectrum S are best explained [1].

We formulate the HomMTM spectral identification problem as the minimum error k-splittable flow (MEkSF) problem on graphs with vertex capacities, in which each path corresponds to a modified proteoform and the flow on the path corresponds to the relative abundance of the proteoform. The goal is to find a k-splittable flow F with a fixed flow value f (F can be decomposed to k or less than k paths) from the source to the sink in a given graph G such that the sum of the errors on the vertices is minimized.

We prove that the MEkSF problem is NP-hard when k is part of the input and propose a polynomial time algorithm for the problem on layered directed graphs when k is a constant. The algorithm consists of two steps: for a given number k, the packing step determines a set of flow value candidates for k flows, and the routing step finds out the paths for the k flow values that minimize the sum of errors on vertices. When $k = 2$, we prove that the number of flow value candidates is limited by $|V|$, which is the number of vertices in the graph, and propose an efficient dynamic programming algorithm for solving the routing problem. The total time complexity of the algorithm is $O(l^4 h |V|)$, where l is the largest number of vertices in a layer and h is the number of layers in the graph.

We tested the algorithm on a data set of the histone H4 protein with $3,254$ top-down tandem mass spectra. The mass spectra were deconvoluted using MS-Deconv [3]. After searching the deconvoluted spectra against the histone H4 sequence, the proposed method identified 625 spectra with at least 10 matched fragment ions, of which 441 were matched to single proteoforms and 184 matched to proteoform pairs. For each identified proteoform pair, we computed the difference between the number of fragment ions matched to the pair and that matched to the higher abundance proteoform only. Compared with the higher abundance proteoform, the proteoform pair increased the number of matched fragment ions by at least 10 for 39 of the 184 proteoform pairs. In addition, we computed the difference between the sum of peak intensities explained by the pair and that by the higher abundance proteoform only. Proteoform pairs increased explained peak intensities by at least 20% for 26 spectra compared with single proteoforms.

We also compared the proposed method with MS-Align-E [2] on the histone H4 data set. MS-Align-E identified from the data set $1,037$ spectra, of which 184 were matched to a proteoform pair by the proposed method. For 43 of the 184 spectra, the proposed method increased the number of matched fragment ions by at least 10 compared with MS-Align-E.

Acknowledgement. The research was supported by the National Institute of General Medical Sciences, National Institutes of Health (NIH) through Grant R01GM118470.

References

1. DiMaggio Jr., P.A., Young, N.L., Baliban, R.C., Garcia, B.A., Floudas, C.A.: A mixed integer linear optimization framework for the identification and quantification of targeted post-translational modifications of highly modified proteins using multiplexed electron transfer dissociation tandem mass spectrometry. Mol. Cell. Proteomics **8**, 2527–2543 (2009)
2. Liu, X., Hengel, S., Wu, S., Tolić, N., Paša-Tolić, L., Pevzner, P.A.: Identification of ultra-modified proteins using top-down tandem mass spectra. J. Proteome Res. **12**, 5830–5838 (2013)
3. Liu, X., Inbar, Y., Dorrestein, P.C., Wynne, C., Edwards, N., Souda, P., Whitelegge, J.P., Bafna, V., Pevzner, P.A.: Deconvolution and database search of complex tandem mass spectra of intact proteins: a combinatorial approach. Mol. Cell. Proteomics **9**, 2772–2782 (2010)
4. Wang, J., Perez-Santiago, J., Katz, J.E., Mallick, P., Bandeira, N.: Peptide identification from mixture tandem mass spectra. Mol. Cell. Proteomics **9**, 1476–1485 (2010)

A New Estimation of Protein-Level False Discovery Rate

Guanying Wu[1], Xiang Wan[2], and Baohua Xu[1]

[1] The Dental Center of China-Japan Friendship Hospital, Beijing, China
[2] Department of Computer Science, Hong Kong Baptist University, Kowloon Tong, Hong Kong

Abstract. In shotgun proteomics, the identification of proteins is a two-stage process: peptide identification and protein inference [1]. In peptide identification, experimental MS/MS spectra are searched against a sequence database to obtain a set of peptide-spectrum matches (PSMs) [2–4]. In protein inference, individual PSMs are assembled to infer the identity of proteins present in the sample [5–7]. Evaluating the statistical significance of the protein identification result is critical to the success of proteomics studies. Controlling the false discovery rate (FDR) is the most common method for assuring the overall quality of the set of identifications. However, the problem of accurate assessment of statistical significance of protein identifications remains an open question [8, 9]. Existing FDR estimation methods either rely on specific assumptions or rely on the two-stage calculation process of first estimating the error rates at the peptide-level, and then combining them somehow at the protein-level. We propose to estimate the FDR in a non-parametric way with less assumptions and to avoid the two-stage calculation process.

We propose a new protein-level FDR estimation framework. The framework contains two major components: the Permutation+BH (Benjamini–Hochberg) FDR estimation method and the logistic regression-based null inference method. In Permutation+BH, the null distribution of a sample is generated by searching data against a large number of permuted random protein database and therefore does not rely on specific assumptions. Then, p-values of proteins are calculated from the null distribution and the BH procedure is applied to the p-values to achieve the relationship of the FDR and the number of protein identifications. The Permutation+BH method generates the null distribution by the permutation method, which is inefficient for online identification. The logistic regression model is proposed to infer the null distribution of a new sample based on existing null distributions obtained from the Permutation+BH method. In our experiment based on three public available datasets, our Permutation+BH method achieves consistently better performance than MAYU, which is chosen as the benchmark FDR calculation method for this study. The null distribution inference result shows that the logistic regression model achieves a reasonable result both in the shape of the null distribution and the corresponding FDR estimation result.

References

1. Nesvizhskii, A.I., Vitek, O., Aebersold, R.: Analysis and validation of proteomic data generated by tandem mass spectrometry. Nat. Meth. **4**, 787–797 (2007)
2. Eng, J.K., McCormack, A.L., Yates III, J.R.: An approach to correlate tandem mass spectra data of peptides with amino acid sequences in a protein database. J. Am. Soc. Mass Spectrom. **5**, 976–989 (1994)
3. Perkins, D.N., Pappin, D.J.C., Creasy, D.M., Cottrell, J.S.: Probability-based protein identification by searching sequence databases using mass spectrometry data. Electrophoresis **20**, 3551–3567 (1999)
4. Craig, R., Beavis, R.C.: TANDEM: matching proteins with tandem mass spectra. Bioinformatics **20**, 1466–1467 (2004)
5. Nesvizhskii, A.I., Keller, A., Kolker, E., Aebersold, R.: A statistical model for identifying proteins by tandem mass spectrometry. Anal. Chem. **75**, 4646–4658 (2003)
6. Bern, M., Goldberg, D.: Improved ranking functions for protein and modification-site identifications. In: International Conference on Research in Computational Molecular Biology, pp. 444–458 (2007)
7. Li, Y., Arnold, R., Li, Y., Radivojac, P., Sheng, Q., Tang, H.: A Bayesian approach to protein inference problem in shotgun proteomics. In International Conference on Research in Computational Molecular Biology, pp. 167–180 (2008)
8. Spirin, V., Shpunt, A., Seebacher, J., Gentzel, M., Shevchenko, A., Gygi, S., Sunyaev, S.: Assigning spectrum-specific p-values to protein identifications by mass spectrometry. Bioinformatics **27**, 1128–1134 (2011)
9. Omenn, G.S., Blackwell, T.W., Fermin, D., Eng, J., Speicher, D.W., Hanash, S.M.: Challenges in deriving high-confidence protein identifications from data gathered by a HUPO plasma proteome collaborative study. Nat. Biotechnol. **24**, 333–338 (2006)

A Generalized Approach to Predicting Virus-Host Protein-Protein Interactions

Xiang Zhou, Byungkyu Park, Daesik Choi, and Kyungsook Han

Department of Computer Science and Engineering,
Inha University, Incheon, South Korea
{jusang486,anrgid6893}@gmail.com
{bpark,khan}@inha.ac.kr

Many computational methods have been developed to predict PPIs, but most of them are intended for PPIs within a same species rather than for PPIs across different species. Motivated by the recent increase in data of virus-host PPIs, a few computational methods have been developed to predict virus-host PPIs, but most of them cannot be applied to new viruses or new hosts that have no known PPIs to the methods. A recent SVM model called DeNovo [1] is perhaps the only one that can predict PPIs of new viruses with a shared host. Protein sequence similarity between different types of viruses or hosts is relatively low, so predicting virus-host PPIs for new viruses or hosts is quite challenging.

We obtained all known PPIs between virus and host from four databases, APID, IntAct, Mentha and UniProt, which use same protein identifiers. As of December 2016, there were a total of 12,157 PPIs between 29 hosts and 332 viruses. For negative data, we obtained protein sequences of major hosts (human, non-human animal, plant, and bacteria) from UniProt, and removed those with a sequence similarity higher than 80% to any positive data.

We constructed several datasets to examine the applicability of our prediction method to new viruses and hosts.

1. Training (TR) and test (TS) sets for assessing the applicability to new viruses
   TR1: 10,955 PPIs between human and any virus except H1N1
   TR2: 11,341 PPIs between human and any virus except Ebola virus
   TR3: 11,617 PPIs between any host and any virus except H1N1
   TR4: 12,007 PPIs between any host and any virus except Ebola virus
   TS1: 381 PPIs between human and H1N1 virus
   TS2: 150 PPIs between human and Ebola virus
2. Training (TR) and test (TS) sets for assessing the applicability to new hosts
   TR5: 11,491 PPIs between human and any virus
   TS5.1: 488 PPIs between non-human animal and any virus
   TS5.2: 17 PPIs between plant and any virus
   TS5.3: 143 PPIs between bacteria and any virus

We built a support vector machine (SVM) model using LIBSVM with the radial basis function as a kernel. The SVM model uses several features of protein sequences: the relative frequency of amino acid triplets (RFAT), frequency difference of amino acid triplets (FDAT), amino acid composition (AC), and transition, distribution and composition of amino acid groups. The first three features (RFAT, FDAT and AC) are improved features developed in our previous study of single host-virus PPIs [2], and the last three features (transition, distribution and composition) were developed by You et al. [3] for PPIs in a single species.

The SVM model was evaluated in several ways: 10-fold cross validation on several datasets with different ratios of positive to negative data instances and independent testing on new viruses and hosts. In the 10-fold cross validation on three datasets of different ratios of positive to negative data (1:1, 1:2 and 1:3), the best performance (sensitivity = 85%, specificity = 96%, accuracy = 86%, PPV = 86%, NPV = 85%, MCC = 0.71, and AUC = 0.93) was observed in the balanced dataset with 1:1 ratio of positive to negative data. As expected, running the SVM model on unbalanced datasets resulted in lower performances than running it on the balanced dataset.

The model was tested on new viruses using 2 independent datasets of PPIs of H1N1 and Ebola virus, which were not used in training the model. Proteins of H1N1 virus have an average sequence similarity of 9.6% to those of other viruses, and proteins of Ebola virus have a sequence similarity of 10.9% to other viruses. Despite such a low sequence similarity of proteins in test datasets to those in training datasets, the model showed a relatively high performance in independent testing (in datasets TR1-TS1, TR2-TS2, TR3-TS1 and TR4-TS2, it showed accuracies of 78%, 78%, 77% and 82%, respectively).

Likewise, we tested the model on new hosts. A model trained with human-virus PPIs (TR5) was tested on PPIs of viruses with non-human, which include non-human animal (TS5.1), plant (TS5.2) and bacteria (TS5.3). The average sequence similarity of human proteins to non-human animal, plant, and bacteria is lower than 10.7%, but the model showed accuracies of 66%, 68% and 67% in test sets of non-human animal, plant, and bacteria, respectively.

In this study, we developed a general method for predicting PPIs between any virus and any host. In independent testing of the model on new viruses and hosts, it showed a high performance comparable to the best performance of other methods for PPIs between a specific virus and its host. This method will be useful in finding potential PPIs of a new virus or host, for which little information is available. The program and data are available at http://bclab.inha.ac.kr/VirusHostPPI.

References

1. Eid, F.E., ElHefnawi, M., Heath, L.S.: DeNovo: virus-host sequence-based protein-protein interaction prediction. Bioinformatics **32**, 1144–1150 (2016)
2. Kim, B., Alguwaizani, S., Zhou, X., Huang, D.-S., Park, B., Han, K.: An improved method for predicting interactions between virus and human proteins. J. Bioinform. Comput. Biol. **15**(1), 1650024 (2016)
3. You, Z.H., Chan, K.C.C., Hu, P.W.: Predicting protein-protein interactions from primary protein sequences using a novel multi-scale local feature representation scheme and the random forest. PLoS One **10**(5), e0125811 (2015)

Deep Learning for Classifying Maize Seeds in Double Haploid Induction Process

Balaji Veeramani, John W. Raymond, and Pritam Chanda

Dow AgroSciences LLC, Indianapolis, IN, USA
{bveeramani,jwraymond,pchanda}@dow.com

1 Introduction

In industrial agricultural breeding, double haploid based generation of inbred maize lines has accelerated the time to market of commercial seed varieties [5]. Traditionally, haploid corn seeds are manually discriminated from the diploid seeds using visual indications of the molecular marker system that is selectively expressed in the embryo region of the diploid seeds. In the industrial scale, there have been two notable automation efforts based on the R1-*nj* marker system [2, 4]. However due to the extensive phenotypic variation of the marker expression [1] and heterogeneity arising from image acquisition in the field, developing computer vision methods to classify seed images is challenging, and approaches robust in recovering haploids are lacking.

2 Results and Discussion

Convolutional neural networks (CNN) have been used successfully for traffic sign recognition, face verification and with autonomous driving vehicles [3]. In this work, we investigate, to our knowledge for the first time, the application of a convolutional network to sort maize haploid seeds from diploids using thousands of images of corn seeds (see Fig. 1). We obtained 4731 corn seed RGB images consisting of 952 haploid and 3779 diploid seeds from several different proprietary maize inbred lines. We train our network using the image dataset that was randomly split into 4021 training (809 haploid and 3212 diploid seeds) and 710 test (143 haploids and 567 diploids) images with 20% haploids in both sets. The training images were further divided into 5-folds to assess its performance under random data splits on unseen data.

We demonstrate deep convolutional networks perform significantly better as compared to several other classifiers that use seed texture, color, and shape features (see Table 1). On the test data set, our network achieved the highest classification accuracy (0.968) among all methods used in our experiments. We looked into the

Balaji Veeramani—Authors acknowledge help/feedback by W. Edsall, S. Cryer, B. John, K. Koehler, G. Tragesser, G. Temnykh, E. Frederickson, and P. Setlur.
Balaji Veeramani and Pritam Chanda contributed equally.

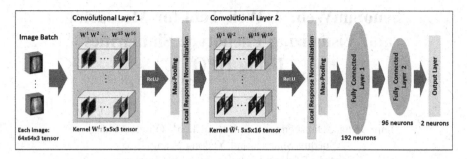

Fig. 1. Convolutional neural network architecture schematic for haploid seed sorting. Input images of the corn seeds are convolved with 16 filter kernels in each convolutional layer, followed by two fully connected layers and an output layer.

misclassifications of our method and the best performing comparative method (SVM) to gain insight into its ability to classify haploid and diploid categories separately in the test dataset. Out of the 567 diploids and 143 haploids in the test dataset, CNN misclassifies 12 haploids as diploids, and 11 diploids as haploids. However, the SVM has a higher tendency to classify haploids as diploids. It classifies 66 haploids as diploids (and 22 diploids as haploids), possibly reflecting dataset class distribution.

Visualizations of the neuronal activations in the convolutional layers indicate the network derives features that are discriminative of embryo regions between haploids and diploids (results not shown here). With the advent of technological advances in agriculture, convolutional networks and other deep learning techniques hold promise for several applications within the agricultural industry.

Table 1. Classification accuracies comparing CNN and other classifiers using texture features (values within brackets indicate results using all features; CV:Cross Validation)

	CNN	SVM	Random forest	Logistic regression
CV	0.961	0.857 (0.836)	0.840 (0.823)	0.749 (0.777)
Train	1.000	0.911 (0.994)	1.000 (0.997)	0.751 (0.786)
Test	0.968	0.876 (0.839)	0.845 (0.824)	0.775 (0.772)

References

1. Kebede, A.Z., Dhillon, B.S., Schipprack, W., Araus, J.L., Bänziger, M., Semagn, K., Alvarado, G., Melchinger, A.E.: Effect of source germplasm and season on the in vivo haploid induction rate in tropical maize. Euphytica **180**(2), 219–226 (2011)
2. Koehler, K.L., Tragesser, G., Swanson, M.: Apparatus and method for sorting plant material. US Patent 9,156,064 (2015)
3. LeCun, Y., Bengio, Y., Hinton, G.: Deep learning. Nature **521**(7553), 436–444 (2015)
4. Modiano, S.H., Deppermann, K.L., Crain, J., Eathington, S.R., Graham, M.: Seed sorter. US Patent 8,189,901 (2012)
5. Prasanna, B., Chaikam, V., Mahuku, G.: Doubled haploid technology in maize breeding: theory and practice. CIMMYT (2012)

PhenoSimWeb: A Web Tool for Measuring and Visualizing Phenotype Similarities Using HPO

Jiajie Peng[1], Hansheng Xue[2], Bolin Chen[1], Qinghua Jiang[3],
Xuequn Shang[1], and Yadong Wang[2,4]
[1] School of Computer Science and Technology,
Northwestern Polytechnical University, Xi'an, China
[2] School of Computer Science and Technology,
Harbin Institute of Technology, Shenzhen, China
[3] School of Life Science and Technology,
Harbin Institute of Technology, Harbin, China
[4] School of Computer Science and Technology,
Harbin Institute of Technology, Harbin, China

The Human Phenotype Ontology (HPO) was constructed by Robinson *et al.* in 2008, which is one of the most widely used bioinformatics resources [7]. The unified and structured vocabulary of HPO helps to display the phenotypic characteristics, constructs a directly acyclic graph (DAG), and provides a convenient way to study the phenotype similarity.

In recent years, various HPO-based semantic similarity measurements have been proposed to measure the phenotype similarity. Most of these methods are based on the Information Content (IC), including Resnik [6], Schlicker measure [8] and Phenomizer [3]. Besides, PhenomeNet [2] and OWLSim [9] are further developed to calculate two phenotype sets similarity based on simGIC [5]. HPOSim [1] provides an open source package to measure phenotype similarity, which integrates seven widely used HPO-based similarity measurements.

Most of the aforesaid methods are revised based on GO-based similarity measurements, which mainly consider the annotations and topological informations of phenotype terms and neglect the unique features of HPO. Therefore, we proposed a novel method, termed as *PhenoSim*, to calculate the phenotype similarity [4]. Our method consists of denoising model, which model the noises in the patient phenotype data set, and a novel path-constrained Information Content similarity measurement. The whole process of *PhenoSim* can be grouped into three steps: constructing the phenotype network, reducing noise data in patients' phenotype set using PageRank algorithm, and calculating the phenotype set similarities by a novel path-constrained Information Content.

Furthermore, the existing tools of measuring phenotype similarity mainly have two drawbacks: Firstly, existing tools ignores the importance of phenotype text, which are

---

Jiajie Peng, Hansheng Xue — Equal contributor.

often used to describe the symptoms of patients, and none of them allow phenotype text as input. Secondly, none of existing tools supplies interface to visualize the similarity results instead of listing the final similarity value directly. Thus, it is necessary to develop an easy-to-used web application to allow researchers to type in phenotype text and visualize the final phenotype similarity results.

In this paper, we present a novel web tool termed as *PhenoSimWeb*, which is available at 120.77.47.2:8080, to measure HPO-based phenotype similarities and to visualize the result with an easy-to-use graphical interface. Comparing with the existing tools, *PhenoSimWeb* has the following advantages:

- *PhenoSimWeb* offers researchers a novel phenotype semantic similarity measurement which considers the unique features of HPO.
- *PhenoSimWeb* allows researchers to type in the phenotype text that describes phenotype features.
- *PhenoSimWeb* provides an easy-to-use graphical interface to visualize phenotype semantic similarity association.

Acknowledgement. This work was supported the Fundamental Research Funds for the Central Universities (Grant No. 3102016QD003), National Natural Science Foundation of China (Grant No. 61602386 and 61332014), the High-Tech Research and Development Program of China (Grant No. 2015AA020101, 2015AA020108).

References

1. Deng, Y., Gao, L., Wang, B., Guo, X.: Hposim: an r package for phenotypic similarity measure and enrichment analysis based on the human phenotype ontology. PloS one **10**(2), e0115692 (2015)
2. Hoehndorf, R., Schofield, P.N., Gkoutos, G.V.: Phenomenet: a whole-phenome approach to disease gene discovery. Nucleic Acids Res. **39**(18), e119–e119 (2011)
3. Köhler, S., Schulz, M.H., Krawitz, P., Bauer, S., Dölken, S., Ott, C.E., Mundlos, C., Horn, D., Mundlos, S., Robinson, P.N.: Clinical diagnostics in human genetics with semantic similarity searches in ontologies. Am. J. Hum. Genet. **85**(4), 457–464 (2009)
4. Peng, J., Xue, H., Shao, Y., Shang, X., Wang, Y., Chen, J.: Measuring phenotype semantic similarity using human phenotype ontology. In: BIBM, pp. 763–766 (2016)
5. Pesquita, C., Faria, D., Bastos, H., Falcão, A., Couto, F.: Evaluating go-based semantic similarity measures. In: Proceedings of 10th Annual Bio-Ontologies Meeting. vol. 37, p. 38 (2007)
6. Resnik, P.: Using information content to evaluate semantic similarity in a taxonomy. In: Proceedings of the 14th International Joint Conference on Artificial Intelligence, pp. 448–453 (1995)
7. Robinson, P.N., Köhler, S., Bauer, S., Seelow, D., Horn, D., Mundlos, S.: The human phenotype ontology: a tool for annotating and analyzing human hereditary disease. Am. J. Hum. Genet. **83**(5), 610–615 (2008)
8. Schlicker, A., Domingues, F.S., Rahnenführer, J., Lengauer, T.: A new measure for functional similarity of gene products based on gene ontology. BMC Bioinform. **7**(1), 1 (2006)
9. Washington, N.L., Haendel, M.A., Mungall, C.J., Ashburner, M., Westerfield, M., Lewis, S.E.: Linking human diseases to animal models using ontology-based phenotype annotation. PLoS Biol. **7**(11), e1000247 (2009)

An Improved Approach for Reconstructing Consensus Repeats from Short Sequence Reads

Chong Chu, Jingwen Pei, and Yufeng Wu

Computer Science and Engineering Department,
University of Connecticut, Storrs, CT, USA
{chong.chu,jingwen.pei,yufeng.wu}@uconn.edu

A repeat is a segment of DNA that appears multiple times in the genome in an identical or near-identical form. There are many types of repeats such as transposable elements (TEs), tandem repeats, satellite repeats, and simple repeats. Among them, TEs are perhaps the most well-known one. Even though many computational approaches have been developed for constructing consensus repeats, it is still useful to construct repeats directly from reads for complex genomes. Repeats usually have many copies in the genome. For low divergent and high copy number repeats, it is highly likely that k-mers generated from their copies will be identical at the same position. Thus, repeats can be assembled from these high frequent k-mers. RepARK [3] and the original REPdenovo [1] are developed based on this observation. The original REPdenovo outperforms RepARK because it conducts a second-round assembly: it attempts to assemble short contigs in order to form longer consensus repeats based on the reliable prefix-suffix matches of contigs. However, REPdenovo performs less well for highly divergent or low copy number repeats. One reason is that k-mers originated from high divergent regions of a long repeat usually have low frequency, and thus will be filtered out. This leads to fragmented assembled repeats. Another reason is that variations make it difficult to merge the fragmented contigs to form complete repeats. In Fig. 1 (A) and (B), we show two examples to illustrate the situation described above.

In this paper, we propose an improved method (with pipeline shown in Fig. 1(C)) for reconstructing repeat elements from short reads. Similar to the original REPdenovo, our new method also finds and assembles these highly frequent k-mers to form consensus repeat sequences. There are two main improvements in the improved REPdenovo over the original REPdenovo:

- Our new method uses more repeat-related k-mers for repeat assembly, and can assemble longer consensus repeats. Briefly, with high frequent k-mers used as a "reference", low frequent k-mers originated from high divergent regions will be recruited by a "mapping-based alignment" approach.
- Our new method uses a randomized algorithm to generate more accurate consensus k-mers. This improves the quality of the assembled repeats.

Compared to the original REPdenovo and RepARK, our new method can construct more fully assembled repeats in Repbase on both Human and Arabidopsis data, especially for higher divergent, lower copy number and longer repeats. Figure 1(D) shows the comparison between the constructed repeats of the two versions in Repbase

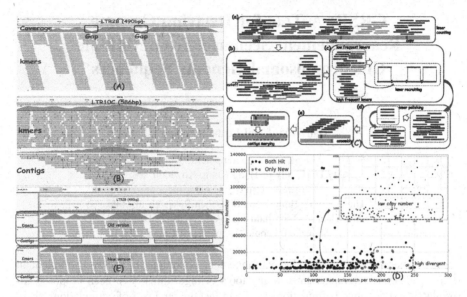

Fig. 1. Observations, pipeline of the method, and results of the improved REPdenovo. (A) Observation one: k-mers from high divergent regions are filtered out and thus form gaps, which leads to fragmented assembled sequences. (B) Observation two: variations make it difficult to assemble long contigs. (C) Pipeline of the improved REPdenovo. (D) Comparison between the original and the improved version of REPdenovo on constructed human repeats in Repbase. (E) One example for comparing the assembly quality on one repeat between the original and the improved REPdenovo.

on Human data. Figure 1(E) illustrates one case that the improved REPdenovo fully construct the repeats while the original REPdenovo fails to. We also apply the new method on Hummingbird data, which has no existing repeat library. Most of the repeats constructed by our new method for Hummingbird can be fully aligned to PacBio long reads. Many of these repeats are long. More than half of the Hummingbird repeats are masked by RepeatMasker, which indicates our assembly works reasonably well. Moreover, many of the assembled repeats are likely to be novel because there are no matches in RepBase. Our new approach has been implemented as part of the REPdenovo software package, which is available for download at https://github.com/Reedwarbler/REPdenovo.

References

1. Chu, C., Nielsen, R., Wu, Y.: Repdenovo: Inferring de novo repeat motifs from short sequence reads. PloS one **11**(3), e0150719 (2016)
2. Jurka, J., Kapitonov, V.V., Pavlicek, A., Klonowski, P., Kohany, O., Walichiewicz, J.: Repbase Update, a database of eukaryotic repetitive elements. Cytogenet Genome Res. **110** (1–4), 462–467 (2005)
3. Koch, P., Platzer, M., Downie, B.R.: RepARK - de novo creation of repeat libraries from whole-genome NGS reads. Nucleic Acids Res. **42**, e80 (2014)

GRSR: A Tool for Deriving Genome Rearrangement Scenarios from Multiple Unichromosomal Genome Sequences

Dan Wang[1] and Lusheng Wang[1,2]

[1] Department of Computer Science,
City University of Hong Kong, Kowloon, HK
cswang1@cityu.edu.hk
[2] University of Hong Kong Shenzhen Research Institute,
Shenzhen Hi-Tech Industrial Park, Nanshan District,
Shenzhen, People's Republic of China

Sorting genomic permutations by rearrangement operations is a classic problem in studying genome rearrangements. Many tools or algorithms have been proposed for sorting signed genomic permutations [1, 2]. In fact, given a pair of permutations, there are often more than one optimal rearrangement scenarios, especially when the rearrangement distance between this permutation pair is large. And sometimes, for the same pair of permutations, the computed rearrangement scenarios using different tools are not consistent. Hence, how to know whether the calculated scenarios are solid and biologically meaningful becomes an essential task. Up to now, several mechanisms for genome rearrangements have been reported [3, 4]. Statistics analyzes showed that breakpoints are often associated with repetitive elements [5, 6]. There was evidence showing that a reversal can be mediated by a pair of inverted repeats (IRs) [7, 8]. Hence, whether there exist repeats at the breakpoints of rearrangement events may give us a clue on whether the calculated rearrangement scenarios are biologically meaningful.

In this paper, we describe a new tool named GRSR for deriving genome rearrangement scenarios from multiple unichromosomal genome sequences and checking whether there are repeats at the breakpoints of each calculated rearrangement event. The input of the GRSR tool is a set of unichromosomal genome sequences and the output is pairwise rearrangement scenario which is a series of transpositions, block interchanges and reversals. Besides, for each calculated rearrangement event, GRSR checks whether there exist repeats which may mediate this rearrangement event.

The GRSR tool is comprised of four primary steps. Firstly, we use Mugsy [9] to conduct a multiple sequence alignment of the input genomes and the alignment result is in an MAF file. Secondly, as transpositions, block interchanges and reversals happen on sequences which are shared by genomes, we extract the coordinates of core blocks (shared by all of the input genomes) from the MAF file. Thirdly, we utilize the coordinates of core blocks to construct synteny blocks using GRIMM [2] and each input genome will be represented by a signed permutation describing the synteny block order on its chromosome. Lastly, we implement a novel method to compute the pairwise rearrangement scenario which is a series of rearrangement events involved in transforming one genome's permutation into another. The computed rearrangement scenarios will only include rearrangement events which happen on a single chromosome, such as transpositions, block interchanges and reversals. Given a pair of signed permutations s and d, the GRSR tool calculate rearrangement scenario from s to d by merging blocks which are on the same order on s and d, then detecting and removing obvious (independent) transpositions and block interchanges and finally sorting permutations s and d by reversals using GRIMM. Once getting a rearrangement event, the GRSR tool will check whether there are repeats at the breakpoints of this event using BLAST [10]. The GRSR tool writes the rearrangement scenarios and whether there are repeats at the breakpoints of each rearrangement event into the *report.txt* file.

We applied the GRSR tool on complete genomes of 28 *Mycobacterium tuberculosis* strains, 24 *Shewanella* strains and 2 *Pseudomonas aeruginosa* strains, respectively. From the results generated by the GRSR tool, we observed that many reversal events were flanked by a pair of inverted repeats so that the two ends of the reversal region remain unchanged before and after the reversal event. We also observed that in other rearrangement operations such transpositions and block interchanges, there exist repeats (not necessarily inverted) at the breakpoints, where the ends remained unchanged before and after the rearrangement operations. In the results for *Pseudomonas aeruginosa* strains, we found an example in which the existence of repeats may explain breakpoint reuse. All the above observations suggest that the conservation of ends could possibly be a popular phenomenon in many types of genome rearrangement events.

References

1. Hannenhalli, S., Pevzner, P.A.: Transforming men into mice (polynomial algorithm for genomic distance problem). In: 36th Annual Symposium on Foundations of Computer Science. Proceedings. IEEE, pp. 581–592 (1995)
2. Tesler, G.: Grimm: genome rearrangements web server. Bioinformatics **18**(3), 492–493 (2002)
3. Darmon, E., Leach, D.R.: Bacterial genome instability. Microbiol. Mol. Biol. Rev. **78**(1), 1–39 (2014)
4. Gray, Y.H.: It takes two transposons to tango: transposable-element-mediated chromosomal rearrangements. Trends Genet. **16**(10), 461–468 (2000)
5. Longo, M.S., Carone, D.M., Green, E.D., O'Neill, M.J., O'Neill, R.J., et al.: Distinct retroelement classes define evolutionary breakpoints demarcating sites of evolutionary novelty. BMC Genomics **10**(1), 334 (2009)
6. Sankoff, D.: The where and wherefore of evolutionary breakpoints. J. Biol. **8**(7), 1 (2009)
7. Small, K., Iber, J., Warren, S.T.: Inversion mediated by inverted repeats. Nat. Genet. **16** (1997)
8. Rajaraman, A., Tannier, E., Chauve, C.: Fpsac: fast phylogenetic scaffolding of ancient contigs. Bioinformatics **29**(23), 2987–2994 (2013)
9. Angiuoli, S.V., Salzberg, S.L.: Mugsy: fast multiple alignment of closely related whole genomes. Bioinformatics **27**(3), 334–342 (2011)
10. Altschul, S.F., Gish, W., Miller, W., Myers, E.W., Lipman, D.J.: Basic local alignment search tool. J. Mol. Biol. **215**(3), 403–410 (1990)

Coestimation of Gene Trees and Reconciliations Under a Duplication-Loss-Coalescence Model

Bo Zhang[1] and Yi-Chieh Wu[2]

[1] Department of Mathematics, Harvey Mudd College,
Claremont, CA, USA
bzhang@hmc.edu
[2] Department of Computer Science, Harvey Mudd College,
Claremont, CA, USA
yjw@cs.hmc.edu

Introduction: Phylogenetic tree reconciliation is fundamental to understanding how genes have evolved within and between species. Given a *gene tree* that depicts how a set of genes has diverged from one another and a *species tree* that depicts how a set of species has speciated, the reconciliation problem proposes a nesting of the gene tree within the species tree and postulates evolutionary events to account for any observed incongruence.

However, within eukaryotes, the most popular reconciliation algorithms consider only a restricted set of evolutionary events, typically modeling only duplications and losses [1, 2] or only coalescences [3, 4]. Recently, the DLCoal model was proposed to unify duplications, losses, and coalescences through an intermediate *locus tree* that describes how new loci are created and destroyed [5]. Here, the locus tree evolves within the species tree according to a duplication-loss model, and the gene tree evolves within the locus tree according to a modified multispecies coalescent model. Two algorithms exist for reconciliations under this model: DLCoalRecon [5], which infers the maximum *a posteriori* reconciliation, and DLCpar [6], which infers a most parsimonious reconciliation. However, both methods assume that the gene tree is known and do not account for errors that may occur during gene tree reconstruction.

To address this challenge, we present DLC-Coestimation, a probabilistic inference method that simultaneously reconstructs the gene tree and reconciles it with the species tree. Given as input a sequence alignment, a species tree, and model parameters including the duplication and loss rate, the population size, and the substitution rate, our algorithm relies on a Bayesian framework to jointly optimize the sequence likelihood and the reconciled tree prior. We show how each term in our inference algorithm corresponds to one component of the underlying generative evolutionary process, and we propose an efficient algorithm for optimizing the overall probability through an iterative hill-climbing procedure combined with Monte Carlo integration.

Results: Our experimental evaluation demonstrates that DLC-Coestimation outperforms existing approaches in ortholog, duplication, and loss inference.

Using a simulated clade of 12 flies, we show that independent reconstruction of the gene tree followed by reconciliation substantially degrades inferences compared to using the true gene tree, even when gene trees are reconstructed with popular top-performing methods. Interestingly, while DLC-Coestimation outperforms DLCoalRecon for every simulation setting, it outperforms DLCpar only for data sets with large amounts of ILS. This finding suggests that our algorithm is better able to handle data sets with low phylogenetic signal, a problem that will become increasingly prevalent as we sequence denser clades.

We also assessed DLC-Coestimation performance on a biological data set of 16 fungi. While all reconciliation methods recover a similar percentage of syntenic orthologs, DLC-Coestimation infers substantially fewer duplications and losses than DLCoalRecon and DLCpar, suggesting that our algorithm is better able to remove spurious duplication and loss events that result from ILS. Furthermore, duplications inferred by DLC-Coestimation are more plausible, with a higher percentage of species overlap post-duplication.

Conclusion: This work demonstrates the utility of coestimation methods for inferences under joint phylogenetic and population genomic models. The DLC-Coestimation software is freely available for download at https://www.cs.hmc.edu/~yjw/software/dlc-coestimation.

References

1. Goodman, M., Czelusniak, J., Moore, G.W., Romero-Herrera, A.E., Matsuda, G.: Fitting the gene lineage into its species lineage, a parsimony strategy illustrated by cladograms constructed from globin sequences. Syst. Zool. **28**(2), 132–163 (1979)
2. Page, R.D.M.: Maps between trees and cladistic analysis of historical associations among genes, organisms, and areas. Syst. Biol. **43**(1), 58–77 (1994)
3. Kingman, J.F.C.: The coalescent. Stoch. Proc. Appl. **13**(3), 235–248 (1982)
4. Pamilo, P., Nei, M.: Relationships between gene trees and species trees. Mol. Biol. Evol. **5**(5), 568–583 (1988)
5. Rasmussen, M.D., Kellis, M.: Unified modeling of gene duplication, loss, and coalescence using a locus tree. Genome Res. 22, 755–765 (2012)
6. Wu, Y.-C., Rasmussen, M.D., Bansal, M.S., Kellis, M.: Most parsimonious reconciliation in the presence of gene duplication, loss, and deep coalescence using labeled coalescent trees. Genome Res. **24**(3), 475–486 (2014)

Reconstruction of Real and Simulated Phylogenies Based on Quartet Plurality Inference

Eliran Avni and Sagi Snir

Department of Evolutionary Biology, University of Haifa, Haifa 31905, Israel
ssagi@research.haifa.ac.il

One of the most fundamental tasks in biology is deciphering the history of life on Earth. To achieve that goal, an important step in many phylogenomic analyses is the reconstruction of a tree of ancestor-descendant relationships, a gene tree, for each family of orthologous genes in a dataset. Such analyses have revealed widespread discordance between gene trees [6]. Apart from statistical errors, various mechanisms may lead to incongruences between gene histories, such as hybridization events, duplications and losses in gene families, incomplete lineage sorting, and most importantly, horizontal genetic transfers [4, 9, 11].

Horizontal gene transfer (HGT) is the non-vertical transfer of genes between contemporaneous organisms (as opposed to the standard vertical transmission between parent and offspring). HGT, which is largely mediated by viruses (bacteriophages), plasmids, transposons and other mobile elements, is particularly common in prokaryotes and has been recognized to play an important role in microbial adaptation, with implications in the study of infectious diseases [13]. Estimates of the fraction of genes that experienced HGT vary widely, some as high as 99% [3, 6]. These have led some researchers to question the meaningfulness of the Tree of Life concept [1, 5, 8, 14]. However, despite HGT, that turns evolution into a network of relationships, there is ample evidence that an underlying species tree signal can still be distilled and separated from non tree-like events [2, 6, 7, 10].

In [12], Roch and Snir investigated the feasibility of reconstructing the phylogeny of a four-taxa set - a quartet - using a simple plurality inference rule. Assuming that HGT events are consistent with a Poisson process of a constant rate, they proved that this reconstruction is achieved with high probability if the number of HGT events per gene is $O(\frac{n}{\log n})$ (where n is the number of species). This implies that the number of HGT events can be almost proportional to the number of gene tree edges without destroying the overall tree signal.

In this work we develop the study of the *quartet plurality rule*, by extending it into a complete tree reconstruction scheme. We first complement [12] by finding a lower bound for the probability of simultaneous correct inference of a multitude of quartets, as a function of the size of the species set, the number of gene trees, and the frequency of HGT events. Since every phylogeny is uniquely determined by its induced quartets, accurate reconstruction of the entire set of quartets implies accurate phylogenetic reconstruction, that can be done in this case in polynomial time. Next, we show via detailed simulations, that even when the number of HGT events is much larger than what the theory of [12] dictates, the plurality inference rule still enables accurate tree reconstruction. In the last part of the paper, we demonstrate that the plurality rule can be a viable tool for real data phylogenetic reconstruction, by applying the above theoretical principles to two sets of prokaryotes. The constructed phylogenies of these two sets are shown to be comparable with (and complementary better than) other suggested evolutionary trees in a number of tests.

Based on our analysis, some interesting questions arise. From a theoretical perspective, our ability to reconstruct accurate phylogenies in practice despite surprisingly high rates of HGT, suggest that the known upper bound for HGT rates that still enable successful tree reconstruction can be further improved. In addition, it is noteworthy that weights were also incorporated in the reconstruction scheme used in this paper. Since only three types of weights were tested, it would be desirable to explore new weighting functions that may be beneficial to the accuracy of tree reconstruction.

References

1. Bapteste, E., Susko, E., Leigh, J., MacLeod, D., Charlebois, R.L., Doolittle, W.F.: Do orthologous gene phylogenies really support tree-thinking? BMC Evol. Biol. **5**, 33 (2005)
2. Beiko, R.G., Harlow, T., Ragan, M.: Highways of gene sharing in prokaryotes. Proc. Natl. Acad. Sci. USA **102**, 14332–14337 (2005)
3. Dagan, T., Martin, W.: The tree of one percent. Genome Biol. **7**(10), 118 (2006)
4. Doolittle, W.F.: Phylogenetic classification and the universal tree. Science **284**(5423), 2124–2129 (1999)
5. Doolittle, W.F., Bapteste, E.: Pattern pluralism and the tree of life hypothesis. Proc. Natl. Acad. Sci. USA **104**, 2043–2049 (2007)
6. Galtier, N. and V. Daubin. Dealing with incongruence in phylogenomic analyses. Philos. Trans. R. Soc. Lond. B Biol. Sci. **363**, 4023–4029 (2008)
7. Ge, F., Wang, L., Kim, J.: The cobweb of life revealed by genome-scale estimates of horizontal gene transfer. PLoS Biol. **3**, e316 (2005)
8. Gogarten, J.P., Townsend, J.P.: Horizontal gene transfer, genome innovation and evolution. Nat. Rev. Micro. **3**(9), 679–687 (2005)
9. Ochman, H., Lawrence, J.G., Groisman, E.A.: Lateral gene transfer and the nature of bacterial innovation. Nature **405**(6784), 299–304 (2000)
10. Koonin, E.V., Puigbó, P., Wolf, Y.I.: Comparison of phylogenetic trees and search for a central trend in the forest of life. J. Comput. Biol. **18**(7), 917–924 (2011)
11. Maddison, W.P.: Gene trees in species trees. System. Biol. **46**(3), 523–536 (1997)
12. Roch, S., Snir, S.: Recovering the tree-like trend of evolution despite extensive lateral genetic transfer: a probabilistic analysis. In: RECOMB, pp. 224–238 (2012)
13. Smets, B.F., Barkay, T.: Horizontal gene transfer: perspectives at a crossroads of scientific disciplines. Nat. Rev. Micro. **3**(9), 675–678 (2005)
14. Zhaxybayeva, O., Lapierre, P., Gogarten, J.: Genome mosaicism and organismal lineages. Trends Genet. **20**, 254–260 (2004)

On the Impact of Uncertain Gene Tree Rooting
on Duplication-Transfer-Loss Reconciliation

Soumya Kundu[1] and Mukul S. Bansal[1,2]

[1] Department of Computer Science and Engineering,
University of Connecticut, Storrs, USA
soumya.kundu@uconn.edu
[2] Institute for Systems Genomics, University of Connecticut, Storrs, USA
mukul.bansal@uconn.edu

Duplication-Transfer-Loss (DTL) reconciliation is one of the most effective techniques for studying the evolution of gene families and inferring evolutionary events. Given the evolutionary tree for a gene family, i.e., a *gene tree*, and the evolutionary tree for the corresponding species, i.e., a *species tree*, DTL reconciliation compares the gene tree with the species tree and reconciles any differences between the two by proposing gene duplication, horizontal gene transfer, and gene loss events. DTL reconciliations are generally computed using a parsimony framework where each evolutionary event is assigned a cost and the goal is to find a reconciliation with minimum total cost [1–3]. The resulting optimization problem is called the *DTL-reconciliation problem*.

The standard formulation of the DTL-reconciliation problem requires the gene tree and the species tree to be rooted. However, while species trees can generally be confidently rooted (using outgroups, for example), gene trees are often difficult to root. As a result, the gene trees used for DTL reconciliation are often unrooted. When provided with an unrooted gene tree, existing DTL-reconciliation algorithms and software first find a root for the unrooted gene tree that yields the minimum reconciliation cost and then use the resulting rooted gene tree for the reconciliation. However, there is a critical flaw in this approach: Many gene trees have multiple optimal roots, and yet, only a single optimal root is randomly chosen to create the rooted gene tree and perform the reconciliation. Here, we perform the first in-depth analysis of the impact of uncertain gene tree rooting on DTL reconciliation and provide the first computational tools to quantify and negate the impact of gene tree rooting uncertainty.

To properly account for rooting uncertainty, we define a *consensus reconciliation*, which summarizes the different reconciliations across all optimal rootings of an unrooted gene tree and makes it possible to identify those aspects of the reconciliation that are conserved across all optimal rootings. We study basic structural properties of consensus reconciliations and analyze a large biological data set of over 4500 gene families from a broadly sampled set of 100 predominantly prokaryotic species [4]. Our analysis focuses on several fundamental aspects of DTL reconciliation with unrooted gene trees including prevalence of multiple optimal rootings, structure of optimal roots in multiply rooted gene trees, impact of gene tree error and evolutionary event costs, information content of consensus reconciliations, and conservation of event and mapping assignments in consensus reconciliations.

Our experimental results show that a large fraction of gene trees have multiple optimal rootings and that gene tree error significantly increases the fraction of multiply rooted gene trees. The prevalence of multiple optimal rootings is also heavily influenced by gene tree size, with smaller gene trees more likely to have multiple optimal roots. An analysis of the placement of optimal roots shows that multiple roots often, but not always, appear clustered together in the same region of the gene tree. This a highly desirable property since it maximizes the information content, or size, of consensus reconciliations and also makes it easier to estimate the "true" root position. A detailed study of the computed consensus reconciliations reveals that most aspects of the reconciliation, i.e., event and mapping assignments, remain conserved across the multiple rootings, showing that unrooted gene trees can be meaningfully reconciled even after accounting for multiple optimal roots. Our analysis also uncovers several interesting patterns in the reconciliations of singly rooted and multiply rooted gene trees.

The results of our experimental analysis have important implications for the application of DTL reconciliation in evolutionary studies, and the techniques introduced in this work make it possible to systematically avoid incorrect evolutionary inferences caused by incorrect or uncertain gene tree rooting. Our tools for computing consensus reconciliations have been implemented into the phylogenetic reconciliation software package RANGER-DTL, freely available from http://compbio.engr.uconn.edu/software/RANGER-DTL/.

References

1. Tofigh, A., Hallett, M.T., Lagergren, J.: Simultaneous identification of duplications and lateral gene transfers. IEEE/ACM Trans. Comput. Biol. Bioinform. **8**(2), 517–535 (2011)
2. Doyon, J.P., Scornavacca, C., Gorbunov, K.Y., Szöllosi, G.J., Ranwez, V., Berry, V.: An efficient algorithm for gene/species trees parsimonious reconciliation with losses, duplications and transfers. In: Tannier, E. (ed.) RECOMB-CG. LNCS, vol. 6398, pp. 93–108. Springer, Berlin (2010)
3. Bansal, M.S., Alm, E.J., Kellis, M.: Efficient algorithms for the reconciliation problem with gene duplication, horizontal transfer and loss. Bioinformatics **28**(12), 283–291(2012)
4. David, L.A., Alm, E.J.: Rapid evolutionary innovation during an archaean genetic expansion. Nature **469**, 93–96 (2011)

Biomedical Event Extraction via Attention-Based Bidirectional Gated Recurrent Unit Networks Utilizing Distributed Representation

Lishuang Li, Jia Wan, Jieqiong Zheng, and Jian Wang

Dalian University of Technology, School of Computer Science and Technology,
Dalian, China
lilishuang314@163.com

The Bacteria Biotope event extraction (BB) task [1] as the one of biomedical event extraction task has been put forward in the BioNLP Shared Task in 2016. The purpose of the BB task is to study the interaction mechanisms of the bacteria with their environment from genetic, phylogenetic and ecology perspectives. The methods based on shallow machine learning methods for BB event extraction need to extract the manual features. However, the construction of complex hand-designed features mainly relies on preferred experience and knowledge. Furthermore, manual efforts may hurt the generalization performance of the system and lead to over-design. Deep learning methods provide an effective way to reduce the number of handcrafted features. But the approaches take all words as equally important and are not able to capture the most important semantic information in a sentence.

In this paper, we propose a novel Bidirectional Gated Recurrent Unit (BGRU) Networks framework based on attention mechanism, using the corpus from the BioNLP'16 Shared Task on BB task. The BGRU networks as a deep learning framework can reduce the number of handcrafted features and the attention mechanism can take advantage of the important information in the sentence. Simultaneously, we employ a biomedical domain-specific word representation training model, which merges relevant biomedical information including stem, chunk, entity and part-of-speech (POS) tags into word embeddings. The system architecture for event extraction based on attention-based BGRU can be summarized in Fig. 1. Firstly, the Shortest Path enclosed Tree (SPT) between two entities is obtained by GENIA Dependency parser (GDEP) [2] and the SPT is extended to the dynamic extended tree (DET) [3], which can accurately encode the input information. Secondly, the DET is mapped to embeddings which are concatenated by the word embeddings, POS embeddings and distance embeddings. Thirdly, a recurrent neural network with attention-based BGRU is established to acquire the hidden layer. Then, the significant information in a sentence is obtained by a weight vector, which could learn word features automatically. Therefore, a sentence feature can be gained by multiplying the weight vector. Lastly, we utilize a softmax function to predict the label for classification.

The experimental results on the BioNLP-ST'16 BB-event corpus show that our attention mechanism and word representation conditioned BGRU can achieve an

F-score of 57.42%. Without using the complex hand-designed features, our system outperforms the previous state-of-the-art BB-event system.

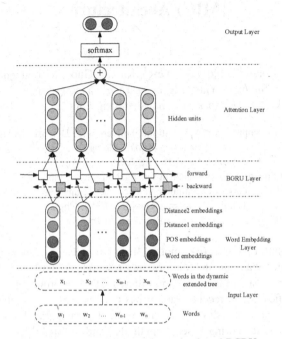

Fig. 1. The architecture of attention-based BGRU

Acknowledgments. The authors gratefully acknowledge the financial support provided by the National Natural Science Foundation of China under No. 61672126, 61173101.

References

1. Deleger, L., Bossy, R., Chaix, E., Ba, M., Ferré, A., Bessieres, P., Nédellec, C.: Overview of the bacteria biotope task at BioNLP shared task. In: Proceedings of the 4th BioNLP Shared Task Workshop, pp. 12–22. The Association for Computational Linguistics, Berlin (2016)
2. Sagae, K., Tsujii, J.I.: Dependency parsing and domain adaptation with LR models and parser ensembles. In: Proceedings of the 2007 Joint Conference on Empirical Methods in Natural Language Processing and Computational Natural Language Learning, pp. 1044–1050. Prague (2007)
3. Li, L., Jin, L., Zheng, J., Zhang, P., Huang, D.: The protein-protein Interaction extraction based on full texts. In: 2014 IEEE International Conference on Bioinformatics and Biomedicine, pp. 497–500. IEEE, Belfast (2014)

Efficient Computation of Motif Discovery on Intel Many Integrated Core (MIC) Architecture

Kaiwen Huang[1], Zhiqiang Zhang[1], Runxin Guo[1], Xiaoyu Zhang[1],
Shunyun Yang[1], Xiangke Liao[1], Yutong Lu[2],
Quan Zou[3], and Shaoliang Peng[1]

[1] School of Computer Science, National University of Defense Technology,
Changsha 410073, China
pengshaoliang@nudt.edu.cn
[2] National Supercomputer Center in Guangzhou, Guangzhou 510275, China
[3] School of Computer Science and Technology,
Tianjin University, Tianjin 300350, China
zouquan@nclab.net

Identifying meaningful patterns (*i.e.*, motifs) from biological sequences is an important problem and a major challenge in bioinformatics research. A motif [1] is a nucleotide or amino-acid sequence pattern that recurs in different DNA or protein sequences and has a biological significance. In recent years, it has emerged a large number of computational algorithms for motif discovery which can be categorized into two groups, including word-based (string-based) methods and probabilistic methods [1]. Word-based methods mostly exhaustive enumerate in their computation and probabilistic methods employ probabilistic sequence models where the model parameters are optimized by maximum-likelihood principle or Bayesian inference. Probabilistic methods have the advantage of few parameters and are more appropriate for finding longer or more general motifs especially for prokaryotes, whose motifs are generally longer than eukaryotes.

MEME (Multiple EM for Motif Elicitation) [2] is one of the currently widely-used algorithms based on maximum-likelihood principle for *de novo* motif discovery [3]. The algorithm consists of two stages: starting point searching and EM. The time complexity of MEME is $O(N^2 \times L^2)$, where N is the number of input sequences and L is the average length of each sequence. However, the high computational cost constrains MEME for handling large datasets [4]. To accelerate motif discovery algorithm, most of previous approaches focus on using parallelization on distributed workstations, Graphics Processing Unit (GPU) and Field Programmable Gate Arrays (FPGA). Farouk *et al.* parallelized the Brute Force algorithm targeted on FPGAs [5]. Marchand *et al.* scaled Dragon Motif Finder (DMF) to IBM Blue Gene/P using mixed-mode MPI-OpenMP programming [6]. mCUDA–MEME is a parallel implementation of MEME running on multiple GPUs using CUDA programming model [7].

Kaiwen Huang, Zhiqiang Zhang, Runxin Guo, Xiaoyu Zhang, Shunyun Yang—These authors contributed equally to this work.

Intel Many Integrated Core (MIC) Architecture [8] is the latest co-processor computer architecture developed by Intel, which combines many Intel processor cores onto a single chip to support the most demanding high-performance computing applications. It is a brand-new many-core architecture that delivers massive thread parallelism, data parallelism, vectorization, and memory bandwidth in a CPU form factor for high throughput workloads.

In this paper, we accelerate MEME algorithm targeted on Intel Many Integrated Core (MIC) Architecture to harness the powerful compute capability of MIC and present a parallel implementation of MEME called MIC-MEME base on hybrid CPU/MIC computing framework. Since the starting point searching stage is the runtime bottleneck of the sequential MEME algorithm, our method focuses on parallelizing the starting point searching method and improving iteration updating strategy of the algorithm. And in EM stage, the M step and E step of EM algorithm are simply parallelized using OpenMP. We also take advantage of the 512 bit vectorization unit to get good performance out of the Intel MIC Architecture.

To evaluate the performance of MIC-MEME, the real datasets with different numbers of sequences and base pairs (bps) were used. MIC-MEME produces the same results as sequential MEME. And it has achieved significant speedups of 26.6 for ZOOPS model and 30.2 for OOPS model on average for the overall runtime when benchmarked on the experimental platform with two Xeon Phi 3120 coprocessors. Furthermore, MIC-MEME shows good scalability with respect to dataset size and the number of MICs. And MIC-MEME has been compared favorably with mCUDA-MEME and BoBro2.0. As the result shows, MIC-MEME is average 2.2 times faster than mCUDA-MEME and MIC-MEME absolutely outperforms BoBro2.0. Comparing with the other methods, we can improve the efficiency of MEME algorithm without losing accuracy and our method which makes full use of computing resources is faster and robustness. With the increase of biological data, we hope the efficient motif discovery of MIC-MEME will be able to help the bioresearch work. Source code can be accessed at https://github.com/hkwkevin28/MIC-MEME.

Acknowledgments. This work was supported by NSFC Grants U1435222, 61625202, 61272056, National Key R&D Program 2016YFC1302500, 2016YFB0200400, and Guangdong Provincial Department of Science and Technology under grant No. 2016B090918122.

References

1. Das, M.K., Dai, H.K.: A survey of DNA motif finding algorithms. BMC Bioinform. **8**, Suppl. 7(7), S21 (2007)
2. Bailey, T.L., Elkan, C., Bailey, T.L., Elkan, C.: Fitting a mixture model by expectation maximization to discover motifs in biopolymers. Proc. Int. Conf. Intell. Syst. Mol. Biol. **2**, 28–36 (1994)
3. Bailey, T.L., et al., MEME: Discovering and analyzing DNA and protein sequence motifs. Nucleic Acids Res. **34**(Web Server issue), 369–373 (2006)

4. Hu, J., Li, B., Kihara, D.: Limitations and potentials of current motif discovery algorithms. Nucleic Acids Res. **33**(33), 4899–4913 (2005)
5. Farouk, Y., Eldeeb, T., Faheem, H.: Massively Parallelized DNA Motif Search on FPGA. InTech. (2011)
6. Marchand, et al.: Highly scalable ab initio genomic motif identification (2011)
7. Liu, Y., Schmidt, B., Maskell, D.L.: An Ultrafast scalable many-core motif discovery algorithm for multiple gpus. In: IEEE International Symposium on Parallel and Distributed Processing Workshops & Phd Forum, pp. 428–434 (2011)
8. Jeffers, J., Reinders, J.: Intel Xeon Phi coprocessor high-performance programming. Morgan Kaufmann Publishers Inc. pp. xvii–xviii (2013)

Predicting Diabetic Retinopathy
and Identifying Interpretable Biomedical
Features Using Machine Learning Algorithms

Hsin-Yi Tsao[1,2], Pei-Ying Chan[3,4], and Emily Chia-Yu Su[1]

[1] Graduate Institute of Biomedical Informatics,
College of Medical Science and Technology,
Taipei Medical University, Taipei, Taiwan
{g658101006,emilysu}@tmu.edu.tw
[2] Division of Endocrinology and Metabolism,
Department of Internal Medicine, Sijhih Cathay General Hospital,
New Taipei City, Taiwan
[3] Department of Occupational Therapy and Healthy Aging Center,
Chang Gung University, Taoyuan, Taiwan
chanp@mail.cgu.edu.tw
[4] Department of Psychiatry, Linkou Chang Gung Memorial Hospital,
Taoyuan, Taiwan

Abstract. Diabetic retinopathy (DR) was found to be a frequent comorbid complication to diabetes. The risk factors of DR were investigated extensively in the past studies, but it remains unknown which risk factors were more associated with the DR than others. If we can detect the DR related risk factors more accurately, we can then exercise early prevention strategies for diabetic retinopathy in the most high-risk population. Thus, using computational approaches to predict diabetes mellitus becomes crucial to support medical decision making.

The purpose of this study is to build a prediction model for the DR in type 2 diabetes mellitus using data mining techniques. First, data consisting of 106 DR and 430 normal patients were collected from the "Diabetes Mellitus Shared Care" database in a private hospital in northern Taiwan. We randomly selected 160 patients were from normal group to combine with DR group, and formed a balanced data set. Ten variables, including systolic blood pressure (SBP), diastolic blood pressure (DPB), body mass index (BMI), age, gender, duration of diabetes, family history of diabetes, self-monitoring blood glucose (SMBG), exercise, and insulin treatment, were extracted. Four machine learning algorithms including support vector machines (SVM), decision trees, artificial neural networks, and logistic regressions, were used to predict diabetic retinopathy.

Among these variables, insulin treatment, SBP, DPB, BMI, age, and duration of diabetes showed significant differences between DR and normal groups. Experimental results demonstrated SVM achieved the best prediction performance with 0.839, 0.795, 0.933, and 0.724 in area under curve, accuracy, sensitivity, and specificity, respectively. The aim of this study is not only to achieve an accurate prediction performance, but also to generate an interpretable model for clinical practice. Table 1 and Fig. 1 demonstrated the interpretable rules generated by logistic regression and decision tree, respectively. Use of insulin and longer duration of DM were major predictors of DR in the decision tree models. If duration of DM increases by 1 year, the odds ratio to have DMR is increased by 9.3%. The odds ratio to have DR is increased by 3.561 times for patients who use insulin compared to patients who do not use insulin. In summary, our method identifies use of insulin and duration of diabetes as novel interpretable features to assist with clinical decisions in identifying the high-risk populations for diabetic retinopathy.

Keywords: Diabetic mellitus retinopathy · Machine learning · Decision support

Table 1. Odds ratio estimates of duration and insulin variables.

Odds ratio estimates		
Effect		Point estimate
Duration		1.093
Insulin	Y vs. N	3.561

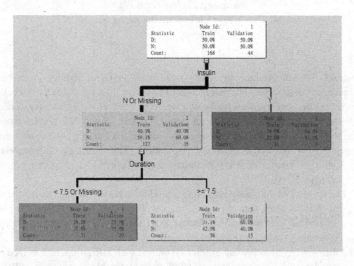

Fig. 1. Interpretable rules for clinical practice generated by decision tress.

Big Data Analysis for Evaluating Bioinvasion Risk

Chenyu Wang[1], Shengling Wang[1], and Liran Ma[2]

[1] College of Information Science and Technology,
Beijing Normal University, Beijing, China
henryascend@gmail.com, wangshengling@bnu.edu.cn
[2] Department of Computer Science, Texas Christian University,
Fort Worth, TX, USA,
l.ma@tcu.edu

Abstract. The global maritime trade makes species get translocated through ballast water and biofouling. We propose a biosecurity triggering mechanism to evaluate the bioinvasion risk of ports. To that aim, we take advantage of big data to compute the invaded risk and construct a species invasion network (SIN). The former is used to evaluate the incoming bioinvasion risk while the latter is employed to estimate the invasion risk spreading capability of a port through s-core decomposition.

1 Introduction

Nowadays, people's daily lives are heavily dependent on global maritime trade. However, marine invasive species and viruses would cause side effects in terms of environment and human health, which lead to huge losses of lives and economy [3].

To address the issue of aquatic bioinvasion, one mainstream countermeasure is to propose suggestions for biomarker identification [1, 2] and bioinvasion management. However, the existing biosecurity suggestions only considered the invaded risk of a port and neglected its role of being a *stepping-stone*.

In this paper, we propose a biosecurity triggering mechanism to address the issues of the existing work. In our biosecurity triggering mechanism, once the bioinvasion risk of a port is larger than a given threshold, biosecurity controls should be triggered. To that aim, we take advantage of the automatic identification system (AIS) data, the ballast water data, and the marine ecoregion data to compute the invasion risk between any two ports, based on which the invaded risk is calculated and a species invasion network (SIN) is constructed. Through s-core decomposition of SIN, the ports whose s-core are higher are identified as the ones transmit bioinvasion risks to others more easily. We found two regions, namely the Western Europe and the Asia-Pacific, which are estimated to be bioinvasion risk intensive regions through our big data analysis.

2 Basis for Our Analysis

For any port j, its invaded risk (i.e. $P_j(Inv)$) is the accumulating invasion risks over all shipping routes passing through it [5], i.e.

$$P_j(Inv) = 1 - \Pi_i[1 - P_{ij}(Inv)] \tag{1}$$

where $P_{ij}(Inv)$ denotes the invasion risk from port i to j.

A SIN can be depicted by a directed graph, namely $S = (V, E, W)$, consisting of a set V of nodes (i.e., ports), a set E of edges (i.e., shipping routes) and the weight $w_{ij} \in W$ ($w_{ij} = P_{ij}(Inv)$) of edge $e_{ij} \in E$ denoting the invasion risk from ports i to j.

According to the description above, both the invaded risk and SIN involve $P_{ij}(Inv)$ ($i, j \in V$). In this paper, we use the model proposed in [5] to calculate $P_{ij}(Inv)$.

To figure out the potential of a port to spread invaded species to others, we need to dig out the transmission power of each node in SIN, which is closely related to the topological property of each port in SIN. We think k-core decomposition is an efficient tool to analyze the structure of complex networks. Larger values of the index k correspond to nodes with larger degree and more central position. According to the algorithm in [4], we can deduce the s-cores of SIN. Seattle, Tokyo and Lima are the top 3 ports ranked by their value of s-shell.

3 Biosecurity Triggering Method

The main idea of the proposed biosecurity triggering method is to trigger bioinvasion treatment according to the bioinvasion risk of each port. As we introduced above, the bioinvasion risk is estimated in light of both the invaded risk of port and its ability of further spreading invaded species. The former is the incoming risk while the latter is the outgoing one. Therefore, we can trigger the corresponding bioinvasion control on a port j based on the following simple criterion:

$$R(j) = \alpha \widetilde{P}_j(Inv) + (1 - \alpha)\widetilde{s}(j) \geq T \tag{2}$$

where $R(j)$ is the bioinvasion risk of port j, and $\widetilde{P}_j(Inv)$ and $\widetilde{s}(j)$ are respectively the normalized $P_j(Inv)$ (the invaded risk of port j calculated using (1)) and the normalized s-shell value of that port; $0 \leq \alpha \leq 1$ is the tradeoff weight. Smaller α means more attention should be paid on the stepping-stone invasion and otherwise, the invaded risk should be obtained more concern. T is the given threshold to help judging whether a bioinvasion treatment should be triggered.

We found two regions, namely the Western Europe and the Asia-Pacific, are bioinvasion risk intensive regions. The result is consistent with the real-world data. Hence, our analysis basically accords with the real-world marine bioinvasion status.

References

1. Cai, Z., Goebel, R., Salavatipour, M.R., Lin, G.: Selecting dissimilar genes for multi-class classification, an application in cancer subtyping. BMC Bioinform. **8**(1), 206 (2007)
2. Cai, Z., Heydari, M., Lin, G.: Iterated local least squares microarray missing value imputation. J. Bioinform. Comput. Biol. **4**(05), 935–957 (2006)
3. Cai, Z., Zhang, T., Wan, X.F.: A computational framework for influenza antigenic cartography. PLoS Comput. Biol. **6**(10), e1000949 (2010)
4. Eidsaa, M., Almaas, E.: S-core network decomposition: A generalization of k-core analysis to weighted networks. Physical Rev. E **88**(6), 062819 (2013)
5. Seebens, H., Gastner, M.T., Blasius, B.: The risk of marine bioinvasion caused by global shipping. Ecology Lett. **16**(6), 782–790 (2013)

Drug Response Prediction Model Using a Component Based Structural Equation Modeling Method

Sungtae Kim[1] and Taesung Park[2]

[1] Interdisciplinary Program in Bioinformatics,
Seoul National University, Seoul, Korea
[2] Department of Statistics, Seoul National University San 56-1,
Sillim-dong, Gwanak-gu, Seoul 151-742, Korea
tspark@stats.snu.ac.kr

The liver is made up of many different types of cells. Mutations in those cells can be developed into several different forms of tumors known as cancers. For this reason, it is hard to expect a single type of liver cancer treatment to have a favorable prognosis for all cancer patients. If we can diagnose and classify the patients who are expected to have good responses to a single therapeutic drug, it will help to reduce the time on choosing an appropriate therapeutic drug for each patient. Therefore, building a decent prediction model became important for an effective treatment. Up to date, several methods such as linear/logistic regression (LR), support vector machine (SVM), random forest (RF) have been used for building prediction models [1–3]. However, occasionally, these methods oversight the biological pathway information with relations between metabolites, proteins, or DNAs.

In this paper, we propose building of prediction model using component based structured equation modeling method which uses the peptide to protein biological structure. Our peptide level data were generated by Multiple Reaction Monitoring (MRM) mass spectrometry for liver cancer patients. MRM is a highly sensitive and selective method for targeted quantitation of peptide abundances in complex biological samples. The advantage of component based structured equation modeling is that it can generate latent variables. These latent variables are not observable but can be inferred from other observed variables. Using latent variables, we can collapse unstructured data into structured data. These latent variables provide more feasible explanation on the results. In our case, multiple peptides can be merged into a protein which is represented as a latent variable. Our proposed schematic model using component based structural equation modeling for MRM data is shown in Fig. 1.

We applied the component based structural equation model to MRM data of liver cancer patients. In our MRM data, there are 124 proteins induced by 231 peptides MRM data. Each protein contains at least one peptides. We identified candidate proteins for a drug Sorafenib response for liver cancer patients. The selected candidate proteins included APOC4, CD163, CD5L, JCHAIN, SERPING1, and RBP4. These proteins were reported as possible cancer biomarkers [4, 5]. Also, CD5L was well known as a liver cancer biomarker [6, 7]. Using these proteins, we evaluated our proposed Sorafeib prediction model by the area under the curve (AUC) score. Also, we compared the performance of our model with generalized linear models with and without ridge penalty. The performance of our model showed a slightly higher AUC score 0.96 compared to 0.949 AUC score of the generalized linear model with ridge penalty.

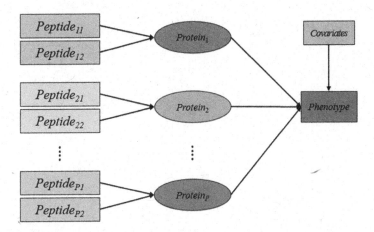

Fig. 1. Proposed component based structural equation model using MRM data.

References

1. Visser, H., le Cessie, S., Vos, K., Breedveld, F.C., Hazes, J.M.: How to diagnose rheumatoid arthritis early: a prediction model for persistent (erosive) arthritis. Arthritis Rheum. **46**(2), 357–365 (2002)
2. Spitz, M.R., Etzel, C.J., Dong, Q., Amos, C.I., Wei, Q., Wu, X., Hong, W.K.: An expanded risk prediction model for lung cancer. Cancer Prev. Res **1**(4), 250–254 (2008)
3. Huang, C.L., Liao, H.C., Chen, M.C.: Prediction model building and feature selection with support vector machines in breast cancer diagnosis. Expert Syst. Appl. **34**(1), 578–587 (2008)
4. Gray, J., Chattopadhyay, D., Beale, G.S., Patman, G.L., Miele, L., King, B.P., Reeves, H.L.: A proteomic strategy to identify novel serum biomarkers for liver cirrhosis and hepatocellular cancer in individuals with fatty liver disease. BMC Cancer **9**(1), 271 (2009)
5. Braconi, C., Meng, F., Swenson, E., Khrapenko, L., Huang, N., Patel, T.: Candidate thera-peutic agents for hepatocellular cancer can be identified from phenotype associated gene expression signatures. Cancer **115**(16), 3738–3748 (2009)
6. Chambers, A.G., Percy, A.J., Simon, R., Borchers, C.H.: MRM for the verification of cancer biomarker proteins: recent applications to human plasma and serum. Expert Rev. Proteomics **11**(2), 137–148 (2014)
7. Rabouhans, J.: A radiologist's guide to the modified Response Evaluation Criteria in Solid Tumours (mRECIST) assessment of therapy for hepatocellular carcinoma. European Congress of Radiology (2011)

Contents

Prediction of Time to Tumor Recurrence in Ovarian Cancer: Comparison of Three Sparse Regression Methods

Mahsa Lotfi[1]($\boxtimes$), Burook Misganaw[2], and Mathukumalli Vidyasagar[1]

[1] University of Texas at Dallas, Richardson, TX 75080, USA
{mahsa.lotfi,m.vidyasagar}@utdallas.edu
[2] Harvard University, Cambridge, MA 02138, USA
burook@g.harvard.edu

Abstract. Ovarian cancer is the most fatal gynecological malignancy among women. Making a reliable prediction of time to tumor recurrence would be a valuable contribution to post-surgery follow-up care. In this paper we study three well-known data sets, known as TCGA, Tothill and Yoshihara, and compare three sparse regression methods, two of which (LASSO and EN) are well-known and the third (CLOT) is from our laboratory. It is established that the three data sets are very different from each other. Therefore a two-stage predictor is built, whereby each test sample is first assigned to the most likely data set and then the corresponding predictor is used. The weighted concordance of each regression method is computed to compare the methods and select the best one. CLOT uses a biomarker panel of 103 genes and achieves a concordance index of 0.7829, which is higher than that achieved by the other two methods.

Keywords: Ovarian cancer · Sparse regression · LASSO · Elastic Net · CLOT · Concordance index

1 Introduction

Ovarian cancer is the fourth most common cause of cancer deaths around the world [1]. It is considered to be highly responsive to the first treatment, but it has very low long-term survival rate because the patient develops drug-resistance [2], which causes the tumor to recur (also called tumor regression). Recent studies have identified several independent prognostic biomarkers such as age at diagnosis, histologic cell-type, stage, histologic grade, FIGO[1] stage, residual tumor size, presence of ascites, albumin, alkaline phosphatase, preoperative serum CA-125, performance stage and other markers [3,4]. Most of the relevant studies in the case of predicting the survival analysis of ovarian cancer have concentrated on molecular markers. However, known ovarian cancer molecular factors are not

---

[1] International Federation of Gynecology and Obstetrics.

© Springer International Publishing AG 2017
Z. Cai et al. (Eds.): ISBRA 2017, LNBI 10330, pp. 1–11, 2017.
DOI: 10.1007/978-3-319-59575-7_1

predictive enough at least for clinical use. Therefore, factors that are independent from clinical parameters are required to achieve better predictions.

Note that the time to tumor recurrence is not always precisely known and depends on the intervals between successive check-ups. In [5], for this reason ovarian cancer patients are grouped into just three categories, namely super responders, medium responders and non-responders, depending on their progression-free survival, which is the same as the time to tumor recurrence. A panel of 25 genes is identified that can classify the patients into one of these three categories.

The present study is more ambitious in that we aim to achieve, not just a coarse classification of patients into three groups, but an actual prediction in terms of the number of days before the tumor recurs. This is achieved by treating the problem as one of sparse regression and not classification. Note that in the world of machine learning, the phrase "regression" basically means "curve-fitting" and has nothing to do with tumor regression. We apply three different sparse regression algorithms, namely the well-known LASSO [6] and Elastic Net [7], together with another one due to this research group called CLOT [8], to predict the time to tumor recurrence.

In [5], the performance of the classification algorithm was evaluated by computing the P-value of the associated 3×3 contingency table. However, when the prediction is a real number rather than just a label, it is more natural to compute the so-called concordance index. The concordance index (C-index hereafter) indicates the probability that a patient that is predicted to be at lower risk of tumor recurrence, survives longer than the other. A C-index of 1 (unachievable in practice or unlikely to occur with real data) indicates perfect performance, while a C-index of 0.5 corresponds to predictions being generated at random. A comprehensive study in [9] shows that most of the currently available predictors for late stage ovarian cancer achieve C-indices of only around 0.6 on independent test sets.

In the present study, we analyzed three different datasets namely the TCGA ovarian cancer dataset, GSE9891 also known as the Tothill dataset and GSE17260 also known as the Yoshihara data set. The usual approach to train and test predictors across data sets is to convert all values to Z-scores. Our analysis indicated that these three data sets are fundamentally different from each other. Specifically, after transformation to Z-scores, for each data set we computed the vector of median values of each gene, that is, a vector dimension 12,229 and for each test sample, we assigned it to the nearest vector of median values in terms of Euclidean distance. In all but three cases (3 out of 282 test samples), the test sample was assigned to the correct data set. Therefore it is unlikely that any predictor based on one of these sets would perform satisfactorily on the other two. This expectation was then verified via numerical computation. Therefore we opted to develop a two-stage prediction process. Three different training sets were identified, one from each data set, and three predictors were developed, one for each data set. The test set consisted of all the remaining samples from all three data sets. To test, a pre-processor assigned each test sample to the most likely data set, and then the corresponding predictor was applied. Through this approach, we are able to achieve a C-index of 0.7829 using the CLOT regressor and a biomarker panel of 103 genes. The CLOT regressor outperformed the well-known LASSO and EN approaches.

2   Materials and Methods

2.1   Datasets

In this study, gene expression datasets were downloaded from the Gene Expression Omnibus (GEO) [10] and The Cancer Genome Atlas (TCGA) websites [11]. More than 80% of ovarian cancer samples used in this study are in stage 3 and grade 3 which indicates that they are mostly extended to the lining of pelvis and lymph nodes and have a tendency to grow quickly [12].

In our project, there are totally three datasets that are used for the training and the validation of the regression models. Details about these datasets are shown in Table 1. It must be noted that samples for which time to tumor recurrence is less than thirty days are removed from the datasets, as these patients were very sick and thus not representative of the overall patient population. In addition, in the dataset GSE9891 (Tothill dataset), there are about 100 samples with time to tumor recurrence value equal to the time of death. This suggests that the clinicians lost track of the patient; therefore these samples were removed from GSE9891 dataset.

According to Table 1, the number of genes is different in different datasets. Therefore, 12249 genes that were common to all three datasets were identified and the rest were removed. Within each dataset, 70% of the samples were used for training and the remaining 30% were kept segregated as test samples. The regression algorithms never see these test samples, so that they constitute an "independent" test set.

Table 1. Details of data sets used

Datasets	Platform	No. of genes	Total samples	Train samples	Test samples	Pairs
TCGA	Affy. HT 133A	13104	512	300	212	22366
GSE9891	Affy. U133 Plus 2.0	19816	169	119	50	1225
GSE17260	Agilent 4112a	20106	105	85	20	190

2.2   Solution Methodology

The problem under study is to predict the time to tumor recurrence in ovarian cancer datasets by using sparse regression methods. Let m denote the number of tumor samples and n the number of genes whose expression levels are measured in each tumor. Let $A \in \mathbb{R}^{m \times n}$ denote the matrix of gene expression values, and $y \in \mathbb{R}^m$ the vector of time to tumor recurrence for each sample. The objective is to find a vector $z \in \mathbb{R}^n$ which is extremely sparse such that Az nearly equals y. In biological data, the number of genes outnumbers the number of samples $(m \ll n)$. Since $m \ll n$, the equation $y = Az$ is under-determined and has infinitely many solutions. The best solution is achieved when the weight vector z has the fewest number of nonzero components, or equivalently, $\|z\|_0$ is minimized while satisfying the equation $y = Az$. However, it is shown in [13] that the minimization of $\|z\|_0$ subject to $y = Az$ is NP-hard. Hence alternate methods based on convex optimization are used instead to find sparse solutions for x. The most popular approach is to define a "regularizer" $R(\cdot)$ that penalizes large vectors z and to minimize

$$\hat{z} = \operatorname*{argmin}_{z} \left(\|y - Az - b\|_2^2 + \lambda R(z) \right) \tag{1}$$

Different choices of the regularizer lead to different solutions. One of the most widely used approaches is known as LASSO, introduced in [6]. Another popular algorithm for sparse regression is the Elastic Net (EN) algorithm introduced in [7]. The Elastic Net algorithm has the "grouping effect" property which means that if two columns of the matrix A are highly correlated, then the corresponding components of $\hat{z}$ are nearly equal. In contrast, in such a situation LASSO chooses one of the highly correlated features and discards the rest. Consequently the final set of features selected is very sensitive to small changes in y. Therefore the Elastic Net approach is particularly useful when there are many correlated predicted variables. However, in compressed sensing applications, where $y = Ax$ and x is a "true but unknown" sparse vector, the LASSO formulation is able to recover x, while EN cannot.

A recent formulation from our research group, known as CLOT is shown to combine the desirable attributes of both LASSO and Elastic Net [8]. In other words, it is shown that CLOT has "grouping effect" and it also achieves robust sparse recovery if the measurement matrix satisfies the Restricted Isometry Property (RIP) [8]. This led us to compare all of these three approaches in the prediction of time to tumor recurrence in Ovarian cancer. The LASSO regularizer is defined by

$$R_{LASSO}(z) = \|z\|_1, \tag{2}$$

The Elastic Net regularizer is defined by

$$R_{EN}(z) = (1 - \mu)\|z\|_1 + \mu\|z\|_2^2, \tag{3}$$

where $0 < \mu < 1$ is an adjustable parameter. The CLOT regularizer is defined by

$$R_{CLOT}(z) = (1 - \mu)\|z\|_1 + \mu\|z\|_2, \tag{4}$$

where $0 < \mu < 1$ is an adjustable parameter.

The Elastic Net and CLOT regularizers are more robust to the variations in the measurement vector than LASSO. Moreover, it is recommended to add a recursive feature elimination step to EN and CLOT algorithms in order to reduce further the dimensionality of the weight vector. Therefore the full algorithm is as described next: Let a_j denote the j-th column of the matrix A and y denote the time to tumor recurrence vector. Then the procedure is as follows:

1. Sort the y vector in an ascending order and sort the rows of A correspondingly.
2. Normalize A by subtracting the mean of each column then dividing by its standard deviation.
3. Run the optimizer (LASSO, EN or CLOT) and obtain a sparse vector z and bias b.
4. Once the predictor is obtained in this manner, validate it on the test set. For each tumor, obtain its predicted time to recurrence by multiplying its gene expression vector by z and adding b. Equation 5 shows how the predicted vector p is formed by the gene expression matrix A for all tumor samples. Note that, since z is sparse, only a few gene expression values are used in the prediction.

$$p = Az + b \qquad (5)$$

5. To assess the performance of the predictor, compute the concordance index of the predictions using the pseudocode in Algorithm 1:

Algorithm 1. Computing the Concordance Index

Assume: p =vector of predicted values, $S = 0$, M = sample size
for $i = 1 : M$ do
   for $j = i + 1 : M$ do
      if $p(j) > p(i)$ then
         $S = S + 1$
      end if
   end for
end for
C-index $= \frac{2S}{M(M-1)}$

The CLOT optimization is carried out in Matlab using the public domain CVX optimization package.

2.3 Parameter Tuning

No matter which regularizer is used for sparse regression, the Lagrangian form of the optimization problem to be solved is

$$\min_z (\|y - Az\|_2^2 + \lambda R(z)) \qquad (6)$$

where λ is a tuning parameter. It is important to notice that high sparsity of z, low error percentage on the training data and high concordance index are the three most crucial criteria for assessing a predictor. The general approach is to divide the training data into two parts via cross-validation and to use prediction error of the cross-validation as a guide to choose the best values for regression parameters [14]. Accordingly, in this paper, 10 fold cross-validation was used and the parameter λ was varied over 50 different values. Then the λ value that led to a good combination of all three parameters (high sparsity, low error and high concordance) was chosen. Our target was to achieve around 10% training error and to find a λ value that led to the fewest number of features for each dataset. As λ increases, fewer features are chosen (higher sparsity), but the error percentage value increases. It must be noted that in order to have completely independent test samples (that were never seen before by the regressors), we applied data splitting method to do the final training and testing.

In the case of EN and CLOT, there are two adjustable parameters, namely $\mu \in [0,1]$ and λ. In principle we could have varied both parameters over a suitable grid. However, to reduce computation time, we fixed λ at a value close to the optimal value generated from applying LASSO (less than 10% training error with the selected λ) and used the same value of λ in EN as well. Then we varied μ over [0,1] to find the best value that gives us the highest concordance index on the training data. Figure 1 shows the concordance indices for different μ values on GSE17260 dataset. The best μ is the one which has the greatest concordance index among all. It must be noted that $\mu \approx 0.2$ has the highest training concordance values in all datasets and 0.2 is chosen as the final value for μ.

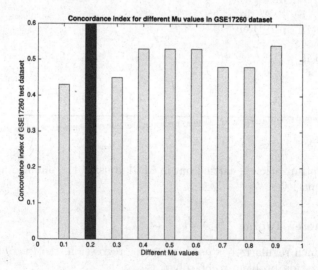

Fig. 1. Concordance indices for different μ values in GSE17260 dataset

3   Experimental Results

The main challenge in applying machine learning techniques to ovarian cancer is that there are very few validation data sets, in contrast with other forms of cancer, such as breast cancer. We have used three datasets, namely TCGA, Tothill (GSE9891) and Yoshihara (GSE17260). These datasets represent patient populations from the USA, Australia and Japan, respectively.

3.1   Training on TCGA and Testing on the Rest

To assess how different these training datasets are from each other, we performed a simple test. Specifically, we computed the median for each feature across all samples within a dataset, leading to three different median vectors of dimension $12{,}229 \times 1$. Then each sample was compared to all three vectors using the Euclidean distance. It was found that all but three samples were closest in Euclidean distance to the median vector of the data set to which they belonged, than to the other two median vectors. This showed that there is near complete separation between the three data sets. Consequently, we believed that it was unlikely that a predictor trained on the TCGA dataset would perform well on the other datasets. Nevertheless, we carried out an exercise of training a regressor using 300 TCGA samples and testing it on 212 TCGA samples and all samples from the Tothill and Yoshihara datasets. Recall that all data vectors were converted to Z-scores as a first step. The results are presented in Table 2, where C denotes the concordance index.

3.2   Multiple Regressor Approach with a Front-End Pre-processor

Since, as expected, the regressors trained using only TCGA samples did not produce good predictions when tested on other datasets, we decided to adopt a different approach. We trained different regressors for each of the three datasets. For testing purposes, all of the test samples were combined into one. Then the expression values of all genes for each test sample were compared to the median gene expression vector from each of the three datasets and the test sample was assigned to the dataset whose median vector was the closest in terms of Euclidean distance. Then the regressor trained on that dataset was applied. It is worth mentioning that each test sample was assigned to the correct dataset by the preprocessing step. This highlights that the three datasets are quite different from each other. In this manner, we developed a multi-regressor approach with a front-end pre-processor and tested each regressor with samples from the same dataset. The results of this experiment are shown in Table 3.

4   Comparison of the Three Methods

According to Table 2, all of the three regression methods, LASSO, EN and CLOT fail to have high C-indices when trained with TCGA and tested with the rest

Table 2. C-indices and number of features in LASSO, EN and CLOT regressors trained with 300 TCGA samples and tested with the rest of the samples

Regressors	No. of features	Training C	TCGA C	GSE9891 C	GSE17260 C
LASSO	260	0.9501	0.5203	0.5421	0.5942
EN	106	0.9324	0.4851	0.5213	0.4623
CLOT	211	0.9334	0.8123	0.4733	0.4512

Table 3. C-indices, number of features and λ in LASSO, EN and CLOT regressors, trained with 70% of each dataset and tested with the remaining part

	Regressor	Dataset	No. of features	λ	Training C	Testing C
LASSO	1	TCGA	260	7.6	0.9501	0.5203
	2	GSE9891	110	8.9	0.9622	0.6392
	3	GSE17260	74	16.6	0.9601	0.5210
EN	1	TCGA	106	10	0.9324	0.4851
	2	GSE9891	241	10	0.9697	0.6122
	3	GSE17260	36	10	0.9151	0.5316
CLOT	1	TCGA	211	10	0.9334	0.8123
	2	GSE9891	127	10	0.9099	0.5112
	3	GSE17260	50	10	0.9332	0.6123

Table 4. Comparison of C-indices and number of features in different methods

Method	Weighted mean (by no. of samples)	Weighted mean (by no. of sample pairs)	Mean of the no. of features
One LASSO reg.	0.5135	0.7167	260
One EN reg.	0.4898	0.4868	106
One CLOT reg.	0.5028	0.7802	65
Three LASSO reg.	0.5262	0.5074	119
Three EN reg.	0.5110	0.4920	128
Three CLOT reg.	**0.7326**	**0.7829**	**103**

(non-TCGA samples). Even testing with TCGA samples only has a high concordance index when CLOT is used (C-index = 0.8123) but using LASSO and EN still generates low concordance indices in this case.

Results of Table 3 indicate that Elastic Net requires fewer features when applied to TCGA and GSE17260 datasets but if GSE9891 is used as the training and test set, EN has the biggest number of features in comparison to LASSO and CLOT. Overall, by considering the results in Table 3, it seems clear that CLOT offers better results, both in terms of high sparsity (small number of

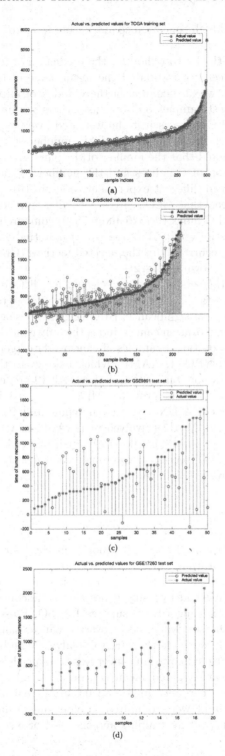

Fig. 2. Actual vs. predicted time values using CLOT for: (a) TCGA training set, (b) TCGA testing set, (c) GSE9891 (Tothill) testing set, (d) GSE17260 (Yoshihara) testing set

features selected) and high concordance index, compared to the other two methods, LASSO and EN.

In order to select the best method for the prediction of time to tumor recurrence in ovarian cancer, we computed the mean value of the final number of features. In addition, we also computed the. C-index of all test samples. However, we compared test samples only against others within the same dataset. Therefore the overall C-index is just the weighted average of the C-indices on individual datasets, where the weight equals the number of test pairs within that dataset. It must be noted that the number of the pairs in each dataset is shown in Table 1. Comparison of C-indices and the number of the features (sparsity in the weight vector) in different experiments done in this paper are shown in Table 4. It can be seen that CLOT gives a smaller number of features (103) and a higher weighted C-index (0.7326 and 0.7829) compared to LASSO (0.5262 and 0.5074) and Elastic Net (0.5110 and 0.4920). Actual versus predicted values of time to tumor recurrence for the selected method (CLOT regression) and different datasets are shown in Fig. 2.

Now we analyze the sets of features selected by all three CLOT regressors across the three data sets TCGA, GSE9891 and GSE17260, to see whether there are any genes of biological significance. BRCA2 gene is recognized as one of the most important genes in ovarian cancer and is the only gene detected by all three CLOT regressors. In addition, examples of cancer-related genes detected by our CLOT regressors are: BRCA1, AAMP (which belongs to the immunoglobulin superfamily and is functional in cell migration), AKT1 (regulating metabolism, proliferation, cell survival and growth), MLH3 (involved in DNA mismatch repair), PMS1 (involved in DNA mismatch repair and PMS1 mutations can cause colorectal cancer), MSH2 (involved in DNA mismatch repair) and NBN (involved in DNA double-strand break, DNA repair and cell cycle control) genes [15]. Therefore the CLOT regressors have succeeded in unearthing many genes that are known to have a role in cancer.

5 Conclusions

In this paper, we introduced a new prognostic method for predicting time to tumor recurrence in ovarian cancer. The recurrence time has generally been assessed on clinical biomarkers therefore novelties in this paper comprise an improved prediction method by using genes as the independent variables. We applied sparse regression algorithms such as LASSO, Elastic Net and CLOT in order to predict the survival time in ovarian cancer samples. Concordance index was computed to evaluate the survival analysis. The best method for the prediction of survival time in ovarian cancer was training three CLOT regressors which achieved a great weighted mean concordance index of 0.7829. This is far higher than the concordance indices of 0.6 or thereabouts achieved in [9]. However, the indices in [9] were computed on independent datasets, whereas we used a part of each dataset for training and the rest for testing.

References

1. Aziz, A.B., Najmi, N.: Is risk malignancy index a useful tool for predicting malignant ovarian masses in developing countries? J. Obstet. Gynecol. Int. (2015). doi:10.1155/2015/951256
2. Hennessey, B.T., Coleman, R.L., Markman, M.: Ovarian cancer. J. Lancet **374**, 1371–1382 (2009). doi:10.1016/S0140-6736(09)61338-6
3. Clark, T.G., Stewart, M.E., Altman, D.G., Gabra, H., Smyth, J.F.: A prognostic model for ovarian cancer. Br. J. Cancer **85**, 944–952 (2001). doi:10.1038/sj.bjc. 6692030
4. Teramukai, S., Ochuau, K., Tada, H., Fukushima, M.: PIEPOC: a new prognostic index for advanced epithelial ovarian cancer? Japan Multinational Trial Organization OC01-01. J. Clin. Oncol. **25**, 3302–3306 (2007). doi:10.1200/JCO.2007.11. 0114
5. Misganaw, B., Ahsen, E., Singh, N., Baggerly, K.A., Unruh, A., White, M.A., Vidyasagar, M.: Optimized prediction of extreme treatment outcomes in ovarian cancer. J. Cancer Inform. **14**, 45–55 (2016). doi:10.4137/CIN.S30803
6. Tibshirani, R.: Regression shrinkage and selection via the lasso. J. Roy. Stat. Soc. **58**, 267–288 (1996). doi:10.1111/j.1467-9868.2011.00771.x
7. Zou, H., Hastie, T.: Regularization and variable selection via the elastic net. J. Roy. Stat. Soc. **67**(2), 301–320 (2005). doi:10.1111/j.1467-9868.2005.00503.x
8. Ahsen, M.E., Challapalli, N., Vidyasagar, M.: Two new approaches to compressed sensing exhibiting both robust sparse recovery and the grouping effect. arXiv. 1410.8229 (2016)
9. Waldron, L., Kains, B.H., Culhane, A.C., Riester, M., Ding, J., Wang, X.V., Ahmadifar, M., Tyekucheva, S., Bernau, C., Risch, T., Ganzfried, B.F., Huttenhower, C., Birrer, M., Parmigiani, G.: Comparative meta-analysis of prognostic gene signatures for late-stage ovarian cancer. J. Natl. Cancer Inst. **106**(5) (2014). doi:10.1093/jnci/dju049
10. Gene Expression Omnibus. https://www.ncbi.nlm.nih.gov/geo
11. The Cancer Genome Atlas. https://gdc-portal.nci.nih.gov
12. Greene, F.L., Page, D.L., Fleming, I.D., Fritz, A.G., Balch, C.M., Haller, D.G., Morrow, M.: AJCC Cancer Staging Manual. Springer, New York (2010)
13. Natarajan, B.K.: Sparse approximate solutions to linear system. SIAM J. Comput. **24**, 227–234 (1995). doi:10.1137/S0097539792240406
14. Friedman, J., Hastie, T., Tibshirani, R.: Regularization paths for generalized linear models via coordinate descent. J. Stat. Softw. **33**(1), 1–22 (2010)
15. The Human Gene Database. https://genecards.org

Histopathological Diagnosis for Viable and Non-viable Tumor Prediction for Osteosarcoma Using Convolutional Neural Network

Rashika Mishra[1(⊠)], Ovidiu Daescu[1], Patrick Leavey[2], Dinesh Rakheja[2], and Anita Sengupta[2]

[1] Department of Computer Science, University of Texas at Dallas, Richardson, TX, USA
{rxm156430,daescu}@utdallas.edu

[2] University of Texas Southwestern Medical Center, Dallas, TX, USA
{patrick.leavey,Dinesh.Rakheja}@UTSouthwestern.edu,
Anita.Sengupta@childrens.com

Abstract. Pathologists often deal with high complexity and sometimes disagreement over Osteosarcoma tumor classification due to cellular heterogeneity in the dataset. Segmentation and classification of histology tissue in H&E stained tumor image datasets is challenging due to intra-class variations and inter-class similarity, crowded context, and noisy data. In recent years, deep learning approaches have led to encouraging results in breast cancer and prostate cancer analysis. In this paper, we propose a Convolutional neural network (CNN) as a tool to improve efficiency and accuracy of Osteosarcoma tumor classification into tumor classes (viable tumor, necrosis) vs non-tumor. The proposed CNN architecture contains five learned layers: three convolutional layers interspersed with max pooling layers for feature extraction and two fully-connected layers with data augmentation strategies to boost performance. We conclude that the use of neural network can assure high accuracy and efficiency in Osteosarcoma classification.

Keywords: Osteosarcoma · Convolutional neural network · Histology image analysis

1 Introduction

Unlike other types of tumor, osteosarcoma has a high degree of heterogeneity, as illustrated in Fig. 1 which makes it difficult in some cases to reach a common diagnostic among pathologists [4,13]. Therefore, automating the analysis of different types of tumor can help to avoid observer bias, reduce diagnosis time, and explore various options for treatment.

Majority of tumor studies rely on Haematoxylin and Eosin (H&E) stain stained images [6], that dye the nuclei blue and background tissues pink in a histology slide. Currently, pathologists must manually evaluate these slides under a

© Springer International Publishing AG 2017
Z. Cai et al. (Eds.): ISBRA 2017, LNBI 10330, pp. 12–23, 2017.
DOI: 10.1007/978-3-319-59575-7_2

microscope to evaluate the extent of tumor and tumor necrosis. A study on renal cell carcinoma [5] found that there was large disagreement between pathologists, on same data samples.

This manual analysis by Pathologists is a labor-intensive process and subject to observer bias. Hence, it is desirable to develop an automatic approach for histopathological slide classification of Osteosarcoma. The whole slide scanning systems provides the opportunity to automate the analysis process. These systems digitize glass slides with the stained tissue at a high resolution (up to 40x). The digital whole slide images (WSIs) allow image processing and analysis techniques by utilizing the morphological and contextual clues present in the WSI as features for tissue classification [7,8].

However, there are several roadblocks towards a fully automatic system. The digital image quality is effected by slide preparation and poor staining response which can cause many tissue and cellular regions to be under-represented.

This diverse cellular morphology resulting in variability in same type of cells (Fig. 1a) and similarity in different cellular structures (Fig. 1b) can make classification of tumor slides challenging. Particularly in Osteosarcoma, both the tumor cells and some types of normal cells (precursor cells) are stained the same blue color but the tumor cells are irregular in shape whereas the precursor cells are more round, close and regular (Fig. 1c). Moreover, each tumor type is significantly different from other types, which makes it difficult to apply one method developed for one tumor type to another tumor type. Osteosarcoma is one such tumor that has a high degree of intra-tumor histological variability and thus methods developed for lung or renal tumor types [5,17] do not work well for it.

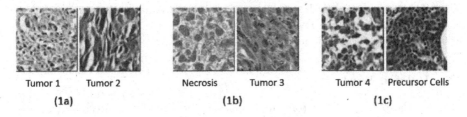

Tumor 1   Tumor 2   Necrosis   Tumor 3   Tumor 4   Precursor Cells

(1a)   (1b)   (1c)

Fig. 1. Examples showing the complexity of dataset. (a) Shows intra class variance for Tumor class. (b) Shows inter class similarity between tumor and necrosis classes. (c) Shows the similarity in color of tumor cells and precursor cells. (Color figure online)

In this paper, we propose a convolution neural network (CNN) architecture to classify the H&E stained histopathology slides of Osteosarcoma. The typical CNN architecture for image processing consists of a series of layers of convolution filters, interspersed with pooling layers. The convolution filters are applied to small patches of the input image to detect and extract image features. Our neural network architecture combines features of AlexNet [9] and LeNet [10] to develop a fast and accurate slide classification system. The proposed system do not require nuclei segmentation which can be a difficult task due to the morphological and system limitations mentioned above. The proposed system works with the annotated image label to generate features at class level. As the paper does not aim to calculate the nuclei properties, we can focus on accurate and efficient class label identification.

Our Contribution. Our main contribution is a new, important, practical and efficient application of CNN, which gives promising results in Osteosarcoma image classification. We developed an efficient CNN architecture used to classify the input images into tumor classes through the use of data augmentation techniques that save time and space. We also provide comparative results of our proposed architecture with existing architectures AlexNet and Lenet to show that the proposed architecture performs better in tumor classification.

1.1 Background

Osteosarcoma is a type of bone cancer. The tumor usually arises in the long bones of the extremities in the metaphyses, next to the growth plates. In order to gauge the extent of treatment response and accurately calculate the percentage of tumor necrosis, it is necessary to consider different histological regions such as clusters of nuclei, fibrous tissues, blood cells, calcified bone segments, marrow cells, adipocytes, osteoblasts, osteoclasts, haemorrhagic tumor, cartilage, precursors, growth plates and osteoid (tumor osteoid and reactive osteoid) with and without cellular material. The goal of this paper is to utilize CNN to identify the four regions of interest (Fig. 2), namely, (1) Viable tumor, (2) Coagulative necrosis, (3) fibrosis or osteoid, and (4) Non tumor (Bone, cartilage) These four regions are used to extract information about the three main classes of interest: viable tumor, necrosis (coagulative necrosis, osteiod, and fibrosis), and other tissue (bone, blood vessels, cartilage, etc.).

1.2 Related Work

Most of the existing work for tumor classification involves thresholding with region growing, k means, otsu, and morphological features like area and shape structures. Arunachalam *et al.* [2] presented multi-level otsu thresholding followed by shape segmentation to identify viable tumor, necrosis and no-tumor regions in osteosarcoma histology slides. Malon *et al.* [12] trained a convolution neural network to classify mitotic and non-mitotic cells using morphological features like color, texture, and shape.

In recent years, machine learning approaches like neural networks have been used for image classification and segmentation but majority of the tumor studies focus on identifying a super-set of features, although not all features are relevant. A recent study on non-small cell lung cancer [17] isolated 9000+ features from images, that consisted of parameters extracted from color, texture, object identification, granularity, density etc. Ciresan *et al.* [3] was the pioneer of utilizing Convolutional Neural Network (CNN) in mitosis counting for primary breast

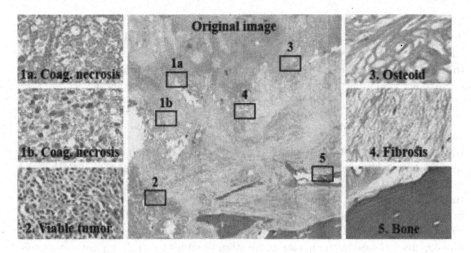

Fig. 2. The figure shows different regions of interest: viable tumor, coagulative necrosis, osteoid, fibrosis, non-tumor (bone) regions in a slide

cancer grading. Litjens *et al.* [11] applied CNN for identifying breast cancer metastases in sentinel lymph nodes and prostate cancer detection.

Many of these methods are focused on nuclei segmentation and not on image classification as tumor or non tumor. Recent studies have proved that deep learning methods are successful in nuclei segmentation and give promising results for image classification. Su *et al.* [16] used a fast scanning deep convolution neural network for region segmentation and classification in breast cancer and Spanhol *et al.* [15] developed on existing AlexNet for different segmentation and classification tasks in breast cancer.

In summary, deep learning algorithms have been successfully implemented in the past for tumor detection in breast cancer and prostate cancer but the work mentioned above is focused on nuclei segmentation whereas evaluation on the classification into tumor classes is limited.

In this paper, we propose a deep learning approach capable of assigning tumor classes (viable tumor, necrosis) vs non-tumor directly to input slides in osteosarcoma, a type of cancer with significantly more variability in tumor description. We extend the successful Alexnet proposed by Krizhevsky (see [9]) and LeNet network architectures introduced by LeCun (see [10]) which uses gradient based learning with back propogation algorithm.

2   Our Approach

2.1   Convolutional Neural Network

Convolutional neural networks (CNNs) are powerful tools in deep learning with high success rate in image classification. The typical CNN architecture for image classification consists of a series of convolution filters paired with pooling layers.

The convolution filters are applied to small patches of the input image which detect increasingly relevant image features like edges or shapes and texture. The output of the CNN is one or more probabilities or class labels. According to Sirinukunwattana *et al.* [14] "Mathematically, CNN can be defined as a feed forward artificial neural network C which is composed of L layers $(C_1, C_2, ..., C_L)$ which maps an input vector x to an output vector y i.e.

$$y = f(x; w_1, w_2, ..., w_L) = f_L(; w_L) \circ f_{L-1}(; w_{L-1}) \circ \cdots \circ f_1(x; w_1) \quad (1)$$

where w_l is the weight and bias vector for the lth layer f_l."

Our approach is conceptually simple. It directly operates on raw RGB data sampled from the source. It is trained to classify patches into three bins: viable tumor, necrosis (coagulative necrosis, osteoid, fibrosis) and non-tumor. Classification in unseen images is done by applying the learned classifier as a sliding window to the data. Because the CNN operates on raw pixel values, no human input is needed beside the initial annotation of slides for training data, a significant advantage over previous attempts [2]. The CNN automatically learns a set of visual features from the training data.

We develop on existing proven networks LeNet and AlexNet because finding a successful network configuration for a given problem can be a difficult challenge given the total number of possible configurations that can be defined. The Lenet architectures [10] have been prototypes for many successful applications in image processing, particularly handwriting recognition and face detection. The data augmentation methods to reduce over-fitting on image data as described by Krizhevsky [9] has been proclaimed for its success rate in various object recognition applications.

2.2   CNN Architecture

CNN Design. Designing the architecture of a neural network is a complex task. We start with a simple 3 layer network [INPUT - CONVOLUTION - MAX POOL - MLP].

(1) INPUT [$128 \times 128 \times 3$] will hold the raw pixel values of the image, i.e. an image of width 128, height 128, and with three color channels R,G,B.

(2) CONVOLUTION layer will compute the output of neurons that are connected to local regions in the input image. Each neuron will compute the dot product between their weights and a small region that they are connected to in the input volume. This may result in volume such as [$124 \times 124 \times 4$] for 4 filters.

(3) MAX POOL layer will down-sample along the spatial dimensions (width, height), resulting in volume [$62 \times 62 \times 4$].

(4) MLP layer will compute the class scores, resulting in volume of size [$1 \times 1 \times 4$], where each of the 4 numbers correspond to a class score for the 4 tumor regions. This simple neural network is not able to identify all the features and the output classification accuracy is very low. This leads to the requirement of increasing the number of hidden layers in the network. But inclusion of many hidden layers can increase the training time and memory requirements making the network impractical. Hence a trade-off is needed between efficiency and

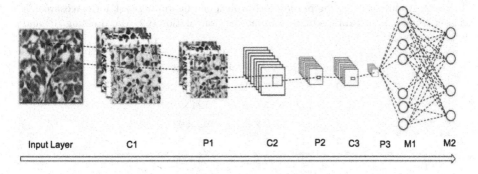

Fig. 3. The figure shows the architecture of a convolution neural network for the classification of osteosarcoma. The different layers in the network are 3 Convolution layer (C), 3 Sub-Sampling layer (P), and 2 fully connected multi-level perceptrons (M).

Table 1. Comparison of accuracy, and running time for 3 different implementation of neural network with different number of hidden layers

Architecture	Accuracy	Running time (in minutes)
3 layer	0.21	3
6 layer	0.86	18
Proposed architecture	0.84	7

accuracy. We worked with different number of hidden layers to define the best output in terms of tumor identification and computational resources needed (see Table 1).

The detailed architecture of the five level CNN for tumor classification is shown in Fig. 3. Our architecture combines the simplicity of Lenet architecture with the data augmentation methods used by AlexNet architecture. The lower 3 layers are comprised of alternating convolution and max-pooling layers. The first convolution layer has filter size 5×5 used to detect low level features like edges which is followed by a max pooling layer of scale 2 to down-sample the data. This data is then sent to second layer of 5×5 filters to detect higher order features like texture and spatial connectivity followed by a max-pooling layer. The last convolution layer uses a filter of size 3×3 and max-pooling size 2 for down- sampling to generate more higher order features. The upper 2 layers are fully-connected multi-level perceptron (MLP) neural network (hidden layer + logistic regression). The second layer of the MLP is the output layer consisting of four neurons (see Table 2). The input to the first MLP layer is the set of all features maps at the layer below and the output is a class probability distribution from the four neurons (p_1, p_2, p_3, p_4) for each image, where p_1, p_2, p_3, p_4 are the probability for viable tumor, coagulative necrosis, osteoid or fibrosis and non-tumor, respectively. The sum of the output probabilities from the MLP is 1, ensured by the use of Softmax algorithm as the activation function in the output

Table 2. Architecture of the proposed convolutional neural network for osteosarcoma classification. The network is built of Input (I), Convolution (C), Max-Pooling (P) and fully connected (M) layers

Layer	Type	Filter size	Output size
0	I		128×128
1	C	5×5	124×124
2	P	2×2	62×62
3	C	5×5	58×58
4	P	2×2	29×29
5	C	3×3	27×27
6	P	2×2	14×14
7	M	32	1×32
8	M	4	1×4

layer of the MLP. The convolution and max pooling layers are feature extractors and the MLP is the classifier.

Data Augmentation. The easiest and most common method to reduce over-fitting of data is to artificially augment the dataset using label-preserving trans-formations. We use two distinct data augmentation techniques both of which allow transformed images to be produced from the original images with very little computation, so the transformed images do not need to be stored on disk. This is a significant saving in both space and time, since WSI images are huge in size and disk read/write is a time consuming process. For this purpose, first we arbitrary rotate the training images by $(0°, 90°, 180°, 270°)$ and flip them along the vertical and horizontal axis to ensure that the network does not learn any rotation dependent features. The second technique for data augmentation alters the intensities of the RGB channels in training images [9]. We perform Principal component analysis (PCA) on the set of RGB pixel values throughout the train-ing set and then, for each training image, we add the following quantity to each RGB image pixel (i.e., $I_{xy} = [I_{xy}^R, I_{xy}^G, I_{xy}^B]^T$): $[p_1, p_2, p_3][\alpha_1\lambda_1, \alpha_2\lambda_2, \alpha_3\lambda_3]^T$, where p_i and λ_i are the i-th eigenvector and eigenvalue of the 3×3 covariance matrix of RGB pixel values, respectively, and α_i is a random variable drawn from a Gaussian with mean 0 and standard deviation 0.1. Data augmentation helps alleviate over-fitting by considerably increasing the amount of training data, removing rotation dependency and making the training images invariant to changes in the color brightness and intensity through PCA.

Initialization and Training. The network is trained with stochastic gradient descent. We initialized all weights with 0 mean by assigning them small, random and unique numbers from 10^{-2} standard deviation Gaussian random numbers, so that each layer calculates unique updates and integrate themselves as different units of the full network.

3 Experimental Setup

3.1 Data

In digital histopathology, the H&E stained microscopic slides are scanned using powerful slidescanner software, such as Aperio, and converted to digital whole slide images (WSIs). Each WSI supports upto 40X magnification, capturing bones, tissues, cellular and sub-cellular structures such as nuclei and cytoplasm. After digitization the digital slides were partitioned into smaller tiles that were evaluated by pathologists to identify patients cases that capture the variability in osteosarcoma. Each case consists of an average of 25 individual svs images representing different sections of the microscopic slide. Three patient cases were identified for training and testing purposes. The dataset used includes three random svs slides from each of the three patient cases. From these 9 svs slides, 81 random tiles of size 1024×1024 that represent different tissue and cellular regions with appearance of both normal and malignant regions were used. For the network to learn the correct representation of tumor, it is important that the training data contain enough information to allow discrimination between the different tissue and cellular structures present in the tiles. As such, the correct resolution used for tile generation was determined through discussions with senior pathologists and was fixed at 20x, which was then used to generate the 81 random tiles.

The pathologists then used an in-house tool that we developed to annotate these 81 tiles as viable tumor, necrosis, non-viable tumor, and non-tumor. As it is difficult to feed 1024×1024 images to the neural network, we extracted small patches from the tiles for training. Patch size was determined through initial trial runs on the network. The 256×256 patches limited the CNN due to memory issues and the 64×64 patch size had very low accuracy. Hence we decided on a 128×128 patch size. This resulted in about 5000 image patches in the dataset. Only 60% patches were used for training, and 20% data was used as validation set, the remaining 20% data was use for test set. Figure 4. Shows some example patches in the training set.

3.2 Implementation

We used existing open source libraries to implement the neural network architecture. The architecture was developed in JAVA using dl4j (deep learning for java) libraries [1]. The training data was fed to the network in batch sizes of 100 to utilize parallelism and improve the network efficiency.

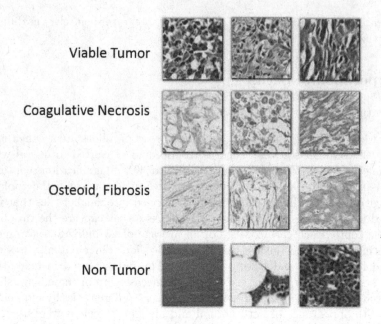

Fig. 4. Example patches of different types of regions found in the dataset.

3.3 Results

Evaluation. The objective of the network was to classify the input images tiles into one of the four regions (viable tumor, coagulative necrosis, osteoid or fibrosis, non-tumor) mentioned before. The output of the neural network is the probability distribution with sum 1. The output class is the class with the highest probability. The regions coagulative necrosis, osteoid and fibrosis fall into class necrosis. The performance of the neural network was monitored by assessing the error rate on the validation set, once the error rate saturated after 10 epochs, training was stopped. The total training time for our implementation of the network was around 7 min.

We evaluate the accuracy of the proposed method quantitatively using accuracy A = (True Positives + True Negatives)/(Total Sample Size), precision P = (True Positives)/(True Positives + False Positives), recall R = (True Positives)/(True Positives + False Negatives), and F1-Score F_1 = (2PR)/(P+R). Our implementation gives F1-score of 0.86 and an accuracy of 0.84.

Comparative Results. The output of a neural network is dependent on the architecture of the network. Different architectures, with different depths and/or numbers of units in the hidden layers result in different output. Shallower networks with fewer number of hidden units are more resistant to over-fitting, require less training data, and train faster per example but can result in loss of precision due to lack of higher order features. A deeper network with more

Table 3. Comparison of accuracy, precision, recall, F1-score and running time for 3 different architectures

Architecture	Accuracy	Precision	Recall	F1-score	Running time (in minutes)
AlexNet	0.73	0.81	0.0.75	0.78	14
LeNet	0.67	0.75	0.67	0.71	5
Proposed architecture	0.84	0.89	0.84	0.86	7

Table 4. Comparison of accuracy of our method with multi-level otsu thresholding

Metric	Multi-level otsu	Proposed neural network
Viable tumor	100	84.5
Necrosis	100	82.6
Non-viable tumor	91	84.9

hidden units may be able to learn patterns from the training data more precisely but could result in over-fitting of the data and loss of efficiency. In this section we present and compare the qualitative output of three architectures: AlexNet, Lenet, and our proposed architecture. We find that the running time of Lenet is fastest but the accuracy and precision of our proposed architecture is better than both AlexNet and Lenet (see Table 3).

We then proceeded to compare these results with a recent study which used color-based multi-level segmentation [2] and found our results to be comparable in both efficiency and accuracy. Arunachalam *et al.* [2] used a multi-level otsu threshold and clustering algorithms to segment out viable-tumor, necrosis, and non-viable tumor regions. The accuracy of the method is around 90% which is close to the accuracy of the neural network (see Table 4).

Results Discrepancy. The method proposed by Arunachalam *et al.* [2] depends on a threshold value which is derived through otsu segmentation, which makes the results biased towards training data. It can be argued that the results are prone to over-fitting and may not generalize well for other datasets whereas the neural network learns the features through the input images and thus can avoid over-fitting, while also becoming better once more data is fed in.

4   Future Work

The architecture of the CNN proposed in this paper was chosen on the basis of datasets and resources available. Justifying any architecture through theory is an ongoing research and is currently done only through experiments and the output results. A deeper network architecture will allow for more variations in the input but will cost more resources. We can continue to explore different architectures and strategies for the training of a neural network by changing the hyper-parameters or pre-processing the input data like using the LAB color space instead of RGB space or by augmenting the results of initial segmentation (otsu segmentation) in the input data. These strategies may improve the output results.

The next step in the development of a fully automatic classification system is to map the output of the CNN to the whole slide images. This can be done by applying the full convolution neural network to generate color coded likelihood maps for the pathologists. This fully automated system can then be used for clinical diagnosis.

5   Conclusion

In this paper, we proposed a deep learning approach using convolutional neural network for tumor classification in osteosarcoma. The proposed method is efficient and accurate and focuses on class level identification instead of nuclei level. The training and evaluation was done on a dataset manually annotated by senior pathologists. As far as the authors are aware, this is the first paper describing the applicability of convolutional neural networks for diagnostic analysis of osteosarcoma. We have shown that the technique has high potential to improve the diagnostic process and be used as a clinical tool in osteosarcoma analysis.

Acknowledgement. This research was partially supported by NSF award IIP14 39718 and CPRIT award RP150164. We would like to thank Harish Arunchalam, Bodgan Armaselu and Dr. Riccardo Ziraldo from our group at UT Dallas, and Dr. Lan Ma, University of Maryland, for their helpful discussions. We also would like to thank John-Paul Bach and Sammy Glick from UT Southwestern Medical Center for their help with the datasets.

References

1. Deep learning libraries. https://deeplearning4j.org/documentation
2. Arunachalam, H.B., Mishra, R., Armaselu, B., Daescu, O., Martinez, M., Leavey, P., Rakheja, D., Cederberg, K., Sengupta, A., Nisuilleabhain, M.: Computer aided image segmentation and classification for viable and non-viable tumor identification in osteosarcoma. In: Pacific Symposium on Biocomputing, vol. 22, p. 195 (2016)
3. Cireşan, D.C., Giusti, A., Gambardella, L.M., Schmidhuber, J.: Mitosis detection in breast cancer histology images with deep neural networks. In: Mori, K., Sakuma, I., Sato, Y., Barillot, C., Navab, N. (eds.) MICCAI 2013. LNCS, vol. 8150, pp. 411–418. Springer, Heidelberg (2013). doi:10.1007/978-3-642-40763-5_51
4. Fischer, A.H., Jacobson, K.A., Rose, J., Zeller, R.: Hematoxylin and eosin staining of tissue and cell sections. Cold Spring Harbor Protoc. **2008**(5), pdb–prot4986 (2008)

5. Fuchs, T.J., Wild, P.J., Moch, H., Buhmann, J.M.: Computational pathology analysis of tissue microarrays predicts survival of renal clear cell carcinoma patients. In: Metaxas, D., Axel, L., Fichtinger, G., Székely, G. (eds.) MICCAI 2008. LNCS, vol. 5242, pp. 1–8. Springer, Heidelberg (2008). doi:10.1007/978-3-540-85990-1_1

6. Goode, A., Gilbert, B., Harkes, J., Jukic, D., Satyanarayanan, M., et al.: Openslide: a vendor-neutral software foundation for digital pathology. J. Pathol. Inform. 4(1), 27 (2013)

7. Irshad, H., Veillard, A., Roux, L., Racoceanu, D.: Methods for nuclei detection, segmentation, and classification in digital histopathology: a review current status and future potential. IEEE Rev. Biomed. Eng. 7, 97–114 (2014)

8. Kothari, S., Phan, J.H., Stokes, T.H., Wang, M.D.: Pathology imaging informatics for quantitative analysis of whole-slide images. J. Am. Med. Inform. Assoc. 20(6), 1099–1108 (2013)

9. Krizhevsky, A., Sutskever, I., Hinton, G.E.: Imagenet classification with deep convolutional neural networks. In: Advances in Neural Information Processing Systems, pp. 1097–1105 (2012)

10. LeCun, Y., Bottou, L., Bengio, Y., Haffner, P.: Gradient-based learning applied to document recognition. Proc. IEEE 86(11), 2278–2324 (1998)

11. Litjens, G., Sánchez, C.I., Timofeeva, N., Hermsen, M., Nagtegaal, I., Kovacs, I., Hulsbergen-Van De Kaa, C., Bult, P., Van Ginneken, B., Van Der Laak, J.: Deep learning as a tool for increased accuracy and efficiency of histopathological diagnosis. Sci. Rep. 6 (2016)

12. Malon, C.D., Cosatto, E., et al.: Classification of mitotic figures with convolutional neural networks and seeded blob features. J. Pathol. Inform. 4(1), 9 (2013)

13. Ottaviani, G., Jaffe, N.: The epidemiology of osteosarcoma. In: Jaffe, N., Bruland, O.S., Bielack, S. (eds.) Pediatric and Adolescent Osteosarcoma, pp. 3–13. Springer, Heidelberg (2009)

14. Sirinukunwattana, K., Raza, S.E.A., Tsang, Y.W., Snead, D.R., Cree, I.A., Rajpoot, N.M.: Locality sensitive deep learning for detection and classification of nuclei in routine colon cancer histology images. IEEE Trans. Med. Imaging 35(5), 1196–1206 (2016)

15. Spanhol, F.A., Oliveira, L.S., Petitjean, C., Heutte, L.: Breast cancer histopathological image classification using convolutional neural networks. In: 2016 International Joint Conference on Neural Networks (IJCNN), pp. 2560–2567. IEEE (2016)

16. Su, H., Liu, F., Xie, Y., Xing, F., Meyyappan, S., Yang, L.: Region segmentation in histopathological breast cancer images using deep convolutional neural network. In: 2015 IEEE 12th International Symposium on Biomedical Imaging (ISBI), pp. 55–58. IEEE (2015)

17. Yu, K.H., Zhang, C., Berry, G.J., Altman, R.B., Ré, C., Rubin, D.L., Snyder, M.: Predicting non-small cell lung cancer prognosis by fully automated microscopic pathology image features. Nat. Commun. 7 (2016)

Relating Diseases Based on Disease Module Theory

Peng Ni[1], Min Li[1(✉)], Ping Zhong[1], Guihua Duan[1], Jianxin Wang[1],
Yaohang Li[2], and FangXiang Wu[1,3]

[1] School of Information Science and Engineering,
Central South University, Changsha 410083, China
limin@mail.csu.edu.cn
[2] Department of Computer Science, Old Dominion University,
Norfolk, VA 23529, USA
[3] Division of Biomedical Engineering and Department of Mechanical
Engineering, University of Saskatchewan, Saskatoon SKS7N5A9, Canada

Abstract. Understanding disease-disease associations can not only help us gain deeper insights into complex diseases, but also lead to improvements in disease diagnosis, drug repositioning and new drug development. Due to the growing body of high-throughput biological data, a number of methods have been proposed for the computation of similarity among diseases during past decades. Recently, the disease module theory has been presented, which states that disease-related genes or proteins tend to interact with each other in the same neighborhood of protein-protein interaction network. In this study, we propose a new method called ModuleSim to measure associations between diseases by using disease-gene association data and protein-protein interaction network data based on disease module theory. By considering the interactions between disease modules and each module's modularity, ModuleSim outperforms other four popular methods for predicting disease-disease similarity.

Keywords: Disease-disease association · Disease module · Protein-protein interaction network

1 Introduction

Quantifying the associations among diseases is now playing an important role in modern biology and medicine, as discovering associations among diseases could be helpful for us to get a deeper knowledge of pathogenic mechanisms of complex diseases. Based on the hypothesis that similar diseases may be caused by the same or similar genes, the measurement of disease-disease associations is widely used in the study of disease gene prediction [1, 2, 33] and drug repositioning [3].

This work is supported by the National Science Fund for Excellent Young Scholars under Grant No. 61622213, the National Natural Science Foundation of China under grant No. 61370024 and No. 61472133, and the Program of Independent Exploration Innovation in Central South University (2016zzts354).

© Springer International Publishing AG 2017
Z. Cai et al. (Eds.): ISBRA 2017, LNBI 10330, pp. 24–33, 2017.
DOI: 10.1007/978-3-319-59575-7_3

A number of approaches measuring disease-disease associations have been proposed during last decade [4–8]. Different approaches measures disease-disease associations from different perspectives by taking advantage of different biological data. These approaches can be broadly grouped into two classes: semantic-based methods and function-based methods [9]. Semantic-based methods take advantage of the structure of disease terminology such as Disease Ontology (DO) [10] and Medical Subject Headings (MeSH) [11] to measure the semantic similarity of diseases [12, 13]. Function-based methods are basically based on the hypothesis that similar diseases may have more same or similar causing genes/gene products [5, 14].

Mathur et al. proposed a method called BOG [15] which calculates disease similarity by comparing the overlapping of disease-related gene sets. Further, Mathur et al. proposed another method called PSB [16] which computes disease similarity based on biological process terms of Gene Ontology (GO) [17] associated with disease-related genes. By exploiting functional associations among disease-related genes based on GO, PSB outperforms BOG. To get a better performance, many other methods take advantage of disease-related genes' interactions in protein-protein interaction networks (PPIN). FunSim [9] measures disease similarity by using a weighted human PPIN in which the weight of each interaction measures the functional association of a gene pair [32]. However, FunSim takes only the first neighbors of each gene into account, rather than making full use of the entire PPIN. Sun et al. [18] applied graphlet theory [19] to calculate gene similarity in PPIN. Then they inferred disease similarity by using disease-related genes' graphlet similarity. Hamaneh et al. [20] proposed a method that first assigns weights to all proteins from a disease to the PPIN and back. Then the method calculates similarity between two diseases as cosine of the angel between their corresponding weight vectors. NetSim [21] uses random walk with restart (RWR) [22] to score the functional relevance between a gene and a disease. The functional relevance scores are then used to measure disease similarity.

Although there have been many methods (such as Sun's method [18], Hamaneh's method [20] and NetSim [21]) which take advantage of PPIN to discover disease-disease associations, these methods rarely consider the modularity of genes related to each disease in PPIN. According to the disease module theory, the disease-related genes or proteins are not scattered randomly in PPIN, but tend to interact with each other, forming one or several connected subgraphs which can be called the disease module [23, 40]. However, as the PPIN and our knowledge of disease-related genes remain incomplete, there also exist lots of disease modules that are not observable in PPIN. In this study, we propose a method to relate diseases based on disease module theory. In this method, we consider the related genes of two diseases as two modules in PPIN. We take advantage of shortest path of each gene pair between the two modules to measure the association of the two modules. Furthermore, for the purpose of overcoming the incompleteness of disease modules, we also take the modularity of each disease module into account. In the comparison with other proposed methods used PPIN, our method shows the best performance.

2   Materials and Methods

2.1   Materials

Disease-Gene Associations: The disease-gene association data are downloaded from two databases: SIDD [25] and DisGeNET [24]. By integrating disease-gene associations from five databases (GeneRIF [34], Online Mendelian Inheritance in Man (OMIM) [35], Comparative Toxicogenomics Database (CTD) [36], Genetic Association Database (GAD) [37], and SpliceDisease [38]), SIDD contains 99658 associations between 2423 diseases and 10527 genes in total (Fig. 1). SIDD uses DOID [10] as the unique identifier for each disease.

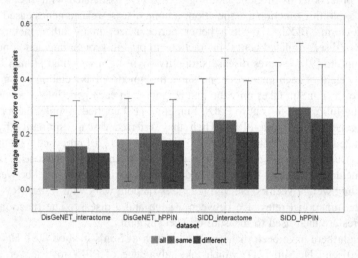

Fig. 1. Evaluation of ModuleSim against DO classification by using different datasets (the barplot shows similarity scores between disease pairs from the same DO categories, compared with those from different DO categories and all disease pairs). Note that two diseases are said to be in the same category if they have at least one common ancestor in the 3rd-level DO categories.

DisGeNET integrates human disease-gene associations from various expert curated databases and text-mining derived associations including Mendelian, complex and environmental diseases [24]. DisGeNET v4.0 contains 429036 associations between 17381 genes and 15,093 diseases. Because of the low reliability of disease-gene associations from literature in DisGeNET, a disease-gene association is adopted only if its DisGeNET score is not less than 0.06 [24]. DisGeNET uses Unified Medical Language System Identifier (UMLS ID) [39] as the unique identifier for each disease. After mapping disease ids from UMLS ID into DOID, in total, we got 1511 diseases, 6929 genes and 20787 associations between them from DisGeNET.

PPIN: Two PPIN datasets were adopted. One is called hPPIN. As Li et al. [21] did, hPPIN was built by integrating four existing protein interaction databases (BioGrid

[26], HPRD [27], IntAct [28], and HomoMINT [29]). In total, hPPIN contains 17506 proteins and 284476 interactions. The other is human interactome which was formed by experimentally documented molecular interactions as Menche et al. [23] did. The interactome integrates protein-protein and regulatory interactions, and metabolic pathway and kinase-substrate interactions. The union of all interactions in the interactome forms a network which contains 13460 proteins and 141296 physical interactions between them.

2.2 Methods

In disease module theory, a disease is considered as a subgraph consisting of genes related to the disease and the interactions between these genes in PPIN [23, 40]. In other words, any perturbation of the nodes in a disease module can be linked to the disease. If genes in two disease modules overlap or stay in the same neighborhood, the perturbations leading to one disease will likely disrupt the other disease modules as well, which results in shared clinical characteristics [23]. However, limited to the fact that our knowledge of disease-related genes and PPIN are still incomplete, lots of disease modules are not observable. Based on disease module theory and the fragmentation of disease modules, we proposed a method called ModuleSim to calculate disease-disease associations. Firstly, we use the length of the shortest path to calculate the strength of two genes' relevance as follows:

$$sim(g_1, g_2) = \begin{cases} 1, & g1 = g2 \\ A * exp^{-b*sp(g_1,g_2)}, & g_1 \in PPIN \, and \, g_2 \in PPIN \\ 0, & else \end{cases} \quad (1)$$

where $sp(g_1,g_2)$ represents the length of the shortest path between node g_1 and node g_2 in PPIN, A and b are two constants. To keep the value of $sim(g_1,g_2)$ within the range [0, 1], we used $A = 1$ and $b = 1$, respectively. A higher $sim(g_1,g_2)$ value represents a closer relationship between g_1 and g_2. Suppose G is a disease module, which means G is a gene set associated with a disease, we then measure a gene's relevance to a disease as follows:

$$F_G(g) = avg\left(\sum_{g_i \in G} sim(g, g_i)\right) \quad (2)$$

As in Eq. (2), the relevance score of a gene g with the disease is calculated as the average transformed distance between g and genes in G.

Suppose $G_1 = \{g_{11}, g_{12}, ..., g_{1m}\}$ is a disease module which contains m genes, $G_2 = \{g_{21}, g_{22}, ..., g_{2n}\}$ is another disease module which contains n genes. The relatedness between the two disease modules is quantified by Eq. (3).

$$spsim(G_1, G_2) = \frac{\sum_{1 \le i \le m} F_{G_2}(g_{1i}) + \sum_{1 \le j \le n} F_{G_1}(g_{2j})}{m+n} \quad (3)$$

Our knowledge of disease-associated genes and PPIN remain incomplete [23]. This is to say, there also exist lots of diseases of whose modularity is not obvious. To overcome the incompleteness of disease modules, we normalize the relatedness score between G_1 and G_2 by dividing the average of relatedness scores of themselves as Eq. (4).

$$ModuleSim(G_1, G_2) = \frac{2 \times spsim(G_1, G_2)}{spsim(G_1, G_1) + spsim(G_2, G_2)} \tag{4}$$

In Eq. (4), *ModuleSim(G_1, G_2)* represents the ModuleSim of disease module G_1 and G_2. A higher ModuleSim value represents a closer connection between G_1 and G_2.

3 Experiments and Results

3.1 Correlation with Disease Classification of DO

The results obtained by ModuleSim were first evaluated against the disease classification of DO. DO is a standardized ontology for human disease concepts with stable identifiers organized by disease etiology [10]. DO (version: releases/2016-05-27) contains 6930 non-obsolete disease terms and 6921 disease terms under the 3rd-level categories. We say that two diseases are in the same class, if they have at least one common ancestor in the 3rd-level DO categories. To investigate the correlation between ModuleSim and the disease classification of DO, we tested whether disease pairs from the same DO classes tends to have higher similarity scores than disease pairs from different DO classes (Fig. 1). Our results show that for all four situations when using different disease-gene association datasets and PPIN datasets, similarity scores of disease pairs from the same classes are higher than those from different classes.

3.2 Evaluation of ModuleSim on the Benchmark Set

We adopted the benchmark set method [9] to evaluate ModuleSim with other methods. 70 disease pairs with high similarity derived from two manually checked datasets by Suthram et al. [30] and Pakhomov et al. [31] were taken as the benchmark set. Receiver operating characteristic (ROC) curves were then drawn with the benchmark set against 100 random sets. Each random set contains 700 randomly selected pairs.

We compared ModuleSim with other four popular methods which are all using disease-gene association data and PPIN data to measure disease-disease associations: Hamaneh [20], FunSim [9], Sun_topo [18], NetSim [21]. As shown in Fig. 2A, when using disease-gene associations from SIDD [25] and hPPIN as the PPIN, the Hamaneh method [20], with an average area under the ROC curve (AUC) of 93.7%, had the worst performance. By considering the functional weights between disease-related genes in PPIN, FunSim [9] got an AUC of 94.4%. NetSim [21] which took the entire interaction network into account by using RWR improved the AUC to 95.1%. By using graphlet theory [19], Sun_topo [18] got a higher AUC of 96.1%. The proposed method, ModuleSim, got the highest AUC of 96.9%. For a further comparison, we also checked how many answer disease pairs out of the top-ranking disease pairs can be found by ranking the benchmark pairs and the random pairs in descending order based on each method. From Fig. 2B we can see that, ModuleSim always find the most answer disease pairs in the top-ranking 150 disease pairs. Furthermore, ModuleSim find all 70 benchmark pairs by using the least top-ranking disease pairs, which showed a quite good performance. For example, "pneumonia" (DOID:552) and "meningitis" (DOID:9471) are two diseases which are validated to have high similarity with each other in the benchmark set. There are only six genes related to "meningitis" based on SIDD [25], which leads to the result that the disease module of "meningitis" is fragmentary. Thus, the average ranking of "pneumonia" and "meningitis" in the 770 disease pairs (70 benchmark pairs and 700 randomly selected pairs) is very low for all five methods, as shown in Table 1. However, by considering the modularity of each diseases, ModuleSim obtained an average ranking of 251 of "pneumonia" and "meningitis", which raised about 100 places compared with Hamaneh and Sun_topo.

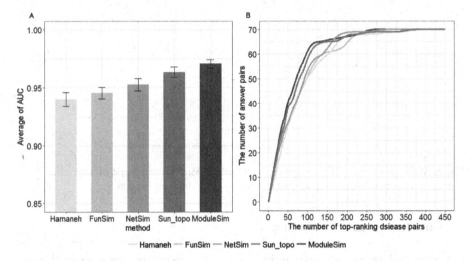

Fig. 2. ModuleSim compared with other four methods on benchmark set by using SIDD [25] and hPPIN [21]. A: average of AUC for 100 permutations. B: the number of answers with varying the number of top-ranking disease pairs.

Table 1. The average ranking of the disease pair ("pneumonia" and "meningitis") in 770 disease pairs, based on the datasets SIDD and hPPIN.

	Hamaneh	FunSim	Sun_topo	NetSim	ModuleSim
Avg ranking	366.45	262.73	354.08	282.04	251.36

Only 55.3% of disease-gene associations in DisGeNET [24] and 11.5% of disease-gene associations in SIDD [25] are shared with each other, which shows that the two databases have a big difference in quantity with each other. Similarly, different PPIN datasets are also very different. The two PPIN datasets (interactome [23] and

hPPIN [21]) used in this paper only have 12560 genes and 90938 interactions in common. To test the influence of different datasets, we further evaluated the five methods by using these two different disease-gene association databases and two different PPIN datasets. As shown in Fig. 3, ModuleSim got the best performance in all four situations, which indicated that ModuleSim have a stable and strong power for discovering disease-disease associations.

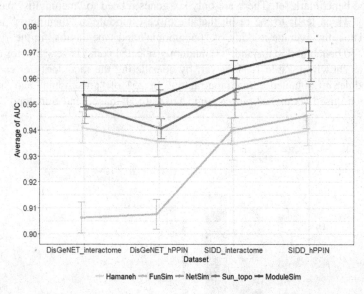

Fig. 3. Average of AUC for 100 permutations when Modulesim compared with other four methods on the benchmark set and random sets by using different datasets.

4   Conclusion and Discussion

It is a big challenge to get a deeper insight into the mechanisms between diseases in modern biology [41, 42]. Measuring disease-disease associations is helpful for us to gain more knowledge about diseases. A number of methods have been proposed for measuring disease-disease associations up to now. The methods which take advantage of disease-gene associations and PPIN have shown a great power to infer disease-disease associations. However, these methods rarely consider the modularity of genes related to each disease in PPIN.

According to the disease module theory, the disease-related genes or proteins are not scattered randomly in PPIN, but tend to interact with each other [23, 40]. In this study, we proposed a method ModuleSim to discovering disease-disease associations based on disease module theory. In the result of ModuleSim, similarity scores of disease pairs from the same DO classes are higher than those from different DO classes. Furthermore, ModuleSim outperformed other four methods (Hamaneh [20], FunSim [9], Sun_topo [18], NetSim [21]) in the evaluation of benchmark set.

ModuleSim considers modularity of each disease module when measuring disease-disease associations. However, our knowledge of disease-related genes and PPIN remains incomplete. Therefore, lots of disease modules remain incomplete. In the future, more disease-gene associations and gene-gene interactions with high quality need to be discovered. In addition, the application of ModuleSim on disease-gene prediction and drug repositioning is worthy of further investigation.

References

1. Vanunu, O., Magger, O., Ruppin, E., et al.: Associating genes and protein complexes with disease via network propagation. PLoS Comput. Biol. **6**(1), e1000641 (2010)
2. Li, M., Zheng, R., Li, Q., et al.: Prioritizing disease genes by using search engine algorithm. Curr. Bioinform. **11**(2), 195–202 (2016)
3. Luo, H., Wang, J., Li, M., et al.: Drug repositioning based on comprehensive similarity measures and Bi-Random walk algorithm. Bioinformatics **32**(17), 2664–2671 (2016)
4. van Driel, M.A., Bruggeman, J., Vriend, G., et al.: A text-mining analysis of the huamn phenome. Eur. J. Hum. Genet. **14**(5), 535–542 (2006)
5. Goh, K.I., Cusick, M.E., Valle, D., et al.: The human disease network. Proc. Natl. Acad. Sci. **104**(21), 8685–8690 (2007)
6. Jung, J., Lee, D.: Inferring disease association using clinical factors in a combinatorial manner and their use in drug repositioning. Bioinformatics **29**(16), 2017–2023 (2013)
7. Sun, K., Buchan, N., Larminie, C., et al.: The integrated disease network. Integr. Biol. **6**(11), 1069–1079 (2014)
8. Frick, J.M., Guha, R., Peryea, T., et al.: Evaluating disease similarity using latent Dirichlet allocation. bioRxiv: 030593 (2015)
9. Cheng, L., Li, J., Ju, P., et al.: SemFunSim: a new method for measuring disease similarity by integrating semantic and gene functional association. PLoS One **9**(6), e99415 (2014)
10. Schriml, L.M., et al.: Disease ontology: a backbone for disease semantic integration. Nucleic Acids Res. **40**(D1), D940–D946 (2012)
11. Lipscomb, C.E.: Medical subject headings (MeSH). Bull. Med. Libr. Assoc. **88**(3), 265 (2000)
12. Yu, G., Wang, L.G., Yan, G.R., et al.: DOSE: an R/Bioconductor package for disease ontology semantic and enrichment analysis. Bioinformatics **31**(4), 608–609 (2015)
13. Wang, D., Wang, J., Lu, M., et al.: Inferring the human microRNA functional similarity and functional network based on microRNA-associated diseases. Bioinformatics **26**(13), 1644–1650 (2010)
14. Zhang, X., Zhang, R., Jiang, Y., et al.: The expanded human disease network combining protein–protein interaction information. Eur. J. Hum. Genet. **19**(7), 783–788 (2011)
15. Mathur, S., Dinakarpandian, D.: Automated ontological gene annotation for computing disease similarity. AMIA Summits Transl. Sci. Proc. **2010**, 12–16 (2010)
16. Mathur, S., Dinakarpandian, D.: Finding disease similarity based on implicit semantic similarity. J. Biomed. Inform. **45**(2), 363–371 (2012)
17. Ashburner, M., Ball, C.A., Blake, J.A., et al.: Gene ontology: tool for the unification of biology. Nat. Genet. **25**(1), 25–29 (2000)
18. Sun, K., Gonçalves, J.P., Larminie, C., et al.: Predicting disease associations via biological network analysis. BMC Bioinform. **15**(1), 1 (2014)

19. Milenkoviæ, T., Pržulj, N.: Uncovering biological network function via graphlet degree signatures. Cancer Inform. **6**, 257 (2008)
20. Hamaneh, M.B., Yu, Y.K.: Relating diseases by integrating gene associations and information flow through protein interaction network. PLoS ONE **9**(10), e110936 (2014)
21. Li, P., Nie, Y., Yu, J.: Fusing literature and full network data improves disease similarity computation. BMC Bioinform. **17**(1), 326 (2016)
22. Köhler, S., Bauer, S., Horn, D., et al.: Walking the interactome for prioritization of candidate disease genes. Am. J. Hum. Genet. **82**(4), 949–958 (2008)
23. Menche, J., Sharma, A., Kitsak, M., et al.: Uncovering disease-disease relationships through the incomplete interactome. Science **347**(6224), 1257601 (2015)
24. Piñero, J., Queralt-Rosinach, N., Bravo, À., et al.: DisGeNET: a discovery platform for the dynamical exploration of human diseases and their genes. Database **2015**, bav028 (2015)
25. Cheng, L., Wang, G., Li, J., et al.: SIDD: a semantically integrated database towards a global view of human disease. PLoS ONE **8**(10), e75504 (2013)
26. Chatr-Aryamontri, A., Breitkreutz, B.J., Heinicke, S., et al.: The BioGRID interaction database: 2013 update. Nucleic Acids Res. **41**(D1), D816–D823 (2013)
27. Prasad, T.S.K., Goel, R., Kandasamy, K., et al.: Human protein reference database—2009 update. Nucleic Acids Res. **37**(suppl 1), D767–D772 (2009)
28. Orchard, S., Ammari, M., Aranda, B., et al.: The MIntAct project—IntAct as a common curation platform for 11 molecular interaction databases. Nucleic Acids Res. **42**(D1), D358–D363 (2013)
29. Persico, M., Ceol, A., Gavrila, C., et al.: HomoMINT: an inferred human network based on orthology mapping of protein interactions discovered in model organisms. BMC Bioinform. **6**(4), 1 (2005)
30. Suthram, S., Dudley, J.T., Chiang, A.P., et al.: Network-based elucidation of human disease similarities reveals common functional modules enriched for pluripotent drug targets. PLoS Comput. Biol. **6**(2), e1000662 (2010)
31. Pakhomov, S., McInnes, B., Adam, T., et al.: Semantic similarity and relatedness between clinical terms: an experimental study. In: AMIA annual symposium proceedings. American Medical Informatics Association, p. 572 (2010)
32. Lee, I., Blom, U.M., Wang, P.I., et al.: Prioritizing candidate disease genes by network-based boosting of genome-wide association data. Genome Res. **21**(7), 1109–1121 (2011)
33. Ni, J., Koyuturk, M., Tong, H., et al.: Disease gene prioritization by integrating tissue-specific molecular networks using a robust multi-network model. BMC Bioinform. **17**(1), 453 (2016)
34. Mitchell, J.A., Aronson, A.R., Mork, J.G., et al.: Gene indexing: characterization and analysis of NLM's GeneRIFs. In: AMIA (2003)
35. Amberger, J.S., Bocchini, C.A., Schiettecatte, F., et al.: OMIM.org: Online Mendelian Inheritance in Man (OMIM®), an online catalog of human genes and genetic disorders. Nucleic Acids Res. **43**(D1), D789–D798 (2015)
36. Davis, A.P., Murphy, C.G., Johnson, R., et al.: The comparative toxicogenomics database: update 2013. Nucleic Acids Res. **41**(D1), D1104–D1114 (2012)
37. Becker, K.G., Barnes, K.C., Bright, T.J., et al.: The genetic association database. Nat. Genet. **36**(5), 431–432 (2004)
38. Wang, J., Zhang, J., Li, K., et al.: SpliceDisease database: linking RNA splicing and disease. Nucleic Acids Res. **40**(D1), D1055–D1059 (2012)
39. Bodenreider, O.: The unified medical language system (UMLS): integrating biomedical terminology. Nucleic Acids Res. **32**(suppl 1), D267–D270 (2004)
40. Barabási, A.L., Gulbahce, N., Loscalzo, J.: Network medicine: a network-based approach to human disease. Nat. Rev. Genet. **12**(1), 56–68 (2011)

41. Guo, X., Zhang, J., Cai, Z., et al.: Searching genome-wide multi-locus associations for multiple diseases based on Bayesian Inference. In: IEEE/ACM transactions on computational biology and bioinformatics (2016)
42. Teng, B., Yang, C., Liu, J., et al.: Exploring the genetic patterns of complex diseases via the integrative genome-wide approach. IEEE/ACM Trans. Comput. Biol. Bioinform. **13**(3), 557–564 (2016)

Reconstructing One-Articulated Networks with Distance Matrices

Kuang-Yu Chang[1], Yun Cui[2], Siu-Ming Yiu[2], and Wing-Kai Hon[1(✉)]

[1] National Tsing Hua University, Hsinchu, Taiwan
wkhon@cs.nthu.edu.tw
[2] The University of Hong Kong, Pokfulam, Hong Kong

Abstract. Given a distance matrix M that represents evolutionary distances between any two species, an edge-weighted phylogenetic network N is said to *satisfy* M if between any pair of species, there exists a path in N with length equal to the corresponding entry in M. In this paper, we consider a special class of networks called *1-articulated network* which is a proper superset of galled trees. We show that if the distance matrix M is derived from an ultrametric 1-articulated network N (i.e., for any species X and Y, the entry $M(X,Y)$ is equal to the *shortest* distance between X and Y in N), we can re-construct an network that satisfies M in $O(n^2)$ time, where n denotes the number of species; furthermore, the reconstructed network is guaranteed to be the *simplest*, in a sense that the number of *hybrid nodes* is minimized. In addition, one may easily index a 1-articulated network N with minimum number of hybrid nodes in $O(n)$ space, such that on given any phylogenetic tree T, we can determine if T *is contained* in N (i.e., if a spanning subtree T' of N is a subdivision of T) in $O(n)$ time.

1 Introduction

It is important to study the evolutionary history and the relationship among a set of related species, especially for viruses and bacteria, in order to trace the origin, understand the infection path, and how they evolved. Rooted (phylogenetic) trees have been the most popular model for decades. However, when there are reticulation events such as hybrid speciations or horizontal gene transfers [14], which are common in viruses and bacteria, rooted trees are not sufficient. Rooted *phylogenetic networks*, which are directed acyclic graphs that may contain vertices with in-degree 2, are more appropriate to capture these evolutionary events. Nodes with in-degree 2 are referred to as *hybrid nodes*.

Models of Networks: There are many different phylogenetic network models proposed in the literature. The simplest model is called *galled trees*. Galled trees have the property that, when we remove the edge orientations (i.e., make them undirected), any biconnected component contains at most 1 hybrid node [8]. There are two well-known generalizations on galled trees. One is *level-k* networks allowing at most k hybrid nodes in each biconnected component [13]. Thus,

Z. Cai et al. (Eds.): ISBRA 2017, LNBI 10330, pp. 34–45, 2017.
DOI: 10.1007/978-3-319-59575-7_4

galled trees are equivalent to level-1 networks. The other is *galled networks* [9]. In contrast to a galled tree, in which all cycles are edge-disjoint, a galled network allows cycles to have shared edges.[1] There are also other generalizations such as *tree-child networks* [4], or other types of networks such as *binary nearly stable networks* [7]. Besides the biological motivation, i.e., different models can capture different types of reticulation events and may be based on different assumptions (i.e., some types of articulation events may be more likely than the others), another reason is that the problem of reconstructing a phylogenetic network with the least possible number of reticulation events is known to be intractable (NP-hard). Thus, it is desirable to have new models with sound biological motivation that capture reticulation events that are not captured by existing models while efficient algorithms exist for solving the reconstruction problem or other related problems.

In this paper, we consider another class of networks called *1-articulated networks*. A 1-articulated phylogenetic network is one where each vertex *corresponds* to at most one hybrid node, where a vertex V is said to correspond to a hybrid node U if there exists a pair of disjoint paths between U and V. This definition of networks is also motivated by a biological observation that articulation events are unlikely to occur multiple times for a species. More importantly, the class of 1-articulated networks properly contains the class of level-1 networks, i.e., galled trees (see Fig. 1 for an example); because of this, 1-articulated networks can capture a larger set of networks that appear in real-life scenarios, and algorithms for 1-articulated networks are directly applicable for level-1 networks. As for the comparison between the classes of 1-articulated networks, tree-child networks and binary nearly stable networks, neither anyone properly contains the other; thus, if we consider all these classes of networks in a reconstruction problem, we would have a higher chance of revealing the true evolutionary history among the species. For 1-articulated networks, we demonstrate that we can derive efficient linear time algorithms (see below for a more detailed description) to reconstruct such a network, or solve the *tree containment problem* (TCP problem). Thanks to the simplicity of the network structure, the TCP problem can be more easily solved when the input network is a 1-articulated network, than the case when the input is a tree-child network or a binary nearly stable network.

The Reconstruction Problem and the TCP Problem: Reconstructing a phylogenetic networks with different inputs (e.g., a set of phylogenetic trees, a set of triplets, and a distance matrix) is a fundamental computational problem in studying phylogenetic networks, and many existing methods are proposed in the literature. Huynh et al. [10] proposed an algorithm RGNet which constructs a galled network in $O(k^2 n^2)$ time with k phylogenetic trees as input. Jansson and Sung [12] presented an algorithm that constructs a galled network given a dense set τ of rooted triplets in $O(|\tau|)$ time. Bryant and Moulton [3] proposed the NeighborNet algorithm, which takes a distance matrix as input (a distance matrix is a symmetric matrix which describes the evolutionary distance

---

[1] In fact, there is no unifying definition for galled networks in the literature. In some papers such as [5,10,12], level-1 networks are also referred as galled networks.

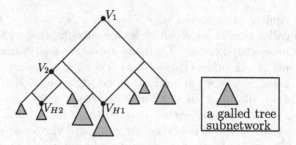

Fig. 1. A galled tree. The split nodes V_1, V_2 correspond to the hybrid nodes V_{H1}, V_{H2}. Since each split node in a galled tree corresponds to exactly one hybrid node on the same galled loop, a galled tree is a 1-articulated network.

between any pair of species), and constructs a phylogenetic network that is planar. Chan et al. [5] considered ultrametric galled network, and proposed an algorithm that runs in $O(n^2 \log n)$ time given an input distance matrix. Bordewich and Tokac [2] presented the NetworkUPGMA algorithm, which constructs a tree-child ultrametric network in $O(n^4)$ time where the input consists of a set of distances for each pair of species. In this paper, we focus on 1-articulated phylogenetic network, and design algorithms to construct such kind of networks when we are given a distance matrix as input. (The formal definition will be given in the next section.) In particular, we present (i) an $O(n^2)$ time (which is linear to the size of input) algorithm for constructing a 1-articulated network satisfying a given *shortest* distance matrix, where all entries are assumed to represent the correct shortest lengths of the corresponding evolutionary paths; (ii) for the general case that the distance matrix may record lengths of non-shortest evolutionary paths, we can still reconstruct the network in $O(n^5)$ time under some reasonable assumptions. Compared to NetworkUPGMA, which requires the distances of all the possible evolutionary paths for each pair of species, our algorithm only requires the distance of the shortest one.

Another important computational problem for phylogenetic networks is called the *tree containment problem* (TCP). Given a phylogenetic network N and a particular phylogenetic tree T, one would like to know if T *is contained* in N (i.e., if a spanning subtree T' of N is a subdivision of T). The answer to this problem provides important evidence whether the phylogenetic network N is consistent with the phylogenetic tree T, where T may capture the evolution history of a subset of species which are known to be more accurate. This *tree containment problem* (TCP) has been discussed in [7] and an $O(n^2)$ time algorithm was given to solve the problem for binary nearly stable networks, and in [11] which shows that the problem is solvable in polynomial time for level-k networks or tree-child networks.

In this paper, we show that one can easily index a 1-articulated network N without skewed loops in $O(n)$ space to solve the TCP problem in linear $O(n)$ time. Also, for general 1-articulated network which may contain skewed loops, we give an algorithm that solves TCP in $O(mn)$ time, where m is the number of nodes in the network.

2   Preliminaries

2.1   Definitions

A *phylogenetic network* is a simple directed acyclic graph in which the following properties are satisfied:

1. The in-degree and out-degree of each node can only be 0, 1, or 2.
2. There is exactly one node, the *root*, with in-degree 0 and out-degree 2.
3. The nodes with in-degree 2 must have out-degree 1. These nodes are referred to as *hybrid nodes*.
4. The nodes with in-degree 1 can have out-degree 0 or 2. The former ones with out-degree 0 are referred to as *leaves*, while the latter ones with out-degree 2 are referred to as *tree nodes*.

In a phylogenetic network for a set S of species, each leaf is labeled by a distinct species in S. A node V is called a *split node* corresponding to a hybrid node U if there exists a pair of edge-disjoint paths, called *merge paths*, from V to U; these two merge paths are said to form a *galled loop* rooted at V. A galled loop rooted at V is said to be *skewed* if one of its merge paths is an edge that links V directly to the corresponding hybrid node U.

A phylogenetic network is *ultrametric* if the weight of each edge is positive real and the sum of edge weights on the directed path from the root to any leaf (i.e., the *distance* from the root to any leaf) is the same. The motivation of defining such networks is based on that the rate of genetic change is constant. [5] by the property, the following lemma must be true:

Lemma 1. *For any vertex V in an ultrametric network, the distances from V to any of its descendant leaves are equal.*

The distance between V and any of its descendant leaves is referred to as the *height* of V.

For species A and B, an *evolutionary path* between A and B in a phylogenetic network N is a simple path which starts from the leaf labeled by A, goes through the edges on a path from V to A in reverse direction, where V is some common ancestor of A and B in N, reaches V, and then goes through the edges on a path from V to B, and finally reaches the leaf labeled by B. By Lemma 1, we have the following corollary:

Corollary 1. *Let V be the highest vertex on an evolutionary path of length d. The height of V is equal to $d/2$.*

One may use an $|S| \times |S|$ matrix M, called a *distance matrix*, to record the evolutionary distance between any two species in S. Precisely, for species A and B, their distance is stored in the entry $M(A, B)$. The matrix M is thus symmetric, and the values on the diagonal, which represent the distance from the species to themselves, shall all be 0. Given a distance matrix M, we say a phylogenetic network N *satisfies* M if for any two species A and B, the length of the *shortest* evolutionary path between A and B in N is equal to $M(A, B)$.

Our Problem: In this paper, we want to determine, for any given input distance matrix M, whether an ultrametric 1-articulated network exists that would satisfy M. Furthermore, if such a network exists, we want to report one whose number of hybrid nodes is minimized. We call this problem a *minimal satisfying network (MSN)* problem, and the network reported (if any) an MSN.

2.2   Properties of an MSN

Suppose that N is an ultrametric 1-articulated network, with the minimum number of hybrid nodes, that satisfies a distance matrix M. In the following, we show some of the important properties about N that will be useful in our algorithm design. Due to space limitations, we defer the proofs of lemmas and theorems to the full paper.

Lemma 2. N *does not contain any skewed galled loop.*

Let $d_{\max}$ be the maximum value in M. For the following lemmas, let V_{root} denote the root of N, and V_L and V_R be the left and right children of V_{root}, respectively. Further, if V_{root} is a split node, let V_H be the hybrid node corresponding to V_{root}. For convenience, we use $\Lambda(V)$ to denote the set of all leaves reachable from a vertex V in N.

Theorem 1. *If* $|S| > 1$, *the height of* V_{root} *is* $d_{\max}/2$.

Assume that $|S| > 1$. Let S_H, S_L, and S_R be sets of species such that

- $S_H = \Lambda(V_H)$ if there exists some hybrid node V_H corresponding to V_{root}. Otherwise, $S_H = \varnothing$.
- $S_L = \Lambda(V_L) \setminus S_H$.
- $S_R = \Lambda(V_R) \setminus S_H$.

We say $(\{S_L, S_R\}, S_H)$ forms a *root partition* of N. See Fig. 2 for an illustration of the possible forms of a root partition.

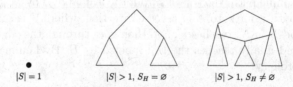

$|S| = 1$ \qquad $|S| > 1, S_H = \varnothing$ \qquad $|S| > 1, S_H \neq \varnothing$

Fig. 2. The possible forms of a root partition. The dot in the first case represents a single leaf, and the triangles represent the subnetworks.

Lemma 3. *For any species* $A, B \in S$, $M(A, B) = d_{\max}$ *if and only if one of them is in* S_L *and the other is in* S_R.

Lemma 4. *If* $S_H \neq \varnothing$, $M(A, X) = M(A, Y)$ *for any species* $A \in (S_L \cup S_R)$ *and* $X, Y \in S_H$.

Lemma 5. *If $S_H \neq \varnothing$,*

$$\max_{X \in S_H} M(Z, X) < \min_{B \in (S_L \cup S_R)} M(Z, B)$$

for any species $Z \in S_H$.

Based on the above lemmas, one may use the following PARTITION procedure to obtain the root partition of N:

1. Pick a species A from S.
2. Find a species B where $M(A, B)$ is maximized. B must be in $S_L \cup S_R$ by Lemmas 3 and 5.
3. By using B and Lemma 3, find S_L (or S_R) and pick a species C from it.
4. Find S_R (or S_L) by using C and Lemma 3.
5. Put the rest of S into S_H.

Indeed, we have the following theorem:

Theorem 2. *Let S_L, S_R, and S_H be the sets returned by the procedure PAR-TITION. Then, for any 1-articulated ultrametric network N' that satisfies M, $(\{S_L, S_R\}, S_H)$ must be the root partition of N'.*

3 The Algorithm

In this section, we will present a recursive algorithm which builds an ultrametric 1-articulated network that satisfies an input distance matrix (whenever such a network exists). Our algorithm consists of the following parts:

1. Prepare data structure for the algorithm.
2. Perform the procedure BUILDNET, which constructs a network N recursively with all the vertices labeled with its height, but the weights of edges are not assigned.
3. Assign weights to edges in N by a graph traversal.
4. Check if N satisfies the distance matrix M. If yes, return N as the output. Else, report that no network satisfying M exists.

3.1 Data Structures

Each species in the input set S is associated with a unique integer between 1 and $|S|$. A species set is represented by a linked list, where the elements are sorted by the integer associated with them. To perform the operations of union, intersection, or difference of two sets, procedures based on the merging process in the merge sort can be used. These procedures can run in $O(|S_1| + |S_2|)$ time where S_1 and S_2 are the input sets of these operations.

For the networks, each vertex is represented by a structure that consists of pointers to its children and parents. The height of each vertex is also stored. The distance matrix is represented by a 2-dimensional array of size $|S| \times |S|$

where each row and each column corresponds to a species in S. An array called *leafIndex* which consists of $|S|$ vertex pointers is normally used as an index to locate a vertex created for a species; however, during the network construction process that involves hybrid nodes, we may select a candidate species Z within a subnetwork (under a hybrid node), and create a temporary leaf for Z to represent this subnetwork, so that *leafIndex*$[Z]$ would be an index from a candidate species Z to its temporary leaf in the above situation. Details on how the array is used will be discussed in the next subsection.

3.2   The Procedure BuildNet

The procedure BUILDNET takes the set of species S and the distance matrix M as input, and outputs an ultrametric 1-articulated network that satisfies M, but with edge weights missing. The basic idea is to partition S into subsets corresponding to different subnetworks, and build each of the subnetworks recursively. The main steps are summarised as follows:

1. If S contains only one species X, return a leaf labeled by X.
2. Partition S into S_L, S_R, and S_H with the PARTITION procedure.
3. If $S_H = \varnothing$:
   Run procedure BUILDNET2:
   – Construct the subnetworks N_L and N_R recursively with S_L and S_R as input, respectively.
   – Connect N_L and N_R.
   – Return the connected network.
4. If $S_H \neq \varnothing$:
   Run procedure BUILDNET3:
   – Construct the subnetwork N_H with S_H as input.
   – Pick a species Z from S_H.
   – Construct the subnetworks N_L and N_R recursively with $S_L \cup \{Z\}$ and $S_R \cup \{Z\}$ as input, respectively.
   – Connect N_L and N_R.
   – Replace the leaf labeled by Z with N_H.
   – Return the connected network.

Depending on the results of PARTITION, BUILDNET invokes different methods for construction:

(i) If the set S_H is empty, then there shall be no hybrid node corresponding to the root. In such a case, the subnetworks rooted at the children of the root are disconnected. Therefore, the subnetworks can just be built recursively with S_L and S_R as input.

(ii) Otherwise, S_H is nonempty. In such a case, we may intuitively construct the desired network by first building the networks for $S_L \cup S_H$ and $S_R \cup S_H$, and merge them. To simplify the task, we pick a species Z from S_H to represent S_H, build the networks N_L for $S_L \cup \{Z\}$ and N_R for $S_R \cup \{Z\}$, merge these two networks, and finally replace Z by the subnetwork N_H for S_H. There are two reasons for picking the species Z. The first is that by Lemma 4, for any particular species $A \in (S_L \cup S_R)$, any species in S_H will have the same distance with A; then, to get the subnetwork for $S_L \cup S_H$, it is the same as to get the subnetwork N_L first, and later replace Z within N_L by the subnetwork N_H. Thus, we may use Z as a representative for N_H, and avoid considering all species in S_H when we build N_L (and N_R). The other reason is that the location of the hybrid node, which links directly to N_H, can be determined easily. Firstly, the species Z has three different roles, namely as a leaf in N_L, a leaf in N_R, and a leaf in N_H; we let V_{Z1}, V_{Z2}, and V_{Z3} to denote the vertices corresponding to Z in these three cases (See Fig. 3). Once N_L and N_R are constructed, we can find the parent V_{PL} of V_{Z1}, the parent V_{PR} of V_{Z2}, and obtain their heights h_{PL} and h_{PR}, respectively. Moreover, once N_H is built, the height h_{HC} of its root V_{HC} is known. Then, we shall create the hybrid node V_H, setting its height to be $(\min(h_{PL}, h_{PR}) + h_{HC})/2$, its parents to be V_{PL} and V_{PR}, and its child to be V_{HC} (See Fig. 4).

If there exists an ultrametric 1-articalted network N' that satisfies M, Theorem 2 implies that the constructed network N would have the same root partition as that of N'. Inductively, this implies that N' and N would have the same structure. Moreover, the height of any corresponding tree nodes in N' and N must be the same. This gives the following lemma.

Lemma 6. *If there exists an ultrametric 1-articulated network N' with minimum number of nodes satisfying M, the network N constructed by our algorithm has the same topology as N'. In addition, any corresponding tree nodes, or leaves, in N' and in N have the same height.*

Thus, if there exists a network N' satisfying M, it remains to show that the height of the hybrid nodes in N are set properly, so that all edges have positive edge weights. This will be the focus of the following subsection.

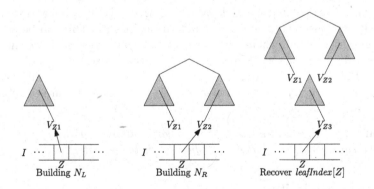

Fig. 3. Maintaining the *leafIndex*

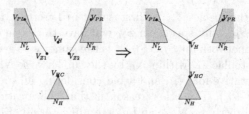

Fig. 4. Connecting the hybrid node

3.3   Assigning Edge Weights

Assigning the edge weights can be easily done by a simple graph traversal. When traversing through an edge (V_1, V_2), assign the edge's weight to be $(h_1 - h_2)$ where h_1 and h_2 are the height of V_1 and V_2 respectively. Based on this assignment, the following lemma is immediate.

Lemma 7. *If there exists an ultrametric 1-articulated network N' satisfying M, then for any vertex V in N, the length of any path from V to any of its descendant leaf is the same. Furthermore, all edge weights are positive. Thus, N is ultrametric.*

Thus, we have the following theorem:

Theorem 3. *If there exists an ultrametric 1-articulated network N' satisfying M, the constructed network N will also be an ultrametric 1-articulated network satisfying M.*

3.4   Verifying the Network

Note that our algorithm may construct some network N even though there does not exist any 1-articulated network satisfying M. Thus, we need an extra step to check if the constructed network N indeed satisfies M. This can be done by another graph traversal, where when visiting a degree-2 vertex V, we check for those pairs A and B of species which take V as their lowest common ancestor, whether $M(A, B)$ is equal to $2 \times height(V)$. In other words, we only need to check for any species $A \in \Lambda(V_L)$ and $B \in \Lambda(V_R)$ where V_L, V_R are the children of V.

Theorem 4. *The procedure CHECKNET reports without reporting failure if and only if the constructed network N satisfies the distance matrix M.*

3.5   Time Complexity

A straightforward analysis shows that Steps 1, 2, and 3 take $O(n)$ time in total, while Step 4 (CHECKNET) takes $O(n^2)$ time. Thus, the overall time is $O(n^2)$.

3.6   Reconstructing Networks with More General Matrix

So far, we have assumed that the entry $M(A, B)$ in the matrix M stores the length of the *shortest* evolutionary path between species A and B. Yet, in real applications, the entry $M(A, B)$ may be storing the length of some evolutionary path between A and B. Now, suppose that there exists a certain ultrametric 1-articulated network N' on the species, and M is a distance matrix such that for any tree node V' in N', and any two species A and B that are reachable from V', we have:

1. $M(A, B)/2 \leq \text{height}(V')$; and
2. $M(A, C) = M(B, C)$ for any C not reachable from V'.

Then, we can still reconstruct an ultrametric 1-articulated network N that satisfies M. To do so, briefly speaking, we perform the same procedures as in the no-skewed-loop case:

1. Find a root partition $(\{S_L, S_R\}, S_H)$ that is consistent with M.
2. Construct the networks for S_L, S_R, S_H, recursively, and connect them to form the complete network.

The difference here (from the previous algorithm) is that there can be at most two possible candidates for the root partition, so that a brute force approach would take $\tilde{O}(2^n)$ time in the worst case. Yet, we can show that if a network that satisfies M exists, then using either root partition would lead to a network that satisfies M, so that we can avoid exponential expansion in the running time. As for finding a root partition, the procedure is more involved, as the entries in M may not be shortest distances. Here, we rely on the properties of M to find the set S_H; this can be reduced to finding a *maximal cluster* problem discussed in [5], and in our case can be solved in $O(n^3)$ time. Furthermore, we can show that our algorithm constructs a network with minimal number of hybrid nodes, and that the total number of nodes is bounded by $O(n^2)$. As we now spend $O(n^3)$ time in each node, the total running time is $O(n^5)$. We defer the details to the full paper.

4   Solving TCP Problem for 1-Articulated Netowrks

In this section, we discuss how to solve the tree containment problem (TCP), where the target is to locate an unweighted tree T within N efficiently (i.e., determine if T is contained in N). We first propose an algorithm for general 1-articulated networks. Then, for a network without skew loops, we show how to build an $O(n)$-space index on the network, such that given any query tree T, the TCP problem can be solved in optimal linear time.

4.1 Algorithm for General 1-Articulated Networks

The algorithm consists of two phases.

Phase 1: First, for each tree node V in T and N, we compute the species set $\Lambda(V)$. This can be done by a post-order graph traversal starting from the root, during which $\Lambda(V)$ can be computed by a union operation of $\Lambda(V_L)$ and $\Lambda(V_R)$, where V_L, V_R are the children of V. If V is a vertex in N, we also compute the set $\Lambda_H(V)$ as $\Lambda(V_L) \cap \Lambda(V_R)$. The set $\Lambda_H(V)$ is non-empty if and only if V is a split node. Moreover, in case V is a split node, $\Lambda_H(V)$ is the set of species which is reachable from the hybrid node U corresponding to V.

Phase 2: Next, we compare T recursively with N. During the process, we keep a list X (initialized as $\varnothing$) which contains the species that are removed from further consideration. Let V_T and V_N be the roots of T and N, respectively. Let V_{TL} and V_{TR} be the children of V_T, and V_{NL} and V_{NR} be the children of V_N. We check if $\{\Lambda(V_{TL}), \Lambda(V_{TR})\}$ equals to $\{(\Lambda(V_{NL}) - X) \setminus \Lambda_H(V_N), \Lambda(V_{NR}) - X\}$ or $\{\Lambda(V_{NL}) - X, (\Lambda(V_{NR}) - X) \setminus \Lambda_H(V_N))\}$. If not, we conclude that the tree T is not found within the network N, and report failure. Otherwise, without loss of generality, asume that it is the former case. Then, we set $X_L = X \cup \Lambda(V_{TR})$ and $X_R = X \cup \Lambda(V_{TL})$. Next, we check, recursively, if the subnetwork rooted at V_{NL} contains the subtree rooted at V_{TL}, and if the subnetwork rooted at V_{NR} contains the subtree rooted at V_{TR}, using X_L and X_R, respectively as X. If both subnetworks contain the desired subtrees, we report T to be found in N.

Time Complexity: Let n be the number of leaves in T or N, and m be the number of vertices in N. The time spent in each node of T or N is $O(n)$, so the overall time is $O(mn)$.

4.2 Indexing 1-Articulated Networks Without Skewed Loops

We sketch the main idea as follows.

Preprocessing: Consider a hybrid node U and its corresponding split node V. Let $([S_L, S_R], S_H)$ be the root partition of the subnetwork N_V of N rooted at V. We may represent V by choosing one leaf from each of the sets S_L, S_R, and S_H. In particular, the leaves from S_L or S_R are chosen such that they are reachable from V without passing through any hybrid node. As for the leaf from S_H, we will choose one that is reachable from U without passing through any hybrid node, if it exists; otherwise, the child of U must be a hybrid node itself, and we pick an arbitrary leaf from S_H. As there are $O(n)$ hybrid nodes in N, the total space includes the representation of all the hybrid nodes takes $O(n)$ extra space, along with the $O(n)$ space to store N itself.

Query Algorithm: Given a rooted tree T, we first construct in $O(n)$ time an $O(n)$-space data structure such that any *lowest common ancestor* (LCA) of any two leaves in T can be reported in $O(1)$ time [1]. Suppose that T can be located within N; then, there will be a unique spanning tree T' of N that is a subdivision of T. Based on the constant-time LCA data structure, we can determine, for each hybrid node U, its parent node P in T' as follows: (i) Take the three leaves $\ell \in S_L, r \in S_R, h \in S_H$ from the representation of U; (ii) Check the LCAs L_1 of (ℓ, r), L_2 of (r, h) and L_3 of (ℓ, h) in T. (iii) Set the parent of U to its left parent (i.e., the one closer to ℓ) if $L_1 = L_2$, to its right parent if $L_1 = L_3$, and remove U if $L_2 = L_3$ (which may happen only if the child of U is a hybrid node). Once each hybrid node has determined its parent, we obtain a spanning tree T' (possibly containing degree-1 internal nodes), and then we can *smooth* the degree-1 nodes of T' in $O(n)$ time, and check if T' is leaf-label-preserving isomorphic to T using another $O(n)$ time (say, by Day's algorithm [6]).

References

1. Bender, M.A., Farach-Colton, M., Pemmasani, G., Skiena, S., Sumazin, P.: Lowest common ancestors in trees and directed acyclic graphs. J. Algorithms **57**(2), 75–94 (2005)
2. Bordewich, M., Tokac, N.: An algorithm for reconstructing ultrametric tree-child networks from inter-taxa distances. DAM **213**, 47–59 (2016)
3. Bryant, D., Moulton, V.: Neighbor-net: an agglomerative method for the construction of phylogenetic networks. Mol. Biol. Evol. **21**(2), 255–265 (2004)
4. Cardona, G., Rossello, F., Valiente, G.: Comparison of tree-child phylogenetic networks. IEEE/ACM TCBB **6**(4), 552–569 (2009)
5. Chan, H., Jansson, J., Lam, T., Yiu, S.: Reconstructing an ultrametric galled phylogenetic network from a distance matrix. JBCB **4**(4), 807–832 (2006)
6. Day, W.H.E.: Optimal algorithms for comparing trees with labeled leaves. J. Classif. **2**(1), 7–28 (1985)
7. Gambette, P., Gunawan, A.D.M., Labarre, A., Vialette, S., Zhang, L.: Locating a tree in a phylogenetic network in quadratic time. In: Przytycka, T.M. (ed.) RECOMB 2015. LNCS, vol. 9029, pp. 96–107. Springer, Cham (2015). doi:10.1007/978-3-319-16706-0_12
8. Gusfield, D., Eddhu, S., Langley, C.H.: Optimal, efficient reconstruction of phylogenetic networks with constrained recombination. JBCB **2**(1), 173–214 (2004)
9. Huson, D.H., Klöpper, T.H.: Beyond galled trees - decomposition and computation of galled networks. In: Speed, T., Huang, H. (eds.) RECOMB 2007. LNCS, vol. 4453, pp. 211–225. Springer, Heidelberg (2007). doi:10.1007/978-3-540-71681-5_15
10. Huynh, T.N.D., Jansson, J., Nguyen, N.B., Sung, W.-K.: Constructing a smallest refining galled phylogenetic network. In: Miyano, S., Mesirov, J., Kasif, S., Istrail, S., Pevzner, P.A., Waterman, M. (eds.) RECOMB 2005. LNCS, vol. 3500, pp. 265–280. Springer, Heidelberg (2005). doi:10.1007/11415770_20
11. van Iersel, L., Semple, C., Steel, M.: Locating a tree in a phylogenetic network. IPL **110**(23), 1037–1043 (2010)
12. Jansson, J., Sung, W.: Inferring a level-1 phylogenetic network from a dense set of rooted triplets. TCS **363**(1), 60–68 (2006)
13. van Iersel, L., Keijsper, J., Kelk, S., Stougie, L., Hagen, F., Boekhout, T.: Constructing level-2 phylogenetic networks from triplets. IEEE/ACM TCBB **6**(4), 667–681 (2009)
14. Nakhleh, L., Warnow, T., Linder, C.R.: Reconstructing reticulate evolution in species: theory and practice. JCB **12**(6), 796–811 (2005)

Computational Methods for the Prediction of Drug-Target Interactions from Drug Fingerprints and Protein Sequences by Stacked Auto-Encoder Deep Neural Network

Lei Wang[1,4], Zhu-Hong You[2(✉)], Xing Chen[3(✉)], Shi-Xiong Xia[1],
Feng Liu[5], Xin Yan[6], and Yong Zhou[1]

[1] School of Computer Science and Technology,
China University of Mining and Technology, Xuzhou 221116, China
{leiwang,xiasx,yzhou}@cumt.edu.cn
[2] Xinjiang Technical Institutes of Physics and Chemistry,
Chinese Academy of Science, Urumqi 830011, China
zhuhongyou@gmail.com
[3] School of Information and Control Engineering,
China University of Mining and Technology, Xuzhou 221116, China
xingchen@amss.ac.cn
[4] College of Information Science and Engineering, Zaozhuang University,
Zaozhuang 277100, Shandong, China
[5] China National Coal Association, Beijing 100713, China
lf@mtkj.org
[6] School of Foreign Languages, Zaozhuang University,
Zaozhuang 277100, Shandong, China
xinyanuzz@gmail.com

Abstract. Identifying the interaction among drugs and target proteins is an important area of drug research, which provides a broad prospect for low-risk and faster drug development. However, due to the limitations of traditional experiments when revealing drug-protein interactions (DTIs), the screening of targets not only takes a lot of time and money, but also has high false-positive and false-negative rates. Therefore, it is imperative to develop effective automatic computational methods to accurately predict DTIs in the post-genome era. In this paper, we propose a new computational method for predicting DTIs from drug molecular structure and protein sequence by using the stacked auto-encoder of deep learning which can adequately extracts the raw data information. The proposed method has the advantage that it can automatically mine the hidden information from protein sequences and generate highly representative features through iterations of multiple layers. The feature descriptors are then constructed by combining the molecular substructure fingerprint information, and fed into the rotation forest for accurate prediction. The experimental results of 5-fold cross-validation indicate that the proposed method achieves superior performance on golden standard datasets (*enzymes, ion channels, GPCRs and*

L. Wang and Z.-H. You—The authors wish it to be known that, in their opinion, the first two authors should be regarded as joint First Authors.

© Springer International Publishing AG 2017
Z. Cai et al. (Eds.): ISBRA 2017, LNBI 10330, pp. 46–58, 2017.
DOI: 10.1007/978-3-319-59575-7_5

nuclear receptors) with accuracy of 0.9414, 0.9116, 0.8669 and 0.8056, respectively. We further comprehensively explore the performance of the proposed method by comparing it with other feature extraction algorithm, state-of-the-art classifier and other excellent methods on the same dataset. The excellent comparison results demonstrate that the proposed method is highly competitive when predicting drug-target interactions.

Keywords: Drug-target interactions · Position-specific scoring matrix · Stacked auto-encoder · Deep learning

1 Introduction

Drug targets are generally those associated with disease or pathological state of biological molecules, and the identification of them is the basis for drug research and development. Although much progress has been made over the past few decades, drug discovery remains a long and expensive process [1–3]. In addition, new drugs typically reach market needs for ten years, and the number of new molecular entities (NMEs) approved annually by the US Food and Drug Administration (FDA) as new drugs is only about 20 [4]. Therefore, researchers have intensified research into the identification of the relationship among drugs and targets, hoping to accelerate the pace of drug development and shorten the time to market.

Traditional drug discovery primarily followed the idea of 'one drug-one target-one disease' and believes that drugs with high selectivity to be safer and more effective. In accordance with this concept, some effective chemical molecules that affect the specific proteins are identified. This traditional concept, however, only focuses on the individual factors that target drug design in the disease system and ignores the complex interactions among drugs and their target proteins, so this model does not achieve the goal of accelerating new drug discovery [5–7]. Recently, more and more researchers have accepted the idea that the target of drugs is not a single target protein, but multiple target proteins [8–11]. So how to identify the complex interactions among drugs and targets rapidly and accurately has become the key to drug development. Because computational methods have the advantages of short time, low cost, high precision and wide range in exploring potential drug-target interactions, researchers hope to use it to solve this problem.

In recent years, many computational methods have been proposed to extrapolate potential drug-target interactions on a genome-wide scale [12]. Yamanishi et al. integrated the relationship among the pharmacological space, the chemical space and the topology of drug-target interaction networks to predict the associations among drugs and targets, and their experimental results have demonstrated that drug-target interactions are more correlated with pharmacological effect similarity than with chemical structure similarity [13]. Wang et al. employed supervised machine learning methods to predict the relationship among drugs and targets. In order to solve the problem of sample imbalance, they are collecting the positive samples from the database, and the negative samples using the random selection method. [14]. Chen et al. developed a novel method of Network-based Random Walk with Restart on the Heterogeneous network (NRWRH) to predict potential drug-target interactions on a large scale. The excellent experimental results show that the proposed method is able to discover new potential drug-target interactions for drug development [5].

In this paper, based on the hypothesis that the interactions among drugs and target proteins are closely related to the sequence of the target proteins and the molecular structure of the drug compounds, a novel computational method is proposed to infer unknown drug-target interactions on a large scale. The proposed method consists of three steps: first, it converts the sequence of the target protein into a matrix containing biological evolutionary information; and then apply the depth learning algorithm to learn the hidden high-level features; finally, combines these features and drug molecule fingerprint information and fed into the rotation forest classifier, according to the decision tree voting results to select the most probable targets. In the experiment, we make the predictions on the golden standard drug-target interactions datasets involving enzymes, ion channels, GPCRs and nuclear receptors. In addition, we compared other feature extraction method and classifier, and the experimental results show that our approach is a promising method for predicting the mutual relationship of drugs and targets.

2 Materials and Methods

2.1 Drug Molecules Description

A growing number of studies have shown that drugs with similar chemical structure have similar therapeutic functions. So far, several types of descriptors have been designed to represent drugs, including molecular substructure fingerprints, topological, constitutional, quantum chemical properties, and geometrical. Here we use the chemical structure of molecular substructure fingerprints to effectively represent the drug [15]. In this type of representation, each molecular structure is encoded as a fingerprint of a structural key according to a substructure pattern of a predefined dictionary, which is described by a Boolean vector.

In this experiment, we used the chemical structure of the molecular substructure fingerprints from PubChem database. It defines an 881 dimensional binary vector to represent the molecular substructure. Depending on the presence or absence of substructures, the corresponding bits of the vector are encoded as 1 or 0.

2.2 Position-Specific Scoring Matrix

The Position-Specific Scoring Matrix (PSSM) is introduced by Gribskov et al. for detecting distantly related protein [16]. The structure of PSSM is a matrix of M rows and 20 columns, where row represents the total number of amino acids in the protein and column represents the 20 naive amino acids. Suppose $R = \{\varrho_{(i,j)}: i = 1 \cdots M$ and $j = 1 \cdots 20\}$ and each matrix is represented as follows:

$$R = \begin{bmatrix} \varrho_{1,1} & \varrho_{1,2} & \cdots & \varrho_{1,20} \\ \varrho_{2,1} & \varrho_{2,2} & \cdots & \varrho_{2,20} \\ \vdots & \vdots & \vdots & \vdots \\ \varrho_{M,1} & \varrho_{M,2} & \cdots & \varrho_{M,20} \end{bmatrix} \tag{1}$$

where $\varrho_{i,j}$ in the i row of PSSM mean that the probability of the ith residue being mutated into type j of 20 native amino acids during the procession of evolutionary in the protein from multiple sequence alignments. In order to obtain highly homologous sequences, we set the number of iterations to 3, the value of e-value to 0.001, and other parameters to the default values.

2.3 Stacked Auto-Encoder

Stacked Auto-Encoder (SAE) is a popular depth learning model, which uses auto-encoders as building blocks to create deep neural network [17]. The auto-encoder (AE) can be considered as a special neural network with one input layer, one hidden layer and one output layer, as shown in Fig. 1.

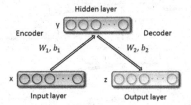

Fig. 1. Structure of auto-encoder

Given a training sample X, the autocoder first encodes the input $X \in R^{d_0}$ into the hidden representation $Y \in R^{d_1}$ by the mapping f_c:

$$Y = f_c(X) = S_c(W_1^T X + b_1) \tag{2}$$

where S_c is the activation function of the encoder, and its input is called the activation of the hidden layer. W_1 and b_1 is the parameter set with a weight matrix $W_1 \in R^{d_0 \times d_1}$ and a bias vector $b_1 \in r^{d_1}$. In the second step, the decoder maps the representation of the hidden layer Y to the output layer $Z \in R^{d_0}$ by the mapping function f_d.

$$Z = f_d(Y) = S_d(W_2^T Y + b_2) \tag{3}$$

where S_d is the activation function of the decoder, W_2 and b_2 is the parameter set with the weight matrix $W_2 \in R^{d_0 \times d_1}$ and the bias vector $b_2 \in r^{d_0}$. The parameters are learned by back-propagation through the minimizing the loss function $\Theta(X,Z)$ in the formula 4.

$$\Theta(X,Z) = \Theta_r(X,Z) + 0.5\tau \left(\|W_1\|_2^2 + \|W_2\|_2^2 \right) \tag{4}$$

where $\Theta_r(X, Z)$ is the reconstruction error, and τ is the weight decay cost. To minimize reconstruction errors, we need to represent as much of the original input as possible on hidden layer features. In this way, the hidden layer learns the feature information of the original input to the maximum extent.

The combination of multiple auto-encoders together constitutes the stacked auto-encoders, which has the characteristics of deep learning. Figure 2 shows the structure of the stacked auto-encoder with h-level auto-encoders which are trained in the layer-wise and bottom-up manner. The input vector is received at the first level of the auto-coder and sent to its hidden layer after training. The second layer of the auto-encoder receives data from the first layer, and sent to its hidden layer after training. The raw data is transformed from layer to layer up to the top layer. The activation function is usually the sigmoid function or tanh function. After completing these unsupervised features training, the entire neural network can use the tagged data to fine-tune the training parameters. The hidden layer of the highest layer auto-encoder can be used as the feature of the original data extraction by the stacked auto-encoder and can be applied to classifiers. In this paper, we set up a 3 layer auto-encoder, and use the rotation forest as the final classifier.

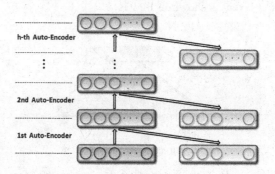

Fig. 2. Structure of stacked auto-encoders

2.4   Rotation Forest Classifier

The rotation forest (RF) proposed by Rodriguez et al. [18] as a popular ensemble classifier has been widely used in various fields. In the execution of RF, the samples are first randomly divided into different subsets; then each subset is rotated to increase diversity using the Principal Component Analysis (PCA); finally, the transformed subsets are fed into different decision trees. The final result of the classification is generated by voting on these decision trees. The steps for rotation forest are as follows

Let M denote the sample set, $X = (x_1, x_2, \ldots, x_n)^T$ be an n × L matrix which is composed of n observation feature vector for each training sample and $Y = (y_1, y_2, \ldots, y_n)^T$ denote the corresponding labels. Therefore, the training samples can be expressed as $\{x_i, y_i\}$, wherein $x_i = (x_{i1}, x_{i2}, \ldots, x_{iL})$ be a L-dimensional feature vector. Suppose that the sample set is randomly divided into K subsets of the same size by an appropriate factor and transformed by PCA. And then all the coefficients of the principal components are rearranged and stored to form a rotation matrix to change the original training set. In this case P decision trees in the forest can be expressed as

$R_1, R_2, \ldots, R_P$, respectively. The preprocessing steps of the training set for a single classifier R_i are shown below:

(1) The sample set M is randomly divided into K (a factor of n) disjoint subsets, and each subset contains the number of features is n/k.

(2) Select the corresponding column of the feature in the subset $M_{i,j}$ to form a new matrix $X_{i,j}$ from the training dataset X. A new training set $X'_{i,j}$ which is extracted from $X_{i,j}$ randomly with 75% of the dataset using bootstrap algorithm. Loop K times in this way, so that each subset is converted

(3) Matrix $X'_{i,j}$ is used as the feature transform by principal component analysis (PCA) technique for producing the coefficient matrix $S_{i,j}$, which jth column coefficient as the characteristic component jth.

(4) A sparse rotation matrix G_i is constructed, and its coefficients which obtained from the matrix $S_{i,j}$ expressed as follows:

$$G_i = \begin{bmatrix} \varkappa_{i,1}^{(1)}, \ldots, \varkappa_{i,1}^{(C_1)} & 0 & \cdots & 0 \\ 0 & \varkappa_{i,2}^{(1)}, \ldots, \varkappa_{i,2}^{(C_2)} & \cdots & 0 \\ \vdots & \vdots & \ddots & \vdots \\ 0 & 0 & \cdots & \varkappa_{i,k}^{(1)}, \ldots, \varkappa_{i,k}^{(C_k)} \end{bmatrix} \quad (5)$$

In the prediction period, provided the test sample x, generated by the classifier R_i of $d_{i,j}(XG_i^\varkappa)$ to determine x belongs to class y_i. And then the class of confidence is calculated by means of the average combination, and the formula is as follows:

$$\theta_j(x) = \frac{1}{p}\sum_{i=1}^{P} d_{i,j}(XG_i^\varkappa) \quad (6)$$

Therefore, the test sample x easily assigned to the classes with the greatest possible. In this experiment, the parameters of the rotation forest are optimized by the grid search method, and finally set K to 52 and L to 5. The flow chart of the proposed method is shown in Fig. 3.

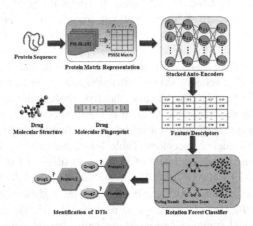

Fig. 3. The flow chart of the proposed method

3 Results and Discussion

3.1 Evaluation Criteria

Evaluation criteria are particularly important for measuring methods. The advantages and disadvantages of this method can be objectively reflected by comparing with other methods under the unified evaluation criteria. The evaluation criteria used in this paper include accuracy (Accu.), sensitivity (Sen.), precision (Prec.), and Matthews correlation coefficient (MCC). They are calculated as:

$$Accu. = \frac{TP + TN}{TP + TN + FP + FN} \tag{7}$$

$$Sen. = \frac{TP}{TP + FN} \tag{8}$$

$$Prec. = \frac{TP}{TP + FP} \tag{9}$$

$$MCC = \frac{TP \times TN - FP \times FN}{\sqrt{(TP + FP)(TP + FN)(TN + FP)(TN + FN)}} \tag{10}$$

where TP (true positive) denotes the number of positive samples correctly identified; FP (false positive) denotes the number of positive samples incorrectly identified; TN (true negative) denotes the number of negative samples correctly identified; FN (false negative) denotes the number of negative samples incorrectly identified. The Receiver Operating Characteristic (ROC) curve [19] is introduced to visually display the performance of classifier.

3.2 Assessment of Prediction Ability

In this paper, we use 5-fold cross-validation to assess the predictive ability of our model in the golden standard datasets involving enzymes, ion channels, GPCRs and nuclear receptors. Cross-validation can not only prevent over-fitting, but also can test the stability of the model. Its implementation steps are: firstly, all the samples are randomly divided into five disjoint subsets of the equal number; then, each time one different subset is used as test set, and the remaining four is used as training set, so that the formation of the five models; finally, the five models are used to predict the classification, and the average value of them is the final result.

The proposed model performs well in the golden standard datasets: enzymes, ion channels, GPCRs and nuclear receptors. Table 1 lists the experimental results on the enzyme dataset, it yielded an accuracy of 0.9414, sensitivity of 0.9555, precision of 0.9293, MCC of 0.8832 and AUC of 0.9425. And their standard deviations are 0.0030, 0.0064, 0.0067, 0.0058 and 0.0022, respectively. The highest accuracy of the five models reached 0.9462, and the lowest also reached 0.9385. Table 2 shows the performance of our model implementation in the icon channel dataset. The accuracy, sensitivity, precision, MCC and AUC of cross-validation are 0.9116, 0.9569, 0.8778, 0.8271 and 0.9107, respectively. The standard deviations for these criteria values are 0.0086, 0.0188, 0.0219, 0.0162 and 0.0074. Table 3 lists the results when the GPCRs dataset is used to predict drug-target interactions. The average accuracy, sensitivity, precision, MCC and AUC are 0.8669, 0.8164, 0.9102, 0.7396 and 0.8743, respectively. The standard deviations for these criteria values are 0.0446, 0.0651, 0.0380, 0.0837 and 0.0417, respectively. The highest accuracy of the five models reached 0.9331. Table 4 summarizes the statistical results of the cross-validation of nuclear receptor dataset. We achieved an accuracy of 0.8056, sensitivity of 0.7627, precision of 0.8410, MCC of 0.6188 and AUC of 0.8176. Their standard deviations are 0.0439, 0.1284, 0.0688, 0.0712 and 0.0676, respectively. Figures 4, 5, 6 and 7 show the ROC curves obtained on the enzymes, ion channels, GPCRs and nuclear receptors datasets by the proposed method.

Table 1. The 5-fold cross-validation results were generated on the *enzyme* dataset by using the proposed method

Test set	Accu. (%)	Sen. (%)	Prec. (%)	MCC (%)	AUC (%)
1	93.93	96.43	91.91	87.97	94.50
2	94.62	95.95	93.59	89.25	94.21
3	93.85	95.43	92.61	87.73	94.03
4	94.19	94.95	93.32	88.39	94.05
5	94.11	94.99	93.22	88.24	94.46
Average	**94.14 ± 0.30**	**95.55 ± 0.64**	**92.93 ± 0.67**	**88.32 ± 0.58**	**94.25 ± 0.22**

Table 2. The 5-fold cross-validation results were generated on the *icon channel* dataset by using the proposed method

Test set	Accu. (%)	Sen. (%)	Prec. (%)	MCC (%)	AUC (%)
1	91.86	96.71	88.55	84.05	90.91
2	90.51	97.59	85.24	81.89	91.28
3	91.53	96.81	86.94	83.59	91.62
4	90.00	93.88	87.07	80.25	89.86
5	91.89	93.46	91.08	83.78	91.69
Average	**91.16 ± 0.86**	**95.69 ± 1.88**	**87.78 ± 2.19**	**82.71 ± 1.62**	**91.07 ± 0.74**

Table 3. The 5-fold cross-validation results were generated on the *GPCR* dataset by using the proposed method

Test set	Accu. (%)	Sen. (%)	Prec. (%)	MCC (%)	AUC (%)
1	86.61	82.68	89.74	73.46	86.56
2	93.31	91.13	94.96	86.66	92.97
3	83.46	79.51	85.09	66.93	83.84
4	81.89	73.05	92.79	66.09	83.39
5	88.19	81.82	92.52	76.68	90.40
Average	**86.69 ± 4.46**	**81.64 ± 6.51**	**91.02 ± 3.80**	**73.96 ± 8.37**	**87.43 ± 4.17**

Table 4. The 5-fold cross-validation results were generated on the nuclear receptor dataset by using the proposed method

Test set	Accu. (%)	Sen. (%)	Prec. (%)	MCC (%)	AUC (%)
1	86.11	87.50	82.35	72.16	86.25
2	83.33	84.62	73.33	65.49	88.29
3	77.78	72.73	88.89	56.98	77.27
4	80.56	80.95	85.00	60.47	84.76
5	75.00	55.56	90.91	54.27	72.22
Average	**80.56 ± 4.39**	**76.27 ± 12.84**	**84.10 ± 6.88**	**61.88 ± 7.12**	**81.76 ± 6.76**

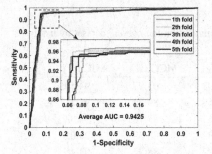

Fig. 4. The ROC curves were generated on the *enzyme* dataset by using the proposed method

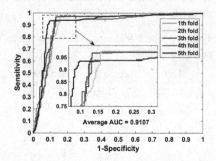

Fig. 5. The ROC curves were generated on the *icon channel* dataset by using the proposed method

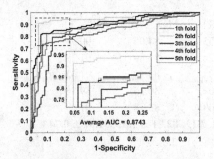

Fig. 6. The ROC curves were generated on the *GPCR* dataset by using the proposed method

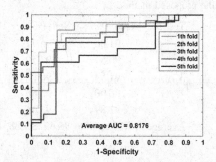

Fig. 7. The ROC curves were generated on the *nuclear receptor* dataset by using the proposed method

3.3 Comparison Between RF Classifier and Support Vector Machine Classifier

Support Vector Machine (SVM) is a supervised learning algorithm, which has outstanding performance on regression tasks and two-class classification problems

Table 5. Comparison of cross-validation results between the RF classifier and the SVM classifier on the *enzyme* dataset

Test set	Accu. (%)	Sen. (%)	Prec. (%)	MCC (%)	AUC (%)
1	86.41	93.21	82.19	73.47	92.85
2	87.61	92.07	84.78	75.47	93.12
3	84.70	91.37	80.84	69.97	92.00
4	86.92	91.46	83.47	74.21	92.25
5	87.12	93.26	82.82	74.85	91.80
Average	**86.55 ± 1.12**	**92.28 ± 0.92**	**82.82 ± 1.47**	**73.95 ± 2.16**	**92.40 ± 0.56**
Our method	**94.14 ± 0.30**	**95.55 ± 0.64**	**92.93 ± 0.67**	**88.32 ± 0.58**	**94.25 ± 0.22**

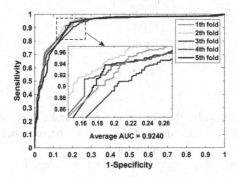

Fig. 8. The ROC curves were generated on the *enzyme* dataset by using the SVM classifier

[20–22]. In this section, we use the same features to compare the proposed classifier with the state-of-the-art SVM classifier. After the parameters of the SVM are optimized by the grid search method, the parameter c is set to 20, g is set to 800. The results of the comparison between them on the enzyme dataset are summarized in Table 5. Our method has achieved good results on all evaluation criteria including accuracy, sensitivity, precision, MCC and AUC. They increased by 7.59%, 3.27%, 10.11%, 14.37% and 1.85%, respectively. While the standard deviations decreased by 0.82%, 0.28%, 0.80%, 1.58% and 0.34%, respectively (Fig. 8).

3.4   Comparison with State-of-the-Art Methods

To test the robustness and reliability of the proposed method, we compared it with state-of-the-art methods on the golden standard datasets. We collected the AUC values generated by the four methods on the enzymes, ion channels, GPCRs and nuclear receptors datasets. As shown in Table 6, our method performs best on the enzymes, ion channels and GPCRs datasets with AUC values of 0.9425, 0.9107, and 0.8743, respectively. On the nuclear receptors dataset, the highest value obtained by the NetCBP method is 0.856, but our method also achieved 0.8176 results, which is only

3.84% lower than it, also reached the average level. The results of the comparison show that the stacked auto-encoder combined with the rotation forest classifier can improve the prediction ability on the golden standard datasets (Table 6).

Table 6. State-of-the-art methods and our method obtained the AUC values on the golden standard datasets

Method	Enzymes	Icon channels	GPCRs	Nuclear receptors
KBMF2 K [23]	0.832	0.799	0.857	0.824
DBSI [24]	0.8075	0.8029	0.8022	0.7578
NetCBP [4]	0.8251	0.8034	0.8235	0.8394
Yamanishi [13]	0.821	0.692	0.811	0.814
WNN-GIP [14]	0.861	0.775	0.872	0.839
SIMCOMP [25]	0.863	0.776	0.867	**0.856**
NLCS [25]	0.837	0.753	0.853	0.815
Our method	**0.9425**	**0.9107**	**0.8743**	0.8176

4   Conclusion

In this paper, based on the idea that the relationship among drugs and targets is largely influenced by the chemical structure of the drug and the sequence information of the protein, we propose a novel computational method to infer potential unknown drug-target interactions on a genome-wide scale by integrating protein amino acid sequence and drug molecular structure. To extract more representative features, we use deep learning technology to learn the protein sequence that is converted into the matrix containing biological evolutionary information. And then combine with the molecular fingerprint information to form the feature descriptor sent to the rotation forest for classification. The proposed method is applied to four classes of target proteins, including enzymes, ion channels, GPCRs and nuclear receptors. To evaluate the performance of our method, we experimented with different feature extraction method, classifier, and compared with other methods. Excellent experimental results show that our method has a prominent ability in mining the hidden interactions among drugs and targets. We have reasons to believe that the proposed method will play an important role in promoting the research and development of drugs. In future work, we plan to integrate more biology knowledge, using more advanced machine learning methods to improve the ability to predict.

Acknowledgements. This work is supported by the Fundamental Research Funds for the Central Universities (2017XKQY083).

References

1. Dickson, M., Gagnon, J.P.: Key factors in the rising cost of new drug discovery and development. Nat. Rev. Drug Discov. **3**, 417–429 (2004)

2. Paul, S.M., Mytelka, D.S., Dunwiddie, C.T., et al.: How to improve R&D productivity: the pharmaceutical industry's grand challenge. Nat. Rev. Drug Discov. **9**, 203–214 (2010)
3. Kola, I., Landis, J.: Can the pharmaceutical industry reduce attrition rates? Nat. Rev. Drug Discov. **3**, 711–715 (2004)
4. Chen, H., Zhang, Z.: A semi-supervised method for drug-target interaction prediction with consistency in networks. PLoS ONE **8**, 5 (2013)
5. Chen, X., Liu, M.X., Yan, G.Y.: Drug-target interaction prediction by random walk on the heterogeneous network. Mol. BioSyst. **8**, 1970–1978 (2012)
6. Iskar, M., Zeller, G., Zhao, X.-M., et al.: Drug discovery in the age of systems biology: the rise of computational approaches for data integration. Curr. Opin. Biotechnol. **23**, 609–616 (2012)
7. Gao, Z.G., Wang, L., Xia, S.X., et al.: Ens-PPI: a novel ensemble classifier for predicting the interactions of proteins using autocovariance transformation from PSSM. In: Biomed Research International, p. 8 (2016)
8. Hopkins, A.L.: Network pharmacology: the next paradigm in drug discovery. Nat. Chem. Biol. **4**, 682–690 (2008)
9. Yang, K., Bai, H.J., Qi, O.Y., et al.: Finding multiple target optimal intervention in disease-related molecular network. Mol. Syst. Biol. **4**, 13 (2008)
10. Xie, L., Xie, L., Kinnings, S.L., et al.: Novel computational approaches to polypharma-cology as a means to define responses to individual drugs. Annu. Rev. Pharmacol. Toxicol. **52**, 361 (2012)
11. Chen, X., Ren, B., Chen, M., et al.: NLLSS: predicting synergistic drug combinations based on semi-supervised learning. PLoS Comput. Biol. **12**, 7 (2016)
12. Chen, X., Yan, C.C., Zhang, X., et al.: Drug–target interaction prediction: databases, web servers and computational models. Briefings Bioinform. **17**, 696–712 (2016)
13. Yamanishi, Y., Kotera, M., Kanehisa, M., et al.: Drug-target interaction prediction from chemical, genomic and pharmacological data in an integrated framework. Bioinformatics **26**, i246–i254 (2010)
14. Wang, Y.-C., Yang, Z.-X., Wang, Y., et al.: Computationally probing drug-protein interactions via support vector machine. Lett. Drug Des. Discov. **7**, 370–378 (2010)
15. Shen, J., Cheng, F.X., Xu, Y., et al.: Estimation of ADME properties with substructure pattern recognition. J. Chem. Inf. Model. **50**, 1034–1041 (2010)
16. Gribskov, M., McLachlan, A.D., Eisenberg, D.: Profile analysis: detection of distantly related proteins. Proc. Natl. Acad. Sci. U.S.A. **84**, 4355–4358 (1987)
17. Bengio, Y., Lamblin, P., Popovici, D., et al.: Greedy layer-wise training of deep networks. Adv. Neural. Inf. Process. Syst. **19**, 153 (2007)
18. Rodriguez, J.J., Kuncheva, L.I.: Rotation forest: a new classifier ensemble method. IEEE Trans. Pattern Anal. Mach. Intell. **28**, 1619–1630 (2006)
19. Zweig, M.H., Campbell, G.: Receiver-operating characteristic (ROC) plots: a fundamental evaluation tool in clinical medicine. Clin. Chem. **39**, 561–577 (1993)
20. Cao, D.-S., Liang, Y.-Z., Xu, Q.-S., et al.: Exploring nonlinear relationships in chemical data using kernel-based methods. Chemometr. Intell. Lab. Syst. **107**, 106–115 (2011)
21. Smola, A.J., Scholkopf, B.: A tutorial on support vector regression. Stat. Comput. **14**, 199–222 (2004)
22. Cao, D.-S., Xu, Q.-S., Liang, Y.-Z., et al.: Prediction of aqueous solubility of druglike organic compounds using partial least squares, back-propagation network and support vector machine. J. Chemometr. **24**, 584–595 (2010)

23. Gonen, M.: Predicting drug-target interactions from chemical and genomic kernels using Bayesian matrix factorization. Bioinformatics **28**, 2304–2310 (2012)
24. Cheng, F., Liu, C., Jiang, J., et al.: Prediction of drug-target interactions and drug repositioning via network-based inference. PLoS Comput. Biol. **8**, 5 (2012)
25. Öztürk, H., Ozkirimli, E., Özgür, A.: A comparative study of SMILES-based compound similarity functions for drug-target interaction prediction. BMC Bioinform. **17**, 1–11 (2016)

Analysis of Paired miRNA-mRNA Microarray Expression Data Using a Stepwise Multiple Linear Regression Model

Yiqian Zhou[1], Rehman Qureshi[2], and Ahmet Sacan[3(✉)]

[1] Pure Storage, 650 Castro Street, Suite #260, Mountain View, CA 94041, USA
[2] Bioinformatics Facility, Wistar Institute,
3601 Spruce Street, Philadelphia, PA 19104, USA
[3] Biomedical Engineering, Drexel University,
3120 Market Street, Philadelphia, PA 19104, USA
ahmet.sacan@drexel.edu

Abstract. MicroRNAs are small endogenous RNAs that play important roles in gene regulation. With the accumulation of expression data, numerous approaches have been proposed to infer miRNA-mRNA regulation from paired miRNA-mRNA expression profiles. These mainly focus on discovering and validating the structure of regulatory networks, but do not address the prediction and simulation tasks. Furthermore, functional annotation of miRNAs relies on miRNA target prediction, which is problematic since miRNA-gene interactions are highly tissue-specific. Thus a different approach to functional annotation of miRNA-mRNA regulation that can generate context-specific expression levels is needed. In this study, we analyzed paired miRNA-mRNA expressions from breast cancer studies. The expression of mRNAs is modeled as a multiple linear function of the expression of miRNAs and the parameters are estimated using stepwise multiple linear regression (SMLR). We demonstrate that the SMLR model can predict mRNA expression patterns from miRNA expressions alone and that the predicted gene expression levels preserve differentially regulated gene sets, as well as the functional categories of these genes. We show that our quantitative approach can determine affected biological activities better than the traditional target-prediction based methods.

Keywords: Micro-RNA · Gene expression · Co-expression · Stepwise multiple linear regression

1 Introduction

MicroRNAs (miRNAs) are small (~ 22 nucleotides) non-coding endogenous RNAs that play important roles in gene regulation by targeting the messenger RNA (mRNA) of protein-coding genes [1]. In most cases, though not always [2], miRNAs act to repress the expression of their target gene [3, 4]. miRNAs guide the repression by either degrading the mRNA molecules, decreasing the translational efficiency, or both. When a miRNA and its target mRNA are highly complementary, the pairing is extensive and the miRNA directs the cleavage of the mRNA, which is the predominant mode of

© Springer International Publishing AG 2017
Z. Cai et al. (Eds.): ISBRA 2017, LNBI 10330, pp. 59–70, 2017.
DOI: 10.1007/978-3-319-59575-7_6

miRNA-guided repression in plants. In animals, extensive miRNA-mRNA complementary pairing and the consequent cleavage of mRNA is less prevalent. Nevertheless, recent studies indicate that target mRNA degradation provides a major contribution to translational repression in animals [5, 6].

miRNAs participate in a wide range of biological processes, affecting the expression of over 60% of mammalian genes [7]. Over the past decade, it has become clear that miRNAs contribute to almost all known physiological and pathological processes, cancer being of particular interest. Since dysregulation of miRNAs is closely linked with dysregulation of oncogenes and tumor suppressors, studying the biological processes of miRNAs provides unique opportunities for the development of miRNA-based diagnostics and treatment of cancer [8, 9].

To understand the functions of miRNAs, a central goal and major challenge is to determine their target mRNAs. There are many experimental techniques for target identification of miRNAs of interest [10]. These experimentally identified miRNA-mRNA interactions are collected in several repositories, such as TarBase [11] and miRTarBase [12]. So far thousands of miRNAs have been identified in animals and plants, but only a small fraction of targets for these miRNAs have been validated experimentally, because of the low efficiency and high cost of experimental validation. Sequence-based computational methods have been developed to fill this gap by generating putative lists of miRNA-mRNA pairs, which have greatly reduced the number of interactions researchers need to validate experimentally. Widely used miRNA target prediction methods include TargetScan [7], miRanda [13], PicTar [14], TargetScanS [15], and DIANA-microT [16].

Currently, reliable prediction of miRNA-mRNA interactions remains a challenge. Predictions based solely on sequence information have high false positive rates [17]. In order to improve the performance, novel integrative approaches that combine sequence based predictions and miRNA experimental data are needed. Genome-wide mRNA expression measurement has become an indispensable tool in molecular biology. Similarly, technological advances have spawned a multitude of miRNA profiling platforms [18]. They together provide paired miRNA-mRNA expression profiles that enable researchers to pinpoint important miRNAs and their roles in particular biological processes.

Several methods that incorporate these high throughput data have been developed to find miRNA-mRNA regulatory pairs, including those based on correlation [19–22] or mutual information [23]. The findings from gene-expression analysis can be integrated with those from sequence-based methods by intersection [24] or weighted sum [20]. These simple approaches are efficient in extracting potential interactions from big datasets but they only consider independent pairwise miRNA-mRNA associations. Since a mRNA can be targeted by several miRNAs and its expression profile is affected by multiple miRNAs at the same time, multiple linear regression models have been proposed [25, 26]. When the data is co-linear or the number of samples is less than the number of regulators, the linear model is underdetermined and optimal solution is unattainable. This can be circumvented by introducing penalty terms to the system, such as $L_1 - norm$, $L_2 - norm$, or combination of both, of the coefficients of regulators [27]. In addition to regression-based approaches, several Bayesian models have been developed, inferring the posterior probability of real miRNA-mRNA interactions based on the expression data, such as implemented in GenmiR++ [28] and its variations [29–31]. Bayesian network structure learning has also been proposed [32], in which regulatory relationships are represented as a graph and the graph that is best supported by the expression data is sought after.

The approaches proposed so far have focused on inference and validation of the "structure" of the miRNA-mRNA regulatory networks from the paired miRNA-mRNA expression data. Although knowing which genes are targeted by which miRNAs is of great value, it is not sufficient for determining whether a gene would be differentially expressed in a particular cellular context.

We have previously shown that a simple linear model is able to quantitatively predict and simulate gene expression levels in time-series data [33]. In this study, we investigate the application of a similar linear model for quantitative estimation of mRNA expression levels from miRNA data. The present study is unique in its focus on explicit quantitative modeling of gene expression levels, rather than just identifying miRNA targets.

2 Methods

We infer miRNA-mRNA regulatory interactions by analyzing paired miRNA-mRNA expression data using stepwise multiple linear regression (SMLR) [33]. Suppose there are M mRNAs and N miRNAs of interest; the expression level of each mRNA is modeled as a linear function of the expression levels of the miRNAs:

$$y_i = \beta_{i0} + \sum_{j=1}^{N} \beta_{ij} x_j + \varepsilon_i \tag{1}$$

where y_i and x_j are variables representing the expression of mRNA i and miRNA j respectively ($i = 1, 2, \ldots, M$ and $j = 1, 2, \ldots, N$); ε_i is the error term; and β_{i0} is a constant term representing the baseline mRNA expression. The β_{ij} term characterizes the regulatory effect of miRNA j on mRNA i. We identify the coefficient weights β_{ij} using stepwise multiple linear regression with a forward selection strategy, as described in our previous study [33]. Briefly, the predictors for a given gene y_i are identified starting with the inclusion of the constant term. In each forward selection step, individual predictor variables are considered for addition based on their statistical significance in the regression fitting. The p-value of an F-statistic for each variable is calculated to determine whether to include or exclude that variable in the model, using the null hypothesis that its weight coefficient is zero.

Suppose there are L samples; we can denote the expression of mRNA i and miRNA j across samples as row vectors: $y_i = [y_{i1}, y_{i2}, \ldots, y_{iL}]$ and $x_j = [x_{j1}, x_{j2}, \ldots, x_{jL}]$. More compactly, let $X = [1; x_1; x_2; \ldots; x_N]$ and $Y = [y_1; y_2; \ldots; y_M]$, with each row representing a mRNA or miRNA and each column representing a sample. If the data is already normalized, the constant term 1 in X can be dropped, leaving X and Y with dimensions of M-by-L and N-by-L, respectively, and representing the experimental data of M miRNAs and N mRNAs across L samples. Let $\beta_i = [\beta_{i0}, \beta_{i1}, \beta_{i2}, \ldots, \beta_{iN}]$ and $B = [\beta_1; \beta_2; \beta_3, \ldots, \beta_M]$. Then the SMLR model can be written in a simple matrix form:

$$Y = B * X \tag{2}$$

The coefficient matrix B is M-by-N, which represents miRNA-mRNA regulatory interactions from M miRNAs and N mRNAs. Note that the coefficient matrix B is sparse, since the coefficients of insignificant interactions are set to zero.

Before estimating the interaction coefficients from training data and predicting gene expression levels, we need to perform necessary data pre-processing. Since we want to have a general model that works for expression datasets from different platforms and given the fact that most expression data available on Gene Expression Omnibus (GEO) database have already been normalized based on different assumptions regarding the specific platform, we avoid extra normalization across each sample unless necessary. First, we remove probes (genes) that have more than 3 missing data points and impute the missing value of the rest using the k-nearest-neighbor method with k = 3. Next, we center and scale the expression of each probe (gene) to have a mean value of zero and a standard deviation of one. This transformation does not alter the correlation between genes or the results of t-test for samples from different subgroups. Data preprocessing ensures that expression levels from different samples are on the same scale and that our predicted values can be directly compared with those from the real data. After preprocessing, we estimate the interaction coefficients B using stepwise multiple linear regression [33].

We evaluate the accuracy of the model predictions on both the training and independent testing datasets. In particular, we focus on how well the predictions preserve the differential expression profiles, as the list of differentially expressed genes is one of the most important outcomes from microarray studies. For both the real and predicted data, we perform Student's t-test to identify the genes that are significantly differentially expressed between experimental groups and analyze the overlap between the lists of genes generated from the real and predicted data.

A common downstream task in differential expression studies is the enrichment of differentially expressed genes into functional categories [34]. Here, we propose to use the mRNA levels estimated from our SMLR model for downstream functional annotation tasks. Considering any negative coefficient in the matrix B to indicate a targeting interaction, we evaluate the ability of our approach to discover mRNA targets and compare its performance to the TargetScan target prediction method [15] and to a negative correlation method where negatively correlated miRNA-mRNA are assumed to be targeting interactions (Pearson $p < 0.01$). Note that our method does not distinguish direct interactions from transitive ones or from those arising from co-regulation. Regardless of the source of the coefficients, our approach generates estimates of mRNA expression values, just as if they were obtained from a microarray gene expression experiment study. Once we obtain these estimated gene expression levels, we calculate a predicted list of differentially expressed genes and then perform gene set enrichment analysis using the DAVID web service [35]. Functional annotation is performed against OMIM, GO terms, BBID pathway, and KEGG pathway databases. We evaluate the performance on the functional enrichment task by comparing the resulting functional categories with those obtained from the real mRNA data and those obtained using target prediction methods.

In the following section, we first illustrate the application of SMLR to predict gene expression levels and functional categories, using a breast cancer expression profiling dataset. We then evaluate the ability of the model coefficients estimated from one dataset to generalize to another dataset generated from different experimental platforms. We compare the gene lists and functional categories predicted from miRNA data to those obtained from the real data and from TargetScan.

3   Results

In order to evaluate the ability of the SMLR model to predict gene expression levels from miRNA data, we first used the dataset available from a paired miRNA-mRNA study [36, 37], in which miRNA and mRNA profiles were obtained from the same primary breast cancer carcinomas (GSE19536, GSE22220), where the TP53 mutational and estrogen receptor (ER) status of each sample are also available. These samples are part of a larger cohort from the Oslo region [38].

After preprocessing, we obtained normalized expression profiles for 489 miRNAs and 40996 genes. We then performed leave-one-out-cross-validation (LOOCV) to evaluate the model, where we set aside one of the samples as the test sample and calculated the interaction coefficients from the remaining 100 training samples. The resulting model is then applied to the miRNA profiles from the training samples and the test sample separately. This procedure is repeated with each sample in the dataset used as the test sample.

Hierarchical clustering of the 1000 most differentially expressed mRNAs in the real data is shown in Fig. 1 (left). For comparison, a heatmap of the predicted expression levels are shown side-by-side (Fig. 1, right) with the same row and column arrangements. The predicted data displays surprisingly similar expression patterns, supporting

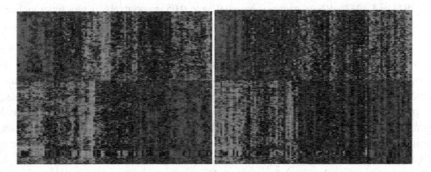

Fig. 1. Hierarchical clustering of mRNA expression. Left: Hierarchical clustering of the 1000 most differentially expressed mRNAs from the GSE19536 dataset. Right: expression levels of the same mRNAs predicted from the paired miRNA expression data, using SMLR with leave-one-out-cross-validation strategy. Rows are mRNA probes and columns are samples. Predicted data is shown with the same row and column arrangement as the real data. Root mean squared error (RMSE) of all predicted values was 1.11.

the idea that the miRNA expression alone provides a good summary of the gene expression state of the cell.

In order to further evaluate the reliability and usefulness of the gene expressions predicted from miRNA data, we examined whether the predicted values can identify a similar set of differentially expressed mRNAs. A two-sampled t-test on predicted gene expression data was performed between the ER-positive and ER-negative subgroups of samples. The p-values of the t-test are compared to those obtained from the original gene expression data (See Fig. 2-Left). These two set of p-values are highly correlated ($r = 0.77$). The mRNAs that are differentially expressed in the real data were likely to be found differentially expressed in the predicted data as well.

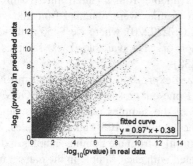 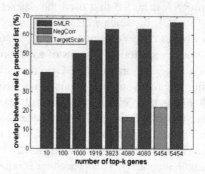

Fig. 2. Left: Comparison of differentially expressed mRNAs identified from the real and predicted expression data. Each point represents a mRNA, where the x and y axes show the $-\log_{10}$ transformed p-values obtained from an unpaired t-test in real and predicted data, respectively, comparing ER-positive and ER-negative breast cancer samples. The least-square fitted line is shown in red. **Right: Amount of overlap between the lists of differentially expressed genes in real and predicted data.** Percentage overlap between the most differentially expressed gene sets obtained from real and predicted data is shown. Each bar shows gene sets obtained with either a top-k or p-value criteria. After false discovery rate (FDR) correction, there were 1923 and 3942 mRNAs with p-value <0.01 and 0.05, respectively. (Color figure online)

Genome-wide microarray analysis is often used to prioritize a set of genes for follow-up wet-lab experimentation; such as reporter assays to confirm transcription, measurement of protein levels by northern blots, or knock-out experiments to evaluate phenotypic outcomes resulting from the absence of a gene. As such, it is important that our predictions preserve the ranking of the differentially expressed genes. Figure 2-Right shows the overlap between the top-k most differentially expressed gene sets obtained from the real and predicted data. The figure also shows the amount of overlap for gene sets obtained with the commonly used p-value thresholds of 0.01 and 0.05. At different top-k or p-value cut-offs, about half of the genes from the predicted gene set are in common with the real gene set.

Considering the noisy nature of gene expression data and the biological complexity of the rules governing translation of mRNAs to different protein isoforms, differential expression detected in microarray experiments is not conclusive for similar expression of the encoded proteins or for regulation of a particular phenotype the genes are involved in. Gene set enrichment is commonly utilized to find biological functions affected by the concerted changes in a set of genes.

For a miRNA study, the functional annotations of miRNAs of interest can be obtained by enrichment analysis with a set of their target mRNAs. Traditionally, the set of miRNAs of interest are selected according to their differential expression patterns and their targets are selected from sequence-based target prediction algorithms or from experimentally validated targets. All targets of differentially expressed miRNAs are then (falsely) assumed to also be differentially regulated, even though these target genes are also targeted by other non-differentially expressed miRNAs. This is an unrealistic assumption that results in thousands of genes, limiting the statistical power of the enrichment analysis. This is demonstrated in Fig. 2-Right, where we compare the accuracy of the genes assumed to be differentially regulated from negative correlation and TargetScan predictions (17% and 22%, respectively) with those obtained from our method (63% and 67% for the same number of genes). Compared to context-agnostic target-prediction methods, we more effectively utilize the cellular context available from the state of all miRNAs in determining whether a gene is differentially expressed.

We performed functional annotation of the gene lists using DAVID [35]. For real data, which is used as the ground truth, and for SMLR, we used differentially expressed genes (p < 0.01) in real and predicted expression data, respectively. For other methods, the gene lists were formed by combining all of the targets of differentially expressed miRNAs (p < 0.01). Overlap of the functional annotation terms obtained from different methods with those generated from the real data are shown in Fig. 3-Right. Top-3 functional categories enriched from the real data were: Phosphoprotein, Alternative Splicing, and Splice Variant. SMLR was able to generate the same three terms in its top-3; whereas TargetScan and negative correlation only ranked only one of them in

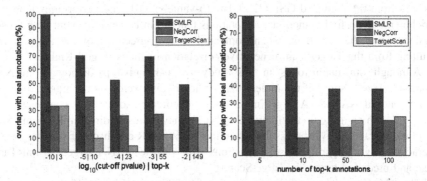

Fig. 3. Left: Functional enrichment from different methods. Percent overlap of functional annotations obtained from different methods with those obtained from real data are shown. At each p-value cutoff from SMLR, the same number of top-k annotations from each method are compared. Full list of enriched terms is available in the supplementary data. **Right: Comparison of functional enrichment in GSE19536 dataset.** SMLR is trained using GSE22220 dataset and differentially expressed genes from the predicted GSE19536 data are used for gene set enrichment. Negative correlation and TargetScan methods use all the predicted targets of differentially expressed miRNAs in GSE19536.

their top-3 lists. For the top-10 functional annotations obtained from each method, 70% were in common between results from real data and SMLR prediction, sharing similar rankings in statistical significance; while 40% and 10% were in common for negative correlation and TargetScan methods. These results support the claim that gene expression values predicted from miRNAs alone can capture the affected biological processes and that the functional annotations from estimated mRNA values are more accurate than those from collection of predicted targets.

The results above were obtained by leave-one-out cross-validation within a single experimental study, where each miRNA to mRNA mapping in a test sample was done using a model trained on the rest of the samples. Here we also evaluate the cross-database performance of SMLR by applying the model trained from one study to a dataset from an independent experimental study. Specifically, we train a model on GSE22220 dataset [36] and test its prediction performance on GSE19536 dataset [37]. Since miRNA-mRNA interactions are highly tissue-specific and development-specific, we focus on datasets from the same cancer type here. Although both datasets were from breast cancer samples, they used different microarray platforms for mRNA and miRNA profiling.

In order to perform a cross-database application of the model, we first find the mRNAs and miRNAs that are in common between the two studies. Since the studies use different microarray platforms with different probe IDs, we convert the mRNA probe IDs to their GeneBank accession numbers and the miRNA probe IDs to their miRBase IDs. This results in 14873 mRNAs and 232 miRNAs that are in common between the two studies.

The comparison of the heat maps generated from real and predicted data illustrates that SMLR is able to predict the overall expression profiles that reflect the ER status of the samples (See Fig. 4, top row). We observe the same behavior when the training and test datasets were switched (Fig. 4, bottom row). Taking the differentially expressed mRNAs from the predicted GSE22220 data (p-value < 0.01) and performing gene set enrichment, again finds functional annotations that are in better agreement with those obtained from the real data, when compared to the agreement of the annotations resulting from the TargetScan or negative correlation methods (Fig. 3-Right).

Although our main focus in this study is quantitative prediction of mRNA expression levels, some of the underlying predictors discovered by our model may be from direct miRNA-mRNA target interactions. Specifically, some of the coefficients w_i in Eq. "1" (which make up the matrix B in Eq. 2) may represent direct miRNA-mRNA targeting interactions. We assess the extent in which SMLR can discover such targeting interactions by comparing these interactions with known miRNA targets in miRTar-Base and predicted targets in TargetScan.

The SMLR model was trained on both GSE22220 and GSE19536 datasets combined and the miRNA-mRNA pairs in the model with negative coefficients, representing a potential targeting effect, were collected. Here, we consider only the 248 miRNAs for which there was at least one such targeting interaction. There were on the average 8 experimentally validated targets for each of these miRNAs, listed in miRTarBase. TargetScan had an average of 341 predicted targets per miRNA. Considering miRTarBase as the ground truth, the accuracy of miRNA-mRNA target pairs predicted by SMLR was 0.10% (41 correct out of 40,633 predictions), whereas TargetScan had

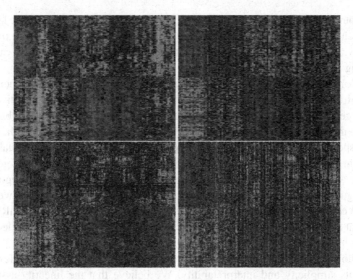

Fig. 4. Hierarchical clustering of true (left) and cross-database predicted (right) mRNA expression. Top: SMLR is trained with GSE22220 dataset and tested on GSE19536 (RMSE = 1.02). Bottom: SMLR is trained with GSE19536 dataset and tested on GSE22220 (RMSE = 1.26). Top 1000 most differentially expressed mRNAs with respect to ER-status are shown. Hierarchical clustering is only done on the real data (left); and the same row-column ordering is used to display the predicted data (right).

an accuracy of 1.12% (944 out of 84,489 predictions) and the negative correlation method had an accuracy of 0.05% (222 out of 428,048 predictions).

Although SMLR had a lower accuracy than TargetScan, we must note that the coverage of miRTarBase is currently very limited. Consequently, these accuracy measures are sensitive to availability of further experimentally validated target data. Furthermore, whereas SMLR finds interactions specific to the datasets it is trained with, namely the breast cancer samples, miRTarBase dataset and TargetScan predictions do not provide any context-specific information for their target interactions. Regardless of these drawbacks in the analysis, combining the predictions from SMLR and TargetScan, by intersecting their miRNA-mRNA target pair lists, achieves an accuracy of 2.17% (23 correct out of 1,060 common predictions), which is better than application of either method alone.

4   Discussion and Conclusion

In this study, we took a radically different approach to miRNA-mRNA interactions and used a multiple linear regression model to directly estimate the mRNA expression levels from miRNA data. Whereas traditional methods try to determine targets of individual miRNAs and rely on these target lists for downstream functional analysis, we estimate mRNA levels from the cellular context captured by the collection of miRNAs. The benefits and opportunities provided by our approach are tremendous. For instance, our approach makes it possible to computationally predict mRNA levels for media, such as serum, where miRNAs are relatively stable and easy to extract and measure with current experimental techniques but mRNAs are less stable and more challenging to measure.

Traditionally, after identifying differentially regulated miRNAs, researchers would sift through hundreds or thousands of targets of these miRNAs and subjectively pick several targets of interest for further experimental validation, e.g., to test for binding of miRNA to mRNA or for differential regulation of the mRNA. Not only are these target lists non-specific to the tissue type, developmental stage, or environmental factors involved in an experimental study; they also ignore the fact that these genes are targeted by multiple miRNAs, some of which may not be differentially regulated or may be regulated in different directions. In our approach on the other hand, we build a model in a cell-type specific manner, connecting multiple miRNAs to each mRNA. We believe that a prioritization of the target genes based on estimated expression levels will result in a higher positive rate in validation experiments.

Our choice of the SMLR model for prediction of mRNA expression levels was based on its simplicity and interpretability. We believe that the linearity assumption used in SMLR provides an appropriate trade-off between the power and generality of the model and the number of parameters that can be correctly estimated from the currently available datasets. Furthermore, the interactions obtained from linear models were previously found to be better than those generated from Bayesian models and Neural Networks [33].

In this study, we mainly focused on breast cancer datasets and demonstrated that a model trained in one experimental platform can be successfully applied to miRNA data from an independent laboratory using different experimental platforms. Although it is possible to apply a model trained on one tissue type to miRNA data from another tissue type; the predicted gene expression values would not be as accurate as restricted the predictions to the same tissue and comparable experimental conditions. For example, applying the model trained on the breast cancer dataset GSE22220 to predict gene expression values from miRNA data in a prostate cancer study GSE20161 resulted in a mean squared error of 1.35, about 33% higher than the error when it was applied to another breast cancer dataset GSE19536. In our future work, we will build a repository of models for different tissue types and experimental conditions of interest. The limiting factor for building such a repository will be the availability of high quality paired miRNA and mRNA data collected from the same samples.

Although our main focus was not identification of the direct miRNA-mRNA targeting interactions, we show that the interactions with negative coefficients in our model can be indicative of direct regulation. Note that the targets from our model were generated only from the two breast cancer studies. We expect that a large scale modeling from all publicly available paired miRNA-mRNA datasets will provide target predictions that are in better agreement with experimentally validated targets. Motivated by the observation that targeting interactions obtained from two breast cancer datasets can improve the accuracy of TargetScan predictions, we expect that our approach will provide a means of improving sequence-based target predictions in a context-specific manner.

References

1. Bartel, D.P.: MicroRNAs: genomics, biogenesis, mechanism, and function. Cell **116**(2), 281–297 (2004)
2. Vasudevan, S., Tong, Y., Steitz, J.A.: Switching from repression to activation: micrornas can up-regulate translation. Science **318**(5858), 1931–1934 (2007)
3. Hobert, O.: Gene regulation by transcription factors and microRNAs. Science **319**(5871), 1785–1786 (2008)
4. Fabian, M.R., Sonenberg, N., Filipowicz, W.: Regulation of mRNA translation and stability by microRNAs. Ann. Rev. Biochem. **79**(1), 351–379 (2010)
5. Huntzinger, E., Izaurralde, E.: Gene silencing by microRNAs: contributions of translational repression and mRNA decay. Nat. Rev. Genet. **12**(2), 99–110 (2011)
6. Guo, H., et al.: Mammalian microRNAs predominantly act to decrease target mRNA levels. Nature **466**(7308), 835–840 (2010)
7. Friedman, R.C., et al.: Most mammalian mRNAs are conserved targets of microRNAs. Genome Res. **19**(1), 92–105 (2009)
8. Croce, C.M.: Causes and consequences of microRNA dysregulation in cancer. Nat. Rev. Genet. **10**(10), 704–714 (2009)
9. Lujambio, A., Lowe, S.W.: The microcosmos of cancer. Nature **482**(7385), 347–355 (2012)
10. Ørom, U.A., Lund, A.H.: Experimental identification of microRNA targets. Gene **451**(1–2), 1–5 (2010)
11. Vergoulis, T., et al.: TarBase 6.0: capturing the exponential growth of miRNA targets with experimental support. Nucleic Acids Res. **40**, D222–D229 (2011)
12. Hsu, S.D., et al.: miRTarBase update 2014: an information resource for experimentally validated miRNA-target interactions. Nucleic Acids Res. **42**(Database issue), D78–D85 (2014)
13. John, B., et al.: Human microRNA targets. PLoS Biol. **2**(11), e363 (2004)
14. Krek, A., et al.: Combinatorial microRNA target predictions. Nat. Genet. **37**(5), 495–500 (2005)
15. Lewis, B.P., Burge, C.B., Bartel, D.P.: Conserved seed pairing, often flanked by adenosines, indicates that thousands of human genes are microRNA targets. Cell **120**(1), 15–20 (2005)
16. Maragkakis, M., et al.: DIANA-microT web server: elucidating microRNA functions through target prediction. Nucleic Acids Res. **37**(Web Server issue), W273–W276 (2009)
17. Sethupathy, P., Megraw, M., Hatzigeorgiou, A.G.: A guide through present computational approaches for the identification of mammalian microRNA targets. Nat. Methods **3**(11), 881–886 (2006)
18. Pritchard, C.C., Cheng, H.H., Tewari, M.: MicroRNA profiling: approaches and considerations. Nat. Rev. Genet. **13**(5), 358–369 (2012)
19. Nam, S., et al.: miRGator: an integrated system for functional annotation of microRNAs. Nucleic Acids Res. **36**(Suppl. 1), D159–D164 (2008)
20. Huang, G.T., Athanassiou, C., Benos, P.V.: mirConnX: condition-specific mRNA-microRNA network integrator. Nucleic Acids Res. **39**(Suppl. 2), W416–W423 (2011)
21. Ritchie, W., Flamant, S., Rasko, J.E.J.: mimiRNA: a microRNA expression profiler and classification resource designed to identify functional correlations between microRNAs and their targets. Bioinformatics **26**(2), 223–227 (2010)
22. Peng, X., et al.: Computational identification of hepatitis C virus associated microRNA-mRNA regulatory modules in human livers. BMC Genom. **10**(1), 373 (2009)

23. Sales, G., et al.: MAGIA, a web-based tool for miRNA and genes integrated analysis. Nucleic Acids Res. **38**(Suppl. 2), W352–W359 (2010)
24. Nam, S., et al.: MicroRNA and mRNA integrated analysis (MMIA): a web tool for examining biological functions of microRNA expression. Nucleic Acids Res. **37**(Suppl. 2), W356–W362 (2009)
25. Kim, S., Choi, M., Cho, K.H.: Identifying the target mRNAs of microRNAs in colorectal cancer. Comput. Biol. Chem. **33**(1), 94–99 (2009)
26. Wang, H., Li, W.H.: Increasing MicroRNA target prediction confidence by the relative R(2) method. J. Theoret. Biol. **259**(4), 793–798 (2009)
27. Beck, D., et al.: Integrative analysis of next generation sequencing for small non-coding RNAs and transcriptional regulation in myelodysplastic syndromes. BMC Med. Genom. **4**(1), 19 (2011)
28. Huang, J.C., Morris, Q.D., Frey, B.J.: Bayesian inference of MicroRNA targets from sequence and expression data. J. Comput. Biol. **14**(5), 550–563 (2007)
29. Huang, J.C., Frey, B.J., Morris, Q.D.: Comparing sequence and expression for predicting microRNA targets using GenMiR3. In: Pacific Symposium on Biocomputing, pp. 52–63 (2008)
30. Su, N., et al.: Predicting microRNA targets by integrating sequence and expression data in cancer. In: 2011 IEEE International Conference on Systems Biology (ISB) (2011)
31. Stingo, F.C., et al.: A Bayesian graphical modeling approach to microRNA regulatory network inference. Ann. Appl. Stat. **4**(4), 2024–2048 (2010)
32. Liu, B., et al.: Exploring complex miRNA-mRNA interactions with Bayesian networks by splitting-averaging strategy. BMC Bioinform. **10**(1), 408 (2009)
33. Zhou, Y., Qureshi, R., Sacan, A.: Data simulation and regulatory network reconstruction from time-series microarray data using stepwise multiple linear regression. Netw. Model. Anal. Health Inform. Bioinform. **1**(1–2), 3–17 (2012)
34. Liu, B., Li, J., Cairns, M.J.: Identifying miRNAs, targets and functions. Briefings Bioinform. **15**(1), 1–19 (2014)
35. da Huang, W., Sherman, B.T., Lempicki, R.A.: Systematic and integrative analysis of large gene lists using DAVID bioinformatics resources. Nat. Protoc. **4**(1), 44–57 (2009)
36. Enerly, E., et al.: miRNA-mRNA integrated analysis reveals roles for miRNAs in primary breast tumors. PLoS ONE **6**(2), e16915 (2011)
37. Buffa, F.M., et al.: microRNA-associated progression pathways and potential therapeutic targets identified by integrated mRNA and microRNA expression profiling in breast cancer. Cancer Res. **71**(17), 5635–5645 (2011)
38. Naume, B., et al.: Detection of isolated tumor cells in bone marrow in early-stage breast carcinoma patients: comparison with preoperative clinical parameters and primary tumor characteristics. Clin. Cancer Res. **7**(12), 4122–4129 (2001)

IsoTree: De Novo Transcriptome Assembly from RNA-Seq Reads
(Extended Abstract)

Jin Zhao[1], Haodi Feng[1(✉)], Daming Zhu[1], Chi Zhang[2], and Ying Xu[3]

[1] School of Computer Science and Technology, Shandong University,
Shun Hua Road, 250101 Jinan, Shandong, China
fenghaodi@sdu.edu.cn
[2] Department of Medical and Molecular Genetics,
Indiana University, Bloomington, USA
[3] Department of Biochemistry and Molecular Biology,
University of Georgia, Athens, USA

Abstract. High-throughput sequencing of mRNA has made the deep and efficient probing of transcriptomes more affordable. However, the vast amounts of short RNA-seq reads make de novo transcriptome assembly an algorithmic challenge. In this work, we present IsoTree, a novel framework for transcripts reconstruction in the absence of reference genomes. Unlike most of de novo assembly methods that build de Bruijn graph or splicing graph by connecting k-mers which are sets of overlapping substrings generated from reads, IsoTree constructs splicing graph by connecting reads directly. For each splicing graph, IsoTree applies an iterative scheme of mixed integer linear program to build a prefix tree, called isoform tree. Each path from the root node of the isoform tree to a leaf node represents a plausible transcript candidate which will be pruned based on the information of pair-end reads. Experiments showed that IsoTree performs better in recall on both pair-end reads and single-end reads and in precision on pair-end reads compared to other leading transcript assembly programs including Cufflinks, StringTie and Bin-Packer.

1 Introduction

Alternative splicing occurs as a normal phenomenon in eukaryotes, where it greatly increases the diversity of proteins that can be encoded by the genome [1]. A recent study estimated that more than 95% of all multi-exon genes are alternative spliced [2]. Besides that, numerous researches have revealed that a great deal of human diseases, especially cancer, are related to abnormal splicing [3,4]. Advances in RNA-seq have opened the way to efficient probing of full-length transcriptome. RNA-seq technology can generate hundreds of millions of short reads (50–250 bp) from expressed transcripts (complete and contiguous mRNA

This work is supported by National Natural Science Foundation of China under No. 61672325 and No. 61472222.

Z. Cai et al. (Eds.): ISBRA 2017, LNBI 10330, pp. 71–83, 2017.
DOI: 10.1007/978-3-319-59575-7_7

sequence from the transcription start site to the transcription end, for multiple alternatively spliced isoforms). Despite the opportunities for transcriptome assembly, great challenge emerges of how to subtly recover as many expressed transcripts as possible at lowest cost from massive short reads. Many obstacles remain in the trancriptome reconstruction problem such as sequencing bias or sequencing error, variable sequence coverage, alternative transcripts from the same locus sharing the same exons, and the existence of paralogs genes. A successful method should address these issues, and apply a suitable data structure to accommodate multiple transcripts per locus.

A growing number of strategies have been developed to solve the transcriptome assembly problem based on RNA-seq. They can be generally divided into two categories, genome-based and de novo assembly approaches. Genome-based approaches, such as StringTie [5], Cufflinks [6], Scripture [7], Bayesember [8], IsoInfer [9], IsoLasso [10], Traph [11], iReckon [12], CIDANE [13], and TransComb [14], usually first align the reads to a reference genome with alignment tools such as TopHat [15], TopHat2 [16], GSNAP [17], STAR [18], and SpliceMap [19], and then merge the sequences from different loci according to overlapping alignments and splicing junctions to build a graph representing all possible isoform transcripts. Finally, different models are adopted to recover the full-length transcripts from the graph. For example, Cufflinks applies the minimum-cost path cover model, StringTie employs a network flow algorithm originally developed in optimization theory, and Traph uses minimum-cost flow model combined with a greedy algorithm. However, the reference genome especially a cancer genome is not always available. In these situations, de novo assembly is required. In theory, a de novo assembler can reconstruct transcripts even on regions that are missing a reference.

The field of de novo assembly developed from pioneering work on de Bruijn graphs [20,21], in which a vertex is a k-mer and an edge exists between two vertices u and v if and only if u and v appear consecutively in a read. Simple paths in such graphs usually represent fragments of transcripts. However, de Bruijn graph may be very tanglesome and therefore hard to deal with. The splicing graph emerges at the right moment, which is more tractable than de Bruijn graph. A splicing graph of a locus is a directed acyclic graph, whose vertices represent exons while edges correspond to splicing junctions. To summarize, Trinity [22], ABySS [21], and IDBA-trande [23] take the advantage of de Bruijn graph approach, while Oases [24], Bridger [25], and BinPacker [26] apply the splicing graph strategy.

Trinity [22] plays a milestone role in de novo transcriptome assembly. It assembles transcripts by first extending contigs greedily, then building de Bruijn graphs from these contigs, and then extracting sufficiently covered paths from these graphs to construct splicing variants based on a brute-force enumeration strategy. Binpacker is a recently developed method, which searches for an optimal edge-path-cover over the splicing graph by iteratively solving a series of bin packing problems. In this scheme, it uses a heuristic algorithm to update trajectories of items (edge-path-cover), and the iteration will terminate within $O(|V|)$ times (where V is the node set of the splicing graph).

So far, all the existing de novo assemblers usually extend the contigs by connecting *k-mers* (a set of overlapping *kbp* substring rising from each read sequence). Given that larger k values tend to perform better on higher expressed transcripts or on longer transcripts while smaller values are more suitable for reconstructing lower expressed transcripts and shorter transcripts, some assemblers apply a multiple-k strategy, such as ABySS [21], Oases-M [24], and IDBA-Tran [23]. However, most of existing $k - mer$ connection strategies usually could not make full use of the information of nucleotides sequence order in each read.

In this paper, we present IsoTree, a de novo transcriptome assembler. The central idea behind IsoTree is to subtly extract transcript-representing paths from the splicing graph. The splicing graph is constructed with contigs that are extended by reads directly. Each vertex as well as each edge in the splicing graph is weighted by reads per base. IsoTree converts the splicing graph to a prefix tree by calling $|V|$ times mixed integer linear program modeled with the objective to seek as few transcripts in the prediction as possible under the coverage constraints(see Methods for details).

We tested the performance of IsoTree compared to other leading transcript assembly programs including Cufflinks, StringTie, and BinPacker on the pair-end reads sets and single-end reads sets. We chose these assemblers for comparison because Cufflinks and StringTie performed comparably best among all the published genome-based assemblers that could run properly while BinPacker performed best among all the published de novo assemblers according to our initial tests. Actually, Bridger as an early work of the authors of BinPacker outperformed all the other de novo competitors except for BinPacker.

We employ blast+ [27] to evaluate the performance of each assembler. Our experiments demonstrated the superior performance of IsoTree in both recall and precision, where recall is defined as the fraction of assembled full-length transcripts out of all reference transcripts in the experiments and the precision is defined as the ratio of assembled full-length transcripts over all assembled transcripts.

2   Method

Splicing graph is originally put forward by Heber et al. in 2002 [28]. IsoTree modified Bridger's splicing graph construction method [25]. Specifically, IsoTree extends the contigs by reads directly while Bridger extends them by *k-mers*. Theoretically, each splicing graph constructed by IsoTree corresponds to an expressed gene: the nodes represent exons, the edges represent splicing junctions, and some paths correspond to isoforms generated by the gene. IsoTree applies a heuristic algorithm to convert the splicing graph into an isoform tree with each path from the root node to a leaf node representing a transcript. The expression level of each transcript is related to the last vertex weight in the corresponding isoform path. The general work flow of IsoTree algorithm is given in Fig. 1.

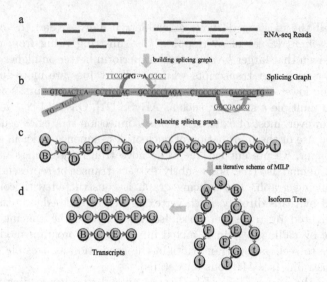

Fig. 1. General work flow of IsoTree. (a) Input single-end or pair-end reads. (b) Splicing graph construction (c) Topological ordering of splicing graph and balancing splicing graph. (d) Constructing isoform tree based on an iterative scheme of mixed integer linear program and recovering transcripts.

2.1   Constructing Splicing Graph

IsoTree applies a hash table of *k-mers* to quickly determine reads that contain a same substring. It decomposes each *Lbp* length read sequence into $L - k + 1$ overlapping *k-mers*. For each *k-mer*, the hash table takes the *k-mer* sequence as key and the set of reads that contain this *k-mer* as value. The *k-mer* sequence is stored as a 64-bit unsigned integer with 2-bit nucleotide encoding. In the process of building a hash table, if the *k-mer* composed by the first k nucleotides of a read appears at the first time, the read is seen as a seed read. The likely error *k-mer* is pruned following the criteria used by Trinity [22]. While pruning *k-mer*, the reads containing the wrong *k-mer* will also be deleted.

(1) Select an unused seed read as the main contig of the initial splicing graph.
(2) Extend the main contig in two directions by repeatedly selecting an unused read with the highest priority in the candidate read set. A candidate read must have x ($a \leq x \leq b$, default $a = L - 1, b = k$) overlaps with the current contig terminus, and its priority is defined as x. The candidate reads with priority x can be found in a linear time according to two *k-mers* in the current contig terminus. Set $k\text{-}mer_1$ as k suffix (or prefix) of current contig and R_1 as the set of reads that contain $k\text{-}mer_1$. In contrast, $k\text{-}mer_2$ is set as $x - k$ to x suffix (or prefix) of current contig and R_2 represents the set of reads that contain $k\text{-}mer_2$. The algorithm to get the candidate reads from sets R_1 and R_2 is described as Algorithm 1. When a contig cannot be extended, it is used as the trunk of a splicing graph to be constructed.

In k-mer extending strategy, the contig is extended by repeatedly selecting the most frequent k-mer that overlaps with the current contig terminus by its $k - 1$-character prefix while neglecting the importance of the reads that support the connection. A less frequent k-mer with more reads supporting the connection may be more reliable than a more frequent k-mer with less reads supporting the extension. In our model, if a read spans both the current contig terminus and a k-mer that intends to be connected to the contig, we regard the read as supporting the connection and extend the contig through this read.

Algorithm 1. Algorithm to get candidate reads

1: int $s_1 = 0$;
2: int $s_2 = 0$;
3: **while** $s_1 < |R_1|$ and $s_2 < |R_2|$ **do**
4:     **if** $R_1[s_1] == R_2[s_2]$ **then**
5:         **if** $R_1[s_1].substr(L - x, x) == contig.substr(0, x)$ or $R_1[s_1].substr(0, x) == contig.substr(contig_1 en - x, x)$ **then**
6:             put $R_1[s_1]$ to candidate read set
7:         **end if**
8:         $s_1 = s_1 + 1$
9:         $s_2 = s_2 + 1$
10:     **else**
11:         **if** $R_1[s_1] > R_2[s_2]$ **then**
12:             $s_2 = s_2 + 1$
13:         **else**
14:             $s_1 = s_1 + 1$
15:         **end if**
16:     **end if**
17: **end while**

---

(3) For each read in the current splicing graph, check if it has an alternative extension that has not been used. Such read is called a junction read. Once we find a junction read, we keep extending it until encountering an already used read or we can make no further extension by using steps (1) to (2). If the former occurs, then a new junction read is identified, and the current splicing graph is updated by merging their matched x nucleotides. Otherwise, the following criteria are used to check if this potential branch can be added to the current splicing graph: (a) the branch is long enough (≥ 80 bp) to be an exon; (b) the branch is not similar with the corresponding part of the trunk; (c) there are at least two single-end reads or pair-end reads supporting this branch. Repeat step (3) until no junction read exists. Now, a splicing graph is constructed.

(4) The graph is trimmed with the similar idea in Trinity: (a) for each edge, there is a minimal number of reads (default 2) matched on each side of the junction; (b) the coverage of each edge must exceed 0.04 times the average coverage of two flanking nodes (twice the sequencing error rate in a read,

the upper bound is about 2%); (c) if there is a node with several outgoing edges, each coverage of them should be more than 5% of total outgoing edge coverage; (d) any outgoing edge coverage should be more than 2% of total incoming edge coverage. Edges that do not meet any one of these criteria are removed. For each isolated vertex in the current splicing graph, it is removed while its sequence length is shorter than the minimum transcript length.

(5) Select a new unused seed read as a new seed, repeat steps (1) to (4) until all seed reads have been used.

2.2 Balancing Splicing Graph

In order to facilitate building isoform tree steps, IsoTree adds a source node s and a sink node t into the splicing graph connects s to all the nodes without in-coming edges, and connects all the nodes without outgoing edges to t. We assign the weight of the new edge connecting s and v to be the sum of weights of the edges leaving v. The new edges entering t can be weighted similarly. Considering that exons are linearly arranged in a gene, IsoTree topologically orders the vertices.

Considering that if an exon is both the end part of a transcript and a middle part of another transcript, its incoming total weights and outgoing total weights may differ greatly. In this case, we balanced this type of exon node by the following rules. Let $G(V, E)$ represent the splicing graph. For each vertex v ($v \epsilon V - \{s, t\}$), check the edges incident with it by the following conditions:

$$(1 - \varepsilon)W_{in}(v) \le (1 + \varepsilon)W_{out}(v), \tag{1}$$

$$(1 + \varepsilon)W_{in}(v) \ge (1 - \varepsilon)W_{out}(v), \tag{2}$$

where ε is an empirical value (default 0.3), $W_{in}(v)$ and $W_{out}(v)$ represent the weight sum of all the edges entering vertex v and the weight sum of all the edges leaving v, respectively. If the edges incident with v can not meet the above conditions at the same time, it means that there is a huge gap between the incoming weights and outgoing weights of v and there needs a balancing deal. The approach to balance vertex v is as follows:

if $W_{in}(v) > W_{out}(v)$, then set $E = E \bigcup (v, t), W(v, t) = W_{in}(v) - W_{out}(v)$;
if $W_{in}(v) < W_{out}(v)$, then set $E = E \bigcup (s, v), W(s, v) = W_{out}(v) - W_{in}(v)$;

where $W(v, t)$ is the weight of edge (v, t).

2.3 Building Isoform Tree and Recover Transcripts

IsoTree iteratively calls a variant of mixed integer linear program model to comb all the transcripts encoded in a splicing graph to a prefix tree, called isoform tree (Fig. 2).

IsoTree first sets vertex s in splicing graph $G(V, E)$ as the root node of isoform tree T, and set $v = s$. Each vertex u ($u \epsilon V, (v, u) \epsilon E$) in graph G is set as a child

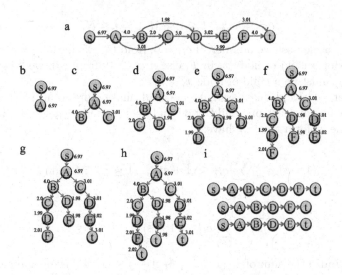

Fig. 2. An example to build isoform tree.

node of v in Tree T, with weight $W_T(u) = W(v, u)$. Then, IsoTree deals each topological ordered vertex of splicing graph G iteratively:

(1) For splicing graph G, set v as v_R (where v_R is the right node of v in topological order), and for the edges $(v, y_j) \epsilon E$ ($1 \leq j \leq N$, N is the total number of out-going edges of vertex v), denote an N-dimensional vector β with component β_j representing the weight of edge (v, y_j).

(2) For isoform tree T, search v in leaf nodes (it is obvious that v must be in leaf nodes), and mark the leaf nodes that represents v as $x_1, x_2, \cdots, x_M$ (M is the sum of leaf nodes that represent v). Denote the weight of x_i ($1 \leq i \leq M$) as α_i. Obviously, α_i must be the weight of an incoming edge of v in splicing graph G or the splitting weight of an incoming edge.

(3) Expand each leaf node x_i ($1 \leq i \leq M$) by y_j according to α and β. IsoTree formalizes it into an optimization problem of how to assign $\alpha_1, \alpha_2, \cdots, \alpha_M$ to $\beta_1, \beta_2, \cdots, \beta_N$. Given $M \times N$ matrix C as the assignments matrix, with component c_{ij} representing the value of α_i assigned to β_j ($1 \leq i \leq M, 1 \leq j \leq N$). If $c_{ij} > 0$, node x_i in tree T will have a child node y_j and the weight of the child is set as $\beta_j c_{ij} / \sum_{t=1}^{M} c_{tj}$. The value of each c_{ij} must satisfy the following constraints:

$$0 \leq c_{ij} \leq (1 + \varepsilon)\alpha_i \quad 1 \leq i \leq M, 1 \leq j \leq N \tag{3}$$

$$0 \leq c_{ij} \leq (1 + \varepsilon)\beta_i \quad 1 \leq i \leq M, 1 \leq j \leq N \tag{4}$$

Here, we introduce a binary integer variable z_{ij} specifying whether a child is added by the following constraints:

$$z_{ij} \leq \lambda c_{ij} \quad 1 \leq i \leq M, 1 \leq j \leq N, \tag{5}$$

$$c_{ij} \leq \lambda z_{ij} \quad 1 \leq i \leq M, 1 \leq j \leq N, \tag{6}$$

where λ is a large positive number. If $c_{ij} = 0$, from (5), we have $z_{ij} = 0$ and thus node y_j is not a child of node x_i in T, else if $c_{ij} > 0$, from (6), we have $z_{ij} = 1$ and node y_j must be a child of node x_i.

In order to avoid bias assignments and make sure that each node in $x_1, \cdots, x_M$ has at least one child node and each node in $y_1, \cdots, y_N$ has been added to the isoform tree, we have:

$$(1 - \varepsilon)\alpha_i \leq \sum_{j=1}^{N} c_{ij} \leq (1 + \varepsilon)\alpha_i \quad 1 \leq i \leq M, \text{ and} \tag{7}$$

$$(1 - \varepsilon)\beta_j \leq \sum_{i=1}^{M} c_{ij} \leq (1 + \varepsilon)\beta_j \quad 1 \leq j \leq N. \tag{8}$$

We minimize the sum of $z_{ij}(1 \leq i \leq M, 1 \leq j \leq N)$ following the parsimony principle to seek as few transcripts in the prediction as possible.

$$minf = \sum_{i=1}^{M} \sum_{j=1}^{N} z_{ij} \tag{9}$$

Equations (3)–(9) form a mixed integer linear program (MILP). The isoform tree is built after call $|V|$ times MIP for each vertex in splicing graph $G(V, E)$. Each path from the root node to a leaf node in isoform tree represents a potential transcript, and the weight of leaf node can be seen as an approximation of the transcript expression level. If pair-end information is available, IsoTree will map the reads to potential transcripts, and a transcript will be discarded if the number of reads with both ends mapped to it is significantly lower than the total number of reads mapped to it.

3   Results

We ran and tested IsoTree with other assemblers: Cufflinks (version 2.1.1), StringTie (version 1.3.1), and BinPacker (version 1.0) both on single-end and pair-end RNA-seq datasets. We tested IsoTree in two versions: IsoTreeI and IsoTreeII. The difference between these two versions is that IsoTreeI constructs splicing graph by the method mentioned above while IsoTreeII built splicing graphs without branch extend steps (we notice that although branching may introduce more transcript candidates, the added candidates may occupy some portion of a real transcript and thus influence the results). These experiments are conducted on a desktop computer with 4 Gb of RAM and Intel Core i5-2400 CPU processor. In the following, we will give a detailed analysis of our experiments.

In this paper, a full-length reconstructed transcript is defined as an assembled transcript whose full-length covers a referenced transcript with at least 95%

sequence identity (where identity is a ratio between the matched length and the length of referenced transcript) and at most 0.5% indels [26]. We use blast+ [27] to align the assembled transcripts to referenced transcripts. We applied recall (the ratio between the number of full-length reconstructed transcripts and the number of reference transcripts) and precision (the ratio between the number of full-length reconstructed transcripts and the number of assembled transcripts) as the measures of prediction quality.

Mimicking the characteristic of real RNA-seq data, we generated 0.1 million pair-end reads of length 75 bp and 0.2 million single-end reads of length 75 bp from 100 isoform transcripts originated from 41 different genes in chromosome 1 using FluxSimulator [29]. We solved our MILP model by Lingo. Lingo is a mathematical modeling language designed for formulating and solving optimization problems, and it adopts the branch-and-bound method to solve the mixed integer linear programming problems. Since the transcript sequence only contains 'A', 'C', 'G', and 'T' four kinds of nucleotides, in most cases the number of outgoing (or incoming) edges of an exon node is less than 3. Besides, the sum of nodes in a splicing graph corresponding to one locus is less than 10 in most of cases. Hence, the variables in our MILP model are usually less than 20 and the MILP can be solved by Lingo in a very short time. The running times of IsoTreeI and IsoTreeII on the pair-end reads datasets mentioned above are about 105 s and 10 s respectively. This is because IsoTreeI spends most of its time on checking and extending branches. (For large real data, branch-and-bound is usually too time-consuming. In this case, we change the MILP to a program with linear constraints while the objective function as the product of the variables. We then solve the nonlinear program by a list of heuristics. We run the algorithm on a server with 256 Gb of RAM and E5-2620V3*2 CPU processor. For the details of the model with its solution as well as the implementation with analysis for large real data, please refer to our later journal version.)

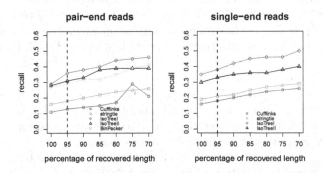

Fig. 3. Comparison of recall for pair-end and single end reads.

For the recall measure, IsoTree outperforms all the other compared assemblers both on single-end and pair-end datasets (Fig. 3). For the data set of pair-end reads, IsoTreeI reconstructed transcripts with a recall value of 36%, a more than 24.1% increase over the recall achieved by BinPacker (29%), StringTie (18%), and Cufflinks (13%). Unfortunately BinPacker does not work on single-end datasets. In this case, IsoTree plays an obvious superior over the other remaining compared assemblers.

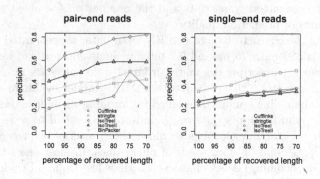

Fig. 4. Comparison of precision for pair-end and single end reads.

For the precision measure, IsoTree has the highest precision on pair-end dataset (Fig. 4). For pair-end dataset, IsoTreeI, IsoTreeII, Cufflinks, StringTie, and Binpacker recovered 36, 31, 13, 18, and 29 full-length reconstructed transcripts from 56, 66, 57, 59, and 77 candidates, respectively. It is worth mentioning that, though IsoTreeI reports the minimum number of candidate transcripts, but it produces the maximum number of full-length reconstructed transcripts on the pair-end dataset. The precision of IsoTreeI on pair-end dataset is 64.3%, best over all assemblers. The precision of BinPacker is 33.7%, higher than all the other assemblers except IsoTree. For single-end dataset, the precision of StringTie is 37.5%, highest among all the assemblers including IsoTreeI (27.7%), IsoTreeII (28%), Cufflinks (25.3%). However, IsoTree recovered more full-length reconstructed transcripts than other assemblers on single-end dataset. IsoTreeI recovered 38 full-length reconstructed transcripts while IsoTreeII, Cufflinks, and StringTie recovered 33, 18, and 21 respectively. We speculate that the reason that IsoTree performs better on pair-end data than on single-end data is that IsoTree prunes the candidate transcripts by pair-end information. We can also see that except for single-end dataset IsoTreeII performs slightly better than IsoTreeI on precision, i.e., 28% vs.27.7%, IsoTreeI always outperforms IsoTreeII. This strongly supports the necessity for branch extension.

From Fig. 5, we conclude that IsoTree performs best on both recall and precision, followed by BinPacker, on pair-end dataset. For single-end dataset, StringTie is superior over IsoTree on precision with 33.9% increase, while IsoTree is superior over StringTie on recall with more than 54.4% increase.

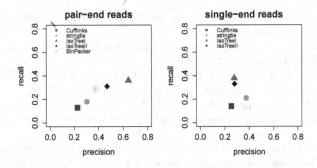

Fig. 5. Transcriptome assemblers' accuracies in detecting expressed transcripts from data set of single-end reads and pair-end reads.

4 Conclusion

We present a novel de novo method IsoTree for transcriptome reconstruction from single-end or pair-end RNA-seq reads. We constructed the splicing by drawing the advantages of Trinity and adding our own innovation of expanding the contig by reads directly. We applied the mixed integer linear program model subtly to build the isoform tree which could express the potential transcripts in a gene. In addition, the process of pruning transcripts with help of pair-end reads information has greatly improved the precision. The experiments shows that IsoTree always holds the best recall among all the compared assemblers on data sets of both single-end and pair-end reads, and its precision is also the highest on pair-end data.

References

1. Chen, M., Manley, J.L.: Mechanisms of alternative splicing regulation: insights from molecular and genomics approaches. Nat. Rev. Mol. Cell Biol. **10**(11), 741–754 (2009)
2. Wang, E.T., Sandberg, R., Luo, S., et al.: Alternative isoform regulation in human tissue transcriptomes. Nature **456**(7221), 470–476 (2008)
3. Faustino, N.A., Cooper, T.A.: Pre-mRNA splicing and human disease. Genes Dev. **17**(4), 419–437 (2003)
4. Sveen, A., Kilpinen, S., Ruusulehto, A., et al.: Aberrant RNA splicing in cancer; expression changes and driver mutations of splicing factor genes. Oncogene **35**, 2413–2427 (2015)
5. Pertea, M., Pertea, G.M., Antonescu, C.M., et al.: StringTie enables improved reconstruction of a transcriptome from RNA-seq reads. Nat. Biotechnol. **33**(3), 290–295 (2015)
6. Trapnell, C., Williams, B.A., Pertea, G., et al.: Transcript assembly and abundance estimation from RNA-Seq reveals throusands of new transcripts and switching among isoforms. Nat. Biotechnol. **28**(5), 511–515 (2010)
7. Guttman, M., Garber, M., Levin, J.Z., et al.: Ab initio reconstruction of cell type-specific transcriptomes in mouse reveals the conserved multi-exonic structure of lincRNAs. Nat. Biotechnol. **28**(5), 503–510 (2010)

8. Maretty, L., Sibbesen, J.A., Krogh, A.: Bayesian transcriptome assembly. Genome Biol. **15**(10), 501 (2014)
9. Feng, J., Li, W., Jiang, T.: Inference of isoforms from short sequence reads. J. Comput. Biol. **18**(3), 305–321 (2011)
10. Li, W., Jiang, T.: IsoLasso: a LASSO regression approach to RNA-Seq based transcriptome assembly. J. Comput. Biol. **18**(11), 1693–1707 (2011)
11. Tomescu, A.I., Kuosmanen, A., Rizzi, R., et al.: A novel min-cost flow method for estimating transcript expression with RNA-Seq. BMC Bioinform. **14**(5), S15 (2013)
12. Mezlini, A.M., Smith, E.J.M., Fiume, M., et al.: iReckon: simultaneous isoform discovery and abundance estimation from RNA-seq data. Genome Res. **23**(3), 519–529 (2013)
13. Canzar, S., Andreotti, S., Weese, D., Reinert, K., Klau, G.W.: CIDANE: comprehensive isoform discovery and abundance estimation. Genome Biol. **17**(1), 16 (2016)
14. Liu, J., Yu, T., Jiang, T., et al.: TransComb: genome-guided transcriptome assembly via combing junctions in splicing graphs. Genome Biol. **17**(1), 213 (2016)
15. Trapnell, C., Pachter, L., Salzberg, S.L.: TopHat: discovering splice junctions with RNA-Seq. Bioinformatics **25**(9), 1105–1111 (2009)
16. Kim, D., Pertea, G., Trapnell, C., et al.: TopHat2: accurate alignment of transcriptomes in the presence of insertions, deletions and gene fusions. Genome Biol. **14**(4), R36 (2013)
17. Wu, T.D., Nacu, S.: Fast and SNP-tolerant detection of complex variants and splicing in short reads. Bioinformatics **26**(7), 873–881 (2010)
18. Dobin, A., Davis, C.A., Schlesinger, F., et al.: STAR: ultrafast universal RNA-seq aligner. Bioinformatics **29**(1), 15–21 (2013)
19. Au, K.F., Jiang, H., Lin, L., et al.: Detection of splice junctions from paired-end RNA-seq data by SpliceMap. Nucleic Acids Res. **38**(14), 4570–4578 (2010)
20. Zerbino, D.R., Birney, E.: Velvet: algorithms for de novo short read assembly using de Bruijn graphs. Genome Res. **18**(5), 821–829 (2008)
21. Simpson, J.T., Wong, K., Jackman, S.D., et al.: ABySS: a parallel assembler for short read sequence data. Genome Res. **19**(6), 1117–1123 (2009)
22. Grabherr, M.G., Haas, B.J., Yassour, M., et al.: Full-length transcriptome assembly from RNA-Seq data without a reference genome. Nat. Biotechnol. **29**(7), 644–652 (2011)
23. Peng, Y., Leung, H.C.M., Yiu, S.M., et al.: IDBA-tran: a more robust de novo de Bruijn graph assembler for transcriptomes with uneven expression levels. Bioinformatics **29**(13), i326–i334 (2013)
24. Schulz, M.H., Zerbino, D.R., Vingron, M., et al.: Oases: robust de novo RNA-seq assembly across the dynamic range of expression levels. Bioinformatics **28**(8), 1086–1092 (2012)
25. Chang, Z., Li, G., Liu, J., et al.: Bridger: a new framework for de novo transcriptome assembly using RNA-seq data. Genome Biol. **16**(1), 30 (2015)
26. Liu, J., Li, G., Chang, Z., et al.: BinPacker: packing-based de novo transcriptome assembly from RNA-seq data. PLoS Comput. Biol. **12**(2), e1004772 (2016)
27. Camacho, C., Coulouris, G., Avagyan, V., et al.: BLAST+: architecture and applications. BMC Bioinform. **10**(1), 421 (2009)

28. Heber, S., Alekseyev, M., Sze, S.H., et al.: Splicing graphs and EST assembly problem. Bioinformatics **18**(suppl 1), S181–S188 (2002)
29. Griebel, T., Zacher, B., Ribeca, P., et al.: Modelling and simulating generic RNA-Seq experiments with the flux simulator. Nucleic Acids Res. **40**(20), 10073–10083 (2012)

Unfolding the Protein Surface for Pattern Matching

Heng Yang[1], Chunyu Zhao[2], and Ahmet Sacan[3(✉)]

[1] Facebook, 1 Hacker Way, Menlo Park, CA 94025, USA
[2] Department of the Biomedical and Health Informatics,
The Children's Hospital of Philadelphia, Philadelphia, PA 19104, USA
[3] Biomedical Engineering, Drexel University,
3120 Market St, Philadelphia, PA 19104, USA
`ahmet.sacan@drexel.edu`

Abstract. Protein 3-D structural data is a valuable resource in computational biology, and the comparison and interpretation of protein structural patterns have remained scientific and computational challenges. We introduce a novel representation of 3-D protein surface patches as 2-D images, obtained using dimension reduction. We utilize image registration to compare these surface patches and infer protein function and binding based on surface similarity. Our surface representation can capture various structural and physicochemical properties, including curvature, electrostatic potential, hydrophobicity, and evolutionary conservation. The results we present support the use of surface images as a new type of family-specific signatures in functional annotation and drug-binding tasks. We demonstrate the ability of our method to detect local surface similarities between proteins and to correctly identify functional classification of proteins.

Keywords: Protein structure · Ligand binding sites · Image processing · Template matching · Dimension reduction

1 Introduction

Determining the functions and interactions of individual proteins is essential for understanding their contribution to the behavior of the cell and the organism as a whole and creates tremendous therapeutic opportunities for treating diseases. Availability of large scale genomic and proteomic data has invited development of automated computational methods for functional annotation of proteins. Traditionally, sequence analysis has been the main source of information, where pairwise and multiple alignments and statistical and machine learning methods have been utilized for classification of proteins into known functional families [1]. However, sequence alone becomes insufficient for making functional inferences for distantly related proteins or those proteins that have discovered the same biomolecular function through convergent evolution. Protein structure is regarded as a stronger determinant of function than sequence alone, placing structure under greater evolutionary pressure, and making structural similarities between homologous proteins detectable even under low sequence similarity conditions.

© Springer International Publishing AG 2017
Z. Cai et al. (Eds.): ISBRA 2017, LNBI 10330, pp. 84–95, 2017.
DOI: 10.1007/978-3-319-59575-7_8

Corroborating the importance of structural data, protein structure initiatives have been established with the goal of expanding the repository of experimentally determined protein structures. As many as 26% of the structures resulting from these Structural Genomics initiatives have unknown or putative functions [2]. Consequently, numerous approaches have been developed for comparing and data mining protein structures, with the hopes of finding functionally relevant similarities among both well-studied and less characterized proteins.

Structure alignment methods make up a majority of the available protein structure comparison approaches. In structure alignment, one seeks to find correspondences between the residues of the proteins being compared and also a translation/rotation matrix that best superposes these corresponding residues. Finding an optimum structural alignment is computationally difficult and available methods employ heuristics to find near-optimal solutions within practical execution times. Available methods are based on distance matrices, common subgraph searches, geometric hashing techniques, and genetic algorithms. An important drawback of structure alignment methods has been their prudent use of geometric information, and only recently methods have been proposed to additionally utilize biochemical and evolutionary information [3].

Structure alignment methods generally represent each amino acid as a single point in space, often using the coordinates of its alpha carbon atom. While this simplification is sufficient for fold recognition purposes where the focus is on categorization of the overall shape of protein domains, it may fail to detect important local arrangements of amino acid side chain atoms. Furthermore, global structural similarity does not necessitate the same enzymatic activity or binding interactions. TIM barrel family of proteins provide an extreme example of this, where proteins sharing the same structural fold can have diverse functions. Ser-His-Asp catalytic triad of serine protease family provide an example from the other extreme, where due to chemical constraints of enzymatic activity, proteins share highly conserved arrangement of active site residues, while having dissimilar global structures.

The need to identify conserved local arrangements of a few amino acids, regardless of the overall fold, has motivated development of a new class of methods for discovery and search of structural patterns, such as LFM-pro [4] and PROMOTIF [5]. These methods try to find spatial configuration of amino acid residues with well-conserved inter-residue distances; and in line with their focus on function rather than structure, they often utilize functional side chain atoms instead of backbone atoms.

It has been observed that proteins with similar active sites have similar functions and that active sites are usually located within pockets formed on the protein surface. These observations have prompted focus on analysis of surface pockets for identification of ligand binding and protein function. Surface pockets have been defined as regions of favorable interaction energies [6] or from purely geometric characteristics [7]. Consequently, a class of structure comparison methods have been developed to compare these surface regions.

The methods that make use of only geometric information for comparison of proteins surfaces include those that summarize the shape by descriptors such as Zernike moments [8] and distance-based features [9] and those that represent surfaces as point clouds [10]. The methods based on shape descriptors generally solve the global structure similarity problem, whereas the point cloud methods try to detect local residue or atomic arrangements. While existing approaches have been useful in comparing and clustering known protein active sites, they are limited in their ability to locate functional sites in new proteins.

In this study, we describe a novel representation and comparison method for protein surface analysis that is able to capture various surface features in a computationally efficient manner. Specifically, we unfold protein surfaces into two dimensional images and perform comparisons using these images. The two dimensional image representation allows the use of fast image registration methods, as opposed to the more demanding graph matching methods required for other surface representations.

Unlike other approaches that try to find equivalences between residues or atoms from the proteins compared, our focus is on the surface itself without enforcing a one-to-one correspondence of residues. This has the potential to identify similar spatial environments created by different number of contributing residues. While other approaches require a priori delineation of functional regions, our method is also able to perform comparison of protein surfaces when such information is not available from one or both of the proteins being compared. Furthermore, whereas other methods utilize mainly the geometrical information, with some methods additionally enforcing residue or functional atom identity, our image representation allows representation of arbitrary surface features, such as hydrophobicity, evolutionary conservation, and electrostatic potential.

2   Methods

The molecular surface of a protein is defined as the set of points traced by the inward-facing part of a hypothetical probe that is rolling on the protein. We calculate the molecular surface using the MSMS program [11] available from the Vasco package [12], using a probe radius of 1.4 Angstrom, and a surface density of one vertex per Angstrom-square. We exclude the water molecules and hetero atoms and consider each peptide chain entry as a separate unit. The generated molecular surface is represented as a point cloud and a corresponding triangular mesh.

2.1   Unfolding the Protein Surface

Once the molecular surface is obtained, we "unfold" it by mapping each surface point to a point in 2-D space using dimension reduction methods. While the idea of mapping protein surfaces to a 2-D space is not new, previous approaches have only explored simple spherical and elliptical projections [13, 14]. However, proteins have more complex shapes than these idealized geometries, and can contain voids and protrusions.

Using dimension reduction methods has allowed us to explicitly optimize the 2-D mapping for its ability to preserve the inter-point distances, neighborhood relationships, and the surface area of the triangular mesh. We have previously demonstrated that dimension reduction methods can generate 2-D representations that surpass simple projections in mapping accuracy [15]. Among the dimension reduction methods investigated, ISOMAP [16] was found to provide an ideal performance-speed tradeoff

[15] and was chosen for this study. In order to preserve the intrinsic geometry of the manifold, the nonlinear dimension reduction method ISOMAP exploits the geodesic distance, instead of straight-line Euclidean distance.

The process of mapping a surface mesh onto a 2-D surface is known in the computer vision domain as "mesh parameterization" [17, 18]. Since it is not possible to unfold a closed shape while preserving distance, neighborhood, and area relationships equally for all points, mesh parameterization methods often perform segmentation of the 3-D shape and consider mapping each segment separately. Additionally, the points that are located in the center of a segment are the most accurately represented, and the error in mapping increases as one moves toward the edges of the mapped mesh structure where the original surface mesh becomes more distorted. Based on these observations, we either generate a single segmentation around a known region of interest to maximize the fidelity of this region; or generate multiple overlapping segmentations from the entire surface, where a different part of the protein surface is best preserved in each of the segments.

When a particular region of the protein such as a set of residues responsible for enzymatic activity or a pocket forming a ligand binding site is of interest, we first identify the surface points that are closest to these residues. This active site surface patch is then extended with other points that have a geodesic distance less than 30 Å from the active site surface points. This ensures that the active site is captured in its entirety within a single segment and that the active site region is mapped at the center of the 2-D image with higher fidelity than it would have had if it were on a segmentation boundary. The top row in Fig. 1 shows segmentation using a region of interest, whereas the bottom row illustrates the general segmentation into many 2-D images.

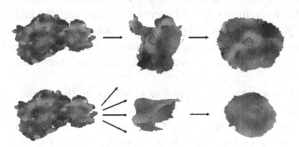

Fig. 1. Segmentation of surface of the protein PDB: 1q8y. Top: A single segment centered on the binding site region is obtained. Left to right images show the overall 3-D structure, the segmented surface section, and the corresponding mapped 2-D image. The binding site is outlined with a red polygon. Bottom: General segmentation into multiple surface patches. Only one of the many sections is shown in the middle, and its corresponding 2-D image is shown on the right. (Color figure online)

The general segmentation procedure repeatedly selects a point on the protein surface to serve as the center of a segment. The points that are within $r_{section}$ of this center are used to define the new segment. Based on typical sizes of active sites, we use $r_{section} = 15$ Å. The points that are within $r_{exclusion}$ are excluded from being used as a new cluster center. In effect, $r_{exclusion}$ determines the amount of overlap between different segments and consequently the number of segments. We use $r_{exclusion} = 10$ Å to obtain 50–100 segments per protein. We pre-calculate all-pairs shortest paths among surface points and use it as an approximation of the geodesic distances.

2.2 Enrichment of Surface with Additional Features

Geometric features of the protein surface are important functional determinants and have been used by some methods as the sole source of information for comparing protein surface regions. However, it is well-known that biochemical properties play an essential role in determining binding interactions and enzymatic activity [19]. These biochemical properties are often captured in the form of amino acid or atom types when a residue-wise or atom-wise matching is performed between surfaces [20].

Without loss of generality, we consider the following properties for each surface point: hydrophobicity (H), electrostatic potential (E), curvature (V), and evolutionary conservation (C). These properties are commonly used in studying protein folding, protein-protein interactions, protein-ligand binding, and enzymatic activity. At each surface point, we calculate the hydrophobicity values using Vasco [12], the Gaussian curvature using surface triangulation [21], and the electrostatic potential using DelPhi [22]. In order to calculate conservation values, a PSI-BLAST search of the protein sequence against NCBI non-redundant protein sequence database is carried out. Multiple sequence alignments are generated using MUSCLE [23], and the conservation scores are derived using the method described in STACCATO [24].

For visual purposes and implementation convenience, we enrich the protein with three properties at a time, where the red, green, and blue channels encode the electrostatic potential, hydrophobicity, and curvature values. Regions high or low in these properties are still visually discernible when the three channels are combined into a single colored surface. Since our image-based implementation is limited to representing at most three features at a time, we consider different combinations of available features.

When the protein surface is mapped to a 2-D mesh, the features associated with each point are also carried over. We then convert the 2-D mapping to an image where each pixel takes on the average values of the features for the points that map into that pixel location. When an active site or another surface region of interest is defined, the points in the image corresponding to that site are used to generate a minimum bounding box enclosing all such points. Although the active site can be more precisely defined by a polygon mask, for computational simplicity we represent the active site region using the smallest rectangle that encloses these points.

2.3 Template Matching

The utility of the 2D representation of the protein surface and its features depends on our ability to compare these representations and find similarities and differences that correlate with functionalities of these proteins. Template matching is an image processing method that finds the region of a *target* image most similar to a *template,* and provides a measure of similarity between these images. Available template matching methods can be broadly categorized into feature-based and area-based approaches. Feature-based approaches [25] are successful in detecting similarities when the images have sharp features, such as edges and corners. Area-based methods are more appropriate for images that do not have such strong features but contain regions unique in their color composition and pixel intensity patterns. The protein surface images we generate are smooth and lack sharp features, hence we utilize area-based template matching methods in this study. Specifically, we use the square root difference measure (SQDIFF) between registered pixel values to evaluate the similarity of two sub-images.

Using a sliding window, we consider each local region from the template protein and search for the most similar local region in the target protein. We rely on the efficient image registration implementation available from OpenCV [26]. We repeat the search for different rotations of the template window, at 1° rotational resolution. In the case where a binding site of the template protein is being searched within the target protein, we use the bounding box of the active site as the only sub-window, rather than searching for all sub-windows of the template.

In the following section, we evaluate the result of the template matching in finding functionally related regions from the target protein and also its ability to differentiate functional categories of multiple proteins.

3   Experiments and Results

The applications of our surface representation and comparison approach can be classified based on whether or not a region of interest is *a priori* defined in one or both of the proteins being compared. In the following subsections, we present case studies that represent different types of applications of our approach.

3.1   Comparison of Known Binding Sites

We use proteins from the Aldehyde dehydrogenase (ALDH) superfamily to demonstrate the application of our approach to comparison of known active sites. ALDH superfamily of enzymes play a crucial role in aldehyde detoxification by catalyzing aldehydes to carboxylic acids. ALDH active sites have been highly conserved over evolution, and share a number of conserved residues for catalysis, including Cys-302, Glu-268, and Asn-169 [27].

Two proteins are chosen from the superfamily: rat liver ALDH3 (PDB: 1ad3) and sheep ALDH1 (PDB: 1bxs). These two proteins share 29% sequence identity, but their binding sites for the ligand NAD are highly conserved. The binding site surface of each protein is segmented and the 2-D binding site images are computed using the EHC color code (electrostatic potential, hydrophobicity, conservation).

Figure (2). The bounding box enclosing the 1ad3 binding site points in 2D is used as the template and searched in the target image of 1bxs. By finding the location in the

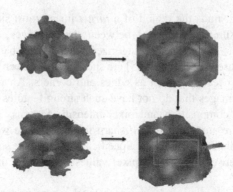

Fig. 2. Template matching of binding site sections between protein 1ad3 and 1bxs. Left: 3-D structures of protein 1ad3 and 1bxs, with their binding sites outlined with a red polygon. Right: the mapped sections as 2-D images. The bounding box encloses all of the points that were within the binding site in 3D. Image registration searches for the 1ad3 binding site within 1bxs image and the located target region is shown with a green rectangle. The amount of overlap between the bounding box enclosing 1bxs binding site and the target bounding box identified from image registration is 76.5%. (Color figure online)

target image that has the minimum square difference in pixel values compared with the template bounding box, the image registration method correctly identifies the binding site region of 1ad3. The overlap between the correct binding site and the one predicted from image registration is 76.5%. The binding regions in both of these proteins have a characteristic positive charge (red color channel) and high conservation (blue color channel).

3.2   Locating a Known Binding Site in an Unknown Protein

In a second case study, we take a binding site section extracted from one protein to search against all sections segmented from another protein. This task represents functional annotation of a new protein using a previously characterized binding site. For this case study, we use two serine proteinase proteins: human trypsin 1 (PDB: 1trn) and bovine trypsin (PDB: 2ptn). The catalytic residues of are extracted from the Catalytic Site Atlas database [28].

Serine proteinases form a classic example of convergent evolution where the catalytic triad responsible for the enzymatic activity has a very well-defined spatial configuration. Human and bovine trypsin proteins share 38% sequence similarity and are highly similar in their active site, including the catalytic triad residues serine, histidine and aspartate.

In this case study, we extract and map the active site region of human trypsin 1 protein (1trn) and use the bounding box enclosing this active site region as the template image. We assume the active site of the bovine trypsin protein (2ptn) is unknown and segment this protein using general segmentation, which results in 62 overlapping segments. The template image from 1trn is then searched on each of the 62 images and

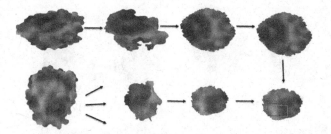

Fig. 3. Searching for a binding site on target protein surface. Top: The active site of human trypsin 1 (PDB: 1trn) is extracted and mapped using electrostatic potential, hydrophobicity, and conservation (EHC) color channels. Bottom: The surface of bovine trypsin (PDB: 2ptn) is segmented using the general segmentation method, resulting in 62 overlapping segments. The template binding site image from 1trn is searched against each of these 62 segments and the match with the smallest SQDIFF measure is reported as a hit (only the segment resulting in smallest SQDIFF is depicted here).

the segment resulting in the smallest SQDIFF is assumed to contain the target binding site. Figure 3 demonstrates this segmentation and search process.

The template search correctly identifies the segment containing the binding site from among the 62 candidate segments and locates the binding site within the target image segment. The overlap between the predicted binding site region and the correct binding site enclosing the catalytic triad is 66%.

3.3 Different Feature Enrichment Combinations

The success of the template matching in identifying similar binding site regions depends on the features being represented in the 2D images. For some protein families, conservation may be sufficient to locate an active site that has a specific spatial arrangement of highly conserved residues; whereas for other protein families additional features may be required to more accurately locate an active site.

In this case study, we investigated the effect of using different feature combinations on correctly locating binding sites of proteins from the Ras superfamily. Human Rac1 protein (PDB: 1mh1) is used as a template to search for the binding site of HRas (PDB: 4g3x). Rac1 is a member of Rho family and downstream effector of Ras. HRas is a member of Ras that operates as molecular switch on the inner surface of the plasma membrane. These proteins have 31% sequence identity and share a common GNP binding sites, but they function on different targets. General segmentation of 4g3x surface resulted in 53 overlapping segments. The binding site template extracted from 1mh1 is used to search against each of these segments using different feature combinations (See Fig. 4).

For the first three feature combinations, template matching was able to correctly identify the segment of 4g3x containing the binding site from among the 53 segments. However, the EHC feature combination, which omits the curvature information performs significantly worse than other color combinations in accurately locating the

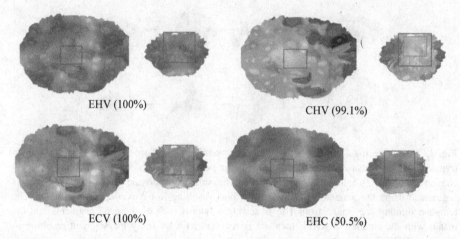

EHV (100%)   CHV (99.1%)

ECV (100%)   EHC (50.5%)

Fig. 4. Searching for 1mh1 binding site in 4g3x segments using different feature combinations. Left: binding site images of protein 1mh1, with each row showing a different feature combination. The red bounding box encloses the surface points of the binding site residues. Right: the detected binding site hits from 4g3x, where the red and green bounding boxes show the correct and predicted binding sites (amount of overlap shown under each color code). The correct segment was identified in all feature combinations except for EHC. Only the segment of 4g3x that contain the binding site are shown here. (Color figure online)

binding site, indicating curvature (V) to be an important geometric characteristic of this active site. We expect different feature combinations to be appropriate for different protein families. Note that we only investigated an unweighted SQDIFF measure for image registration in this study; a more general image representation with multiple color channels and a corresponding weighted SQDIFF measure are left as future work and are expected to provide higher accuracy in locating the binding sites.

3.4   Functional Annotation Using Database Search

While the previous case studies investigated pairwise comparison of proteins to locate a binding site of a target protein, here we investigate the ability of our approach to identify the functional category of a target protein by comparison to a number of available protein families. For this purpose, we constructed a benchmark dataset from Metapocket [29], which contains 198 drug-target complexes. The proteins in Metapocket are non-redundant, with at most 40% pairwise sequence identity.

For a database retrieval task, we select only the ligands that contain at least 10 proteins in the dataset, resulting in the three ligand groups: Adenine (ADE, DB00173) with 10 proteins, Glutathione (GSH, DB00143) with 10 proteins, and Pyridoxal Phosphate (PLP, DB00114) with 15 proteins. The surface points within 30 Å from the ligands are considered as the binding site. The binding sites from each of the 35 proteins are mapped into 2D images and a binding site database of these template images is constructed for each feature combination.

The proteins that bind to Adenine are used as query and searched against the binding sites in the database (excluding the binding site generated from the query protein itself). The query protein is segmented using general segmentation, and the segment with the smallest SQDIFF to a template is used for scoring for that template. As a result, each query protein is associated with a numerical vector of 34 distance scores. These scores are sorted and at different score thresholds, the true and false positive rates are calculated. True positive rate is defined as the proportion of ADE binding sites that are correctly predicted as such, and the false positive rate is the proportion of other binding sites that are incorrectly predicted as an ADE binding site. The receiver operating curve (ROC) showing the true and false positive rates at different score thresholds is shown in Fig. 5.

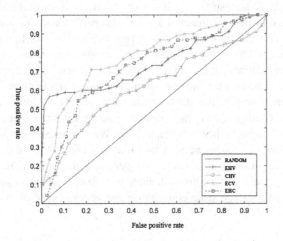

Fig. 5. Receiver Operating Curves (ROC) for retrieval of Adenine binding sites using SQDIFF scores. The areas under the curves for EHV, CHV, ECV, and EHC color combinations are 76%, 63%, 79%, and 74%, respectively.

The database retrieval performance were similar for EHV, ECV, and EHC feature combinations, and the CHV feature combination performed worse than others. Unlike the previous case studies that compared evolutionarily related proteins from the same protein family or superfamily, the proteins in the Metapocket database are categorized based on the ligands they bind to, without enforcing any homology relationship. The Adenine-binding proteins in this dataset have an average pairwise sequence identity of 15%. Consequently, conservation alone is not sufficient to characterize and relate these Adenine binding proteins and the accuracy of database retrieval is not adversely affected when conservation feature is excluded. On the contrary, excluding the electrostatic potential feature significantly reduces the accuracy of identifying Adenine-binding sites. This is in line with the observation that electrostatics is an important contributing factor for binding Adenine and is sufficient for distinguishing between Adenine and Guanine binding sites [30].

4 Conclusion

In this paper, we reported a novel representation of protein surfaces as feature-enriched 2D images and a corresponding image registration algorithm to compare these images in order to find similar surface patches. The case studies demonstrate that our approach can successfully locate active sites on protein surfaces and that this ability can be utilized in functional annotation of protein structures. Note that our surface representation does not try to correspond individual residues from proteins and consequently our approach is not directly comparable to traditional protein structure comparison methods that focus on residue-by-residue correspondences.

Although our approach can be used to represent other features of interest, we considered electrostatic potential, evolutionary conservation, hydrophobicity, and curvature properties at protein surface points. For visual and computational convenience, we only used three of these features at a time. The performance of image registration depended on the features used and different protein families and ligand binding sites were best characterized by different feature combinations. Our future work will include encoding all of the features within a single high-dimensional image and using weighted SQDIFF measure to better tune the contribution of individual features.

Because different protein functions are characterized by different physicochemical and evolutionary characteristics, the contributions of different features need to be adjusted for each protein family and ligand type. We also expect the functional annotations to not reflect a uniform characterization of these features. We therefore propose re-categorization of binding sites using unsupervised clustering methods. A protein active site database can then be compiled, along with feature weights optimized for each family, and utilized for large scale comparison and annotation of protein structures.

References

1. Najmanovich, R., Kurbatova, N., Thornton, J.: Detection of 3D atomic similarities and their use in the discrimination of small molecule protein-binding sites. Bioinformatics **24**(16), i105–i111 (2008)
2. Chruszcz, M., et al.: Unmet challenges of structural genomics. Curr. Opin. Struct. Biol. **20**(5), 587–597 (2010)
3. Zhao, C., Sacan, A.: UniAlign: protein structure alignment meets evolution. Bioinformatics **31**(19), 3139–3146 (2015)
4. Sacan, A., et al.: LFM-Pro: a tool for detecting significant local structural sites in proteins. Bioinformatics **23**(6), 709–716 (2007)
5. Hutchinson, E.G., Thornton, J.M.: PROMOTIF–a program to identify and analyze structural motifs in proteins. Protein Sci. **5**(2), 212–220 (1996)
6. Laurie, A.T., Jackson, R.M.: Q-SiteFinder: an energy-based method for the prediction of protein-ligand binding sites. Bioinformatics **21**(9), 1908–1916 (2005)
7. Hendlich, M., Rippmann, F., Barnickel, G.: LIGSITE: automatic and efficient detection of potential small molecule-binding sites in proteins. J. Mol. Graph Model **15**(6), 359–363 (1997)

8. Sael, L., et al.: Rapid comparison of properties on protein surface. Proteins **73**(1), 1–10 (2008)
9. Das, S., Kokardekar, A., Breneman, C.M.: Rapid comparison of protein binding site surfaces with property encoded shape distributions. J. Chem. Inf. Model. **49**(12), 2863–2872 (2009)
10. Kinoshita, K., Nakamura, H.: Identification of protein biochemical functions by similarity search using the molecular surface database eF-site. Protein Sci. **12**(8), 1589–1595 (2003)
11. Connolly, M.L.: The molecular surface package. J. Mol. Graph. **11**(2), 139–141 (1993)
12. Steinkellner, G., et al.: VASCo: computation and visualization of annotated protein surface contacts. BMC Bioinform. **10**, 32 (2009)
13. Fanning, D.W., Smith, J.A., Rose, G.D.: Molecular cartography of globular proteins with application to antigenic sites. Biopolymers **25**(5), 863–883 (1986)
14. Pawlowski, K., Godzik, A.: Surface map comparison: studying function diversity of homologous proteins. J. Mol. Biol. **309**(3), 793–806 (2001)
15. Yang, H., Qureshi, R., Sacan, A.: Protein surface representation and analysis by dimension reduction. Proteome Sci. **10**(Suppl. 1), S1 (2012)
16. Tenenbaum, J.B., de Silva, V., Langford, J.C.: A global geometric framework for nonlinear dimensionality reduction. Science **290**(5500), 2319–2323 (2000)
17. Sheffer, A., Praun, E., Rose, K.: Mesh parameterization methods and their applications. Found. Trends. Comput. Graph. Vis. **2**(2), 105–171 (2006)
18. Levy, B., et al.: Least squares conformal maps for automatic texture atlas generation. In: Proceedings of the 29th Annual Conference on Computer Graphics and Interactive Techniques, pp. 362–371. ACM, San Antonio (2002)
19. Bertolazzi, P., Guerra, C., Liuzzi, G.: A global optimization algorithm for protein surface alignment. BMC Bioinform. **11**, 488 (2010)
20. Angaran, S., et al.: MolLoc: a web tool for the local structural alignment of molecular surfaces. Nucleic Acids Res. **37**(Web Server issue), W565–W570 (2009)
21. Dong, C.-S., Wang, G.-Z.: Curvatures estimation on triangular mesh. J. Zhejiang Univ. Sci. **6**, 128–136 (2005)
22. Li, L., et al.: DelPhi: a comprehensive suite for DelPhi software and associated resources. BMC Biophys. **5**, 9 (2012)
23. Edgar, R.C.: MUSCLE: a multiple sequence alignment method with reduced time and space complexity. BMC Bioinform. **5**, 113 (2004)
24. Shatsky, M., Nussinov, R., Wolfson, H.J.: Optimization of multiple-sequence alignment based on multiple-structure alignment. Proteins **62**(1), 209–217 (2006)
25. Zhao, W., et al.: Face recognition: a literature survey. ACM Comput. Surv. (CSUR) **35**(4), 399–458 (2003)
26. Bradski, G., Kaehler, A.: Learning OpenCV: Computer Vision with the OpenCV Library. O'reilly, Sebastopol (2008)
27. Marchitti, S.A., et al.: Non-P450 aldehyde oxidizing enzymes: the aldehyde dehydrogenase superfamily. Expert Opin. Drug Metab. Toxicol. **4**(6), 697–720 (2008)
28. Porter, C.T., Bartlett, G.J., Thornton, J.M.: The catalytic site atlas: a resource of catalytic sites and residues identified in enzymes using structural data. Nucleic Acids Res. **32** (Database issue), D129–D133 (2004)
29. Huang, B.: MetaPocket: a meta approach to improve protein ligand binding site prediction. OMICS **13**(4), 325–330 (2009)
30. Basu, G., et al.: Electrostatic potential of nucleotide-free protein is sufficient for discrimination between adenine and guanine-specific binding sites. J. Mol. Biol. **342**(3), 1053–1066 (2004)

Estimation of Rates of Reactions Triggered by Electron Transfer in Top-Down Mass Spectrometry

Michał Aleksander Ciach[1]([✉]), Mateusz Krzysztof Łącki[1]([✉]),
Błażej Miasojedow[1], Frederik Lermyte[2,3], Dirk Valkenborg[3,4],
Frank Sobott[2], and Anna Gambin[1]

[1] Faculty of Mathematics, Informatics and Mechanics,
University of Warsaw, Warsaw, Poland
m_ciach@student.uw.edu.pl, mateusz.lacki@biol.uw.edu.pl
[2] Biomolecular and Analytical Mass Spectrometry Group,
Department of Chemistry, University of Antwerp, Antwerp, Belgium
[3] Centre for Proteomics, University of Antwerp, Antwerp, Belgium
[4] Interuniversity Institute for Biostatistics and Statistical Bioinformatics,
Hasselt University, Hasselt, Belgium

Abstract. Electron transfer dissociation (ETD) is a versatile technique used in mass spectrometry for the high-throughput characterization of proteins. It consists of several competing reactions triggered by the transfer of an electron from its anion source to the sample cations. One can retrieve relative quantities of the products from mass spectra.

We present a method to analyze these results from the perspective of the reaction kinetics. A formal mathematical model of the ETD process is introduced and parametrized by intensities of the occurring reactions. Also, we introduce a method to estimate the reaction intensities by solving a nonlinear optimization problem. The presented method proves highly robust to noise on *in silico* generated data. Moreover, the presented model can explain a considerable amount of experimental results obtained under various experimental settings.

1   Introduction

Motivation. One of the principal fragmentation methods used in top-down mass spectrometry is Electron Transfer Dissociation, ETD, which is based on the interaction of a multiply charged, non-radical protein/peptide cation and a radical reagent anion [1,2]. However, while this method is becoming ever more ubiquitous in MS-based proteomics analyses, important questions remain regarding the precise reaction mechanism, and which level(s) of protein structure can be probed using ETD [3,4]. Therefore, shedding more light on the nature of ETD can lead to optimization of instrumental settings and improvement of the identification of peptide sequences and post-translational modifications.

There are several other fragmentation techniques, most importantly the Collision-Induced Dissociation, CID, where the cleavage is induced by colliding ions with non-reactive gas molecules [5]. A major disadvantage of the CID is that

Z. Cai et al. (Eds.): ISBRA 2017, LNBI 10330, pp. 96–107, 2017.
DOI: 10.1007/978-3-319-59575-7_9

it often leads to loss of posttranslational modifications, particularly phosphorylation [6]. Electron Transfer Dissociation has also been found to provide more uniform fragmentation than CID, which preferentially cleaves the weakest bonds [2,6]. However, a notable amount of work has been devoted to analyzing and mathematically modelling the CID process [7–9], while ETD has received less attention.

The fragmentation in ETD is induced by transfer of an electron from a radical anion to the sample peptide/protein cation and results in cleavage of one of the peptide bonds $N-C_\alpha$ in the proteins backbone. The sample cations are positively charged during the electrospray ionization (ESI) step [10], leading to the formation of $[M + nH]^{n+}$ ions, i.e. adding both charge and mass to the analyte molecule M.

Anions and cations may interact in several ways (see Fig. 1):

1. in the *default* reaction, ETD, the cation accepts the electron and, after rapid neutralization of charge and a series of electron rearrangements, a backbone $N-C_\alpha$ bond breaks leading to the formation of so-called c and z fragments where the c-ion contains the N-terminus and the z-ion the C-terminus.
2. the cation accepts the electron but no fragmentation is observed, and the electron stays on the analyte. As no dissociation occurs, it is called ETnoD.
3. one of the cations' protons is transferred to the anion; a situation referred to as a proton transfer reaction, or PTR.

$$\textbf{ETD} \quad [M + nH]^{n+} + A^{\bullet-} \longrightarrow [c + xH]^{x+} + [z + (n\text{-}x)H]^{(n-x-1)\bullet} + A$$

$$\textbf{PTR} \quad [M + nH]^{n+} + A^{\bullet-} \longrightarrow [M + (n\text{-}1)\,H]^{(n-1)+} + AH$$

$$\textbf{ETnoD} \quad [M + nH]^{n+} + A^{\bullet-} \longrightarrow [M + nH]^{(n-1)+\bullet} + A$$

Fig. 1. Studied chemical reactions.

The appearance of the ETnoD fragments in the experimental data can be traced to the folding of proteins: although backbone cleavage occurs, noncovalent interactions keep the resulting fragments from separating, see [11,12]. The ETnoD can also be caused by accommodation of an electron, e.g. in an aromatic side chain. It is assumed that, regardless of the precise reaction mechanism, the electron obtained by ETnoD causes neutralization of one ESI-generated proton [13], referred to as the *quenched proton* further on. In all of the reactions described above, one charge is neutralized. The mass of electrons can be neglected, falling beyond the resolving power of most modern instruments.

Cations can undergo several reaction events, being approached multiple times by different anions. However, the so-called internal fragments of proteins, i.e. resulting from two backbone cleavage events, are usually not observed suggesting, that double ETD scarcely ever occurs. On the other hand, there is a lot of evidence of multiple ETnoD and PTR occurring on one analyte molecule [14]. Note that only ions and not neutral molecules are observed in the mass spectrometer. The isotope distributions of reaction products show considerable overlap, especially for large molecules, as illustrated in Fig. 2.

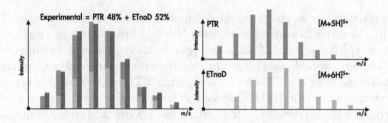

Fig. 2. The deconvolution of isotopic structures performed by MASSTODON: the mixture of two distributions is represented by a convex combination of two theoretical isotopic patterns.

The peptide bond cleavage induced by ETD is believed to be fairly uniform [15]. A notable exception from this rule is the peptide bond of proline: due to the ring structure of this amino acid, the c- and z-ions are held together even after the $N-C_\alpha$ bond cleavage. A specific type of $N-C_\alpha$ bond cleavage occurs on the N-terminus, leading to a loss of one ammonia molecule. The precise mechanism of this reaction is not known. However, in the current work we assume this reaction to be an ETD, and the ammonia molecule is treated as a c fragment. Therefore, the number of possible ETD cleavage sites is equal to the number of amino acids other than proline in the protein/peptide sequence.

Related Research. Various approaches have been taken to model different protein fragmentation techniques [2,16–18]. A similar approach was presented by Zhang [7,8] who studies CID fragmentation using a kinetic model. In [19], Zhang adapts the model to ETD mass spectra. The model relies upon 280 parameters and its derivation is grounded in the theory of statistical mechanics. The model was fitted to a training data set consisting of more than 7000 ETD spectra simultaneously.

A notable amount of literature has been built up around the idea of purely data-driven prediction of the intensity of peptides in tandem MS experiments [20–22]. A more exploratory approach targeted at studying fragmentation patterns was taken by Li et al. [15]. That said, the above approaches have been applied mainly to study CID.

To the best of our knowledge, none of the existing approaches allows estimating reaction intensities directly from a single mass spectrum.

Our Contribution. We propose a formal model of the electron-driven reactions occurring inside the mass spectrometer. We follow a modeling strategy first developed by Gambin and Kluge [23] to study the degradation of proteins induced by various peptidases. The solution to the problem of ETD reaction can be obtained conceptually in the same way: the stochastic description based on Markov Jump Process, MJP, can be transformed to a populational description of a large number of molecules based on a system of Ordinary Differential Equations, ODEs. Given the intensities of transitions in the process, one can solve the ODEs with a recursive algorithm to obtain the expected number of molecules. The space of possible intensitities has to be searched for the best possible set of parameters by some optimization algorithm.

We study mass spectra gathered in controlled experiments, obtained for highly purified compounds. The identity of the precursor ion and all fragments obtained given a set of possible reactions is known and the quantities of these fragments can be established using our in-house developed identification tool called MASSTODON [13,24]. Given a mass spectrum, MASSTODON outputs a list of reaction products together with their estimated intensities (that are usually assumed to be proportional to the actual number of ions). MASSTODON merges peaks that can be traced to originate from different isotopologues of the same molecule and also deconvolves isotope clusters into their sources or origin, see Fig. 2. The obtained information is more compact, but the observed products can rarely be attributed to only one specific reaction.

Here, we propose a method that can further reduce the dimensionality of the data obtained by MASSTODON. It represents MASSTODON's outcomes in terms of (relative) reaction rates – the key notion in the theory of reaction kinetics.

The model we propose lets us express the mass spectrum in terms of parameters such as the total intensity of reactions and the probabilities of the three studied reactions: ETD, PTR, and ETnoD. A process described by a handful of parameters can be easily visualized and thus easily understood. Also, the comparison of different spectra, e.g. coming from different instrument settings, is highly simplified.

Organization of the Paper. First, we introduce the theoretical considerations behind our model. Then, we describe the procedures used to obtain our data sets (experimental and *in silico*). Then, we assess the performance of the model. Finally, we discuss existing problems and possible extensions.

2   Formal Model of the ETD Reaction

Following the ideas outlined in [23], we model ETD as a continuous time Markov Jump Process, MJP, which is a well-established approach to modeling chemical reactions. To describe the state space for our model we introduce the *reaction graph*: a bipartite directed graph with two types of nodes which we call *molecular species* and *reactions*. The molecular species correspond to cations that are substrates or products of the studied reactions, see Fig. 3. Each molecular species u can be uniquely described by (1) the sequence of amino acids s, (2) the charge of the cation q, and (3) the number of quenched protons g, $u = (s, q, g)$. All molecules that cannot be observed, e.g. the internal fragments or ions in which all charges have been neutralized, are merged into one molecular species called the *cemetery*. We assume to know only the numbers of protons and not their positioning. We also assume that ESI-generated protons can only be attached to basic amino acids: lysine, arginine, and histidine.

The molecular species are occupied by tokens that represent their numbers. Denote the number of tokens at place u by x_u. The state x of the MJP is defined as a collection of all such counts at a given moment in time so that $x = (x_u)$. From a state x, the system can evolve to another state following one of the possible reactions, see Fig. 3(B).

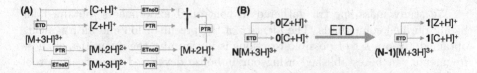

Fig. 3. A fragment of the reaction graph for triply charged precursor. The *molecular species* are depicted in black and the *reactions* in orange. The dagger symbolizes the *cemetery*. The reaction graph serves as a board for *tokens* that represent the number of molecules of a given species. Counts of tokens are plotted in red in panel (B). During each reaction a token disappears on the substrate side and product tokens appear: one in case of ETnoD and PTR, two in case of ETD, as seen in (B). (Color figure online)

We investigate L fragmenting reactions, where L equals the number of amino acids of the protein being studied minus the number of prolines (which ETD cannot fragment). We also consider two reactions corresponding to ETnoD and PTR, which convert one substrate into one product.

We model the time of experiment as an interval $[0, 1]$, where 0 is the beginning of the process and 1 corresponds to the moment of observation. The probability that the process ends up in state x at a given time t is denoted by p_x^t. The derivative of this quantity follows the master equation,

$$\dot{p}_x^t = \sum_{y \neq x} p_y^t Q_{yx} - p_x^t \sum_{y \neq x} Q_{xy},$$

Above, Q_{xy} is the intensity of the reaction leading from state x to state y. The intensity is zero if y cannot be obtained from x. Otherwise, the intensity is proportional to the number of the substrate molecules, $Q_{xy} = c_R x_{s_R}$. Here, c_R (described later on) is the rate of reaction R and s_R is its substrate species.

The average numbers of cations at that place u is $E_u^t = \sum_x x_u p_x^t$. At $t = 0$, the process is deterministic: all tokens can be found only in the precursor node with maximal charge state, denoted q_0. We call this state the *root*, r. Thus, $E_r^0 = N$ and $E_u^0 = 0$ if $u \neq r$. Differentiating the expressions for averages we arrive at

$$\dot{E}_u^t = \sum_x x_u \sum_{y \neq x} p_y^t Q_{yx} - \sum_x x_u p_x^t \sum_{y \neq x} Q_{xy}.$$

We rewrite the above in terms of reactions R and their substrate states to get

$$\dot{E}_u^t = \sum_R c_R \sum_{x : \exists_y x = Ry} x_u (x_{s_R} + 1) p_{R^{-1}x}^t - \sum_R c_R \sum_x x_u x_{s_R} p_x^t,$$

where Ry denotes the state obtained from y after reaction R, and $R^{-1}x$ is the substrate state of x given R. Note that the minuend enumerates only states x for which $R^{-1}x$ is properly defined. We can rephrase the sum in terms of the source states y and then retag them to x,

$$\sum_{x : \exists_y x = Ry} x_u (x_{s_R} + 1) p_{R^{-1}x}^t = \sum_y (Ry)_u y_{s_R} p_y^t = \sum_x (Rx)_u x_{s_R} p_x^t.$$

Thus, $\dot{E}_u^t = \sum_R c_R \sum_x [(Rx)_u - x_u] x_{s_R} p_x^t$. Denote $[(Rx)_u - x_u]$ by K_R. It equals one if place u is a product of reaction R, minus one for substrate, and zero otherwise. It is a value dependent on R and not on particular state x. It results in

$$\dot{E}_u^t = \sum_R c_R K_R \sum_x x_{s_R} p_x^t = \sum_R c_R K_R E_{s_R}^t = \sum_{R:u \in P_R} c_R E_{s_R}^t - E_u^t \sum_{R:u=s_R} c_R,$$

where P_R are the products of reaction R. It follows that the average inflow of molecules in place u is proportional to the average numbers of the parent molecules of u. Their proportionality constants are equal to reaction rates. All reactions are enumerated in the outflow, while some might lead directly to the *cemetery*, e.g. in the case of the second fragmentation. This technical assumption is also needed to correctly solve the resulting ODEs, as described later on.

The last formula can be rewritten as $\dot{E}_u^t = \sum_{v \to u} \lambda_{vu} E_v^t - \lambda_{uu} E_u^t$. This underlines the dependence of the average amount of u upon respective average levels of species it originates from, v. We call v a parent of u. The presented system of ODEs is recursive and can be solved from the root r downwards. For the root, $\dot{E}_r^t = -\lambda_{rr} E_r^t$. The function $E_r^t = N e^{-\lambda_{rr} t}$ solves this ODE. Knowing solutions for all the ancestors of u, we explicitly solve the corresponding ODE. First, consider the ODE of any child u of the root species r, $\dot{E}_u^t = \lambda_{ru} E_r^t - \lambda_{uu} E_u^t$. Applying the integrating factor $e^{\lambda_{uu} t}$ (provided $\lambda_{rr} \neq \lambda_{uu}$) one obtains

$$E_u^t = e^{-\lambda_{uu} t} \int_0^t e^{\lambda_{uu} s} \lambda_{ru} E_r^s ds = \frac{N \lambda_{ru}}{\lambda_{rr} - \lambda_{uu}} (e^{-\lambda_{uu} t} - e^{-\lambda_{rr} t}).$$

In general, instead of $\lambda_{ru} E_r^t$, one has to consider a linear combination of solutions to the ODEs of u's ancestors, $f_{>u}$. Then, $E_u^t = e^{-\lambda_{uu} t} \int_0^t e^{\lambda_{uu} s} f_{>u}(s) ds$. E_u^t can be shown to be equal to $\sum_{v>u} b_{vu} e^{-\lambda_{vv} t} - b_{uu} e^{-\lambda_{uu} t}$, with parameters $b_{vu} = \frac{1}{\lambda_{uu} - \lambda_{vv}} \sum_{w:v \geq w \to u} \lambda_{wu} b_{vw}$ and $b_{uu} = -\sum_v b_{vu}$.

The above is true only if $\lambda_{vv} \neq \lambda_{uu}$ for all parents v of u. These inequalities are satisfied if we make a natural assumption that the intensities can be factorized so that $\lambda_{uv} = I q_i^2 P_{R_{uv}}$, where I is the overall intensity of all reactions, q_u is the charge of molecules in place u, and $P_{R_{uv}}$ is the probability of reaction R_{uv}, where u is the substrate and v one of the products. Then $\lambda_{vv} - \lambda_{uu} = I \sum_R P_R(q_v^2 - q_u^2) > 0$, because u has a lower charge state than any of its ancestors (as all reactions reduce the charge state by at least one). The quadratic dependence on the charge, q_u^2, can be motivated by theoretical considerations [25].

For ETnoD and PTR respectively, $P_{R_u} = P_{\text{PTR}}$ and $P_{R_u} = P_{\text{ETnoD}}$, which are both parameters of the model. The case of ETD is more complex, as it can cleave different bonds. We denote the probability of the cleavage of the lth bond by P_{ETD_l}—another parameter of the model. Additionally, one has to distribute the $q-1$ protons and g quenched protons between both the c and z fragments, so that $q_c + q_z = q - 1$ and $g_c + g_z = g$. The division of remaining $q - 1$ protons and g quenched protons depends on the available number of basic amino-acids on both fragments, denoted by B_c and B_z. It is assumed to occur with probability

$$P_l(q_c, g_c) = \frac{\binom{B_c}{q_c}\binom{B_z}{q_z}}{\binom{B_c+B_z}{q-1}} \frac{\binom{B_c-q_c}{g_c}\binom{B_z-q_z}{g_z}}{\binom{B_c+B_z-q+1}{g}},$$

which corresponds to placing the q_c charges on the basic sites of the fragment peptide followed by placing g_c quenched protons on the remaining free basic sites. Placing quenched protons first would result in the same formula.

To conclude, the probability of the ETD on the lth amino acid with a given proton distribution among fragments equals $P_{R_u} = P_{\text{ETD}_l} P_l(q_c, g_c)$. The probability of observing any ETD reaction, P_{ETD}, can be obtained by summing P_{ETD_l} over possible cleavage sites. The model is parametrized by the total reaction intensity I, two reaction probabilities ($P_{\text{PTR}}, P_{\text{ETnoD}}$) and L different ETD reaction probabilities P_{ETD_l}.

For a given set of model parameters θ we can calculate the average number of cations for all molecular species u at observation time, $E_u^1(\theta)$. On the other hand, MASSTODON provides the estimates y_u of the percentual content of u in the experimental mass spectrum. For a given θ, we measure the difference between these quantities using the logarithm of their euclidean distance. The error minimizer, θ^*, is our estimate of the true reaction intensities. To get it we use the BFGS algorithm that evaluates all $E_u^1(\theta)$ for all species u and compares them to respective y_u and updates θ iteratively until reaching convergence. The cost of evaluating E_u^t is $\mathcal{O}(Lq_0^5)$. To see this, note that there are $\mathcal{O}(Lq_0^2)$ nodes in the reaction graph, each with links to $\mathcal{O}(q_0^2)$ parents. Moreover, reaction intensities can be obtained in constant time, except for ETD which is obtainable in $\mathcal{O}(q_0)$ because of the $P_l(q_c, g_c)$ term.

3 Validation and Results

Numerical Simulations. Numerical simulations of ETD process were performed to assess the quality of the fitting procedure under fully controlled conditions. The simulation was performed as follows: we start with a given number of substance P (amino acid sequence RPKPQQFFGLM) precursor cations. We simulate the electrospray ionization by placing a given number of protons on randomly chosen basic amino acids. Then, we simulate the Markov Jump Process using standard simulation techniques [26], noting that our process can be simulated as if the cations reacted independently of each other. Ions that find themselves in the same state at the end of the simulation are aggregated. The resulting counts of ions simulate results obtainable with MASSTODON.

We also analyze the robustness of the fitting procedure to noisy or missing data. The random noise is modeled by adding gaussian noise to the counts, with zero mean and standard deviation expressed as a given percentage of the count. Missing data is modeled by randomly removing a given proportion of the peaks. Finally, the counts obtained in this way are normalized to sum to one. Altogether, the simulation was repeated 100 times for 20 different values of data distortion parameters, see Fig. 4.

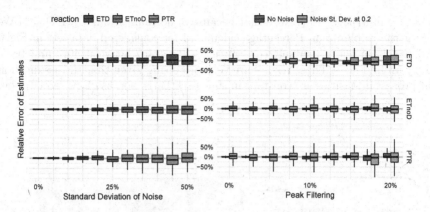

Fig. 4. Relative errors of the fitting procedure on in silico Substance P data. The known true values of parameters are respectively $P_{ETD} = 30\%, P_{ETnoD} = 25\%, P_{PTR} = 45\%$. Cleavage probabilities were assumed to be uniform (proline being the obvious exception). Each boxplot summarizes the results of 100 independent simulations: whiskers denote the first and ninth decile and the box lids - the first and third quartiles. The left panel presents the response of the relative error of the estimates to the increasing amount of noise in the intensities reported by MASSTODON. On the right panel, we study the impact of the random removal of information on the molecular species, both in noiseless conditions (in gold) and with a modest amount of noise (standard deviation set to 20% of the intensity of the simulated molecule).

Experimental Data. Mass spectra have been acquired for purified Substance P. The precise experimental setting is described in detail in [27].

We have tested the model and the fitting procedure on both simulated and experimental data to assess their robustness and to estimate real reaction intensities.

Fitting to Simulated Data. The fitting procedure turned out to be fairly robust toward moderate noise and missing data, see Fig. 4. The results of the fitting procedure are unbiased. On noiseless data and data with a moderate amount of noise (up to 50% of variation in simulated intensities), the model was able to predict the reaction intensities with very high accuracy (only after introducing more than 25% of peak variation do the estimates start to surpass the limit of 50% relative error in more than 20% of cases).

Fitting to Experimental Data. The model has been fitted to 53 substance P spectra, obtained at various travelling-wave height/velocity combinations (design of the instrument and physical meaning of these parameters are described in detail in [27]). After fitting the model to the data, the validity of the model was further investigated by computing the percentage of the experimental spectrum accounted for by the theoretically predicted spectrum. We call this value the *Explanation Percentage* (EP) and define it to be the common part of the theoretical and experimental spectrum. Since both spectra are normalized so that they sum to one, the Explanation Percentage can be expressed in a simple formula,

$EP = \sum_u \min\{y_u, e_u^{\text{norm}}\}$. Note that because of normalization, $0 \leq EP \leq 1$. We present the Explanation Percentage calculated for considered data sets in Fig. 5 (two panels in the upper left corner): the values are between 40% and 98%, mostly around 80%. These results are very promising given that the assumption that process intensities are constant in time is rather strict, as discussed later on.

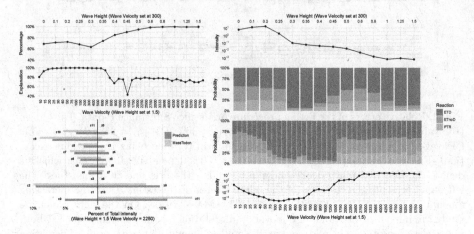

Fig. 5. Results of fitting to experimental data preprocessed by the MASSTODON software. Two plots in the top left corner report the Explanation Percentages calculated for the input data. Below, we present results of fitting the model to the fragments obtained at Wave Height = 1.5 and Wave Velocity = 2250. Intensities of the c fragments are plotted to the left of the black division line. The intensities of their corresponding z fragments are plotted to the right. The figure aggregates results for different charges and quenched protons. The right side of the panel presents estimates of intensities (top and bottom) and estimates of reaction probabilities (middle).

The predicted total intensity of all reactions, I, can be found between 10^{-3} and 10, see Fig. 5. In regions of low reaction intensity, the explanation percentage approaches 100%; however, in these conditions mass spectra contain mostly unreacted precursors and so the fitting is relatively easy to perform. In regions of high reaction intensity (wave height between 0 and 0.3) the spectra are much more informative and even then the model can explain around 70% of the input information. Similar results are obtained for different values of wave velocity. In the regions of high intensity (wave velocity above 1750) the model explains around 75% of the input.

In the bottom left corner of Fig. 5 we present the comparison of peak heights obtained with MASSTODON and those fitted by described procedure. The fragments have been aggregated over protons and quenched protons to simplify the plot. Note that in the input data there are more c fragments than z fragments, with the exception of the z_{11} fragment, corresponding to the loss of NH_3 from the N-terminus. This lack of symmetry is stronger than that expected in the current setting but likely related to the electrostatic repulsion and the asymmetric distribution of basic sites within the substance P sequence, as all the basic residues are located in the proximity of the N-terminus.

4   Discussion and Conclusions

We present a kinetic model of electron transfer driven reactions. The obtained results are promising for future work, as the model can explain around 80% of the observed intensities of the molecular species. The model is based on stochastic foundations and so the estimated parameters have a probabilistic interpretation, such as the probability of a given cleavage or reaction.

Due to its simplicity, the model described here can be used in further fundamental research into the ETD mechanism, as a discrepancies between experimental observations and the model predictions is expected to have a relatively straightforward physical interpretation. For instance, the underestimation of the asymmetry of corresponding c and z fragment intensity in the current results might indicate that a more sophisticated model of protonation sites should be used (e.g. one that accounts for electrostatic repulsion, see [28]). Similarly, using the MassTodon software, it has been recently shown [24] that the observed ratio of PTR to ETnoD depends on protein conformation for intermediate charge states of ubiquitin and, thus, on the reaction history. A more detailed analysis could be easily performed (and similar dependencies thus revealed) using ETDetective.

As mentioned in the Introduction, our kinetic model is somewhat similar to that of Zhang [19]. However, there are many differences in the conceptual approach to the problem. The earlier model is derived from first principles of statistical physics, whereas that proposed by us is much more phenomenological: the physical content is reflected only in the construction of possible states and enumeration of ways of how one state can be modified into another state. Knowing this, we cast the problem into the well-studied setting of continuous time Markov Jump Processes. Because of that, the number of parameters that describe our model is fairly limited, in contrast to the approach described in [19]. Another difference is that we do not estimate any parameters common to more than one dataset: we can fully estimate our model based on one spectrum alone. This allows us to compare reaction intensities for different experimental conditions. As shown in the Results section, this allows us to compare many mass spectra acquired under different experimental conditions and summarize the results in a convenient way using just a few dimensions. Note that in precisely the same way one can compare spectra acquired using different instruments, which is important to properly design the experiment, see [13].

A natural way for this work to proceed is to explain the influence of the instrumental settings and experimental conditions on the reaction intensity and cleavage preferences. This can be investigated using statistical methodology, like the generalized linear models, Dirichlet regression in particular.

The code used in this work is available at https://github.com/mciach/ETDetective under the 2-clause BSD license.

Acknowledgements. This work was partially supported by the National Science Centre grants number 2013/09/B/ST6/01575, 2014/12/W/ST5/00592 and 2015/17/N/ST6/03565, the SBO grant *InSPECtor* (120025) of the Flemish agency for Innovation by Science and Technology (IWT). The authors thank the Research Foundation – Flanders (FWO) for funding a Ph.D. fellowship (F.L.). The Synapt G2 mass spectrometer is funded by a grant from the Hercules Foundation – Flanders.

References

1. Syka, J.E.P., Coon, J.J., Schroeder, M.J., Shabanowitz, J., Hunt, D.F.: Peptide and protein sequence analysis by electron transfer dissociation mass spectrometry. Proc. Natl. Acad. Sci. USA **101**(26), 9528–9533 (2004)
2. Zhurov, K.O., Fornelli, L., Wodrich, M.D., Laskay, Ü.A., Tsybin, Y.O.: Principles of electron capture and transfer dissociation mass spectrometry applied to peptide and protein structure analysis. Chem. Soc. Rev. **42**(12), 5014–5030 (2013)
3. Sohn, C.H., Chung, C.K., Yin, S., Ramachandran, P., Loo, J.A., Beauchamp, J.L.: Probing the mechanism of electron capture and electron transfer dissociation using tags with variable electron affinity. J. Am. Chem. Soc. **131**(15), 5444–5459 (2009)
4. Sohn, C.H., Yin, S., Peng, I., Loo, J.A., Beauchamp, J.L.: Investigation of the mechanism of electron capture and electron transfer dissociation of peptides with a covalently attached free radical hydrogen atom scavenger. Int. J. Mass Spectrom. **390**, 49–55 (2015)
5. Mitchell Wells, J., McLuckey, S.A.: Collision induced dissociation (CID) of peptides and proteins. In: Methods in Enzymology, pp. 148–185 (2005)
6. Kim, M.S., Pandey, A.: Electron transfer dissociation mass spectrometry in proteomics. Proteomics **12**(4–5), 530–542 (2012)
7. Zhang, Z.: Prediction of low-energy collision-induced dissociation spectra of peptides. Anal. Chem. **76**(14), 3908–3922 (2004)
8. Zhang, Z.: Prediction of low-energy collision-induced dissociation spectra of peptides with three or more charges. Anal. Chem. **77**(19), 6364–6373 (2005)
9. Wysocki, V.H., Tsaprailis, G., Smith, L.L., Breci, L.A.: Mobile and localized protons: a framework for understanding peptide dissociation. J. Mass Spectrom. **35**(12), 1399–1406 (2000)
10. Fenn, J., Mann, M., Meng, C., Wong, S., Whitehouse, C.: Electrospray ionization for mass spectrometry of large biomolecules. Science **246**(4926), 64–71 (1989)
11. Lermyte, F., Konijnenberg, A., Williams, J., Brown, J., Valkenborg, D., Sobott, F.: ETD allows for native surface mapping of a 150 kDa noncovalent complex on a commercial Q-TWIMS-TOF instrument. J. Am. Soc. Mass Spectrom. **25**(3), 343–350 (2014)
12. Lermyte, F., Sobott, F.: Electron transfer dissociation provides higher-order structural information of native and partially unfolded protein complexes. Proteomics **15**(16), 2813–2822 (2015)
13. Lermyte, F., Łącki, M.K., Valkenborg, D., Baggerman, G., Gambin, A., Sobott, F.: Understanding reaction pathways in top-down ETD by dissecting isotope distributions: a mammoth task. Int. J. Mass Spectrom. **390**, 146–154 (2015)

14. Lermyte, F., Williams, J.P., Brown, J.M., Martin, E.M., Sobott, F.: Extensive charge reduction and dissociation of intact protein complexes following electron transfer on a quadrupole-ion mobility-time-of-flight MS. J. Am. Soc. Mass Spectrom. **26**(7), 1068–1076 (2015)

15. Li, W., Song, C., Bailey, D.J., Tseng, G.C., Coon, J.J., Wysocki, V.H.: Statistical analysis of electron transfer dissociation pairwise fragmentation patterns. Anal. Chem. **83**(24), 9540–9545 (2011)

16. Breuker, K., Oh, H., Lin, C., Carpenter, B.K., McLafferty, F.W.: Nonergodic and conformational control of the electron capture dissociation of protein cations. Proc. Natl. Acad. Sci. USA **101**(39), 14011–14016 (2004)

17. Simons, J.: Mechanisms for S-S and $N - C_\alpha$ bond cleavage in peptide ECD and ETD mass spectrometry. Chem. Phys. Lett. **484**(4–6), 81–95 (2010)

18. Tureček, F., Julian, R.R.: Peptide radicals and cation radicals in the gas phase. Chem. Rev. **113**(8), 6691–6733 (2013)

19. Zhang, Z.: Prediction of electron-transfer/capture dissociation spectra of peptides. Anal. Chem. **82**(5), 1990–2005 (2010)

20. Elias, J.E., Gibbons, F.D., King, O.D., Roth, F.P., Gygi, S.P.: Intensity-based protein identification by machine learning from a library of tandem mass spectra. Nat. Biotechnol. **22**(2), 214–219 (2004)

21. Arnold, R.J., Jayasankar, N., Aggarwal, D., Tang, H., Radivojac, P.: A machine learning approach to predicting peptide fragmentation spectra. In: Pacific Symposium on Biocomputing, pp. 219–230 (2006)

22. Degroeve, S., Martens, L., Jurisica, I.: MS2PIP: a tool for MS/MS peak intensity prediction. Bioinformatics **29**(24), 3199–3203 (2013)

23. Gambin, A., Kluge, B.: Modeling proteolysis from mass spectrometry proteomic data. Fund. Inform. **103**(1–4), 89–104 (2010)

24. Lermyte, F., Łącki, M.K., Valkenborg, D., Gambin, A., Sobott, F.: Conformational space and stability of ETD charge reduction products of ubiquitin. J. Am. Soc. Mass Spectrom. **28**(1), 69–76 (2017)

25. McLuckey, S.A., Stephenson, J.L.: Ion/ion chemistry of high-mass multiply charged ions. Mass Spectrom. Rev. **17**(6), 369–407 (1999)

26. Gillespie, D.T.: Exact stochastic simulation of coupled chemical reactions. J. Phys. Chem. **81**(25), 2340–2361 (1977)

27. Lermyte, F., Verschueren, T., Brown, J.M., Williams, J.P., Valkenborg, D., Sobott, F.: Characterization of top-down ETD in a travelling-wave ion guide. Methods **89**, 22–29 (2015)

28. Morrison, L.J., Brodbelt, J.S.: Charge site assignment in native proteins by ultraviolet photodissociation (UVPD) mass spectrometry. Analyst **141**(1), 166–176 (2016)

Construction of Protein Backbone Fragments Libraries on Large Protein Sets Using a Randomized Spectral Clustering Algorithm

Wessam Elhefnawy[1], Min Li[2], Jianxin Wang[2], and Yaohang Li[1(✉)]

[1] Department of Computer Science,
Old Dominion University, Norfolk, VA 23529, USA
{welhefna, yaohang}@cs.odu.edu
[2] Department of Computer Science,
School of Information Science and Engineering,
Central South University, Changsha, China
{limin, jxwang}@mail.csu.edu.cn

Abstract. The protein fragment libraries play an important role in a wide variety of structural biology applications. In this work, we present the use of a spectral clustering algorithm to analyze the fixed-length protein backbone fragment sets derived from the continuously growing Protein Data Bank (PDB) to construct libraries of protein fragments. Incorporating the rank-revealing randomized singular value decomposition algorithm into spectral clustering to fast approximate the dominant eigenvectors of the fragment affinity matrix enables the clustering algorithm to handle large-scale fragment sample sets. Compared to the popularly used protein fragment libraries developed by Kolodny et al., the fragments in our new libraries exhibit better representability across diverse protein structures in PDB. Moreover, using much larger fragment sample sets, libraries of longer fragments with length up to 20 residues are also generated. Our fragment libraries can be found at http://hpcr.cs.odu.edu/frag/.

1 Introduction

Local interactions play an important role in stabilizing a protein structural conformation [1, 2]. Therefore, in any short amino acid sequence segment, the molecular interactions constrain the structure into limited number of possible conformations. These conformations are often modeled as protein fragments, which are distributed across many protein structures from different families and allow the study of local interactions in isolation from the protein context. Libraries of these protein fragment have applications in a wide variety of structural biology problems. Giving a few examples, *de novo* protein structure predictions rely on accurate fragment libraries to generate good structural models [3, 4]; in homology-based modeling, protein loop fragments can be used to generate scoring functions [5, 24] and build up loop structures [6, 7]; fragment patterns are also helpful in interpreting experimental electron-density map for protein structure determination [8, 9]; and furthermore, using fragment libraries to represent protein structural features enables effective protein databases search and data mining [10, 11].

© Springer International Publishing AG 2017
Z. Cai et al. (Eds.): ISBRA 2017, LNBI 10330, pp. 108–119, 2017.
DOI: 10.1007/978-3-319-59575-7_10

The protein fragments are often constructed from the protein structures stored in the Protein Data Bank (PDB). Typically, the construction process [12] involves steps of protein selection, measuring fragment similarities, clustering, refinement, and extraction. In the selection step, a subset of high resolution protein chains covering diverse structural conformations is chosen and for each structure, every structural segment containing fixed-length, consecutive residues is taken as a fragment sample. An affinity matrix is built where each entry measures pair-wise similarity among these fragment samples. Then, a clustering algorithm is carried out on the affinity matrix to group the fragment samples with high similarity into k clusters. The refinement step is taken to filter, merge, and optimize these clusters. Finally, the most representative fragment from each cluster is selected to make up the fragment libraries.

One of the well-known protein fragment libraries is developed by Kolodny et al. [12] in 2003. These fragment libraries contain backbone fragments ranging from 5–7 residues, which are generated from a set of 145 protein chains. The sizes of the fragment sample sets are less than 10,000. These libraries are later extended to include fragments of length 5–12 in 2009 [10]. Nevertheless, in recent years, the number of experimentally determined protein structures steadily grows in PDB, with over 10,000 per year. By Feb. 10, 2017, 117,479 protein structure entries are recorded in PDB. The continuously increasing number of experimentally determined protein structures provides rich information sources that enable us to gain important insights into protein structures and their relationship to sequences and functions in a scale and at a level that has never been possible before. Moreover, the protein structure universe tends to be complete with high percentage [13]. All these motivate us to take advantage of the large-scale protein structure information available in PDB to generate high-quality protein backbone fragments. However, one of the main challenges is the clustering algorithms. The popularly used k-means algorithms often have difficulty to scale to handle very large fragment sample sets. Also, k-means algorithms are unable to find clusters with concave boundaries. Moreover, the k-means algorithms are sensitive to the initial randomly picked cluster centers [12], which often generate different results with different initial settings.

In this paper, we present the use of a spectral clustering algorithm to cluster large-scale protein fragment sample sets generated from a large number of protein structures covering diverse conformations in the protein structure universe. Spectral clustering is a graph-cut based algorithm aims at extracting the global patterns of the fragment sample sets [14]. Compared to the commonly used clustering algorithms such as k-means, the graph-cut based clustering algorithms such as normalized cuts have the advantages of generating stable clusters with non-convex boundaries, achieving theoretical optimum, and providing an efficient way to estimate the number of clusters. A rank-revealing randomized singular value decomposition (R^3SVD) technique [19] is employed to fast approximate the dominant eigenvectors of the fragment affinity matrices, which allows the spectral clustering method to scale up to large fragment sample sets. With fragment sample sets of significantly larger sizes, we are able to generate new protein backbone fragment libraries up to length 20. Finally, we analyze the new fragment libraries we generate and compare their representability with the existing ones [10, 12].

The rest of the paper is organized as follows. In Sect. 2, the randomized spectral clustering method to cluster the large-scale fragment sample sets is described and discussed. We analyze our fragment libraries in Sect. 3. Finally, Sect. 4 summarize our work and future research directions.

2 Methods

2.1 Datasets

We use the Protein Sequence Culling Server (PISCES) [15] to extract a non-redundant and non-homologous set of protein chains from PDB. This set contains 2,491 high resolution protein chains with at most 20% sequence identity, 1.6 A resolution cut-off, and 0.25 R-factor. Approximately 70% (1,757) of protein chains are selected as the training set to generate the fragment libraries and the remaining 30% (734) are designated as the testing set for validation.

For each chain in the training set, we use a fixed-length sliding window to consecutively segment the protein sequence into overlapping fragments. We use sliding window sizes ranging from 7 to 20 residues to generate fragment samples from 7–20 in length. Fragments with gaps are excluded. A reduced fragment representation is employed such that each residue in a fragment sample is encoded by the spatial coordinates of heavy backbone atoms while side chains are removed. Residue identities in each fragment are also ignored. Table 1 lists the total numbers of the generated protein fragment samples for each length.

Table 1. Total numbers of fragment samples with respect to fragment lengths in the training data set

Length	# of samples	Length	# of samples
7	490044	14	461217
8	485766	15	457295
9	481540	16	453421
10	477375	17	449583
11	473266	18	445785
12	469210	19	442018
13	465188	20	438295

2.2 Fragment Affinity Matrices

For a given pair of fragments f_i and f_j of the same length, we calculate the Root Mean Square Deviation (RMSD) of the corresponding Cα atoms to measure the distance score between these two fragments. An undirected, weighted fragment affinity graph $G = (V, E, a)$ is created where $f_i \in V$, $(f_i, f_j) \in E$ if the RMSD value between fragments f_i and f_j is within the RMSD cutoff τ, and the corresponding connection affinity $a(f_i, f_j)$ is calculated by applying the Gaussian kernel to convert the RMSD value into the affinity score such that

$$a(f_i,f_j) = \begin{cases} \exp\left(-\dfrac{rmsd(f_i,f_j)}{\sigma^2}\right), & rmsd(f_i,f_j) < \tau \\ 0, & rmsd(f_i,f_j) \geq \tau \end{cases},$$

where σ^2 is the standard deviation of the RMSD distribution of the fragment sample set and τ varies with respect to fragment length. Then, the fragment affinity graph G is converted into a fragment affinity matrix A, where $A_{ij} = a(f_i,f_j)$. Due to the positive RMSD values and the commutative property of RMSD calculation, A is symmetric positive definite (SPD). Moreover, A is sparse when an efficient RMSD cutoff is applied.

2.3 Spectral Clustering

Spectral clustering [16] is a graph-based clustering technique [14] that can be views as finding partitions of a graph that minimizes the graph cut property. The fundamental idea of spectral clustering [17] is to make use of the spectrum (eigenvalues/eigenvectors) of the affinity matrix with respect to graph G to perform dimensionality reduction before clustering in lower dimensions. Starting from the fragment affinity matrix A, a diagonal degree matrix D is defined as

$$D_{ii} = \sum_{j=1}^{n} A_{ij}.$$

Then, a normalized Laplacian matrix is obtained such that

$$L = D^{-1/2}AD^{-1/2}.$$

Afterwards, the largest eigenvector of L is used to bipartition the graph $G = (V, E, a)$ into two complementary partitions S and $\overline{S}$, where $S, \overline{S} \subseteq V, S + \overline{S} = V$, and $S \cap \overline{S} = \emptyset$. Define the normalized cut property $ncut(S, \overline{S})$ of G as

$$ncut(S, \overline{S}) = \frac{w(S, \overline{S})}{w(S, V)} + \frac{w(S, \overline{S})}{w(\overline{S}, V)},$$

where $w(.)$ is the weight function summing all weights between two partitions. According to the theoretical analysis of spectral clustering in [14], $ncut(S, \overline{S})$, which measures the balanced similarity between S and $\overline{S}$, is minimized.

Unlike the classical clustering techniques such as the k-means approaches, the spectral clustering method is able to produce clusters with concave cluster boundaries due to the nonlinear separation hyper surfaces obtained. As a result, spectral clustering does not need any priori information on the shapes of the clusters. Moreover, spectral clustering often yields more robust clustering results because it does not rely on the initial, randomly selected cluster centers.

2.4 Randomized Singular Value Decomposition

The most computationally costly operation in the spectral clustering method described above is the calculation of the bipartitioning eigenvector from the large fragment affinity matrix to bipartition G as well as the subsequent big subgraphs, particularly when the training set includes a large number of fragment samples. Fortunately, we only need the dominant eigenvectors and thus there is no need to calculate the whole spectrum information from the affinity matrix [18]. Notice that because the normalized Laplacian matrix is SPD, its eigenvalue decomposition and SVD coincide. Therefore, we adopt a rank-revealing randomized singular value decomposition (R3SVD) algorithm [19] to fast approximate the dominant eigenvector of the normalized Laplace matrix.

The R^{3}SVD algorithm includes four major steps: Gaussian sampling, QB decomposition, error estimation, and SVD. First of all, given an $n \times n$ Laplacian matrix L, an $n \times k$ Gaussian matrix Ω is randomly generated and an $n \times k$ matrix Y is obtained by projecting L onto Ω such that

$$Y = L^q \Omega,$$

where k is the guessed rank and q is the number of power iterations. Here, we adopt $q = 2$ as recommended by [20]. Then, a QB decomposition is generated by

$$[Q, R] = qr(Y)$$

$$B = Q^T L$$

where $qr(Y)$ is a QR decomposition of Y and Q^T denotes the transpose of Q. Then $QB \approx L$ is a k-rank approximation of L. The error percentage of the QB decomposition can be computationally efficiently estimated by calculating the squares of the Frobenius norms of L and B such that

$$\frac{||L - QB||_F^2}{||L||_F^2} = \frac{||L||_F^2 - ||B||_F^2}{||L||_F^2}.$$

The mathematical proof of this property can be found in [20]. Our empirical results show that when the error percentage is less than 20%, the QB decomposition can lead to a high-quality approximation of the dominant eigenvector of L. If the error percentage is over 20%, the Gaussian sampling step is repeated with an increased k value until the error percentage drops below 20%. Due to the fact that there are limited number of independent factors that determine the formations of structures of short protein fragments, the Laplacian matrix L is of low rank. Typically, $k < 100$ can capture most of the actions of L. Afterwards, the low-rank approximated SVD of L, $U_L \Sigma_L V_L^T$, is obtained by carrying out SVD on the "short-and-wide" matrix B by

$$[U, \Sigma, V] = svd(B)$$

$$U_L = Q^T U, \Sigma_L = \Sigma, V_L = V.$$

Finally, the dominant eigenvector of L can be extracted from the obtained approximated SVD.

The R^3SVD algorithm is able to adaptively estimate the appropriate rank of the approximated SVD to calculate the dominant eigenvector of L. In the randomized algorithm, most numerical linear algebraic operations are carried out on "skinny" block matrices, which are both computation and memory efficient. This allows the spectral clustering method to scale up to handle the large sample sets in this study with close to half million protein fragments.

2.5   Generation of Fragment Libraries

We adopt an iterative bipartitioning approach to generate the fragment libraries given a predefined RMSD clustering cutoff. Starting from the graph G generated from all fragments in the training set, we bipartition it into two complementary subgraphs using the spectral clustering approach described above. The bipartitioning process is repeated on the subgraphs until all RMSD values between fragments in the subgraph fall below a RMSD cutoff. By applying the spectral clustering and iterative bipartitioning techniques, we are able to generate protein backbone fragment clusters of different lengths (7–20 residues) as well as under different RMSD cutoffs. For specific length and RMSD cutoff, these clusters are ranked according to their size, i.e., the number of their member fragments.

Instead of calculating the member-wise mean which may generate unrealistic structural conformation, given a cluster and its RMSD cutoff, the centroid of the cluster is determined by the fragment that has most friend fragments in the cluster within the RMSD cutoff. The conformation of the fragment corresponding to the centroid of the cluster is then deposit into the fragment library. The fragments of the same length are ranked in the fragment library according to the sizes of the clusters that they are generated from.

3   Results

3.1   Analysis of Fragment Libraries

In this work, we have generated protein backbone fragment libraries of length ranging from 7 to 20 under Cα RMSD cutoffs from 0.5 A to 3.0 A. We use the testing set described in Sect. 2.1 to validate our fragment libraries. Figure 1 shows the percentage of chains in the testing set (overall 734 protein chains) that a fragment can be found for the top-100 fragments in the fragment libraries of length 10–20 with 1.0 A cutoff. We define a fragment in the fragment library is found in a protein chain if there exists a same length segment in this chain whose Cα RMSD is less than 1.0 A. One can find that the top 100 fragments in these libraries of different lengths are well represented in the testing set and the ranks of the fragments are mostly preserved. Even the 100-ranked 20-residue fragment can be found in about 5% of the protein chains in the testing set.

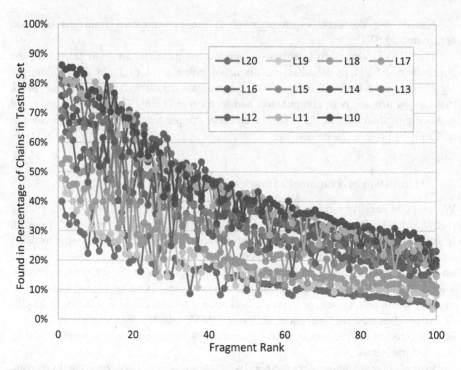

Fig. 1. The percentage of chains that a fragment can be found in the testing set (overall 733 protein chains) for the top-100 fragments in the fragment libraries of length 10–20, 1.0 A cutoff.

3.2 Comparison with Existing Fragment Libraries

We compare the representability of the top-200 12-residue fragments in the fragment library generated by our randomized spectral clustering method with that of the top-200 ones from fragment libraries developed by Kolodny et al. [10, 12] in the testing set. Both fragment libraries have minimum 0.9 A RMSD cutoff. The comparisons measured by the number of fragments matched and percentage of chains found are shown in Figs. 2 and 3, respectively. Two fragments are considered a match if their RMSD is less than 1.0 A. If a fragment in the fragment library can match with a segment in the protein chain, we consider this fragment is found in the protein chain. From Figs. 2 and 3, one can find that except for the first several fragments that are α-helix like, the rest of the top-ranked fragments in our 12-residue fragment library can find significantly more matches, both in terms of the number of fragments and the number of chains, in the testing set. The similar observations are also found in fragment libraries of other lengths.

The conformations of the top-200 12-residue fragments generated by randomized spectral clustering are displayed in Fig. 4. New fragments that have RMSD over 1 A to all top-200 12-residue fragments in the fragment libraries developed by Kolodny et al. are found and are highlighted. As shown in Figs. 2 and 3, these fragments are well representative in the protein chains in the testing set. This indicates that with a much

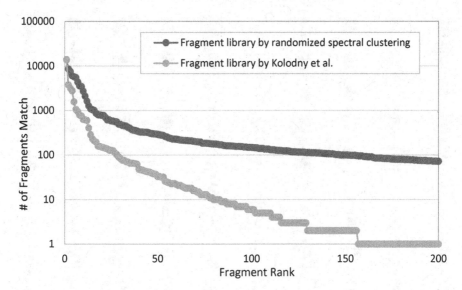

Fig. 2. Comparison of the numbers of chains in the testing set that matches the fragments in the 12-residue fragment library by randomized spectral clustering and the fragments in the fragment library developed by Kolodny et al. [10, 12]. The testing set contains 138,004 overall fragments from 734 protein chains. Both fragment libraries have minimum 0.9 A RMSD cutoff.

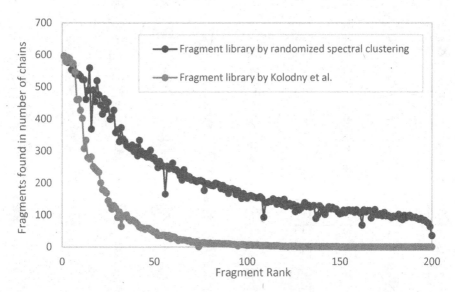

Fig. 3. Comparison of the numbers of chains in the testing set that match each one of the top-200 fragments in the 12-residue fragment library by randomized spectral clustering and the fragments in the fragment library developed by Kolodny et al. [10, 12]. The testing set contains 734 protein chains.

Fig. 4. Top-200 12-residue fragments generated by randomized spectral clustering. The highlighted fragments in orange, green, and blue indicate that the RMSD distances between these fragments and the top-200 12-residue ones in the fragment libraries developed by Kolodny et al. [10, 12] are at least 1 A, 2 A, and 3 A, respectively. (Color figure online)

larger scale fragment data set and a more powerful clustering algorithm, we are able to find more representative backbone fragment conformations across different protein structural conformations.

3.3 Libraries of Long Fragments

For longer protein fragments, the degree of freedom is higher. When the fragment sample set was small, the libraries for long fragments were difficult to obtain because they are sparsely distributed in the high dimensional space. Generated from significantly more structures deposited in the PDB, the large-scale fragment sample sets also allow us to derive libraries for longer fragments. Figure 5 displays the top-100 20-residue fragments in the fragment library with 3.0 A RMSD cutoff generated by our

Fig. 5. Top-100 20-residue fragments with 3.0 A RMSD cutoff generated by randomized spectral clustering.

randomized spectral clustering method. Many super secondary structure motifs [22], such as β-hairpins, short β-sheets, helix-loop-helix, and helix-turn-helix, are found and listed.

4 Conclusions

In this paper, we revisit the problem of constructing representative protein backbone fragment libraries from PDB. The continuous growth of experimentally determined protein structures in PDB provides rich information to construct high quality protein backbone fragments libraries that contain common fragments shared across diverse protein structural conformations. We used the spectral clustering method to bipartition the fragment samples to obtain stable fragment clusters potentially with non-convex boundaries. A rank-revealing randomized SVD algorithm is employed to enable the spectral clustering method to scale up to handle large-scale datasets with nearly half million fragment samples. Compared to the existing protein backbone fragment libraries [10, 12], our new fragment libraries exhibit better representability across diverse protein structures. Libraries for long fragments up to 20 residues are also generated.

Our fragment libraries are deposited at http://hpcr.cs.odu.edu/frag/. Our future work will include using these new fragment libraries for investigating interactions between fragments [23], studying motif formations in protein families, and *de novo* protein structure design.

Acknowledgements. Y. Li acknowledges support from National Science Foundation through Grant No. CCF-1066471. W. Elhefnawy acknowledges support from Old Dominion University Modeling and Simulation Fellowship.

References

1. Munoz, V., Serrano, L.: Local versus nonlocal interactions in protein folding and stability – an experimentalist's point of view. Fold. Des. **1**(4), R71–R77 (1996)
2. Chikenji, G., Fujitsuka, Y., Takada, S.: Shaping up the protein folding funnel by local interaction: lesson from a structure prediction study. Proc. Natl. Acad. Sci. **103**(9), 3141–3146 (2006)
3. Simons, K.T., Kooperberg, C., Huang, E., Baker, D.: Assembly of protein tertiary structures from fragments with similar local sequences using simulated annealing and Bayesian Scoring functions. J. Mol. Biol. **268**, 209–225 (1997)
4. de Oliveira, S.H.P., Shi, J., Deane, C.M.: Building a better fragment library for de novo protein structure prediction. PLoS ONE **10**(4), e0123998 (2015)
5. Rata, I., Li, Y., Jakobsson, E.: Backbone Statistical Potential from Local Sequence-Structure Interactions in Protein Loops. J. Phys. Chem. B **114**(5), 1859–1869 (2010)
6. Li, Y., Rata, I., Jakobsson, E.: Sampling multiple scoring functions can improve protein loop structure prediction accuracy. J. Chem. Inf. Model. **51**(7), 1656–1666 (2011)
7. Li, Y.: Conformational sampling in template-free protein loop structure modeling: an overview. Comput. Struct. Biotechnol. J. **5**(6), e201302003 (2013)

8. Di Maio, F., Shavlik, J., Phillips, G.: A probabilistic approach to protein backbone tracing in electron density maps. Bioinformatics **22**(14), 81–89 (2006)
9. Terwiliger, T.C.: Automated main-chain model building by template matching and iterative fragment extension. Acta Crystallogr. D Biol. Crystallogr. **59**(1), 38–44 (2003)
10. Budowski-Tal, I., Nov, Y., Kolodny, R.: FragBag, an accurate representation of protein structure, retrieves structural neighbors from the entire PDB quickly and accurately. Proc. Natl. Acad. Sci. **107**, 3481–3486 (2010)
11. Keasar, C., Kolodny, R.: Using protein fragments for searching and data-mining protein databases. In: Proceedings of AAAI workshop of Artificial Intelligence and Robotics Methods in Computational Biology (2013)
12. Kolodny, R., Koehl, P., Guibas, L., Levitt, M.: Small Libraries of Protein Fragments Model Native Protein Structures Accurately. J. Mol. Biol. **323**, 297–307 (2005)
13. Denise, C.: Structural GENOMICS exploring the 3D protein landscape. Simbios (2010)
14. Shi, J., Malik, J.: Normalized cuts and image segmentation. IEEE Trans. Pattern Anal. Mach. Intell. **22**(8), 888–905 (2000)
15. Wang, G.L., Dunbrack, R.L.: PISCES: a protein sequence culling server. Bioinformatics **19**, 1589–1591 (2003)
16. von Luxburg, U.: A tutorial on spectral clustering. Stat. Comput. **17**(4), 395–416 (2007)
17. Ng, A.Y., Jordan, M.I., Weiss, Y.: On spectral clustering: analysis and an algorithm. Adv. Neural. Inf. Process. Syst. **14**, 849–856 (2001)
18. Ji, H., Weinberg, S., Li, Y.: A revisit of block power methods for finite state markov chain applications. arXiv:1610.08881 (2016)
19. Ji, H., Yu, W., Li, Y.: A rank revealing randomized singular value decomposition (R^3SVD) algorithm for low-rank matrix approximations. arXiv:1605.08134 (2016)
20. Gu, Y., Yu, W., Li, Y.: Efficient randomized algorithms for adaptive low-rank factorizations of large matrices. arXiv:1606.09402 (2016)
21. Halko, N., Martinsson, P.G., Tropp, J.A.: Finding structure with randomness: probabilistic algorithms for constructing approximate matrix decompositions. SIAM Rev. **53**(2), 217–288 (2009)
22. Chiang, Y.S., Gelfand, T.I., Kister, A.E., Gelfand, I.M.: New classification of supersecondary structures of sandwich-like proteins uncovers strict patterns of strand assemblage. Proteins **68**(4), 915–921 (2007)
23. Elhefnawy, W., Chen, L., Han, Y., Li, Y.: ICOSA: a distance-dependent, orientation-specific coarse-grain contact potential for protein structure modeling. J. Mol. Biol. **427**(15), 2562–2576 (2015)
24. Li, Y., Liu, H., Rata, I., Jakobsson, E.: Building a knowledge-based statistical potential by capturing high-order inter-residue interactions and its applications in protein secondary structure assessment. J. Chem. Inf. Model. **53**(2), 500–508 (2013)

Mapping Paratope and Epitope Residues of Antibody Pembrolizumab via Molecular Dynamics Simulation

Wenping Liu[1] and Guangjian Liu[2(✉)]

[1] School of Bioscience and Bioengineering,
South China University of Technology, Guangzhou 510006, China
liuwenp@mail3.sysu.edu.cn
[2] Division of Birth Cohort Study,
Guangzhou Women and Children's Medical Center,
Guangzhou Medical University, Guangzhou 510623, China
liugjcn@163.com

Abstract. Blocking the programmed death receptor 1 (PD-1)/programmed death ligand 1 protein (PD-L1) interaction has come up as a promising cancer immunotherapy. Pembrolizumab is a therapeutic monoclonal antibody targeting PD-1 and received widespread attention. However, the messages for the paratope and epitope residues of pembrolizumab are insufficient. Here molecular dynamics (MD) simulation was used to map epitope on PD-1 to paratope residues on pembrolizumab. A total of twenty-nine key residues were predicted in the PD-1/pembrolizumab interaction. Of the fourteen epitope residues, three (i.e., ASN66, LYS78 and ALA132 on PD-1) were found to play critical roles in the interaction of PD-1 and PD-L1. Therefore, pembrolizumab prevents PD-L1 from interacting with PD-1 through steric hindrance, and the key residues sorted out here were potential hotspots for the optimization of pembrolizumab.

Keywords: Epitope · Paratope · Molecular dynamics simulation · PD-1 · Pembrolizumab · PD-L1

1 Introduction

Cancer immunotherapy has great achievements in cancer treatment in past few years. It comprises a variety of treatment approaches and the blockade of the interaction between programmed death receptor 1 (PD-1) and its ligand programmed death ligand 1 protein (PD-L1) is the most promising [1, 2]. PD-1 is a 288 amino acid type I transmembrane protein expressed on tumor and immune cells and PD-L1 is a 290 amino acid type I transmembrane protein belonging to the B7 family [3, 4]. several therapeutic monoclonal antibodies targeting PD-1 or PD-L1 have come up, such as MDX-1106, MK3475, CT-011, AMP-224 and MDX-1105 [5].

Pembrolizumab, also known as KEYTRUDA, is a humanized IgG4 antibody blocking PD-1. It was approved for the treatment of advanced melanoma and metastatic non-small-cell lung cancer by FDA in 2015 [6]. Mapping paratope to epitope of

Z. Cai et al. (Eds.): ISBRA 2017, LNBI 10330, pp. 120–127, 2017.
DOI: 10.1007/978-3-319-59575-7_11

pembrolizumab is an essential step to increase its efficacy to better meet the clinical demands [7]. The PD-1/pembrolizumab complex was crystallized in 2016 and provided us with the information of the interface at atomic level [8]. However, protein-protein interaction is a dynamic process, and that conformational transformation is missed in frozen structure, which might results in loss of key residues [9, 10]. Besides, the blocking mechanism of pembrolizumab is not clear.

Here, molecular dynamic (MD) simulations were used to map paratope to epitope residues of pembrolizumab, as well as the key residues in the interface of PD-1/PD-L1 complex. Altogether fourteen epitope and fifteen paratope residues were sorted out and three epitope residues (i.e., ASN66, LYS78 and ALA132 on PD-1) were also found to be critical for the binding between PD-1 and PD-L1. Therefore, pembrolizumab prevents PD-L1 from interacting with PD-1 through steric hindrance, and the key residues sorted out here were potential hotspots for the optimization of pembrolizumab.

2   Materials and Methods

2.1   MD Simulations

The crystal structures of the PD-1/pembrolizumab and PD-1/PD-L1 complexes were downloaded from Protein Data Bank with accession code of 5GGS and 4ZQK. The VMD program was used for modeling [11]. Residues 85 to 92 of PD-1 were missing in the crystal structure and modeled by the SWISS-MODEL server [12, 13]. Two complexes were first solvated with TIP3P water molecules in rectangular boxes, respectively. Then, Na^+ and Cl^- ions were added to neutralize two systems at a 150 mM salt concentration.

The NAMD 2.11 program [14] with CHARMM36 all-atom force field [15, 16] were used for the simulations. Two systems were energy-minimized for 5,000 steps with all protein atoms fixed and for another 5,000 steps with all atoms free. After that, equilibrium of 20 ns was performed thrice (I, II and III) for each complex, respectively, during which the temperature was held at 310 K using Langevin dynamics and the pressure was held at 1 atmosphere by the Langevin piston method.

2.2   Survival Ratio of H-Bonds or Salt Bridges

The H-bonds and/or salt bridges across the complex interface were detected through the VMD software. An H-bond was defined if the donor–acceptor distance and bonding angle were smaller than 3.5 Å and 30°, respectively. But only the bond-length cutoff of 4 Å was applied to examine the salt bridges in binding site. The survival ratio of an H-bond (or salt bridge) was defined as the percentage of bond survival time. The maximum value of the survival ratios with the initial equilibrated conformations I, II, and III was considered the survival ratio of a bond (Tables 2 and 4).

3 Results

3.1 Mapping Paratope to Epitope Residues of Pembrolizumab/PD-1 Complex via Molecular Dynamics Simulation

Mapping paratope to epitope is an essential step to increase the efficacy of antibody pembrolizumab. The crystal structure of the pembrolizumab/PD-1 complex provides us that information at atomic level, which could be downloaded from the PDB database with accession code 5GGS. H-bonds and salt bridges at the complex interface were believed to mainly contribute to the binding and they were analyzed with VMD to obtain information on paratope to epitope residues [8]. The crystal structure of pembrolizumab/PD-1 complex had five H-bonds and two salt bridges in its interface, involving four residues (i.e., ASN66, ASP85, LYS131 and ALA132) on PD-1 and three residues (i.e., H99ARG, H102ARG, and L59GLU) on pembrolizumab (Table 1).

Table 1. Interaction residue between PD-1 and pembrolizumab in the crystal structure

Bond no.	Bond type[*]	PD-1		Pembrolizumab[#]	
		Residue	Atom	Residue	Atom
1	H	ASN66	ND2	H102ARG	O
2	H	ASP85	OD1	H99ARG	NH1
3	H	ASP85	OD2	H99ARG	NH2
4	H	ALA132	O	H102ARG	NH1
5	S	ASP85		H99ARG	
6	S	LYS131		L59GLU	

[*]H denotes the hydrogen bond and S denotes the salt bridge.
[#]The name of the residues with H or L indicating that the residues are on the heavy or the light chain of pembrolizumab, respectively, and with the number indicating the position of the residue.

Above analysis provided us binding residues in the crystal structure. But protein-protein interaction is a dynamic progress [9] and protein flexibility also plays a significant role in predicting locations of interacting interface [17]. That information was lost in static structure analysis. Therefore, MD simulations were performed on pembrolizumab/PD-1 complex because it is a useful tool to studying the dynamics of proteins at atomic level [18]. System equilibrium process of 20 ns was conducted thrice after energy minimization of 10,000 steps for pembrolizumab/PD-1 complex.

Table 2 showed all detected bonds in equilibriums with survival ratios above 0.2 (Methods, Table 2). Altogether twenty-five bonds were detected in simulations, involving fourteen residues (i.e., SER62, PHE63, ASN66, THR76, ASP77, LYS78, GLU84, ASP85, ARG86, SER87, GLY90, LEU128, LYS131 and ALA132) on PD-1 and fifteen residues (i.e., H35TYR, H54SER, H55ASN, H58THR, H99ARG, H101TYR, H102ARG, H104ASP, H108ASP, L32SER, L34TYR, L36TYR, L53TYR, L57TYR and L59GLU) on pembrolizumab (Fig. 1a and b). Compared with the results of structure analysis, additional ten residues (i.e., SER62, PHE63, THR76, ASP77, LYS78, GLU84, ARG86, SER87, GLY90 and LEU128) on PD-1 and twelve residues (i.e., H35TYR, H54SER, H55ASN, H58THR, H101TYR, H104ASP, H108ASP, L32SER, L34TYR, L36TYR, L53TYR and L57TYR) on pembrolizumab were sorted out. It implied that PD-1/pembrolizumab complex had conformational transitions in water and new interaction residues in the interface were found.

Table 2. Survival ratios of bonds detected from MD simulations for pembrolizumab/PD-1 complex

Bond no.	Bond type	PD-1		Pembrolizumab[#]		Survival ratio			
		Residue	Atom	Residue	Atom	I	II	III	Max
1	S	LYS131		L59GLU		0.99	0.99	0.98	0.99
2	H	ASP85	OD2	H99ARG	NH2	0.99	0.98	0.99	0.99
3	H	ASP85	OD1	H99ARG	NH1	0.96	0.95	0.97	0.97
4	S	ASP85		H99ARG		0.96	0.95	0.95	0.96
5	H	LEU128	O	L53TYR	OH	0.93	0.88	0.83	0.93
6	H	SER87	O	H35TYR	OH	0.84	0.89	0.87	0.89
7	H	SER62	OG	L57TYR	OH	0.71	0.80	0.52	0.80
8	H	LYS131	O	H102ARG	NH2	0.28	0.73	0.67	0.73
9	H	SER87	OG	H99ARG	NH1	0.68	0.59	0.71	0.71
10	H	LYS78	NZ	H101TYR	O	0.71	0.58	0.66	0.71
11	H	SER87	OG	H104ASP	OD2	0.19	0.67	0.50	0.67
12	H	SER87	OG	H104ASP	OD1	0.62	0.14	0.36	0.62
13	H	ASP77	OD1	H54SER	OG	0.13	0.17	0.50	0.50
14	H	ASN66	ND2	H102ARG	O	0.15	0.45	0.50	0.50
15	H	LYS131	NZ	L59GLU	OE2	0.42	0.43	0.33	0.43
16	H	GLY90	N	H58THR	O	0.37	0.22	0.42	0.42
17	H	LYS131	NZ	L59GLU	OE1	0.35	0.31	0.41	0.41
18	H	ASP77	OD2	H55ASN	ND2	0.40	0.02	0.03	0.4
19	H	THR76	O	H101TYR	OH	0.00	0.35	0.18	0.35
20	H	LYS131	NZ	H108ASP	OD2	0.19	0.31	0.19	0.31
21	H	LYS131	NZ	H108ASP	OD1	0.27	0.00	0.10	0.27
22	H	GLU84	O	L36TYR	OH	0.21	0.24	0.00	0.24
23	H	ALA132	O	H102ARG	NH1	0.00	0.23	0.13	0.23
24	H	PHE63	O	L34TYR	OH	0.07	0.22	0.10	0.22
25	H	ARG86	NH2	L32SER	OG	0.02	0.09	0.21	0.21

The headings I, II, and III denote three equilibrations. H denotes the hydrogen bond and S denotes the salt bridge in Column 2.

[#]The name of the residues with H or L indicating that the residues are on the heavy or the light chain of pembrolizumab, respectively, and with the number indicating the position of the residue.

a
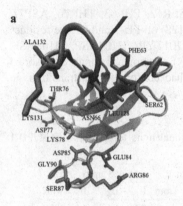

b
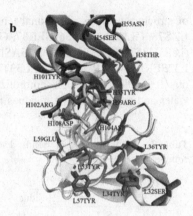

Fig. 1. The predicted key residues on the PD-1/pembrolizumab complex interface: (a) the key residues on PD-1 and (b) the key residues on pembrolizumab are shown in red licorice. PD-1 is shown in silver, the heavy chain of pembrolizumab is shown in green and the light chain of pembrolizumab is shown in yellow. (Color figure online)

3.2   Identify Key Residues in the Interface of PD-1/PD-L1 Complex

Although paratope to epitope residues were identified via MD simulations, we also need the information of the key residues in the interface of PD-1/PD-L1 complex to better understand how pembrolizumab blocks the interaction between PD-1 and PD-L1. The crystal structure of PD-1/PD-L1 was obtained from the PDB database with accession code 4ZQK. H-bonds and salt bridges was also suggested to mainly donate to the binding and they were detected through VMD software [19]. Five H-bonds and one salt bridge, involving four residues (i.e., ASN66, GLN75, ALA132, and GLU136) on PD-1 and four residues (i.e., ASP26, GLN66, ALA121, and ARG125) on PD-L1 (Table 3) were sorted out. But those interactions might not be stable when the complex was in solution.

Table 3. Residue interactions between PD-1 and PD-L1 in crystal structure

No.	Hydrogen bond				Salt bridge	
	PD-1		PD-L1		PD-1	PD-L1
	Residue	Atom	Residue	Atom	Residue	Residue
1	ASN66	ND2	ALA121	O	GLU136	ARG125
2	GLN75	OE1	ARG125	N		
3	GLN75	NE2	ASP26	OD1		
4	GLN75	NE2	ARG125	O		
5	ALA132	N	GLN66	OE1		

Therefore, system equilibrium process of 20 ns was performed thrice after energy minimization of 10,000 steps for PD-1/PD-L1 complex. Stable bonds with survival ratios above 0.2 were listed in Table 4. Altogether seventeen bonds were detected in simulations, involving six residues (i.e., ASN66, TYR68, GLN75, LYS78, ALA132, GLU136) on PD-1 and seven residues (i.e., ASP26, GLN66, ARG113, ALA121, ASP122, TYR123, ARG125) on PD-L1. Additional two residues (i.e., TYR68, LYS78) on PD-1 and three residues (i.e., ARG113, ASP122, TYR123) on PD-L1 were sorted out compared with the results of crystal structure analysis, which again proved the flexibility of interface (Fig. 2a and b).

Table 4. Survival ratios of bonds detected from MD simulations for PD-L1/PD-1 complex

Bond no.	Bond type	PD-1		PD-L1		Survival ratio			
		Residue	Atom	Residue	Atom	I	II	III	Max
1	H	GLU136	OE2	TYR123	OH	0.54	0.97	0.97	0.97
2	H	TYR68	OH	ASP122	OD1	0.40	0.95	0.95	0.95
3	S	GLU136		ARG125		0.73	0.91	0.94	0.94
4	H	GLU136	OE1	ARG125	NH1	0.45	0.91	0.92	0.92
5	H	LYS78	NZ	ASP122	OD1	0.41	0.58	0.91	0.91
6	H	GLU136	OE2	ARG125	NH2	0.66	0.91	0.89	0.91
7	S	LYS78		ASP122		0.83	0.52	0.78	0.83
8	H	ALA132	N	GLN66	OE1	0.47	0.66	0.75	0.75
9	H	GLN75	OE1	ARG125	N	0.38	0.61	0.73	0.73
10	H	GLN75	NE2	ARG125	O	0.66	0.68	0.64	0.68
11	H	TYR68	OH	TYR123	N	0.12	0.51	0.51	0.51
12	H	ASN66	ND2	ALA121	O	0.25	0.49	0.37	0.49
13	H	LYS78	NZ	ASP122	OD2	0.43			0.43
14	H	GLU136	OE1	TYR123	OH	0.43			0.43
15	H	GLU136	OE2	ARG113	NH2	0.3			0.3
16	H	GLU136	OE1	ARG125	NE	0.25			0.25
17	H	GLN75	NE2	ASP26	OD1	0.23		0.01	0.23

The headings I, II, and III denote three equilibrations. H denotes the hydrogen bond and S denotes the salt bridge in Column 2.

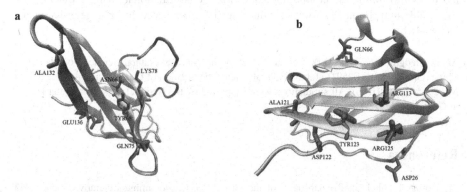

Fig. 2. The predicted key residues on the PD-1/PD-L1 complex interface: (a) the key residues on PD-1 and (b) the key residues on PD-L1 are shown in red licorice. PD-1 is shown in silver, whereas PD-L1 is shown in cyan. (Color figure online)

3.3   Pembrolizumab Prevents PD-L1 from Interacting with PD-1 Through Steric Hindrance

Our results showed that fourteen residues (i.e., SER62, PHE63, ASN66, THR76, ASP77, LYS78, GLU84, ASP85, ARG86, SER87, GLY90, LEU128, LYS131 and ALA132) on PD-1 were responsible for the binding between PD-1 and antibody pembrolizumab while six residues (i.e., ASN66, TYR68, GLN75, LYS78, ALA132, GLU136) on PD-1 were critical for its binding with PD-L1. Three of them (ASN66, LYS78, ALA132) were the same, indicating that pembrolizumab could occupy about 50% of the binding site of PD-L1. Therefore, pembrolizumab prevents PD-L1 from interacting with PD-1 through steric hindrance.

4   Discussion

Mapping paratope to epitope residues is an essential step in therapeutic antibody design and optimization. Here we used MD simulations to find the paratope and epitope residues of antibody pembrolizumab and altogether twenty-nine residues were sorted out. The key residues in the interface of PD-1/PD-L1 complex were also predicted by MD simulations to further study the block mechanism of pembrolizumab. Our residues showed that three of fourteen epitopes (ASN66, LYS78 and ALA132) also played important roles in the recognition of PD-1 and PD-L1. Therefore, pembrolizumab prevents PD-L1 from interacting with PD-1 through steric hindrance, and the left three residues on PD-1(TYR68, GLN75, GLU136) that critical for the binding with PD-L1 were potential hotspots for the optimization of pembrolizumab.

Protein flexibility was demonstrated to play important roles in protein-protein recognition. From the crystal structure of PD-1/pembrolizumab and PD-1/PD-L1 complexes, we only got seven and eight key residues. But additional twenty-two and five residues were sorted out for two complexes through MD simulations, respectively. The reason lies in that molecular recognition and drug binding are dynamic processes and the conformation transforming is missed in crystal structure [9]. MD simulation is a useful tool to mimic the atomic fluctuations and conformational changes of biomolecules, although the newly found residues need further tested by mutagenesis experiments [18].

Acknowledgments. This work was supported by the National Natural Science Foundation of China (Grant No. 31500591) and the Natural Science Foundation of Guangdong Province (Grant No. 2015A030310106). All simulations were supported by the National Super Computer Center in Guangzhou.

References

1. Couzin-Frankel, J.: Breakthrough of the year 2013. Cancer immunotherapy. Science **342** (6165), 1432–1433 (2013). doi:10.1126/science.342.6165.1432
2. Pardoll, D.M.: The blockade of immune checkpoints in cancer immunotherapy. Nat. Rev. Cancer **12**(4), 252–264 (2012). doi:10.1038/nrc3239

3. Freeman, G.J., Long, A.J., Iwai, Y., Bourque, K., Chernova, T., Nishimura, H., Fitz, L.J., Malenkovich, N., Okazaki, T., Byrne, M.C., Horton, H.F., Fouser, L., Carter, L., Ling, V., Bowman, M.R., Carreno, B.M., Collins, M., Wood, C.R., Honjo, T.: Engagement of the PD-1 immunoinhibitory receptor by a novel B7 family member leads to negative regulation of lymphocyte activation. J. Exp. Med. **192**(7), 1027–1034 (2000)

4. Keir, M.E., Butte, M.J., Freeman, G.J., Sharpe, A.H.: PD-1 and its ligands in tolerance and immunity. Ann. Rev. Immunol. **26**, 677–704 (2008). doi:10.1146/annurev.immunol.26. 021607.090331

5. Kono, K.: Current status of cancer immunotherapy. J. Stem Cells Regen. Med. **10**(1), 8–13 (2014).

6. Callahan, M.K., Postow, M.A., Wolchok, J.D.: Targeting T cell co-receptors for cancer therapy. Immunity **44**(5), 1069–1078 (2016). doi:10.1016/j.immuni.2016.04.023

7. Gershoni, J.M., Roitburd-Berman, A., Siman-Tov, D.D., Tarnovitski Freund, N., Weiss, Y.: Epitope mapping: the first step in developing epitope-based vaccines. BioDrugs **21**(3), 145–156 (2007)

8. Lee, J.Y., Lee, H.T., Shin, W., Chae, J., Choi, J., Kim, S.H., Lim, H., Won Heo, T., Park, K.Y., Lee, Y.J., Ryu, S.E., Son, J.Y., Lee, J.U., Heo, Y.S.: Structural basis of checkpoint blockade by monoclonal antibodies in cancer immunotherapy. Nat. Commun. **7**, Article No. 13354 (2016). doi:10.1038/ncomms13354

9. Moroni, E., Paladino, A., Colombo, G.: The dynamics of drug discovery. Curr. Top. Med. Chem. **15**(20), 2043–2055 (2015)

10. Durrant, J.D., McCammon, J.A.: Molecular dynamics simulations and drug discovery. BMC Biol. **9**, 71 (2011). doi:10.1186/1741-7007-9-71

11. Humphrey, W., Dalke, A., Schulten, K.: VMD: visual molecular dynamics. J. Mol. Graph. **14**(1), 27–38 (1996)

12. Bordoli, L., Kiefer, F., Arnold, K., Benkert, P., Battey, J., Schwede, T.: Protein structure homology modeling using SWISS-MODEL workspace. Nat. Protoc. **4**(1), 1–13 (2009)

13. Biasini, M., Bienert, S., Waterhouse, A., Arnold, K., Studer, G., Schmidt, T., Kiefer, F., Cassarino, T.G., Bertoni, M., Bordoli, L.: SWISS-MODEL: modelling protein tertiary and quaternary structure using evolutionary information. Nucleic Acids Res. **42**(w1), 252–258 (2014)

14. Phillips, J.C., Braun, R., Wang, W., Gumbart, J., Tajkhorshid, E., Villa, E., Chipot, C., Skeel, R.D., Kale, L., Schulten, K.: Scalable molecular dynamics with NAMD. J. Comput. Chem. **26**(16), 1781–1802 (2005). doi:10.1002/jcc.20289

15. MacKerell Jr., A.D., Feig, M., Brooks III, C.L.: Improved treatment of the protein backbone in empirical force fields. J. Am. Chem. Soc. **126**(3), 698–699 (2004). doi:10.1021/ja036959e

16. MacKerell, A.D., Bashford, D., Bellott, M., Dunbrack, R.L., Evanseck, J.D., Field, M.J., Fischer, S., Gao, J., Guo, H., Ha, S., Joseph-McCarthy, D., Kuchnir, L., Kuczera, K., Lau, F.T., Mattos, C., Michnick, S., Ngo, T., Nguyen, D.T., Prodhom, B., Reiher, W.E., Roux, B., Schlenkrich, M., Smith, J.C., Stote, R., Straub, J., Watanabe, M., Wiorkiewicz-Kuczera, J., Yin, D., Karplus, M.: All-atom empirical potential for molecular modeling and dynamics studies of proteins. J. Phys. Chem. B **102**(18), 3586–3616 (1998). doi:10.1021/jp973084f

17. Lexa, K.W., Carlson, H.A.: Protein flexibility in docking and surface mapping. Q. Rev. Biophys. **45**(3), 301–343 (2012). doi:10.1017/S0033583512000066

18. Adcock, S.A., McCammon, J.A.: Molecular dynamics: survey of methods for simulating the activity of proteins. Chem. Rev. **106**(5), 1589–1615 (2006). doi:10.1021/cr040426m

19. Zak, K.M., Kitel, R., Przetocka, S., Golik, P., Guzik, K., Musielak, B., Domling, A., Dubin, G., Holak, T.A.: Structure of the complex of human programmed death 1, PD-1, and its ligand PD-L1. Structure **23**(12), 2341–2348 (2015). doi:10.1016/j.str.2015.09.010

A New 2-Approximation Algorithm
for rSPR Distance

Zhi-Zhong Chen[1]([⊠]), Youta Harada[1], and Lusheng Wang[2]

[1] Division of Information System Design, Tokyo Denki University, Tokyo, Japan
zzchen@mail.dendai.ac.jp
[2] Department of Computer Science, City University of Hong Kong,
Kowloon Tong, HK, China

Abstract. Due to hybridization events in evolution, studying two
different genes of a set of species may yield two related but different
phylogenetic trees for the set of species. In this case, we want to measure the dissimilarity of the two trees. The rooted subtree prune and
regraft (rSPR) distance of the two trees has been used for this purpose. The problem of computing the rSPR distance of two given trees
has many applications but is NP-hard. The previously best approximation algorithm for rSPR distance achieves a ratio of 2 in *polynomial*
time and its analysis is based on the duality theory of linear programming. In this paper, we present a *cubic*-time approximation algorithm
for rSPR distance that achieves a ratio of 2. Our algorithm is based on
the notion of *key* and several structural lemmas; its analysis is purely
combinatorial and explicitly uses a search tree for computing rSPR distance exactly. Our experimental results show that the algorithm can be
implemented into a program which outputs significantly better lower
and upper bounds on the rSPR distance of the two given trees than the
previous best.

Keywords: Phylogenetic tree · rSPR distance · Approximation algorithm · Fixed-parameter algorithm

1   Introduction

When studying the evolutionary history of a set X of existing species, one can
obtain a phylogenetic tree T_1 with leaf set X with high confidence by looking at a
segment of sequences or a set of genes [13,14]. When looking at another segment
of sequences, a different phylogenetic tree T_2 with leaf set X can be obtained
with high confidence, too. In this case, we want to measure the dissimilarity of T_1
and T_2. The rooted subtree prune and regraft (rSPR) distance between T_1 and
T_2 has been used for this purpose [12]. It can be defined as the minimum number
of edges that should be deleted from each of T_1 and T_2 in order to transform
them into *essentially identical* rooted forests F_1 and F_2. Roughly speaking, F_1
and F_2 are *essentially identical* if they become identical forests (called *agreement*

© Springer International Publishing AG 2017
Z. Cai et al. (Eds.): ISBRA 2017, LNBI 10330, pp. 128–139, 2017.
DOI: 10.1007/978-3-319-59575-7_12

forests of T_1 and T_2) after repeatedly contracting an edge (p, c) in each of them such that c is the unique child of p (until no such edge exists).

The rSPR distance is an important metric that often helps us discover reticulation events. In particular, it provides a lower bound on the number of reticulation events [1,2], and has been regularly used to model reticulate evolution [15,16].

Unfortunately, it is NP-hard to compute the rSPR distance of two given phylogenetic trees [5,12]. This has motivated researchers to design approximation algorithms for the problem [3,4,12,17]. Hein *et al.* [12] were the first to come up with an approximation algorithm. They also introduced the important notion of maximum agreement forest (MAF) of two phylogenetic trees. Their algorithm was correctly analyzed by Bonet *et al.* [3]. Rodrigues *et al.* [17] modified Hein *et al.*'s algorithm so that it achieves an approximation ratio of 3 and runs in quadratic time. Whidden *et al.* [21] came up with a very simple approximation algorithm that runs in linear time and achieves an approximation ratio of 3. Although the ratio 3 is achieved by a very simple algorithm in [21], no polynomial-time approximation algorithm had been designed to achieve a better ratio than 3 before Shi *et al.* [11] presented a polynomial-time approximation algorithm that achieves a ratio of 2.5. Schalekamp *et al.* [18] presented a *polynomial*-time 2-approximation algorithm for the same problem. However, they use an LP-model of the problem and apply the duality theory of linear programming in the analysis of their algorithm. Hence, their analysis is not intuitively understandable. Moreover, they did not give an explicit upper bound on the running time of their algorithm. Unaware of Schalekamp *et al.*'s work [18], we [9] presented a *quadratic*-time $\frac{7}{3}$-approximation algorithm for the problem; the algorithm is relatively simpler and its analysis is purely combinatorial.

In certain real applications, the rSPR distance between two given phylogenetic trees is small enough to be computed exactly within reasonable amount of time. This has motivated researchers to take the rSPR distance as a parameter and design fixed-parameter algorithms for computing the rSPR distance of two given phylogenetic trees [5,8,19–21]. These algorithms are basically based on the branch-and-bound approach and use the output of an approximation algorithm (for rSPR distance) to decide if a branch of the search tree should be cut. Thus, better approximation algorithms for rSPR distance also lead to faster exact algorithms for rSPR distance. It is worth noting that approximation algorithms for rSPR distance can also be used to speed up the computation of hybridization number and the construction of minimum hybridization networks [6,7].

In this paper, we sketch how to improve our $\frac{7}{3}$-approximation algorithm in [9] to a new 2-approximation algorithm. Our algorithm proceeds in stages until the input trees T_1 and T_2 become identical forests. Roughly speaking, in each stage, our algorithm carefully chooses a dangling subforest S of T_1 and uses S to carefully choose and remove a set B of edges from T_2. B has a crucial property that the removal of the edges of B decreases the rSPR distance of T_1 and T_2 by at least $\frac{1}{2}|B|$. Because of this property, our algorithm achieves a ratio of 2. As in [9], the search of S and B in our algorithm is based on our original notion of *key.* However, unlike the algorithm in [9], the subforest S in our new algorithm is not bounded from above by a constant. This difference is crucial, because the small bounded size of S in [9] makes for a tedious case-analysis. Fortunately, we can prove a number of structural lemmas which enable us to construct B systematically and hence avoid complicated case-analysis. Our analysis of the algorithm explicitly uses a search tree (for computing the rSPR distance of two given trees exactly) as a tool, in order to show that it achieves a ratio of 2. To our knowledge, we were the first to use a search tree explicitly for this purpose.

Unfortunately, our new algorithm and its analysis are so complicated that it is impossible to include the details here. In this paper, we only sketch the algorithm and the details can be found in [10]. We also implement our algorithm. Like Schalekamp *et al.*'s implementation of their 2-approximation algorithm for rSPR distance [18], our implementation can output both a lower bound and an upper bound on the rSPR distance of the input trees. Our experimental results show that our implementation gives much better lower and upper bounds than Schalekamp *et al.*'s.

The remainder of this paper is organized as follows. Section 2 reviews the rSPR distance problem and states our main theorems. Section 3 first gives the basic definitions that will be used thereafter, then shows how to build a search tree for computing the rSPR distance exactly, further defines the important notion of *key*, and finally sketches how to compute a good key or cut. The final section states our experimental results.

2   The rSPR Distance Problem

A *phylogenetic forest* is a rooted forest F in which each vertex has at most two children, each root has zero or two children, and the leaves are distinctively labeled but the non-leaves are unlabeled. A non-leaf v of F is *unifurcate* (respectively, *bifurcate*) if the number of children of v in F is 1 (respectively, 2). F is a *phylogeny* if it is connected and has no unifurcate vertices. Figure 1 shows two phylogenies T and F.

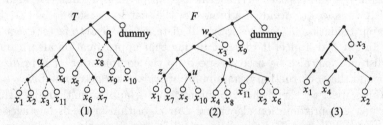

Fig. 1. (1) A phylogeny T, (2) another phylogeny F, (3) $F{\uparrow}_{\{x_1,\dots,x_4\}}$.

For a phylogenetic forest F and a set or sequence U of vertices, $\ell_F(U)$ denotes the lowest common ancestor (LCA) of the vertices in U if the vertices in U are in the same connected component of F, while $\ell_F(U)$ is undefined otherwise. Let C be a set of edges in F. $F - C$ denotes the forest obtained from F by deleting the edges in C. $F - C$ may not be phylogenetic, because it may have unlabeled leaves or unifurcate roots. $F \ominus C$ denotes the phylogenetic forest obtained from $F - C$ by first removing all vertices without labeled descendants and then repeatedly removing a unifurcate root until no root is unifurcate. Note that both $F - C$ and $F \ominus C$ are subgraphs of F. C is a *cut* of F if each connected component of $F - C$ has a labeled leaf. If in addition every leaf of $F - C$ is labeled, then C is a *canonical cut* of F. For example, if F is as in Fig. 2(1), then the 4 dashed edges in F form a canonical cut of F. It is known that if C is a set of edges in F, then F has a canonical cut C' such that $F \ominus C = F \ominus C'$ [4].

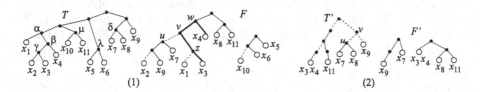

Fig. 2. (1) A TF-pair (T, F) and (2) an induced sub-TF pair (T', F') of (T, F).

Let F_1 and F_2 be two phylogenetic forests with the same set of leaf-labels. A leaf x_1 of F_1 is *agreed with* a leaf x_2 of F_2 if the labels of x_1 and x_2 are the same. We can extend this agreement between the leaves of F_1 and F_2 to (some of) their bifurcate non-leaves recursively as follows. Suppose that two non-roots v_1 and v_1' in F_1 are agreed with two non-roots v_2 and v_2' in F_2, respectively. Further assume that $\ell_{F_i}(v_i, v_i')$ is defined and bifurcate for each $i \in \{1, 2\}$. Then, $\ell_{F_1}(v_1, v_1')$ in F_1 is *agreed with* $\ell_{F_2}(v_2, v_2')$ in F_2 if for each $i \in \{1, 2\}$, every vertex of the path between v_i and v_i' in F_i other than v_i, v_i', and $\ell_{F_i}(v_i, v_i')$ is unifurcate. This finishes the extension. A vertex of F_1 (respectively, F_2) is *agreed* if it is agreed with a vertex of F_2 (respectively, F_1). F_1 and F_2 are *identical* if the roots of F_1 are agreed and so are the roots of F_2.

The rSPR Distance Problem: Given a pair (T, F) of phylogenies with the same set of leaf-labels, find a cut C_T in T and a *smallest* cut C_F in F such that $T \ominus C_T$ and $F \ominus C_F$ are identical.

For example, if T and F in Fig. 1 are the input to the rSPR distance problem, then the dashed edges in T and those in F together form a possible output. In the above definition, we require that the size of C_F be minimized; indeed, it is equivalent to require that the size of C_T be minimized because the output C_T and C_F have the same size.

To solve the rSPR distance problem, it is more convenient to relax the problem by only requiring that T be a phylogenetic tree (i.e., a connected phylogenetic forest) and F be a phylogenetic forest. Hereafter, we assume that the problem has been relaxed in this way. Then, we refer to each input (T, F) to the problem as a *tree-forest (TF) pair*. In the sequel, we assume that a TF-pair (T, F) always satisfies that no leaf of F is a root of F. This assumption does not lose generality, because we can remove x from both T and F if x is both a leaf and a root of F. We also emphasize that for each TF-pair (T, F), T and F have the same set of leaf-labels. The size of the output C_F is the *rSPR distance* of T and F, and is denoted by $d(T, F)$. It is worth pointing out that to compute $d(T, F)$, it is required in the literature that we preprocess each of T and F by first adding a new root and a *dummy* leaf and further making the old root and the *dummy* be the children of the new root. However, the common *dummy* in the modified T and F can be viewed as an ordinary labeled leaf and hence we do not have to explicitly mention the *dummy* when describing an algorithm.

To compute $d(T, F)$ for a given TF-pair (T, F), it is unnecessary to compute both a cut C_T in T and a cut C_F in F. Indeed, it suffices to compute only C_F, because a cut in F forces a cut in T. To make this clear, we define the *sub-TF pair of (T, F) induced by a (possibly empty) cut C of F* to be the TF pair (T', F') obtained as follows.

1. Initially, $T' = T$ and $F' = F \ominus C$.
2. While F' has a connected component K whose root is agreed with a vertex r in T', delete K from F' and delete all descendants of r (including r) from T'.
3. While T' has a non-leaf agreed with a non-leaf of F', first find a non-leaf u in T' such that u is agreed with a non-leaf v of F' but no proper ancestor of u in T' is agreed with a non-leaf of F', next modify T' (respectively, F') by contracting the subtree rooted at u (respectively, v) into a single leaf $\tilde{u}$ (respectively, $\tilde{v}$), and finally assign the same new label to $\tilde{u}$ and $\tilde{v}$.

We can view T' (respectively, F') as a subgraph of T (respectively, F), by viewing $\tilde{u}$ (respectively, $\tilde{v}$) as u (respectively, v). For example, if C consists of the 4 dashed edges in Fig. 2(1), then the sub-TF pair induced by C is as in Fig. 2(2).

Let $\perp$ denote the empty forest. If $(T', F') = (\perp, \perp)$, then C is an *agreement cut* of (T, F) and $F \ominus C$ is an *agreement forest* of (T, F). If in addition, C is canonical, then C is a *canonical agreement cut* of (T, F). The smallest size of an agreement cut of (T, F) is actually $d(T, F)$.

To compute an approximation of $d(T, F)$, our idea is to look at a local structure of T and F and find a cut within the structure. A cut C of F is *good* if $d(T, F \ominus C) \leq d(T, F) - \frac{1}{2}|C|$. Theorem 1 is hard to prove and its proof is detailed in [10]. Section 3 outlines the proof. Theorem 2 follows from Theorem 1.

Theorem 1. [10] *Given a TF-pair (T, F), we can find a good cut of F in quadratic time.*

Theorem 2. *Given a TF-pair (T, F), we can compute an integer d in cubic time such that $d \leq 2d(T, F)$ and there is an agreement cut of (T, F) with size d.*

3 Finding a Good Cut

3.1 Definitions and Notations

Throughout this subsection, let F be a phylogenetic forest. We view each vertex v of F as an ancestor and descendant of itself. For brevity, we refer to a connected component of F simply as a *component* of F. We use $L(F)$ to denote the set of leaves in F, and use $|F|$ to denote the number of components in F. A *dangling subtree* of F is the subtree rooted at a vertex of F. If a vertex v of F has a bifurcate proper-ancestor, then $e_F(v)$ denotes the edge whose tail is the lowest bifurcate proper-ancestor of v in F; otherwise, $e_F(v)$ is undefined. For example, if F is as in Fig. 1(3), then $e_F(x_1)$ is the bold edge. Also, if (T, F) is a TF-pair, then $e_F(x)$ is defined for all $x \in L(F)$ because no leaf of F is a root of F.

Let u_1 and u_2 be two vertices in the same component of F. If u_1 and u_2 have the same parent in F, then they are *siblings* in F. We use $u_1 \sim_F u_2$ to denote the path between u_1 and u_2 in F. Note that $u_1 \sim_F u_2$ is not a directed path if $\ell_F(u_1, u_2) \neq u_1$ and $\ell_F(u_1, u_2) \neq u_2$. For convenience, we still view each edge of $u_1 \sim_F u_2$ as a directed edge (whose direction is the same as in F) although $u_1 \sim_F u_2$ itself may not be a directed path. Each vertex of $u_1 \sim_F u_2$ other than u_1 and u_2 is an *inner vertex* of $u_1 \sim_F u_2$. A *dangling edge between u_1 and u_2* in F is an edge in F but not in $u_1 \sim u_2$ whose tail is an inner vertex of $u_1 \sim_F u_2$. $D_F(u_1, u_2)$ denotes the set of dangling edges between u_1 and u_2 in F. Moreover, if $\ell_F(u_1, u_2) \notin \{u_1, u_2\}$, then $D_F^+(u_1, u_2)$ denotes the set consisting of the edges in $D_F(u_1, u_2)$ and all defined $e_F(v)$ such that v is a vertex of $u_1 \sim_F u_2$ but $e_F(v) \neq e_F(u_i)$ for each $i \in \{1, 2\}$ with $u_i \in L(F)$; otherwise $D_F^+(u_1, u_2) = \emptyset$. For convenience, if w_1 and w_2 are two vertices in different components in F, we define $D_F(w_1, w_2) = \emptyset$ and $D_F^+(w_1, w_2) = \emptyset$. For example, in Fig. 1(2), $D_F(x_1, x_9) = \{e_F(x_7), e_F(u), e_F(v), e_F(x_3)\}$, while in Fig. 2(2), $D_{T'}^+(u, x_{11})$ consists of the five dashed edges. For each $e \in D_F(u_1, u_2)$, the subtree of F rooted at the head of e is a *dangling subtree between u_1 and u_2* in F. If $\ell_F(u_1, u_2) \notin \{u_1, u_2\}$, and u_i is not unifurcate in F but each inner vertex of $\ell_F(u_1, u_2) \sim_F u_i$ is unifurcate in F for each $i \in \{1, 2\}$, then u_1 and u_2 are *semi-siblings* in F and the *semi-children* of $\ell_F(u_1, u_2)$ in F, and $\ell_F(u_1, u_2)$ is the *semi-parent* of u_1 and u_2 in F. For example, if F is the tree in Fig. 1(3), then x_2 and x_4 are semi-siblings and their semi-parent is v.

Let X be a subset of $L(F)$, and v be a vertex of F. A descendant x of v in F is an X-*descendant* of v if $x \in X$. $X^F(v)$ denotes the set of X-descendants of v in F. If $X^F(v) \neq \emptyset$, then v is X-*inclusive*; otherwise, v is X-*exclusive*. If v is bifurcate and both children of v are X-inclusive, v is X-*bifurcate*. Similarly, if exactly one child of v in F is X-inclusive, then v is X-*unifurcate*. An edge of F is X-*inclusive* (respectively, X-*exclusive*) if its head is X-inclusive (respectively, X-exclusive). For an X-bifurcate v in F, an X-*semi-child* of v in F is a vertex u such that each edge in $D_F(v, u)$ is X-exclusive and either $u \in X$ or u is X-bifurcate; we also call v the X-*semi-parent* of u in F; note that v has exactly two X-semi-children in F and we call them X-*semi-siblings*. In particular, when $X = L(F)$, X-semi-parent, X-semi-children, and X-semi-siblings become semi-parent, semi-children, and semi-siblings, respectively. For example, if F is as in Fig. 2(1) and $X = \{x_1, \ldots, x_4\}$, then u is X-unifurcate, but v is X-bifurcate and its X-semi-children are x_2 and z. $F \upharpoonright_X$ denotes the phylogenetic forest obtained from F by removing all X-exclusive vertices and all vertices without X-bifurcate ancestors. See Fig. 1 for an example.

For two phylogenetic forests F_1 and F_2 with the same set of leaf labels, we always view two leaves of F_1 and F_2 with the same label as the same vertex although they are in different forests.

3.2 Search Trees

A simple way to compute $d(T, F)$ for a TF-pair (T, F) is to build a *search tree* Γ as follows. The root of Γ is $(\emptyset, \emptyset)$. In general, each node of Γ is a pair (C_T, C_F) satisfying the following conditions:

- C_T and C_F are canonical cuts of T and F, respectively.
- Each root of $T \ominus C_T$ except one is agreed.
- If (C_T, C_F) is left as a leaf in Γ, then each root of $T \ominus C_T$ is agreed.

Now, suppose that a node (C_T, C_F) of Γ has been constructed but should not be left as a leaf in Γ. For convenience, let $T' = T \ominus C_T$ and $F' = F \ominus C_F$. To construct the children of (C_T, C_F) in Γ, we first select a pair (u_1, u_2) of semi-siblings in T' such that $\ell_T(u_1, u_2)$ is still not agreed but u_i is agreed with a vertex v_i in $F \ominus C_F$ for each $i \in \{1, 2\}$. The children of (C_T, C_F) are then constructed by distinguishing three cases as follows:

Case 1: v_1 or v_2 is a root of F'. If v_1 is a root of F', then $(C_T \cup \{e_{T'}(u_1)\}, C_F)$ is the only child of (C_T, C_F) in Γ; otherwise, $(C_T \cup \{e_{T'}(u_2)\}, C_F)$ is the only child of (C_T, C_F) in Γ.

Case 2: v_1 and v_2 fall into different components of F' but Case 1 does not occur. In this case, (C_T, C_F) has two children in Γ, where for each $i \in \{1, 2\}$, the i-th child of (C_T, C_F) in Γ is $(C_T \cup \{e_{T'}(u_i)\}, C_F \cup \{e_{F'}(v_i)\})$.

Case 3: v_1 and v_2 fall into the same component of F'. In this case, (C_T, C_F) has three children in Γ. The first two are constructed as in Case 2. The third child is $(C_T, C_F \cup D_{F'}(v_1, v_2))$.

This finishes the construction of Γ (see Fig. 3 for an example). The path from the root of Γ to a leaf is a *root-leaf path* in Γ. Let P be a root-leaf path in Γ. We use $C_T(P)$ (respectively, $C(P)$) to denote the canonical cut of T (respectively, F) contained in the leaf of P. Clearly, $C(P)$ is a canonical agreement cut of (T, F).

(T, F) may have multiple search trees. Nonetheless, it is known that for each search tree Γ of (T, F), $d(T, F) = \min_P |C(P)|$, where P ranges over all root-leaf paths in Γ [20]. Basically, this is true because the root-leaf paths in a search tree represents an exhaustive search of a smallest agreement cut of (T, F).

3.3 Keys

Throughout this subsection, let (T, F) be a TF pair. Instead of cuts, we consider a more useful notion of *key*. Intuitively speaking, a key contains not only a cut B within a local structure of F but also possibly two leaves in $F \ominus B$ to be merged into a single leaf. Formally, a *key* of (T, F) is a triple $\kappa = (X, B, R)$ satisfying the following conditions:

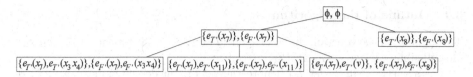

Fig. 3. A search tree for the TF-pair (T', F') in Fig. 2(2).

1. X is a set of leaves in T such that each component of $T{\uparrow}_X$ is a dangling subtree of T.
2. $B \subseteq E_X$ is a cut of F, where E_X is the set of all defined $e_F(v)$ such that v is a vertex of $F{\uparrow}_X$. Moreover, either $\{e_F(x) \mid x \in X\} \subseteq B$ or $|\{e_F(x) \mid x \in X\} \setminus B| = 2$.
3. If $\{e_F(x) \mid x \in X\} \subseteq B$, then $R = \emptyset$; otherwise, for the two vertices x_1 and x_2 in X with $\{e_F(x_1), e_F(x_2)\} = \{e_F(x) \mid x \in X\} \setminus B$, we have that x_1 and x_2 are semi-siblings in $F \ominus B$, R is the edge set of $x_1 \sim_F x_2$, and $B \cap R = \emptyset$. (*Comment:* By Condition 1, x_1 and x_2 are semi-siblings in $T \ominus \{e_T(x) \mid x \in X \setminus \{x_1, x_2\}\}$ as well. So, when we compute the sub-TF pair of (T, F) induced by B, x_1 and x_2 will be merged into a single leaf.)

For example, if (T, F) is as in Fig. 2(1), then $\kappa_e = (X, B, R)$ is a key of (T, F), where $X = \{x_2, x_3, x_4\}$, $B = \{e_F(x_1), e_F(x_2), e_F(u)\}$, and R is the edge set of $x_3 \sim_F x_4$.

If $R = \emptyset$, then κ is *normal* and we simply write $\kappa = (X, B)$ instead of $\kappa = (X, B, R)$; otherwise, it is *abnormal*. In essence, only normal keys were considered in [11].

In the sequel, let $\kappa = (X, B, R)$ be a key of (T, F). The *size* of κ is $|B|$ and is also denoted by $|\kappa|$. The *sub-TF pair of* (T, F) *induced by* κ is the sub-TF pair of (T, F) induced by B. Let P be a root-leaf path in a search tree of (T, F), and $M = C(P) \setminus R$. An edge $e \in B$ is *free* with respect to (w.r.t.) P if $e \in M$ or the leaf descendants of the head of e in $F - (M \cup B)$ are all unlabeled. We use $f_e(\kappa, P)$ to denote the set of edges in B that are free *w.r.t.* P. A component K in $F - (M \cup (B \setminus f_e(\kappa, P)))$ is *free* if the leaves of K are all unlabeled and there is at least one edge $e \in B \setminus f_e(\kappa, P)$ whose tail is a leaf of K. We use $f_c(\kappa, P)$ to denote the set of free components in $F - (M \cup (B \setminus f_e(\kappa, P)))$. The *lower bound* achieved by κ w.r.t. P is $b(\kappa, P) = |f_e(\kappa, P)| + |f_c(\kappa, P)| + |C(P) \cap R|$.

The *lower bound* achieved by κ is $b(\kappa) = \max_{\Gamma} \min_P b(\kappa, P)$, where Γ ranges over all search trees of (T, F) and P ranges over all root-leaf paths in Γ. We call $b(\kappa)$ the *lower bound* achieved by κ. The next lemma shows why we can call $b(\kappa)$ a lower bound.

Lemma 1. [10] *For a key* $\kappa = (X, B, R)$ *of* (T, F), $d(T, F) - d(T', F') \geq b(\kappa)$.

A key κ of (T, F) is *good* if $|\kappa| \leq 2b(\kappa)$, while κ is *fair* if $|\kappa| \leq 2b(\kappa) + 1$. If $\kappa = (X, B, R)$ is a good key of (T, F), then by Lemma 1, B is a good cut of F. So, in order to find a good cut of F, it suffices to find a good key of (T, F).

3.4   Outline of the Algorithm

Throughout this subsection, fix a TF-pair (T, F) such that for each pair (x_1, x_2) of semi-sibling leaves in T, $D_F(x_1, x_2) \neq \emptyset$.

For a vertex β in T, let L_β denote the set of leaf descendants of β in T. β is *consistent* with F if either β is a leaf, or β is a bifurcate vertex in T such that $F{\uparrow}_{L_\beta}$ is a tree and the root in $T{\uparrow}_{L_\beta}$ is agreed with the root in $F{\uparrow}_{L_\beta}$. For example, if (T, F) is as in Fig. 2, then β is consistent with F but α is not.

Let A be a (possibly empty) set of edges in F, and X be a subset of $L(T)$. An *X-path* in $F - A$ is a directed path q to an $x \in X$ in $F - A$ such that each vertex of q other than x is either unifurcate or X-bifurcate in $F - A$. For each vertex v of $F - A$, let $N_{A,X}(v)$ denote the number of X-paths starting at v in $F - A$. When $A = \emptyset$, we write $N_X(v)$ instead of $N_{A,X}(v)$.

Example: Let T and F be as in Fig. 2(1), and $X = \{x_2, x_3, x_4\}$. Then, $w \sim_F x_4$ is an X-path in F but $v \sim_F x_3$ is not; indeed, $N_X(v) = 0$, $N_X(w) = 1$. However, if $A = \{e_F(x_1)\}$, then $v \sim_F x_3$ is an X-path in $F - A$, $N_{A,X}(v) = 1$, and $N_{A,X}(w) = 2$.

Our algorithm finds good cuts within several types of local structures, called *stoppers* for (T, F), defined as follows. Let β be a vertex in T. β is a *close stopper* for (T, F) if it is consistent with F, $N_{L_\beta}(\ell_F(L_\beta)) \geq 2$, and $N_{L_\beta}(v) \leq 1$ for all proper descendants v of $\ell_F(L_\beta)$ in F. β is a *semi-close stopper* if it contains two vertices x_1 and x_2 (called the *anchors* of β) such that $\ell_F(x_1, x_2) = \ell_F(L_\beta)$, $|A| \leq 2$, and β becomes a close stopper for $(T, F \ominus A)$, where A consists of X-exclusive edges in $D_F(x_1, x_2)$. For example, if T and F are as in Fig. 2(1), then β is not a close stopper but is a semi-close stopper for (T, F), while γ is neither.

β is a *root stopper* for (T, F) if it is consistent with F, no descendant of β in T is a semi-close stopper for (T, F), and $\ell_F(L_\beta)$ is a root in F, For example, if T and F are as in Fig. 2(1), then δ is a root stopper for (T, F), but λ is not because it is a semi-close stopper.

β is a *disconnected stopper* for (T, F) if $\ell_F(L_\beta)$ is undefined, no descendant of β in T is a semi-close stopper or a root stopper for (T, F), and both semi children of β in T are consistent with F. For example, if T and F are as in Fig. 2(1), then μ is a disconnected stopper for (T, F).

β is an *overlapping stopper* for (T, F) if $\ell_F(L_\beta)$ is defined, no descendant of β in T is a semi-close stopper, a root stopper, or a disconnected stopper for (T, F), and both semi-children λ_1 and λ_2 of β in T are consistent with F but β is not. For example, if T and F are as in Fig. 1(1), then both α and β are overlapping stoppers for (T, F). The next lemma is easy to prove.

Lemma 2. *There always exists a semi-close, root, disconnected, or overlapping stopper for (T, F).*

Now, our algorithm for finding a good cut of F proceeds as follows.

1. Find a bifurcate vertex β in T such that β is not consistent with F but both semi-children of β are.

2. If some proper descendant γ of β in T is a semi-close stopper β for (T, F), then use γ to find a good abnormal key $\kappa = (L_\gamma, B, R)$ of (T, F) and return B.

3. If some semi-child γ of β in T is a root stopper for (T, F), then use γ to find a good normal key $\kappa = (L_\gamma, B)$ of (T, F) and return B.

4. If β is a disconnected stopper for (T, F), then use β to find a good normal key $\kappa = (L_\beta, B)$ of (T, F) and return B.

5. Use β to find and return a good cut C. (*Comment:* β is an overlapping stopper for (T, F).)

Step 1 can be easily done in quadratic time. However, the other four steps are very complicated and their details can be found in [10]. Roughly speaking, to perform the four steps, we process the vertices of the subtree of T rooted at β in a bottom-up fashion as follows. First, we construct a fair normal key $\kappa_x = (\{x\}, \{e_F(x)\})$ of (T, F) for each $x \in L_\beta$. Once we have constructed fair normal keys $\kappa_{\alpha_1} = (L_{\alpha_1}, B_{\alpha_1})$ and $\kappa_{\alpha_2} = (L_{\alpha_2}, B_{\alpha_2})$ of (T, F) for two semi-siblings α_1 and α_2 in T, we then try to combine κ_{α_1} and κ_{α_2} into a fair normal key $\kappa_\alpha = (L_\alpha, B_\alpha)$ of (T, F) for the semi-parent α of α_1 and α_2 in T. Indeed, if α is a root or disconnected stopper for (T, F), then κ_α will be good, basically because $|\kappa_\alpha| = |\kappa_{\alpha_1}| + |\kappa_{\alpha_1}|$ and $b(\kappa_\alpha) \geq b(\kappa_{\alpha_1}) + b(\kappa_{\alpha_2}) + 1$. However, we may fail to construct κ_α when α is a semi-close or overlapping stopper for (T, F). In case α is a semi-close stopper for (T, F), we instead use the anchors x_1 and x_2 of α to construct a good abnormal key $\kappa_\alpha = (L_\alpha, B_\alpha, R)$ of (T, F) for α, where R is the edge set of $x_1 \sim_F x_2$ and B_α is the union of $D_F(x_1, x_2)$ and all B_γ such that for some $\ddot{e} = (\ddot{v}, \ddot{u}) \in D_F(x_1, x_2)$, γ is the L_α-semi-child of $\ddot{v}$ that is a descendant of $\ddot{u}$ in F. This abnormal κ_α is good, basically because B_α contains neither $e_F(x_1)$ nor $e_F(x_2)$ and in turn x_1 and x_2 will be merged into a single leaf in $F \ominus B_\alpha$.

The most difficult case is when α is an overlapping stopper for (T, F). This case is split into three subcases which are handled separately. In two of the subcases, there exist $x_1 \in L_{\alpha_1}$ and $x_2 \in L_{\alpha_2}$ such that x_1 and x_2 are semi-siblings in F; we construct a cut C of F with $\{e_F(x) \mid x \in L_\alpha\} \subseteq C$ and $d(T, F \ominus C) \leq d(T, F) - \frac{|C|-1}{2}$. A crucial point is that we can merge x_1 and x_2 into a single leaf in $F \ominus C$ and hence can modify C so that $|C|$ decreases by 1 but $d(T, F \ominus C)$ remains the same. So, the modified C is a good cut of F. In the other subcase, such x_1 and x_2 do not exist and we can find $\hat{u}$ in F such that for some $i \in \{1, 2\}$, (1) $\hat{u}$ is X_i-inclusive but X_{3-i}-exclusive in F, (2) $\hat{u}$ has at least two leaf descendants in F, and (3) the edge entering $\hat{u}$ in F belongs to $D_F(x_1, x_2)$ for some vertices x_1 and x_2 in X_{3-i}. We can require that the fair normal key $\kappa_{\alpha_j} = (L_{\alpha_j}, B_{\alpha_j})$ with $j = 3 - i$ satisfy $e_F(\hat{u}) \in B_{\alpha_j}$, and can further prove that $\kappa_\alpha = (L_{\alpha_j} \cup L_\gamma, B_{\alpha_j} \cup B_\gamma)$ is a good normal key of (T, F), where γ is the L_α-semi-child of the parent of $\hat{u}$ in F that is a descendant of $\hat{u}$ in F.

4   Experimental Results

Since the algorithm in [18] has been implemented by its authors, we have also implemented our new algorithm. Both algorithms output not only an upper bound but also a lower bound on the rSPR distance of the two given trees. In order to compare the real performance of the two algorithms, we use the program of [2] to generate three datasets. Each dataset consists of 120 pairs of trees. Each tree in the first dataset has 100 leaves, while each tree in the second and third has 200 leaves. Moreover, to generate a tree pair (T_1, T_2) in the first (respectively, second or third) dataset, we first generate T_1 randomly and then obtain T_2 by applying 50 (respectively, 80 or 100) random rSPR operations on T_1. Our experimental result for the first (respectively, second or third) dataset is shown in the left (respectively, center or right) in Fig. 4. As seen from the figure, our algorithm outputs significantly better lower and upper bounds than the algorithm in [18].

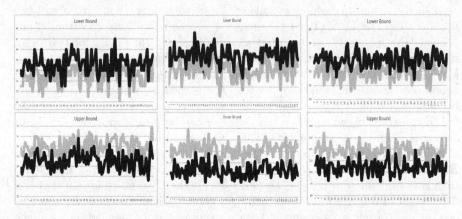

Fig. 4. Comparing the lower and the upper bounds for the three datasets, where our results are shown in black while Schalekamp *et al.*'s in gray.

Acknowledgments. Lusheng Wang was supported by a National Science Foundation of China (NSFC 61373048) and a grant from the Research Grants Council of the Hong Kong Special Administrative Region, China [Project No. CityU 123013].

References

1. Baroni, M., Grunewald, S., Moulton, V., Semple, C.: Bounding the number of hybridisation events for a consistent evolutionary history. J. Math. Biol. **51**, 171–182 (2005)
2. Beiko, R.G., Hamilton, N.: Phylogenetic identification of lateral genetic transfer events. BMC Evol. Biol. **6**, 159–169 (2006)
3. Bonet, M.L., John, K.S., Mahindru, R., Amenta, N.: Approximating subtree distances between phylogenies. J. Comput. Biol. **13**, 1419–1434 (2006)
4. Bordewich, M., McCartin, C., Semple, C.: A 3-approximation algorithm for the subtree distance between phylogenies. J. Discrete Algorithms **6**, 458–471 (2008)

5. Bordewich, M., Semple, C.: On the computational complexity of the rooted subtree prune and regraft distance. Ann. Comb. **8**, 409–423 (2005)
6. Chen, Z.-Z., Wang, L.: FastHN: a fast tool for minimum hybridization networks. BMC Bioinform. **13**, 155 (2012)
7. Chen, Z.-Z., Wang, L.: An ultrafast tool for minimum reticulate networks. J. Comput. Biol. **20**(1), 38–41 (2013)
8. Chen, Z.-Z., Fan, Y., Wang, L.: Faster exact computation of rSPR distance. J. Comb. Optim. **29**(3), 605–635 (2015)
9. Chen, Z.-Z., Machida, •E., Wang, L.: An improved approximation algorithm for rSPR distance. In: Dinh, T.N., Thai, M.T. (eds.) COCOON 2016. LNCS, vol. 9797, pp. 468–479. Springer, Cham (2016). doi:10.1007/978-3-319-42634-1_38
10. Chen, Z.-Z., Machida, E., Wang, L.: A Cubic-Time Approximation Algorithm for rSPR Distance. CoRR, abs/1609.04029, 2016 (2016)
11. Shi, F., Feng Q., You, J., Wang, J.: Improved approximation algorithm for maximum agreement forest of two rooted binary phylogenetic trees. J. Comb. Optim. (2014, to appear)
12. Hein, J., Jing, T., Wang, L., Zhang, K.: On the complexity of comparing evolutionary trees. Discrete Appl. Math. **71**, 153–169 (1996)
13. Ma, B., Wang, L., Zhang, L.: Fitting distances by tree metrics with increment error. J. Comb. Optim. **3**, 213–225 (1999)
14. Ma, B., Zhang, L.: Efficient estimation of the accuracy of the maximum likelihood method for ancestral state reconstruction. J. Comb. Optim. **21**, 409–422 (2011)
15. Maddison, W.P.: Gene trees in species trees. Syst. Biol. **46**, 523–536 (1997)
16. Nakhleh, L., Warnow, T., Lindner, C.R., John, L.S.: Reconstructing reticulate evolution in species - theory and practice. J. Comput. Biol. **12**, 796–811 (2005)
17. Rodrigues, E.M., Sagot, M.-F., Wakabayashi, Y.: The maximum agreement forest problem: approximation algorithms and computational experiments. Theor. Comput. Sci. **374**, 91–110 (2007)
18. Schalekamp, F., van Zuylen, A., van der Ster, S.: A duality based 2-approximation algorithm for maximum agreement forest. Proc. ICALP **70**(1–70), 14 (2016)
19. Wu, Y.: A practical method for exact computation of subtree prune and regraft distance. Bioinformatics **25**(2), 190–196 (2009)
20. Whidden, C., Beiko, R.G., Zeh, N.: Fast FPT algorithms for computing rooted agreement forests: theory and experiments. In: Festa, P. (ed.) SEA 2010. LNCS, vol. 6049, pp. 141–153. Springer, Heidelberg (2010). doi:10.1007/978-3-642-13193-6_13
21. Whidden, C., Zeh, N.: A unifying view on approximation and FPT of agreement forests. LNCS **5724**, 390–401 (2009)

An SIMD Algorithm for Wraparound Tandem Alignment

Joshua Loving[1][(✉)], John P. Scaduto[3], and Gary Benson[1,2]

[1] Bioinformatics Program, Boston University, Boston, MA 02215, USA
jloving@bu.edu
[2] Department of Computer Science, Department of Biology, Boston University,
Boston, MA 02215, USA
[3] Department of Computer Science, Boston College, Chestnut Hill, MA 02467, USA

Abstract. DNA tandem repeats (TRs), and in particular, variable number of tandem repeat (VNTR) loci, can have functional effects on gene regulation and disease mechanisms and are useful for forensics studies. The need to quickly analyze high volumes of sequencing data for TRs and VNTRs has motivated the search for a more efficient sequence alignment algorithm for tandem repeats. Alignment of a pattern to a sequence, which may contain zero or more tandem copies of the pattern, can be accomplished using wraparound dynamic programming (WDP). This paper presents the use of Single Instruction, Multiple Data (SIMD) computer instructions as well as a parallel scan to accelerate WDP, extending earlier SIMD algorithms for global alignment. The SIMD data types and intrinsics store data in 128 bit computer words partitioned into 16 1-byte blocks. Operations are performed on the bytes separately and simultaneously. We allow either single values for match and mismatch, or a substitution scoring scheme that assigns a potentially different substitution weight to every pair of alphabet characters. Additionally, for indels, we allow either a simple linear gap penalty or an affine gap penalty. Benchmarking demonstrated that SIMD tandem alignment runs over 3 times faster than standard wraparound dynamic programming.

1 Introduction

Tandem repeats (TRs), often subclassified as microsatellites and minisatellites, are a common genomic feature [6]. At some TR loci, the number of pattern copies, within the TR array, is variable among members of the population, and these loci are termed variable number of tandem repeats (VNTRs). VNTRs are useful in DNA fingerprinting [14] and bacterial strain identification [9,11, 15,18,27]. They have also been implicated in a large number of neurological diseases, including Fragile-X syndrome [28], Friedreich's ataxia [4], Alzheimer's disease [24], myotonic dystrophy, [10], Huntington's disease [13], and certain psychiatric disorders [5,16,17]. VNTRs are also known to have important effects on chromatin structure [1,25,26,30] and gene expression [29].

© Springer International Publishing AG 2017
Z. Cai et al. (Eds.): ISBRA 2017, LNBI 10330, pp. 140–149, 2017.
DOI: 10.1007/978-3-319-59575-7_13

Our programs, Tandem Repeats Finder [2], which identifies TRs in genomic sequences, and VNTRseek [12], which detects VNTRs in whole genome sequencing data, make extensive use of tandem alignment, *i.e.*, the aligning of a pattern to multiple adjacent copies in a text. The large and growing volume of DNA sequencing data makes methods to speed up tandem alignment, and as a consequence, these and other programs, highly desireable.

Wraparound dynamic programming (WDP, Fig. 1) [8,22], an extension of standard global alignment [23], efficiently solves the tandem alignment problem by using a single copy of the pattern versus the text. Here we present a new method that adapts our bit-parallel alignment techniques [19–21] to WDP.

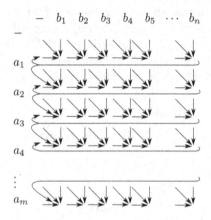

Fig. 1. The WDP scoring matrix (1 copy of the pattern aligned to the text) showing the wraparound computation from the right most cells to the left most cells.

The remainder of this paper is organized as follows. In Sect. 2, we define the problem and give necessary definitions and notation. In Sect. 3, we describe our new algorithm for tandem alignment. In Sect. 4 we give the complexity of both algorithms, and in Sect. 5 we give results of experiments comparing our algorithm with standard WDP alignment.

2 Problem Description

Given:

- a text sequence $a = a_1 a_2 \ldots a_m$ and pattern sequence $b = b_1 b_2 \ldots b_n$, of length m and n respectively,
- a similarity scoring function, S, with substitution score defined by either
  - a single match weight and a single mismatch weight, or
  - a table of integer substitution weights, $subst(x, y)$, one weight for each character pair (x, y) from the alphabet Σ,
- and with indel score defined by either

- a negative, per position integer gap penalty, G, or
- a negative integer gap opening penalty α and a negative, per position gap extension penalty β (affine gap)

Calculate: a global or semi-global alignment score for one copy of a versus an unknown number of tandem copies of b, that is the maximum of the global alignments of a versus b, a versus bb, and so on, using bit operations, addition, and max/min comparisons on computer words of length w.

For semi-global alignment, an initial or final gap in the alignment occurs without penalty. That is, a gap spanning a proper prefix or suffix of a, or a gap spanning a proper prefix of the first copy of b or a proper suffix of the last copy of b, is allowed without penalty.

For the remainder of this abstract, we assume that the problem choices are a per position gap penalty, G, a substitution scoring table $subst()$, and global alignment. All other variants are computed similarly. We do not restrict the size of the alphabet, although the time complexity depends, in part, on the alphabet size as a result of required pre-processing of the $subst()$ table.

2.1 Definitions and Notation

Let S be the recursively defined WDP scoring matrix:

Row zero ($1 \leq j \leq n$):

$$S[0,0] = 0$$
$$S[0,j] = S[0,0] + j \cdot G$$

Column zero ($1 \leq i \leq m$):

$$S[i,0] = S[0,0] + j \cdot G$$

First pass ($i \geq 1, j \geq 1$):

$$S[i,j] = \max \begin{cases} \begin{cases} S[i-1,0] + subst(a_i, b_1) & \backslash\backslash\text{diagonal} \\ S[i-1,n] + subst(a_i, b_1) & \backslash\backslash\text{wraparound diagonal} \\ S[i,0] + G & \backslash\backslash\text{from left} \\ S[i-1,1] + G & \backslash\backslash\text{from above} \end{cases} & \text{if } j = 1 \\[2em] \begin{cases} S[i-1,j-1] + subst(a_i, b_j) & \backslash\backslash\text{diagonal} \\ S[i,j-1] + G & \backslash\backslash\text{from left} \\ S[i-1,j] + G & \backslash\backslash\text{from above} \end{cases} & \text{if } j > 1 \end{cases} \tag{1}$$

Second pass ($i \geq 1, 1 \leq j < n$):

$$S[i,j] = \max \begin{cases} S[i,j] \\ S[i,n] + G & \text{if } j = 1 \\ S[i,j-1] + G & \text{if } j > 1 \end{cases} \tag{2}$$

For our new algorithm, we two define sets of horizontal and vertical score differences for a given row i, one based on S after the first pass and one based on S after the second pass. For row $i > 0$, we have for pass $l \in \{1, 2\}$:

$$\Delta v_l[i, j] = S[i, j] - S[i - 1, j]$$
$$\Delta h_l[i, j] = S[i, j] - S[i, j - 1]$$

For row 0 there is no prior row, so we only have the horizontal differences, which in the case of global alignment reduce to:

$$\forall j > 1, \quad \Delta h_1[0, j] = \Delta h_2[0, j] = G.$$

To simplify algorithmic explanation, for the remainder of this paper we map Δv_l, and Δh_l, $l \in \{1, 2\}$, into new variables ΔV_l and ΔH_l using the formulas:

$$\Delta V_l = \Delta v_l - G, \quad \Delta H_l = G - \Delta h_l$$

For a fixed letter x in sequence a and any position j in pattern b, we define a lower bound on the value of ΔV_1 and ΔV_2, that is the value of ΔV_1 or ΔV_2 from a substitution.

$$L[j] = subst(x, y_j) - 2G.$$

Finally, we define the *second pass sum* SPS_i, as the sum of Δh_2 values in row $i - 1$:

$$SPS_i = \sum_2^n \Delta h_2[i - 1, j].$$

and for row zero, $SPS_0 = 0$.

3   Algorithm

In our bit-parallel approach, we calculate score differences rather than actual scores in the alignment scoring matrix. We start with the ΔH_2 values in row zero, which are known, and proceed row by row to calculate new ΔH_2 values. At the end, we use the ΔH_2 in the final row to calculate the alignment score. Our goal then is to calculate the ΔH_2 values in row i from:

- ΔH_2 values in row $i - 1$,
- SPS_i,
- $L[j]$ values for row character x_i.

The L values are computed in a pre-processing step (outlined in Sect. 4) so that for any given row character x, we have the appropriate L values available.

3.1   Data Structure

Our data structure stores individual ΔH, ΔV, and L values in blocks of k bits within a single computer word of length w. Each computer word holds $W = \lceil w/k \rceil$ values. Due to the limitations of SIMD instructions, 8 bit blocks are used. Within the SIMD words, we use a 'striped' data format.

Definition 1. *'Striped' data format* [7] – *For a given data set of n consecutive values, and SIMD words that store W values, the data is stored in $g = \lceil n/W \rceil$ words such that the zeroth value is in the zeroth position of the zeroth word, the first value in the zeroth position of the first word, ..., the gth value is in the first position of the zeroth word, and so on such that the kth value is in the k modulo g word in the $\lfloor k/g \rfloor$ position (Fig. 2).*

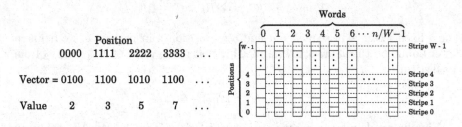

Fig. 2. Left: The representation of consecutive Δ values. Here a block is 4 bits long. Right: An example of how values are stored across SIMD words in the *striped* format.

3.2   Method of Tandem Alignment by Partial Sums

We examine the algorithm for computing the ΔV_l and ΔH_l in a single row. First, we calculate the ΔV_l values, after which we can calculate the ΔH_2 values from the ΔV_2 values in the current row and ΔH_2 values of the previous row. The process is repeated for each successive row. We split the operation on a row into two categories. First, the wraparound cases where the values $\Delta V_l[i, 1]$ depend on the values at position n, then the remaining cases.

Due to the possibility of a wraparound, each row is computed in two passes, the first calculates ΔV_1 and the second calculates ΔV_2. The algorithm for calculating $\Delta V_1[i, j]$ and $\Delta V_2[i, j]$ for $j > 1$ in both passes is very similar to our previously described SIMDParSum algorithm [19]. In order to compute the wraparound values, we introduce the variable SPS_i that contains the sum of Δh_2 values. Note that SPS_i does not need to be fully recomputed for each row (which would be $O(n)$ work), it only requires knowing the previous row's SPS value and a pair of operations as outlined below. (Proofs and additional theorems are omitted due to space constraints.)

Lemma 1. *In row $i > 0$, given the values $\Delta V_2[i, 1]$ and $\Delta V_2[i, n]$, SPS_{i+1} for row $i + 1$ can be calculated as*

$$SPS_{i+1} = SPS_i + \Delta V_2[i, n] - \Delta V_2[i, 1].$$

First pass: Using SPS_i to compute the diagonal wraparound, we compute $\Delta V_1[i, 1]$:

Lemma 2. $\Delta V_1[i, 1]$ *can be computed as:*

$$\Delta V_1[i, 1] = \max \begin{cases} SPS_i + L[j] + G & \backslash\backslash diagonal\ wraparound \\ & substitution \\ subst(a_i, b_1) + \Delta H_2[i - 1, 1] & \backslash\backslash diagonal\ substitution \\ 0 & \backslash\backslash vertical\ gap \end{cases}$$

Given $\Delta V_1[i, 1]$, $\Delta H_2[i - 1, j]$ and L, the remaining values of $\Delta V_1[i, j]$ for $j > 1$ and $\Delta H_1[i, j]$ can be computed according to the following theorems.

Theorem 1. $\forall j > 1$, ΔV_1 *and* ΔH_1 *are computed by the following formulas:*

$$\Delta V_1[i, j] = \max \left(0, \ \max \left(\Delta V_1[i, j - 1], L[j] \right) + \Delta H_2[i - 1, j] \right) \tag{3}$$

$$\Delta H_1[i, j] = \min \left(0, \ \min \left(- L[j], \Delta H_2[i - 1, j] \right) + \Delta V_1[i, j - 1] \right). \tag{4}$$

By application of a parallel scan, [3], these operations can be performed in $O(n/W + \log(W))$ time for all j.

Second pass: We compute $\Delta V_2[i, 1]$ from SPS_i, $\Delta V_1[i, 1]$, and $\Delta V_1[i, n]$.

Lemma 3. *In row $i > 0$, given the values $\Delta V_1[i, 1]$, $\Delta V_1[i, n]$, and SPS_i, $\Delta V_2[i, 1]$ can be calculated as*

$$\Delta V_2[i, 1] = \max(\Delta V_1[i, 1], SPS_i + \Delta V_1[i, n] + G).$$

$\Delta V_2[i, j]$ and $\Delta H_2[i, j]$ can be computed from $\Delta V_2[i, j - 1]$, $\Delta V_2[i, j]$, and $\Delta H_2[i - 1, j]$ for all j just as ΔV_1 and ΔH_1 were.

Theorem 2. $\forall j > 1$, ΔV_2 *and* ΔH_2 *are computed by the following formulas:*

$$\Delta V_2[i, j] = \max \left(0, \ \max \left(\Delta V_2[i, j - 1], L[j] \right) + \Delta H_2[i - 1, j] \right) \tag{5}$$

$$\Delta H_2[i, j] = \min \left(0, \ \min \left(- L[j], \Delta H_2[i - 1, j] \right) + \Delta V_2[i, j - 1] \right). \tag{6}$$

By application of a parallel scan, [3] these operations can be performed in $O(n/W + \log(W))$ time for all j.

4   Complexity and Space

The preprocessing costs when storing a $subst()$ table consist of computing the $L[j]$ values for each possible row character x. The pattern sequence Y is scanned one character, y_j, at a time and for each $x \in \Sigma$, we store $L(x, y_j)$ in the appropriate position of the L variables for x. The time required is $O(|\Sigma| n)$. When using match, mismatch scoring, each alphabet character x has a two-valued $L[j]$ vector. The collection of vectors can be computed more simply in time $O(n + \Sigma)$ (details omitted).

Excluding the pre- and post-processing, the time complexity is

$$O\left(m \left[\frac{n}{W} + \log W \right] \right),$$

where m is the number of rows, and the scans are proportional to $(n/W + \log W)$ as stated in Theorems 1 and 2. (The upsweep and downsweep parts of the scan are linear in the number of words, n/W, and the final step of the upsweep, which occurs in the final word, is logarithmic, $\log W$, in the number of values, W, in that word.) Deciding if a wraparound update of a row is required takes constant time and every wraparound means computing a row twice, which is just a constant multiple of the number of operations and does not change the time complexity. All variations of WDP, described in Sect. 2, have the same complexity as that discussed above.

Post-processing involves retrieving the alignment score from the final ΔH_2 values and intermediate ΔV_2 values. This is an extension of the method described in [20]. The time required is $O(m + n)$, since one operation is required for each row and for each column.

5   Experimental Results

We compared the running time of our tandem alignment algorithm against a standard WDP algorithm, as described in Appendix D of [2]. The algorithms are designated: (1) SIMDTandem (our new algorithm) and (2) WDP (standard algorithm).

For all experiments, we performed 250 thousand alignments, using randomly generated nucleic acid pattern sequences and text sequences built from concatenated pattern copies. The length of the pattern sequence Y (along the top of the alignment scoring matrix which defines the number of columns) was 120. At this length, SIMDTandem uses eight words. Five lengths were used for text sequence X (along the left side of the alignment scoring matrix which defines the number of rows), $|X| = 120, 240, 360, 480, 600$.

All programs were compiled with GCC using optimization level O3 and `march = native` (for SIMD commands) and run on an Intel Core i7-4710HQ CPU 2.50 - 3.5GHz CPU running Ubuntu Linux 14.04. Results are shown in Fig. 3. As can be seen, our new algorithm SIMDTandem is significantly faster than WDP at all but very short sequence X lengths. When X is 5 times the pattern length,

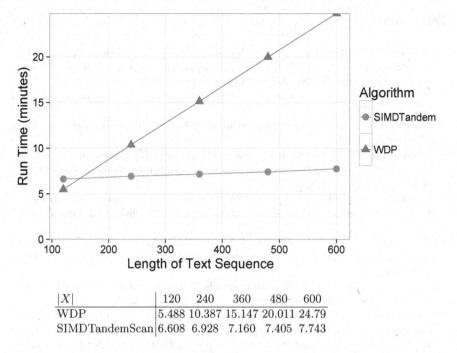

$|X|$	120	240	360	480	600
WDP	5.488	10.387	15.147	20.011	24.79
SIMDTandemScan	6.608	6.928	7.160	7.405	7.743

Fig. 3. Comparison of algorithm run times for 250 thousand alignments. Top: Shown are averages over three trials, with pattern length, $|Y| = 120$, and text length, $|X| =$ integer multiples of the pattern length from 1 (120 characters) to 5 (600 characters). Bottom: Table of run times.

the speedup factor is 3.2. Also apparent from the graph is that the a major cost is the preprocessing since the growth in time for SIMDTandem is very slow with increasing text length.

6   Discussion

We have presented a new algorithm for wraparound tandem alignment. Our algorithm is significantly faster than the standard iterative dynamic programming solution. It illustrates the flexibility of our SIMD algorithm for global alignment and will be useful for the analysis of tandem repeats done in our lab and others. Future work will improve the preprocessing performance of the algorithm. Expected updates to the Intel SSE instruction set will lead to greater flexibility in use of larger register sizes (256 bits and 512 bits) and will further enhance our algorithm's performance.

References

1. Alleman, M., Sidorenko, L., McGinnis, K., Seshadri, V., Dorweiler, J.E., White, J., Sikkink, K., Chandler, V.L.: An RNA-dependent RNA polymerase is required for paramutation in maize. Nature **442**, 295–298 (2006)
2. Benson, G.: Sequence alignment with tandem duplication. J. Comput. Biol. **4**, 351–367 (1997)
3. Blelloch, G.E.: Vector Models for Data-parallel Computing, vol. 356. MIT Press, Cambridge (1990)
4. Campuzano, V., Montermini, L., Molto, M., Pianese, L., Cossee, M.: Friedreich's ataxia: autosomal recessive disease caused by an intronic GAA triplet repeat expansion. Science **271**, 1423–1427 (1996)
5. Clarke, H., Flint, J., Attwood, A., Munafo, M.: Association of the 5-HTTLPR genotype and unipolar depression: a meta-analysis. Psychol. Med. **40**, 1767–1778 (2010)
6. de Koning, A.P., Gu, W., Castoe, T.A., Batzer, M.A., Pollock, D.D.: Repetitive elements may comprise over two-thirds of the human genome. PLoS Genet. **7**(12), e1002384 (2011)
7. Farrar, M.: Striped Smith-Waterman speeds database searches six times over other SIMD implementations. Bioinformatics **23**(2), 156–161 (2007)
8. Fischetti, V.A., Landau, G.M., Schmidt, J.P., Sellers, P.H.: Identifying periodic occurrences of a template with applications to protein structure. In: Apostolico, A., Crochemore, M., Galil, Z., Manber, U. (eds.) CPM 1992. LNCS, vol. 644, pp. 111–120. Springer, Heidelberg (1992). doi:10.1007/3-540-56024-6_9
9. Frothingham, R., Meeker-O'Connell, W.A.: Genetic diversity in the *Mycobacterium tuberculosis* complex based on variable numbers of tandem DNA repeats. Microbiology **144**(5), 1189–1196 (1998)
10. Fu, Y.-H., Pizzuti, A., Fenwick, R., King, J., Rajnarayan, S., Dunne, P., Dubel, J., Nasser, G., Ashizawa, T., DeJong, P., Wieringa, B., Korneluk, R., Perryman, M., Epstein, H., Caskey, C.: An unstable triplet repeat in a gene related to myotonic muscular dystrophy. Science **255**, 1256–1258 (1992)
11. Gascoyne-Binzi, D., Barlow, R., Frothingham, R., Robinson, G., Collyns, T., Gelletlie, R., Hawkey, P.: Rapid identification of laboratory contamination with *Mycobacterium tuberculosis* using variable number tandem repeat analysis. J. Clin. Microbiol. **39**, 69–74 (2001)
12. Gelfand, Y., Hernandez, Y., Loving, J., Benson, G.: VNTRseek - a computational tool to detect tandem repeat variants in high-throughput sequencing data. Nucleic Acids Res. **42**(14), 8884–8894 (2014). http://dx.doi.org/10.1093/nar/gku642
13. Huntington's disease collaborative research group: A novel gene containing a trinucleotide repeat that is expanded and unstable on Huntington's disease chromosomes. Cell **72**, 971–983 (1993)
14. Jobling, M.A., Gill, P.: Encoded evidence: DNA in forensic analysis. Nat. Rev. Genet. **5**(10), 739–751 (2004)
15. Keim, P., Pearson, T., Okinaka, R.: Microbial forensics: DNA fingerprinting of *Bacillus anthracis* (anthrax). Anal. Chem. **80**(13), 4791–4800 (2008). doi:10.1021/ac086131g
16. Lasky-Su, J.A., Faraone, S.V., Glatt, S.J., Tsuang, M.T.: Meta-analysis of the association between two polymorphisms in the serotonin transporter gene and affective disorders. Am. J. Med. Genet. B Neuropsychiatr. Genet. **133B**, 110–115 (2005)

17. Lesch, K.P., Bengel, D., Heils, A., Sabol, S.Z., Greenberg, B.D., Petri, S., Benjamin, J., Muller, C.R., Hamer, D.H., Murphy, D.L.: Association of anxiety-related traits with a polymorphism in the serotonin transporter gene regulatory region. Science **274**, 1527–1531 (1996)

18. Lindstedt, B.-A.: Multiple-locus variable number tandem repeats analysis for genetic fingerprinting of pathogenic bacteria. Electrophoresis **26**(13), 2567–2582 (2005)

19. Loving, J.: Bit-parallel and SIMD alignment algorithms for biological sequence analysis. Ph.D. thesis, Boson University (2017)

20. Loving, J., Hernandez, Y., Benson, G.: BitPAl: a bit-parallel, general integer-scoring sequence alignment algorithm. Bioinformatics **30**(22), 3166–3173 (2014)

21. Loving, J., Becker, E., Benson, G.: Bit-parallel alignment with substitution scoring. In: Proceedings of the 8th International Conference on Bioinformatics and Computational Biology (BICoB), pp. 149–154 (2016)

22. Miller, W., Myers, E.: Approximate matching of regular expressions. Bull. Math. Biol. **51**, 5–37 (1989)

23. Needleman, S., Wunsch, C.: A general method applicable to the search for similarities in the amino acid sequence of two proteins. J. Mol. Biol. **48**, 443–453 (1970)

24. Pritchard, A.L., Pritchard, C.W., Bentham, P., Lendon, C.L.: Role of serotonin transporter polymorphisms in the behavioural and psychological symptoms in probable Alzheimer disease patients. Dement. Geriatr. Cogn. Disord. **24**, 201–206 (2007)

25. Stam, M., Belele, C., Dorweiler, J.E., Chandler, V.L.: Differential chromatin structure within a tandem array 100 kb upstream of the maize b1 locus is associated with paramutation. Genes Dev. **16**, 1906–1918 (2002)

26. Teixeira, F.K., Colot, V.: Repeat elements and the *Arabidopsis* DNA methylation landscape. Heredity **105**, 14–23 (2010). http://dx.doi.org/10.1038/hdy.2010.52

27. Van Belkum, A.: Tracing isolates of bacterial species by multilocus variable number of tandem repeat analysis (MLVA). FEMS Immunol. Med. Microbiol. **49**(1), 22–27 (2007)

28. Verkerk, A., Pieretti, M., Sutcliffe, J., Fu, Y., Kuhl, D., Pizzuti, A., Reiner, O., Richards, S., Victoria, M., Zhang, F., Eussen, B., van Ommen, G., Blonden, A., Riggins, G., Chastain, J., Kunst, C., Galjaard, H., Caskey, C., Nelson, D., Oostra, B., Warren, S.: Identification of a gene (FMR-1) containing a CGG repeat coincident with a breakpoint cluster region exhibiting length variation in fragile X syndrome. Cell **65**, 905–914 (1991)

29. Vinces, M.D., Legendre, M., Caldara, M., Hagihara, M., Verstrepen, K.J.: Unstable tandem repeats in promoters confer transcriptional evolvability. Science **324**, 1213–1216 (2009)

30. Walker, E.L.: Paramutation of the r1 locus of maize is associated with increased cytosine methylation. Genetics **148**, 1973–1981 (1998)

PhAT-QTL: A Phase-Aware Test for QTL Detection

Meena Subramaniam[1,2,3,4,5], Noah Zaitlen[2,5], and Jimmie Ye[3,4,5(✉)]

[1] UCSF Biological and Medical Informatics Graduate Program, San Francisco, USA
[2] UCSF Department of Medicine, San Francisco, USA
[3] UCSF Department of Biostatistics and Epidemiology, San Francisco, USA
jimmie.ye@ucsf.edu
[4] UCSF Department of Bioengineering and Therapeutic Sciences,
San Francisco, USA
[5] UCSF Institute for Human Genetics, San Francisco, USA

Abstract. Next generation sequencing based molecular assays have enabled unprecedented opportunities to quantitatively measure genome function. When combined with dense genetic data, quantitative trait loci (QTL) mapping of molecular traits is a fundamental tool for understanding the genetic basis of gene regulation. However, standard computational approaches for QTL mapping ignore the diploid nature of human genomes, testing for association between genotype and the total counts of sequencing reads mapping to both alleles at each genomic feature. In this work, we develop a new phase-aware test for QTL analysis (PhAT-QTL) leveraging the inherent single nucleotide resolution of sequencing reads to associate the alleles of each marker with the allele-specific counts (ASC) at a genomic feature. Through simulations, we show PhAT-QTL achieves increased power relative to standard genotype-based tests as a function of the number of heterozygotes for a given marker, the noise correlation between haplotypes, and the number of samples with detectable allele-specific counts at a genomic feature. Simulations further show that phasing error and error in quantifying ASC results in a loss of power as opposed to bias. Read simulations on varying haplotype structures (simulated from 1000 Genomes phased genomes) demonstrate that PhAT-QTL is able to detect 20% more QTLs while maintaining the same false positive rate as previous approaches. Applied to RNA-sequencing data, PhAT-QTL achieves similar performance as previous phase-aware methods in detecting *cis* expression QTLs (*cis*-eQTLs) but at a fraction of the computational cost.

Keywords: Statistical genetics · Next-generation sequencing · QTL detection

1 Introduction

Mapping quantitative trait loci (QTLs) of molecular traits is a fundamental tool for identifying and interpreting genetic variants that affect gene regulation.

© Springer International Publishing AG 2017
Z. Cai et al. (Eds.): ISBRA 2017, LNBI 10330, pp. 150–161, 2017.
DOI: 10.1007/978-3-319-59575-7_14

Recently, next generation sequencing has emerged as a compelling approach to quantitatively measure molecular traits, enabling the mapping of QTLs associated with gene expression [1–4] and alternative splicing [5] (both measured by RNA-sequencing), translation [6] (measured by ribosomal footprinting), chromatin accessibility [7] (measured by ATAC-seq), and *cis*-regulatory element activity [8,9] (measured by ChIP-seq). These studies have yielded dense maps of genetic variants associated with variability in gene regulation, shed new light on the genetic architecture of molecular traits in humans, and aided the annotation of disease-associated variants, particularly those located in non-coding regions of the genome.

Despite the initial success of sequencing-based QTL mapping studies, standard methods based on linear regression, associating genotypes with the total counts of sequencing reads mapping to both alleles of each genomic feature (e.g. isoform, region of open chromatin) are underpowered to detect QTLs. This is because these methods were developed for mapping non-sequencing-based molecular quantitative traits (e.g. gene expression measured by microarrays) that do not inherently contain allele-specific information. Phase-aware approaches could significantly improve the power of sequencing-based QTL mapping by leveraging allelic-specific reads in heterozygous individuals, effectively increasing the sample size, and concomitantly, the power of the study.

Recent strategies that utilize allele-specific reads including WASP [10], TreCASE [11] and RASQUAL [12] have shown promise but several challenges remain. One, for genomic features larger than the read length (e.g. isoform), not all methods correctly account for the dependencies between allele-specific and non-allele-specific reads in estimates of allele-specific counts from each chromosome. Two, current methods use generative models that are not easily extended to include additional covariates and variance components for modeling confounding effects from population structure or assay heterogeneity. Finally, current methods are computationally intensive thus prohibiting their application to real world datasets.

Here we introduce a novel computational strategy called phase-aware test for QTL (PhAT-QTL) mapping that utilizes a fast and flexible linear mixed model framework to model allele-specific counts. We implement two versions of the model, PhAT-QTL-joint and PhAT-QTL-meta trading off inference accuracy and computational performance. We also develop a new pipeline that leverages the latest phasing information from large-scale population genetic studies (1000 Genomes) and annotated gene structures to estimate allele-specific counts of gene expression. We demonstrate the performance and computational properties of PhAT-QTL using both extensive simulations and by mapping *cis* expression QTLs (*cis*-eQTLs) in lymphoblastoid cell lines derived from the 1000 Genomes cohort. Because PhAT-QTL uses a fast linear mixed model framework to detect QTLs, it is scalable to thousands of individuals unlike some previous phase-aware methods. PhAT-QTL is freely available for download at https://github. com/meenasub/phatqtl.

2   Methods

The goal of a QTL study is to detect associations between genetic markers and the abundance of genomic features. Consider the illustration of an RNA-sequencing based *cis*-expression QTL (*cis*-eQTL) study presented in Fig. 1. In panel A, the alleles of the tested genetic marker (A/C) and the reads mapping to the exons of a gene are shown for a heterozygous individual in the study. Traditional approaches such as linear regression examine the relationship between the genotype and total read counts (Fig. 1B), which is the sum of reads mapping to both haplotypes. However, each haplotype in a cell can act independently of the other and so this approach does not take full advantage of all available information. Figure 1A shows how a genetic variant in one of the exons (T/G) can be used to assign reads to a haplotype. Namely that all sequencing reads overlapping the variant will either contain a T or G. These reads can be used to estimate allele-specific counts (ASC) of each haplotype, which can then be used to test for association with the noncoding genetic variant A/C. In Fig. 1C, we see a phase-aware analysis in which allele-specific counts are tested for association with the alleles of each haplotype.

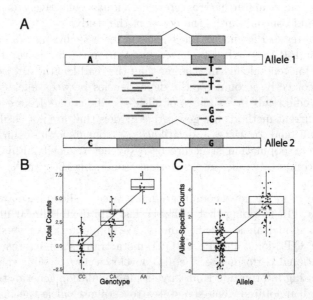

Fig. 1. A. Sequencing reads containing exonic variants are used to estimate the expression coming from each haplotype. B. *cis*-eQTL model for individuals without detectable ASC. C. *cis*-eQTL model for individuals with detectable ASC.

Formally, the standard linear regression method for QTL mapping tests for association between a genotype (sum of alleles at a given marker) and overall abundance of a genomic feature with the underlying model:

$$y = \beta g + \epsilon \tag{1}$$

where y is the total read counts and g is the genotype. When using such a test, the power of the method is determined by the genetic effect size β, the number of individuals N, and the residual noise ϵ. While this model has been successfully employed in multiple QTL studies, it is not optimally powered for sequencing-based QTL studies. In particular, for each individual it combines the reads coming from both haplotypes, which may be differentially regulated in the cell.

We first estimate the allele-specific counts for each individual by aligning reads containing variants overlapping a genomic feature (e.g. exonic variants for gene expression) to a diploid reference genome with phased variant information (Fig. 1A) [13]. The underlying model of PhAT-QTL uses phased haplotypes h_1 and h_2, and the allele-specific counts corresponding to each haplotype y_1 and y_2 to increase the effective sample size.

$$y_{i1} + y_{i2} = y_i \tag{2}$$

$$h_{i1} + h_{i2} = g_i \tag{3}$$

To detect a genetic effect, one possibility is to test for association between each haplotype and its corresponding ASC with standard linear regression. However, shared genetic and environmental factors will induce correlation between allele-specific counts from the two haplotypes for each individual. This violates the independence assumption of linear regression and will result in biased test statistics. To account for this, we propose the following model:

$$y_{i1} = \beta h_{i1} + u_{i1} + \epsilon_{i1} \tag{4}$$

$$y_{i2} = \beta h_{i2} + u_{i2} + \epsilon_{i2} \tag{5}$$

where u_i is the random effect that accounts for the shared noise between ASC from the two haplotypes of the same individual and ϵ_i is the independent residual error. We assume that $cov(u_{i1}, u_{i2}) = \sigma_u^2$ for allele-specific counts coming from the same individual, and $cov(u_i, u_j) = 0$ in all other cases.

To run a statistical test with this model, allele-specific counts from every individual in the cohort are required. While this may be possible with future technological advancements, current allele-specific mapping methods rely on the presence of sequencing reads from heterozygous individuals overlapping SNPs in the feature [13]. When sequencing data are unavailable, for small genes, or genes undergoing strong purifying selection, there may not be any heterozygous individuals for SNPs in the features of interest. We therefore develop a method that can jointly model allele-specific counts from individuals where ASC can be estimated and total counts from individuals where ASC cannot be estimated. In the PhAT-QTL-joint method, we estimate the genetic effect size jointly across individuals with and without ASC, and allow for separate error components for the two groups of individuals,

$$\begin{pmatrix} y_1 \\ y_2 \\ y_{adj} \end{pmatrix} = \beta \begin{pmatrix} h_1 \\ h_2 \\ g \end{pmatrix} + u + \epsilon_{asc} + \epsilon_{noasc} \tag{6}$$

$$y_{adj} = \log_2(y/2) \tag{7}$$

Here $\epsilon_{iasc} \sim N(0, \sigma^2_{\epsilon asc})$ if individual i has detectable ASC, and $\epsilon_{inoasc} \sim N(0, \sigma^2_{\epsilon noasc})$ if there is no detectable ASC. For the individuals without ASC, the total counts y_{adj} is adjusted to be on the same scale as the allele-specific counts. We estimate σ^2_u, $\sigma^2_{\epsilon ase}$, and $\sigma^2_{\epsilon noasc}$ jointly for each gene-SNP pair.

In practice we find that fitting a model with three variance components is slower than a model with two variance components. Additionally, there are faster methods for modeling variance components which can further decrease the overall runtime of the method [14]. For this reason, we sought to develop a method that tests the association in ASC and non-ASC individuals separately. In a meta-analysis version (PhAT-QTL-meta), we first estimate genetic effect sizes from the ASC and linear regression components separately, and then use an inverse-variance weighted meta analysis to estimate a global beta and standard error.

$$\begin{pmatrix} y_1 \\ y_2 \end{pmatrix} = \beta_{asc} \begin{pmatrix} h_1 \\ h_2 \end{pmatrix} + \epsilon_{asc} \tag{8}$$

$$y = \beta_{noasc} g + \epsilon_{noasc} \tag{9}$$

$$\beta_{meta} = (\beta_{asc}(1/SE_{asc}) + \beta_{noasc}(1/SE_{noasc})) / ((1/SE_{asc}) + (1/SE_{noasc})) \tag{10}$$

$$SE_{meta} = \sqrt{1/((1/SE_{asc}) + (1/SE_{noasc}))} \tag{11}$$

We use the average information Restricted Maximum Likelihood (ai-REML) method to estimate all of the variance components in the joint model and the ASC portion of the meta analysis model [15]. Because the meta analysis model estimates one fewer variance component, the method is faster than the joint model.

3 Simulation Framework

3.1 Simulation of Genotypes and Allele-Specific Counts

To compare the power of PhAT-QTL to the standard QTL detection method (linear regression), we simulated QTL effects with simulated genotypes and allele-specific counts. We simulated correlated allele-specific counts according to the model described in Eqs. 4 and 5. Specifically, we simulated haplotypes from a binomial distribution where p was set to the minor allele frequency. The associated allele-specific counts were simulated with a specified β and correlated noise terms were drawn from a multivariate normal distribution. We fixed the number of individuals to $N = 200$, and assessed the power of PhAT-QTL-joint and standard linear regression to detect genetic associations as a function of σ^2_u. Here we define power as the fraction of tests with a p-value <0.05. We show that the power of PhAT-QTL-joint increases as σ^2_u increases, and it outperforms linear regression for a range of minor allele frequencies at a fixed $\beta = 0.3$ (Fig. 2A). Similarly, at a fixed minor allele frequency of 0.15, the power of PhAT-QTL-joint increases relative to linear regression as β increases from 0.2–0.4 (Fig. 2B).

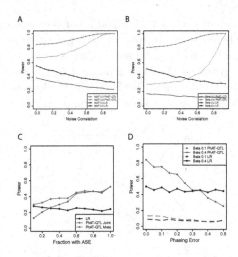

Fig. 2. A. Power as a function of noise correlation (varying MAF), B. Power as a function of noise correlation (varying β), C. Power as a function of ASE fraction, D. Power as a function of phasing error

Notably, in this simulation where each individual has ASC, PhAT-QTL-joint and PhAT-QTL-meta have exactly the same performance.

Next, we incorporated known biases in ASC estimation for genomic features into this simulation framework. Because the number of individuals with detectable ASC varies across genomic features, we simulated different fractions of ASC with $\beta = 0.3$ and minor allele frequency (MAF) = 0.15, and compared the power between the two implementations of PhAT-QTL and linear regression (Fig. 2C). PhAT-QTL-joint and PhAT-QTL-meta matched the performance of linear regression in cases where there is no allelic imbalance (difference in allele-specific counts), and outperformed linear regression when there is any allelic imbalance. We also examined the power of PhAT-QTL as a function of phasing error, which here we define as the expected fraction of individuals in which the causal variant is incorrectly phased with respect to the variants we use to estimate ASC. This would cause the estimation of ASC to be swapped with respect to the causal variant. Although the power of PhAT-QTL-joint decreases as the phasing error increases, we find that PhAT-QTL-joint outperforms linear regression when the phasing error <40% (Fig. 2D). The switch error rate of current phasing methods ranges from 0.28% to 5.57%, suggesting that PhAT-QTL should be robust to most computational phasing errors [16]. In summary, phasing error and ASC quantification error result in decreased power as opposed to bias.

3.2   Simulation of Gene Expression over Real Haplotypes

We next assessed the performance of PhAT-QTL by simulating gene expression over real haplotypes from 1000 Genomes. We simulate allele-specific reads with

respect to a causal variant chosen within 200 kb of the gene in a random subset
of 100 individuals. Allele-specific counts for each gene were simulated according
to the previously described model, and sequencing reads were simulated from the
phased haplotypes in proportion to the simulated ASC values. We then applied
PhAT-QTL-joint, PhAT-QTL-meta, RASQUAL (the leading phase-aware QTL
detection method), and linear regression and compared their performances.

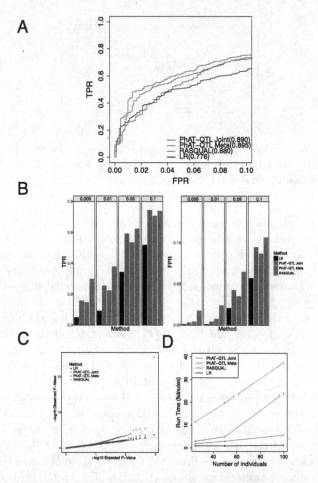

Fig. 3. A. ROC Curves for FPR <0.1 with the AUC scores for each method, B. TPR
and FPR at different FDR cutoffs, C. QQ plot with observed p-values under the null,
D. Run time for each method across different numbers of individuals

For estimating allele-specific counts, simulated reads were aligned using
individual-specific diploid transcriptomes that included exonic variants, and
reads were allocated to each haplotype using RSEM at the transcript level
[17]. For linear regression (LR) and RASQUAL, reads were aligned to the standard Hg19 reference genome, and either total read counts (LR) or total read

counts and allele-specific reads (RASQUAL) were used to test the performance. Adjusted p-values using the qvalue package [18] were used to assess performance.

Out of the 2000 randomly chosen genes in the human genome, 50% of those genes were simulated to have no QTL effect ($\beta = 0$). The other 50% of genes were simulated to have genetic effect sizes inversely proportional to the minor allele frequency of the causal variant, consistent with the distribution of effect sizes in the human genome. Comparing the performance of all four methods using the known true positives, we find that PhAT-QTL-joint (AUC = 0.890) and PhAT-QTL-meta (AUC = 0.895) outperformed both linear regression (AUC = 0.776) and RASQUAL (AUC = 0.880) (Fig. 3A). Furthermore, at a false discovery rate (FDR) of <0.005–0.1, we show that the true positive rate of PhAT-QTL-joint is almost doubled compared to linear regression, detecting a total of 582 eQTLs at an FDR cut-off of 0.05 (Fig. 3B). Although RASQUAL has a higher true positive rate than PhAT-QTL-joint at different FDR cut-offs, it also has a higher false positive rate, thus decreasing its overall performance. While PhAT-QTL-joint and PhAT-QTL-meta have a higher false positive rate than linear regression, we note that the p-values are controlled in the simulated null eQTLs (Fig. 3C).

We also compared the time taken to run all methods across different numbers of individuals, showing that PhAT-QTL-joint and PhAT-QTL-meta are faster than RASQUAL across the range of 25–100 individuals (Fig. 3D). At N = 100 individuals, PhAT-QTL-meta is 6.5x faster than RASQUAL, making it computationally feasible to run on larger cohorts.

4   Detecting *cis*-eQTLs from RNA-seq Data Using PhAT-QTL

In order to compare PhAT-QTL to the leading QTL detection method RASQUAL while accommodating the increased computational burden, we applied PhAT-QTL-joint and PhAT-QTL-meta to a reduced dataset consisting of 50 individual from the GEUVADIS dataset [3].

4.1   Data Acquisition and Processing

We downloaded phased haplotypes and RNA-sequencing data for a random subset of 50 EUR individuals from the GEUVADIS portal [19]. Reads were aligned to individual-specific transcriptome references with the same procedure as previously described. For *cis*-eQTL detection, we filtered out lowly expressed genes by requiring at least 10% of the individuals to have at least 10 transcripts per million (TPM) counts. We used the log2 median-normalized TPM counts as inputs to PhAT-QTL-joint, PhAT-QTL-meta and linear regression, and the raw expected counts as inputs to RASQUAL. For each gene, we tested all variants with MAF >0.05 that satisfy Hardy-Weinberg Equilibrium (HWE) with p-value >0.05 within a 200 kb window of the transcription start site.

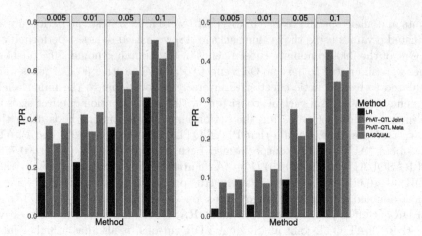

Fig. 4. True positive rates and false positive rates for each method at different FDR cutoffs.

4.2   Comparison of Methods

To assess performance, *cis*-eQTLs were identified as true positives if they were in the GEUVADIS published list of *cis*-eQTLs, and false positives if they were not present in the published list. We then examined true positive rates and false positive rates at different FDR cutoffs (0.005–0.1) for each method (Fig. 4). Consistent with the simulations, the true positive rates of PhAT-QTL-joint and PhAT-QTL-meta are much higher than the true positive rate of linear regression. However, the false positive rates of PhAT-QTL-joint and PhAT-QTL-meta are much higher in the real dataset compared to the simulated dataset, suggesting that there are biases in the allele-specific quantification that are resulting in false *cis*-eQTL discoveries.

5   Discussion

Here we propose a fast linear mixed model framework to detect QTLs from functional genomic sequencing data. By leveraging allele-specific reads, we can increase the power to detect QTLs relative to linear regression and previous phase-aware mapping methods. PhAT-QTL-joint and PhAT-QTL-meta are robust to biases in phasing error and haplotype structure, and have well-calibrated p-values under the null. We eliminate the challenge of mapping bias by aligning all reads to an individual-specific reference that contains variant information overlapping genomic features. In the case where the genomic features are isoforms and genes, we applied an expectation-maximization algorithm to use all allele-specific and non allele-specific reads to obtain robust estimates of allele-specific counts for each isoform or gene. PhAT-QTL-joint and PhAT-QTL-meta have the potential to discover isoform-specific QTLs, which cannot be achieved with methods that model allele-specific counts at each exon and do not account for shared exon usage between isoforms.

Our approach is not without shortcomings. In this work we assume that the relationship between haplotype and log-normalized allele-specific counts is linear. However, raw counts of genomic features from functional genomic sequencing data are over-dispersed, suggesting that a negative binomial model may be more suited to testing the association. In application to real data (GEUVADIS RNA-seq), we observed an inflated type I error rate for all phase-aware methods relative to linear regression at a given false discovery rate (FDR) threshold. Depending on the purpose of the study, researchers may prefer the well-calibrated distribution of linear regression to that of our approach and previous approaches.

By decoupling estimating allele-specific counts from testing the association between a genotype and allele-specific counts, we can apply PhAT-QTL-joint and PhAT-QTL-meta across a number of quantitative traits measured by next generation sequencing (e.g. chromatin accessibility and transcription factor binding). In addition to increasing power, PhAT-QTL-joint and PhAT-QTL-meta can detect associations that are undetectable via linear regression such as auto-regulated genes. In this work we used computational phasing and existing mapping methods to estimate allele-specific counts. Going forward we will explore alternative phasing and mapping methods as well as changes to the underlying statistical model that directly incorporate phasing, mapping, and genotype errors. Additionally, as experimental techniques are being developed to capture haplotype specific reads through linked reads [20] and long-reads [21, 22], we can obtain more accurate estimates of allele-specific counts to further improve the performance of PhAT-QTL-joint and PhAT-QTL-meta.

References

1. Montgomery, S.B., Sammeth, M., Gutierrez-Arcelus, M., Lach, R.P., Ingle, C., Nisbett, J., Guigó, R., Dermitzakis, E.T.: Transcriptome genetics using second generation sequencing in a Caucasian population. Nature **464**, 773–777 (2010)
2. Pickrell, J.K., Marioni, J.C., Pai, A.A., Degner, J.F., Engelhardt, B.E., Nkadori, E., Veyrieras, J.-B., Stephens, M., Gilad, Y., Pritchard, J.K.: Understanding mechanisms underlying human gene expression variation with RNA sequencing. Nature **464**, 768–772 (2010)
3. Lappalainen, T., Sammeth, M., Friedländer, M.R., Hoen, P.A.C.T., Monlong, J., Rivas, M.A., Gonzàlez-Porta, M., Kurbatova, N., Griebel, T., Ferreira, P.G., Barann, M., Wieland, T., Greger, L., van Iterson, M., Almlöf, J., Ribeca, P., Pulyakhina, I., Esser, D., Giger, T., Tikhonov, A., Sultan, M., Bertier, G., MacArthur, D.G., Lek, M., Lizano, E., Buermans, H.P.J., Padioleau, I., Schwarzmayr, T., Karlberg, O., Ongen, H., Kilpinen, H., Beltran, S., Gut, M., Kahlem, K., Amstislavskiy, V., Stegle, O., Pirinen, M., Montgomery, S.B., Donnelly, P., McCarthy, M.I., Flicek, P., Strom, T.M., Consortium, T.G., Lehrach, H., Schreiber, S., Sudbrak, R.: Carracedo, Á., Antonarakis, S.E., Häsler, R., Syvänen, A.-C., van Ommen, G.-J., Brazma, A., Meitinger, T., Rosenstiel, P., Guigó, R., Gut, I.G., Estivill, X., Dermitzakis, E.T.: Transcriptome and genome sequencing uncovers functional variation in humans. Nature. **501**, 506–511 (2013)

4. Battle, A., Mostafavi, S., Zhu, X., Potash, J.B., Weissman, M.M., McCormick, C., Haudenschild, C.D., Beckman, K.B., Shi, J., Mei, R., Urban, A.E., Montgomery, S.B., Levinson, D.F., Koller, D.: Characterizing the genetic basis of transcriptome diversity through RNA-sequencing of 922 individuals. Genome Res. **24**, 14–24 (2014)

5. Li, Y.I., van de Geijn, B., Raj, A., Knowles, D.A., Petti, A.A., Golan, D., Gilad, Y., Pritchard, J.K.: RNA splicing is a primary link between genetic variation and disease. Science **352**, 600–604 (2016)

6. Battle, A., Khan, Z., Wang, S.H., Mitrano, A., Ford, M.J., Pritchard, J.K., Gilad, Y.: Genomic variation. Impact of regulatory variation from RNA to protein. Science **347**, 664–667 (2015)

7. Cheng, C.S., Gate, R.E., Aiden, A.P., Siba, A., Tabaka, M., Lituiev, D., Machol, I., Subramaniam, M., Shammim, M., Hougen, K.L., Wortman, I., Huang, S.-C., Durand, N.C., Feng, T., De Jager, P.L., Chang, H.Y., Lieberman Aiden, E., Benoist, C., Beer, M.A., Ye, C.J., Regev, A.: Genetic determinants of chromatin accessibility and gene regulation in T cell activation across human individuals. bioRxiv 090241 (2016)

8. Grubert, F., Zaugg, J.B., Kasowski, M., Ursu, O., Spacek, D.V., Martin, A.R., Greenside, P., Srivas, R., Phanstiel, D.H., Pekowska, A., Heidari, N., Euskirchen, G., Huber, W., Pritchard, J.K., Bustamante, C.D., Steinmetz, L.M., Kundaje, A., Snyder, M.: Genetic control of chromatin states in humans involves local and distal chromosomal interactions. Cell **162**, 1051–1065 (2015)

9. Waszak, S.M., Delaneau, O., Gschwind, A.R., Kilpinen, H., Raghav, S.K., Witwicki, R.M., Orioli, A., Wiederkehr, M., Panousis, N.I., Yurovsky, A., Romano-Palumbo, L., Planchon, A., Bielser, D., Padioleau, I., Udin, G., Thurnheer, S., Hacker, D., Hernandez, N., Reymond, A., Deplancke, B., Dermitzakis, E.T.: Population variation and genetic control of modular chromatin architecture in humans. Cell **162**, 1039–1050 (2015)

10. van de Geijn, B., McVicker, G., Gilad, Y., Pritchard, J.K.: WASP: allele-specific software for robust molecular quantitative trait locus discovery. Nat. Methods **12**, 1061–1063 (2015)

11. Sun, W.: A statistical framework for eQTL mapping using RNA-seq data. Biometrics **68**, 1–11 (2012)

12. Kumasaka, N., Knights, A.J., Gaffney, D.J.: Fine-mapping cellular QTLs with RASQUAL and ATAC-seq. Nat. Genet.: Nat. Res. **48**, 206–213 (2016)

13. Munger, S.C., Raghupathy, N., Choi, K., Simons, A.K., Gatti, D.M., Hinerfeld, D.A., Svenson, K.L., Keller, M.P., Attie, A.D., Hibbs, M.A., Graber, J.H., Chesler, E.J., Churchill, G.A.: RNA-Seq alignment to individualized genomes improves transcript abundance estimates in multiparent populations. Genetics **198**, 59–73 (2014)

14. Kang, H.M., Zaitlen, N.A., Wade, C.M., Kirby, A., Heckerman, D., Daly, M.J., Eskin, E.: Efficient control of population structure in model organism association mapping. Genetics **178**, 1709–1723 (2008)

15. Gilmour, A.R., Thompson, R., Cullis, B.R.: Average information REML: an efficient algorithm for variance parameter estimation in linear mixed models. Biometrics **51**, 1440–1450 (1995)

16. O'Connell, J., Gurdasani, D., Delaneau, O., Pirastu, N., Ulivi, S., Cocca, M., Traglia, M., Huang, J., Huffman, J.E., Rudan, I., McQuillan, R., Fraser, R.M., Campbell, H., Polasek, O., Asiki, G., Ekoru, K., Hayward, C., Wright, A.F., Vitart, V., Navarro, P., Zagury, J.-F., Wilson, J.F., Toniolo, D., Gasparini, P., Soranzo, N., Sandhu, M.S., Marchini, J.: A general approach for haplotype phasing across the full spectrum of relatedness. PLoS Genet. **10**, e1004234 (2014)

17. Li, B., Dewey, C.N.: RSEM: accurate transcript quantification from RNA-Seq data with or without a reference genome. BMC Bioinform. **12**, 323 (2011)

18. Storey, J.D., Tibshirani, R.: Statistical significance for genomewide studies. Proc. Natl. Acad. Sci. USA **100**, 9440–9445 (2003)

19. GEUVADIS portal. http://www.geuvadis.org/web/geuvadis/rnaseq-project

20. Weisenfeld, N.I., Kumar, V., Shah, P., Church, D., Jaffe, D.B.: Direct determination of diploid genome sequences. bioRxiv 070425 (2016)

21. Jain, M., Olsen, H.E., Paten, B., Akeson, M.: The Oxford Nanopore MinION: delivery of nanopore sequencing to the genomics community. Genome Biol. **17**, 239 (2016)

22. Eid, J., Fehr, A., Gray, J., Luong, K., Lyle, J., Otto, G., Peluso, P., Rank, D., Baybayan, P., Bettman, B., Bibillo, A., Bjornson, K., Chaudhuri, B., Christians, F., Cicero, R., Clark, S., Dalal, R., deWinter, A., Dixon, J., Foquet, M., Gaertner, A., Hardenbol, P., Heiner, C., Hester, K., Holden, D., Kearns, G., Kong, X., Kuse, R., Lacroix, Y., Lin, S., Lundquist, P., Ma, C., Marks, P., Maxham, M., Murphy, D., Park, I., Pham, T., Phillips, M., Roy, J., Sebra, R., Shen, G., Sorenson, J., Tomaney, A., Travers, K., Trulson, M., Vieceli, J., Wegener, J., Wu, D., Yang, A., Zaccarin, D., Zhao, P., Zhong, F., Korlach, J., Turner, S.: Real-time DNA sequencing from single polymerase molecules. Science **323**, 133–138 (2009)

Unbiased Taxonomic Annotation
of Metagenomic Samples

Bruno Fosso[1], Graziano Pesole[1], Francesc Rosselló[2], and Gabriel Valiente[3(✉)]

[1] Institute of Biomembranes and Bioenergetics,
Consiglio Nazionale delle Ricerche, 70126 Bari, Italy
[2] Department of Mathematics and Computer Science,
Research Institute of Health Science,
University of the Balearic Islands, 07122 Palma de Mallorca, Spain
[3] Algorithms, Bioinformatics, Complexity and Formal Methods Research Group,
Technical University of Catalonia, 08034 Barcelona, Spain
valiente@cs.upc.edu

Abstract. The classification of reads from a metagenomic sample using a reference taxonomy is usually based on first mapping the reads to the reference sequences and then, classifying each read at a node under the lowest common ancestor of the candidate sequences in the reference taxonomy with the least classification error. However, this taxonomic annotation can be biased by an imbalanced taxonomy and also by the presence of multiple nodes in the taxonomy with the least classification error for a given read. In this paper, we show that the Rand index is a better indicator of classification error than the often used area under the ROC curve and F-measure for both balanced and imbalanced reference taxonomies, and we also address the second source of bias by reducing the taxonomic annotation problem for a whole metagenomic sample to a set cover problem, for which a logarithmic approximation can be obtained in linear time.

Keywords: Metagenomics · Classification · Taxonomic annotation · Correlation · Set cover

1 Introduction

Next generation sequencing technologies have moved forward the development of metagenomics, a new field of science devoted to the study of microbial communities by the analysis of their genomic content, directly sequenced from the environment [15,20,21]. A sequenced metagenomic sample consists of a large number of relatively short DNA or RNA fragments, called reads, and one of the first steps in the computational analysis of a metagenomic sample is the identification of the organisms present in the sequenced environment and their relative abundance, that is, the classification of the metagenomic sample.

In this paper, we focus on the taxonomic annotation problem, that is, the classification of the reads from a metagenomic sample using a reference taxonomy, for which we adapt some basic notions from statistical classification in

© Springer International Publishing AG 2017
Z. Cai et al. (Eds.): ISBRA 2017, LNBI 10330, pp. 162–173, 2017.
DOI: 10.1007/978-3-319-59575-7_15

	Positive prediction	Negative prediction
Positive class	True Positive (*TP*)	False Negative (*FN*)
Negative class	False Positive (*FP*)	True Negative (*TN*)

Fig. 1. Confusion matrix for a binary classification problem

machine learning. We abstract away from the computational problem of mapping reads to reference sequences, and assume that a set of candidate sequences in a reference taxonomy is given for each read in the metagenomic sample to be classified. These candidate sequences are usually obtained either by sequence composition methods (those reference sequences with oligonucleotide frequencies within a given distance threshold to the oligonucleotide frequencies of the read) or by sequence similarity methods (those reference sequences that the read can be aligned to within a given threshold of sequence similarity, or those reference sequences that the read can be mapped to with at most a given number of mismatches).

In a statistical binary classification problem, the confusion matrix (Fig. 1) shows the number of correctly and incorrectly classified instances of each class. True positives (*TP*) are the correctly classified positive instances, true negatives (*TN*) are the correctly classified negative instances, false positives (*FP*) are the misclassified negative instances, and false negatives (*FN*) are the misclassified positive instances. The *true positive rate*, *sensitivity*, or *recall R* of a classification is the ratio $TPR = TP/(TP + FN)$ of true positives to the total number of positive instances, the *false positive rate* is the ratio $FPR = FP/(FP + TN)$ of false positives to the total number of negative instances, the *true negative rate* or *specificity* is the ratio $TNR = TN/(FP + TN)$ of true negatives to the total number of negative instances, and the *false negative rate* is the ratio $FNR = FN/(TP + FN)$ of false negatives to the total number of positive instances. Further, the *precision* of a classification is the ratio $P = TP/(TP + FP)$ of true positives to the total number of positive predictions. They are usually combined into a single indicator of classification error as either the area under the *ROC* curve $AUC = (TPR - FPR + 1)/2$ or the *F*-measure, which is the harmonic mean $F = 2/(1/P + 1/R)$ of precision and recall [18].

In a metagenomic classification problem, the annotation of a read as coming from a particular sequence in a reference taxonomy often involves solving the ambiguity of multiple candidate sequences, caused among other factors by reads being not long enough to ensure a unique identification of the reference sequences they come from. Reference taxonomies are rooted trees, with the leaves labeled by sequences at the taxonomic rank of species or strain, and these ambiguities are solved by annotating reads as coming from internal nodes, at higher taxonomic ranks in the reference taxonomy. When classifying a read as coming from an internal node in a reference taxonomy (Fig. 2), the leaves under the internal node are true positives if they are labeled by candidate sequences, otherwise they are false positives, and the remaining leaves under the lowest common ancestor (LCA) of the candidate sequences are false negatives if they are labeled

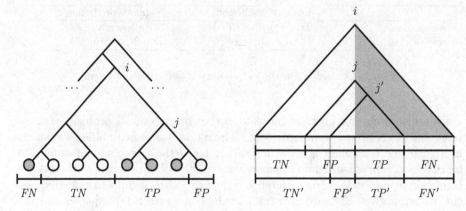

Fig. 2. Classifying a read using a reference taxonomy. The grayed leaves are the candidate sequences for the classification of the read, and node i is their LCA in the reference taxonomy. The taxonomic annotation of the read at node i implies the absence of true negatives and false negatives. With a taxonomic annotation of the read at node j, which is the LCA in the reference taxonomy of the true positives, however, the remaining grayed leaves are the false negatives, the remaining leaves under node j are the false positives, and the still remaining leaves under node i are the true negatives of the metagenomic classification problem

by candidate sequences, otherwise they are true negatives. Annotating a read as coming from the LCA of the candidate sequences in a reference taxonomy [12] maximizes precision, as in that case there are no true negatives and no false negatives, but at the expense of specificity, because the number of false positives in a reference taxonomy can be very large. Annotating a read as coming from an internal node with the largest F-measure value [1,3,8,9] minimizes the classification error as a combination of precision and sensitivity.

However, there are at least two sources of bias in the taxonomic annotation of a metagenomic sample. One the one hand, reference taxonomies are imbalanced, that is, the instances of one class significantly outnumber the instances of the other classes, and this can be observed at any taxonomic rank. For example, the NCBI Taxonomy [5,6], which is the most comprehensive taxonomic reference to date, includes as of 13 March 2017 an imbalanced number of sequences for Bacteria (1,412,065), Eukaryota (685,380), and Archaea (27,322). Within the Bacteria, for example, there is also an imbalanced number of sequences for the Actinobacteria (593,837), Proteobacteria (440,315), Firmicutes (245,632), Bacteroidetes (77,866), Planctomycetes (8,899), Fusobacteria (7,789), and others (37,727). In a statistical binary classification problem, imbalanced datasets result in a good coverage of the positive instances and a frequent misclassification of the negative instances, since most of the standard machine learning algorithms consider a balanced training set [16]. In a metagenomic classification problem, an imbalanced reference taxonomy may also yield an imbalance between the positive and negative classes, because the larger the clade of the LCA in a reference taxonomy of the candidate sequences for a read, the larger the negative class for the classification of the read. In Sect. 2, we show that this is in general not the case, and we also show that the Rand index is a better indicator of classification error than the often used area under the ROC curve and F-measure, when the reference taxonomy is imbalanced and also for balanced reference taxonomies.

Another source of bias in the taxonomic annotation of a metagenomic sample lies in the existence of multiple candidate nodes in a reference taxonomy with the least classification error for a given read, one of which is usually chosen arbitrarily for the taxonomic annotation of the read [1,3]. Instead of breaking ties independently for each read in a metagenomic sample, we show in Sect. 3 that the shift from a one-sequence-read-at-a-time view to a whole-set-of-sequence-reads view yields a better resolution of any remaining ambiguities in the taxonomic annotation of a metagenomic sample.

2   Taxonomic Annotation Using Imbalanced Reference Taxonomies

Recall from Sect. 1 that in a metagenomic classification problem, an imbalanced reference taxonomy yields an imbalance between the positive and negative classes. Let us define the *balance ratio* of a classification problem as the ratio of the size of the positive class to the size of the negative class.

Definition 1. *Let TP, TN, FP, and FN be the number of true positives, false positives, true negatives, and false negatives in a binary classification problem. The balance ratio of the classification problem is* $(TP + FN)/(FP + TN)$.

Recall also from Sect. 1 that the reference taxonomies used in metagenomic classification are highly imbalanced. It turns out that balanced and imbalanced reference taxonomies yield exactly the same metagenomic classification problems, as long as they have the same number of internal nodes. Some evidence supporting this observation follows.

The topology of the most possible balanced binary reference taxonomy is a complete binary tree, as every internal node (and also the root) has two descendant clades of exactly the same size. On the other hand, the topology of the least possible balanced binary reference taxonomy is a degenerate binary tree, as every internal node (and also the root) has one big descendant clade and one small (with only one node) descendant clade.

Now, in a metagenomic classification problem, any subset of the leaves of a reference taxonomy may be labeled by the candidate sequences for the classification of a given read. For a given subset of the leaves of a reference taxonomy, each candidate internal node (at or under the LCA of the subset of the leaves) for the taxonomic annotation of the read yields a certain number of true positives, false positives, true negatives, and false negatives. For example, for the reference taxonomy in Fig. 2, the subset of grayes leaves yields, for the candidate internal node j, a metagenomic classification problem with $TP = 3$, $FP = 1$, $TN = 3$, $FN = 1$ and thus, balance ratio $(3 + 1)/(1 + 3) = 1$. Table 1 shows the distribution of the number of true positives, false positives, true negatives,

Table 1. Distribution of TP, FP, TN, FN values (left) and distribution of $TP + FN$ values (right) in metagenomic classification problems for different taxonomic reference topologies: complete (C) and degenerate (D) binary trees with 8 leaves

TP	FP	TN	FN	C	D		$TP + FN$	Count
0	2	0	6	4	1		1	56
0	2	1	5	24	6		2	196
0	2	2	4	60	15		3	392
0	2	3	3	80	20		4	490
...	...	...	...	...	...		5	392
7	0	1	0	0	1		6	196
7	1	0	0	8	8		7	56
8	0	0	0	1	1		8	7

and false negatives for all subsets of the leaves of a reference taxonomy and for every candidate internal node for the taxonomic annotation of a read having as candidate sequences the subset of the leaves, for both a complete binary tree and a degenerate binary tree with 8 leaves.

The resulting distribution of $TP + FN$ values (Table 1, right) is exactly the same in both cases and thus, a complete binary tree and a degenerate binary tree with the same number of leaves have the same balance ratio. In fact, any two reference taxonomies for the same taxa have the same balance ratio as long as they have the same number of internal nodes, because they yield a metagenomic classification problem for any subset of the leaves and for any candidate internal node, and $TP + FN$ equals the number of leaves in the subset.

Let us assume that the reads in a metagenomic sample to be classified come from known sequences in a reference taxonomy, as it is usually the case in the taxonomic annotation of metagenomic samples, whereas reads coming from novel sequences are annotated by using clustering methods instead. Given a read and a set of candidate sequences in a reference taxonomy, the taxonomic annotation of the read at a certain node in the clade of the LCA in the reference taxonomy of the set of candidate sequences can then be taken to be correct if, and only if, the candidate sequence that the read comes from lies in the clade of the node at which it is annotated.

Based on this observation, we have studied the performance of some of the most often used indicators of classification error: the Yule ϕ [23], also known as Matthews correlation coefficient [17], the area under the ROC curve, the Youden J [22], the F-measure [18], the Jaccard similarity coefficient [13], and the Rand index [19], in the taxonomic annotation of metagenomic samples.

Definition 2. *Let* TP, TN, FP, *and* FN *be the number of true positives, false positives, true negatives, and false negatives in a binary classification problem.*

– *The Yule* ϕ *is given by*

$$\phi = \frac{TP\,TN - FP\,FN}{\sqrt{(TP + FP)(TP + FN)(TN + FP)(TN + FN)}}$$

– *The Youden J is given by*

$$J = \frac{TP\,TN - FP\,FN}{(TP + FN)(FP + TN)}$$

– *The area under the ROC curve is given by*

$$AUC = \frac{1}{2}\left(\frac{TP}{TP + FN} + \frac{TN}{FP + TN}\right)$$

– *The F-measure is given by*

$$F = \frac{2\,TP}{2\,TP + FP + FN}$$

– *The Jaccard similarity coefficient is given by*

$$C = \frac{TP}{TP + FP + FN}$$

– *The Rand index is given by*

$$R = \frac{TP + TN}{TP + FP + TN + FN}$$

If the denominator in any of these formulas is zero, the value of the indicator is arbitrarily set to zero.

We have computed the value of all these indicators of classification error for each possible set of candidate sequences in a reference taxonomy and for each possible candidate node for the taxonomic annotation of a read coming from each of the candidate sequences, for different taxonomic reference topologies: complete binary trees, that have the largest possible balance but yield the least balanced metagenomic classification problems, and degenerate binary trees, that have the smallest possible balance but yield the most balanced metagenomic classification problems. For these classification problems, we have counted the number of times the taxonomic annotation is correct, that is, the number of times a read is annotated to a node in the reference taxonomy whose clade includes the reference sequence that the read comes from.

The results (Table 2) show that the worst indicator of classification error is the Yule ϕ, followed by AUC and the Youden J (which are equivalent, as $J = 2\,AUC - 1$), the F-measure and the Jaccard similarity coefficient C (which are also equivalent, as $C = F/(2 - F)$), and that the Rand index R is the best indicator of classification error for the taxonomic annotation of metagenomic samples. This can be explained by the fact that in a metagenomic classification problem, we focus on the correct classification of a correct taxonomic annotation while in a statistical classification problem in machine learning, where both positive and negative instances are taken into account, correlation measures such as the Yule ϕ (which is equivalent to the Pearson correlation coefficient for binary classification problems) often are the best indicators of classification error.

Table 2. Total number of correct taxonomic annotations under the Yule (ϕ), the area under the ROC curve (A) or the Youden J, the F-measure (F) or the Jaccard similarity coefficient, and the Rand index (R) for reads coming from known sequences, for different taxonomic reference topologies (complete binary tree and degenerate binary tree) with n leaves

Complete binary tree														
n	2	3	4	5	6	7	8	9	10	11	12	13	14	15	16
ϕ	4	14	40	70	262	306	824	1,450	4,318	6,156	17,064	28,158	63,378	118,292	270,448
A	4	14	40	70	262	306	920	1,530	4,726	6,316	22,056	29,528	79,322	138,477	352,496
F	4	12	32	78	220	407	984	2,234	5,188	10,251	24,844	49,019	112,812	235,322	493,856
R	4	12	48	90	344	485	1,544	2,742	8,308	11,845	37,764	54,757	154,012	239,147	672,416
Degenerate binary tree														
n	2	3	4	5	6	7	8	9	10	11	12	13	14	15	16
ϕ	4	14	38	80	203	388	945	1,961	4,344	8,592	20,152	39,474	88,063	183,603	398,700
A	4	14	38	80	211	384	973	1,952	4,628	8,346	22,230	38,088	94,962	188,986	421,697
F	4	12	32	79	195	441	1,024	2,270	5,104	10,994	24,491	51,959	113,305	241,277	518,937
R	4	12	36	89	222	512	1,191	2,652	5,949	12,971	28,459	61,189	132,263	281,547	602,076

Now, the taxonomic annotation of a metagenomic sample involves obtaining the candidate nodes in a reference taxonomy with the least classification error (for a given indicator) for each of the reads in the metagenomic sample. We have proved in [3] that, when the F-measure is taken as indicator, it suffices to consider candidate nodes that are either candidate sequences themselves, or the LCA of two or more candidate sequences in the reference taxonomy. That is, it suffices to consider as candidate nodes the LCA skeleton tree [7] of the set of candidate sequences for a given read.

We prove below that it also suffices to consider the LCA skeleton tree when the Youden J, the area under the ROC curve, or the Jaccard similarity coefficient is taken as indicator of classification error. The proof for the Yule ϕ is left to the reader.

Let T be a reference taxonomy, let M_i be the set of candidate sequences for the classification of read i, and let T_i be the subtree of T rooted at the LCA of M_i. See Fig. 2 for a schematic view.

Definition 3. *A node j in T_i is called* relevant *if it is equal to a candidate sequence in M_i or equal to the LCA of two or more candidate sequences in M_i.*

Also, for every node j in T_i, let $T_{i,j}$ be the subtree of T_i rooted at j, let L_i be the set of all candidate sequences in T_i, and let N_i be the set of all candidate sequences in T_i that do not belong to M_i (hence, $L_i = M_i \cup N_i$). Similarly, let $M_{i,j}$ be the set of all candidate sequences in $T_{i,j}$ that belong to M_i, let $N_{i,j}$ be the set of all candidate sequences in $T_{i,j}$ that do not belong to $M_{i,j}$, and let $L_{i,j} = M_{i,j} \cup N_{i,j}$. Using this notation, for the taxonomic annotation at node j of a read i with candidate sequences M_i (see Fig. 2), the true positives are $TP_{i,j} = M_{i,j}$, the false positives are $FP_{i,j} = N_{i,j}$, the true negatives are $TN_{i,j} = N_i \setminus N_{i,j}$, and the false negatives are $FN_{i,j} = M_i \setminus M_{i,j}$. Let $C_{i,j}$ be the Jaccard correlation coefficient for node j in T_i, that is,

$C_{i,j} = TP_{i,j}/(TP_{i,j} + FP_{i,j} + FN_{i,j})$. Similarly, let $J_{i,j}$ and $A_{i,j}$, and $F_{i,j}$ be the Youden J and the area under the ROC curve for node j in T_i, respectively. We have:

Theorem 1. *For each node j in T_i, there exists a relevant node j' such that $J_{i,j'} \geqslant J_{i,j}$, $A_{i,j'} \geqslant A_{i,j}$, and $C_{i,j'} \geqslant C_{i,j}$.*

Proof. Suppose that j is a node in T_i that is not relevant. Let j' be the LCA of the candidate sequences in $M_{i,j}$. Clearly, j' is relevant and, furthermore, $|M_{i,j}| = |M_{i,j'}|$ while $|N_{i,j}| \geqslant |N_{i,j'}|$ since $T_{i,j'}$ is a subtree of $T_{i,j}$.

Let $TP = |M_{i,j}|$, $FP = |N_{i,j}|$, $FN = |M_i| - |M_{i,j}|$, $TN = |N_i| - |N_{i,j}|$ and, similarly, let $TP' = |M_{i,j'}|$, $FP' = |N_{i,j'}|$, $FN' = |M_i| - |M_{i,j'}|$, $TN' = |N_i| - |N_{i,j'}|$. We have that $TP' = TP$, $FP' \leqslant FP$, $FN' = FN$, $TN' \geqslant TN$, and $TN' + FP' = TN + FP$.

- Youden J: It has to be proved that

$$\frac{TP'\,TN' - FP'\,FN'}{(TP' + FN')(FP' + TN')} \geqslant \frac{TP\,TN - FP\,FN}{(TP + FN)(FP + TN)}$$

We have that $(TP' + FN')(FP' + TN') = (TP + FN)(FP + TN)$. Then, it suffices to prove that $TP'\,TN' - FP'\,FN' \geqslant TP\,TN - FP\,FN$, that is, $TP(TN' - TN) \geqslant FN(FP' - FP)$. But $TP \geqslant 0$, $(TN' - TN) \geqslant 0$, $FN \geqslant 0$, $(FP' - FP) \leqslant 0$ and thus, the inequality follows.
- Area under the ROC curve: It has to be proved that

$$\frac{TP'(FP' + TN') + TN'(TP' + FN')}{(TP' + FN')(FP' + TN')} \geqslant \frac{TP(FP + TN) + TN(TP + FN)}{(TP + FN)(FP + TN)}$$

We have that $(TP' + FN')(FP' + TN') = (TP + FN)(FP + TN)$ and $TP'(FP' + TN') = TP(FP + TN)$. Then, it suffices to prove that $TN'(TP' + FN') \geqslant TN(TP + FN)$. But $TP' = TP$, $FN' = FN$, $TN' \geqslant TN$ and thus, the inequality follows.
- Jaccard similarity coefficient: It has to be proved that

$$\frac{TP'}{TP' + FP' + FN'} \geqslant \frac{TP}{TP + FP + FN}$$

We have that $TP' = TP$, $FP' \leqslant FP$, $FN' = FN$ and thus, the inequality follows.
- Rand index: It has to be proved that

$$\frac{TP' + TN'}{TP' + FP' + TN' + FN'} \geqslant \frac{TP + TN}{TP + FP + TN + FN}$$

We have that $TP' = TP$, $FN' = FN$, $TN' \geqslant TN$, $FP' + TN' = FP + TN$ and thus, the inequality follows.

$\square$

Corollary 1. *The Youden $J_{i,j}$, the area under the ROC curve $A_{i,j}$, the Jaccard correlation coefficient $C_{i,j}$ and the Rand index $R_{i,j}$ only need to be computed for nodes j in T_i that are relevant.*

3   A Set Cover Approach to Taxonomic Annotation

Let us recall from [10] that an instance of the set cover problem is a collection C of subsets of a finite set X whose union is X, and a solution to the set cover problem is a subset $C' \subseteq C$ such that every element in X belongs to at least one member of C'. The set cover problem is NP-complete, but a logarithmic approximation can be computed in linear time [2,14].

Recall also that in a metagenomic classification problem, the are often multiple candidate nodes in a reference taxonomy with the least classification error for a given read. As a set cover problem, the set of elements X is the set of candidate nodes in a reference taxonomy with the least classification error for the reads in a metagenomic sample, and the collection C of subsets of X is the collection of sets of candidate nodes in the reference taxonomy with the least classification error for each read.

The following example is adapted from [4, Sect. 35.3]; see Fig. 3.

Example 1. Consider a metagenomic sample with reads $x_1, \ldots, x_{12}$ and candidate nodes in a reference taxonomy with the least classification error as follows: $\{y_1, y_3\}$ for x_1, $\{y_1, y_4\}$ for x_2, $\{y_1, y_5\}$ for x_3, $\{y_1, y_3\}$ for x_4, $\{y_1, y_2, y_4\}$ for x_5, $\{y_1, y_2, y_5\}$ for x_6, $\{y_3, y_4\}$ for x_7, $\{y_2, y_4\}$ for x_8, $\{y_2, y_5\}$ for x_9, $\{y_3, y_6\}$ for x_{10}, $\{y_4, y_6\}$ for x_{11}, and $\{y_5\}$ for x_{12}. Then, as an instance of the set cover problem, $X = \{x_1, \ldots, x_{12}\}$ and $C = \{y_1 \ldots, y_6\}$, where $y_1 = \{x_1, x_2, x_3, x_4, x_5, x_6\}$, $y_2 = \{x_5, x_6, x_8, x_9\}$, $y_3 = \{x_1, x_4, x_7, x_{10}\}$, $y_4 = \{x_2, x_5, x_7, x_8, x_{11}\}$, $y_5 = \{x_3, x_6, x_9, x_{12}\}$, and $y_6 = \{x_{10}, x_{11}\}$.

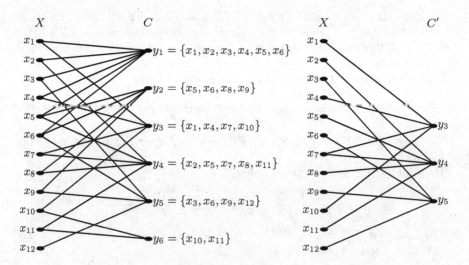

Fig. 3. (left) A metagenomic classification problem viewed as a set cover problem. X is the set of reads from a metagenomic sample, and C is the collection of candidate nodes in the reference taxonomy with the least classification error for some read from the metagenomic sample. (right) The smallest solution to the set cover problem instance.

In a solution C' to a metagenomic classification problem viewed as a set cover problem (X, C), each read in X is annotated to a node in $C' \subseteq C$. Such a taxonomic annotation is not necessarily unique, and there may still be ambiguities in the classification of the metagenomic sample. For the problem instance from Example 1, the smallest solution is $\{y_3, y_4, y_5\}$, which implies the taxonomic annotation of reads x_1, x_4 and x_{10} to node y_3, reads x_2, x_5, x_8 and x_{11} to node y_4, reads x_3, x_6, x_9 and x_{12} to node y_5, and read x_7 to either node y_3 or node y_4 in the reference taxonomy. The greedy algorithm of [14] yields the approximate solutions $\{y_1, y_4, y_5, y_3\}$ and $\{y_1, y_4, y_5, y_6\}$.

The taxonomic annotation of a metagenomic sample can thus be seen as the reduction, and ideally the removal, of ambiguity in the identification of the reads in the metagenomic sample, where a read is ambiguous if it is annotated to more than one node in a reference taxonomy. Viewing the metagenomic classification problem as a set cover problem, an element of X is ambiguous if it belongs to more than one subset of the collection $C' \subseteq C$. The subsets of a set cover overlap on ambiguous elements.

Definition 4. *Let X be a finite set and let C be a collection of subsets of X whose union is X. The overlap of a set cover $C' \subseteq C$ is the total size of the subsets minus the size of X.*

Let the *size* of a set cover be the number of subsets of X that it contains, and let the *total size* of a set cover be the total size of the subsets of X that it contains. This corresponds to set cover problems I and II in [14]. It turns out that a set cover of smallest size does not necessarily have the least overlap, while a set cover of smallest total size always has the least overlap.

Proposition 1. *A set cover with the least number of subsets does not necessarily have the least overlap.*

Proof. Let $X = \{1, \ldots, n\}$ and assume, without loss of generality, that $n = 2k$ for $k \geqslant 3$. Let S be the following collection of subsets of X:

$$\{1, 2\}, \{3, 4\}, \ldots, \{n-1, n\}, \{1, \ldots, n-1\}, \{2, \ldots, n\}$$

The set cover $\{1, \ldots, n-1\}, \{2, \ldots, n\}$ has size 2, which is the smallest possible for S and X, and overlap n. The set cover $\{1, \ldots, n-1\}, \{n-1, n\}$ also has size 2, but it has overlap 1. Same for the set cover $\{1, 2\}, \{2, \ldots, n\}$, and S and X have no other set cover of size 2. However, the set cover $\{1, 2\}, \{3, 4\}, \ldots, \{n-1, n\}$ has size $n/2$ and overlap 0, which is the least possible overlap.

The following result follows directly from Definition 4.

Corollary 2. *A set cover with the least total size of subsets has the least overlap.*

Based on the solution of a set cover problem with the least total size of subsets, the abundance profile of a metagenomic sample is given by the proportion of reads mapped to each node in the set cover, adjusted by a uniform distribution of any still ambiguous reads among all the nodes in the set cover which they are mapped to.

We have implemented the set cover approach to taxonomic annotation in a next release of the TANGO software [1,3], which belongs in the BioMaS [9] and MetaShot [8] pipelines. The new implementation of TANGO consists of a Python script for taxonomic annotation using the NCBI Taxonomy [5,6], based on the ETE Toolkit [11], along with another Python script for resolving any remaining ambiguities by finding an approximate solution to a set cover problem with the least total size of subsets. While the first script processes the input metagenomic sample one-sequence-read-at-a-time, the second script processes the output of the first script for the whole set of reads, and produces both a taxonomic annotation of the reads and an abundance profile of the metagenomic sample.

4   Conclusion

We have addressed two potential sources of bias in the taxonomic annotation of metagenomic samples, which is usually done by first mapping the reads to the reference sequences and then, classifying each read at a node in the clade of the LCA of the candidate sequences in the reference taxonomy with the least classification error. On the one hand, we have shown that the reference taxonomy being balanced or imbalanced does not affect the balance of the metagenomic classification problem, and we also shown that the Rand index is a better indicator of classification error for metagenomic classification problems than the often used area under the ROC curve and F-measure. On the other hand, we have reduced the taxonomic annotation problem for a whole metagenomic sample to a set cover problem, for which a logarithmic approximation can be obtained in linear time, and we have shown that a solution to the set cover problem with the least total size of subsets minimizes the ambiguity in the taxonomic annotation of the reads in a metagenomic sample.

Future work includes extending the computation of balance ratio and total number of correct taxonomic annotations from Sect. 2 to the NCBI Taxonomy, taking ancestry relationships among the nodes in the reference taxonomy into account in the set cover formulation of the taxonomic annotation problem from Sect. 3 and last, but not least, extending the set cover problem formulation of the taxonomic annotation problem to a non-taxonomic metagenomic classification problem, with reference sequences but without a reference taxonomy.

Acknowledgements. Partially supported by Spanish Ministry of Economy and Competitiveness and European Regional Development Fund project DPI2015-67082-P (MINECO/FEDER).

References

1. Alonso, D., Barré, A., Beretta, S., Bonizzoni, P., Nikolski, M., Valiente, G.: Further steps in TANGO: improved taxonomic assignment in metagenomics. Bioinformatics **30**(1), 17–23 (2013)
2. Bar-Yehuda, R., Even, S.: A linear-time approximation algorithm for the weighted vertex cover problem. J. Algorithms **2**(2), 198–203 (1981)
3. Clemente, J.C., Jansson, J., Valiente, G.: Flexible taxonomic assignment of ambiguous sequencing reads. BMC Bioinform. **12**(1), 8 (2011)
4. Cormen, T.H., Leiserson, C.E., Rivest, R.L., Stein, C.: Introduction to Algorithms, 3rd edn. MIT Press, Cambridge (2009)
5. Federhen, S.: The NCBI taxonomy database. Nucleic Acids Res. **40**(D1), D136–D143 (2012)
6. Federhen, S.: Type material in the NCBI taxonomy database. Nucleic Acids Res. **43**(D1), D1086–D1098 (2015)
7. Fischer, J., Huson, D.H.: New common ancestor problems in trees and directed acyclic graphs. Inform. Process. Lett. **110**(8–9), 331–335 (2010)
8. Fosso, B., Santamaria, M., D'Antonio, M., Lovero, D., Corrado, G., Vizza, E., Passero, N., Garbuglia, A.R., Capobianchi, M.R., Crescenzi, M., Valiente, G., Pesole, G.: MetaShot: An accurate workflow for taxon classification of host-associated microbiome from shotgun metagenomic data. Bioinformatics (2017, in press)
9. Fosso, B., Santamaria, M., Marzano, M., Alonso, D., Valiente, G., Donvito, G., Monaco, A., Notarangelo, P., Pesole, G.: BioMaS: a modular pipeline for bioinformatic analysis of metagenomic amplicons. BMC Bioinform. **16**(1), 203 (2015)
10. Garey, M.R., Johnson, D.S.: Computers and Intractability: A Guide to NP-Completeness. Freeman, Dallas (1979)
11. Huerta-Cepas, J., Serra, F., Bork, P.: ETE 3: reconstruction, analysis and visualization of phylogenomic data. Mol. Biol. Evol. **33**(6), 1635–1638 (2016)
12. Huson, D.H., Auch, A., Qi, J., Schuster, S.C.: MEGAN analysis of metagenomic data. Genome Res. **17**(3), 377–386 (2007)
13. Jaccard, P.: Étude comparative de la distribution florale dans une portion des Alpes et du Jura. Bull. Soc. Vaud. Sc. Nat. **37**(142), 547–579 (1901)
14. Johnson, D.S.: Approximation algorithms for combinatorial problems. J. Comput. Syst. Sci. **9**(3), 256–278 (1974)
15. Kunin, V., Copeland, A., Lapidus, A., Mavromatis, K., Hugenholtz, P.: A bioinformatician's guide to metagenomics. Microbiol. Mol. Biol. Rev. **72**(4), 557–578 (2008)
16. López, V., Fernández, A., García, S., Palade, V., Herrera, F.: An insight into classification with imbalanced data: empirical results and current trends on using data intrinsic characteristics. Inform. Sci. **250**(1), 113–141 (2013)
17. Matthews, B.W.: Comparison of the predicted and observed secondary structure of T4 phage lysozyme. Biochim. Biophys. Acta **405**(2), 442–451 (1975)
18. Powers, D.M.W.: Evaluation: from precision, recall and F-measure to ROC, informedness, markedness and correlation. J. Mach. Learn. Tech. **2**(1), 37–63 (2011)
19. Rand, W.M.: Objective criteria for the evaluation of clustering methods. J. Am. Stat. Assoc. **66**(336), 846–850 (1971)
20. Thomas, T., Gilbert, J., Meyer, F.: Metagenomics: a guide from sampling to data analysis. Microb. Inform. Exp. **2**(1), 3 (2012)
21. Wooley, J.C., Godzik, A., Friedberg, I.: A primer on metagenomics. PLoS Comput. Biol. **6**(2), e1000667 (2010)
22. Youden, W.J.: Index for rating diagnostic tests. Cancer **3**(1), 32–35 (1950)
23. Yule, G.U.: On the methods of measuring association between two attributes. J. R. Statist. Soc. **75**(6), 579–642 (1912)

Genetic Algorithm Based Beta-Barrel Detection for Medium Resolution Cryo-EM Density Maps

Albert Ng and Dong Si[✉]

University of Washington, Bothell,
18115 Campus Way NE, Bothell, WA 98011, USA
{alng180, dongsi}@uw.edu

Abstract. Cryo-electron microscopy (Cryo-EM) is a technique that produces three-dimensional density maps of large protein complexes and enables the study of the interactions and structures of those molecules. Identifying the secondary structures (α-helices and β-sheets) located in proteins using density maps is vital in identifying and matching the backbone of the protein with the cryo-EM density map. The β-barrel is a unique β-sheet structure commonly found in proteins, such as membranes and lipocalins. We present a new approach utilizing a genetic algorithm and ray tracing to automatically identify and extract β-barrels from cryo-EM density maps. This approach was tested using ten simulated density maps at 9 Å resolution and six experimental density maps at various resolutions. The results suggest that our approach is capable of performing automatic detection and extraction of the β-barrels from medium resolution cryo-EM density maps.

Keywords: Protein · Secondary structures · Genetic algorithm · Ray tracing · Cryo-electron microscopy · Beta-barrel · Density map · Feature detection · Pattern recognition

1 Introduction

Cryo-electron microscopy (cryo-EM) is an experimental technique that allows for the study of the structure of large molecules and protein complexes [1]. In cryo-EM, the molecule being studied is frozen and millions of two-dimensional images are taken at numerous different angles, and then used to generate a three-dimensional density map of the molecule being studied [2]. Although the technique has improved recently to the point of being able to resolve to near-atomic resolutions, there still exists a significant amount of data that are resolved at medium resolutions between 5–10 Å [3, 4]. These resolutions are relatively too low to be able to differentiate between individual atoms within the protein and an alternative approach must be used to discover the structure and mechanism of the protein.

Although individual atoms cannot be seen at medium resolution cryo-EM data, the secondary structures of a protein are still prominent and visible [5, 6]. Secondary structures, local structures within proteins that help form the overall structure, generally consist of either α-helices or β-sheets. α-helices have a very regular and distinct shape at medium resolutions as they tend to appear as a thick rod of density within the

© Springer International Publishing AG 2017
Z. Cai et al. (Eds.): ISBRA 2017, LNBI 10330, pp. 174–185, 2017.
DOI: 10.1007/978-3-319-59575-7_16

cryo-EM density maps. Because of this, α-helices are much easier to identify and detect. Much work has already been done in attempting to identify α-helices within cryo-EM density maps and are quite successful in doing so [6, 7, 9, 10, 12, 13].

β-sheets, on the other hand, are formed when two or more β-strands line up side by side to form the characteristic sheet-like structure and tend to appear as a thin layer of density within the cryo-EM density maps. Work has been done to attempt to identify the position and location of the individual β-strands within some β-sheets [7, 8, 11, 14–16]. However, as β-strands can line up, twist, and combine to form vastly different geometries, this means that one method is rarely capable predict the position of β-sheets/strands for all possible β-sheet structures (Fig. 1).

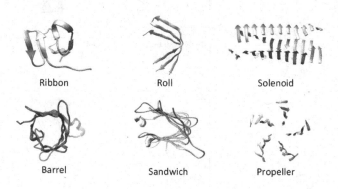

Ribbon   Roll   Solenoid

Barrel   Sandwich   Propeller

Fig. 1. Examples of various β-sheet structures

The β-barrel is a specific type of β-sheet structure where the first β-strand in the sheet is hydrogen bonded to the last β-strand in the β-sheet. This results in a distinct hollow cylindrical structure. Because of this unique cylindrical structure, β-barrels are generally found as part of a cell membrane, in porins, and in lipocalins [17, 18].

In Si [16], a random sample consensus (RANSAC) based approach, *BarrelMiner*, was developed to find and detect β-barrels from cryo-EM density maps. *BarrelMiner* attempts to fit an ideal cylinder template to the β-barrel. However, *BarrelMiner* assumes that the β-barrel's shape is that of an ideal cylinder. In this paper, we propose an alternative approach towards the automatic detection of β-barrels from medium resolution cryo-EM density maps. Like *BarrelMiner*, we attempt to fit an ideal cylinder into the β-barrel region. Instead of RANSAC, we use a genetic algorithm to fit an ideal cylinder inside the β-barrel region. Additionally, we utilize a ray tracing algorithm to attempt to detect the true shape of the β-barrel using the fitted cylinder. Our algorithm can also suggest if there exists a β-barrel in the cryo-EM density map or not.

2   Method

The method described in this paper utilizes a genetic algorithm to fit an ideal cylinder to the center of the β-barrel in a density map. Ray tracing is then applied using the fitted ideal cylinder to attempt to discover the true shape of the β-barrel (Fig. 2).

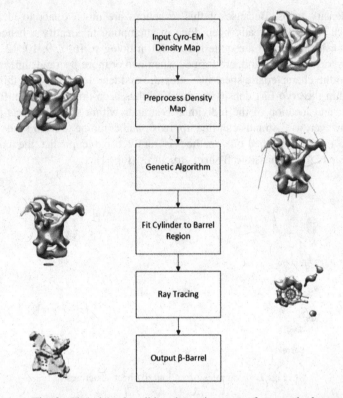

Fig. 2. Flowchart describing the major steps of our method

This approach can be separated into three main components:

Preprocessing Component. The goal of this component is to reduce the size of the search space and to subdivide it into smaller sections.

Genetic Algorithm. Agenetic algorithm is performed to search each section to detect the location of the β-barrel and fit an ideal cylinder to it.

Postprocessing Component. The goal of this component is to utilize ray tracing technique to discover the true shape of the β-barrel.

2.1 Preprocessing

The purpose of preprocessing in our method is to remove as many background noise and non β-barrel voxels to reduce the amount of computation time needed. A global density threshold was selected to filter and remove all voxels in the density map with a lower density than the selected threshold. This should remove the voxels associated with background noise, as these voxels tend to have very small density values. For all density maps tested, it was loaded into Chimera [19] and a global threshold was selected such that most of the non-secondary structure voxels were eliminated, while still retaining the general cylindrical shape of the β-barrel. To further decrease the number of non β-barrel voxels in the density map, the voxels associated with α-helix secondary structures were also removed from the density map. This was done using Gorgon and *SSETracer* [15, 20]. Any voxel found within 2.0 Å of a detected α-helix using *SSETracer* would be removed from the density map. The last step in preprocessing involves subdividing the entire density map into clusters of voxels. A cluster is defined as a group of voxels that are at least 2.0 Å away from another cluster. Once the clustering is completed, all clusters that contain a population below a pre-defined threshold were removed. These removed clusters are assumed to be background noise. The genetic algorithm is then performed on each remaining cluster to attempt to discover the location and shape of β-barrels, if any, located within the cluster.

2.2   Genetic Algorithm

The genetic algorithm is an extremely efficient algorithm at searching and optimizing solutions within a search space by mimicking natural selection [21, 22]. The purpose of the genetic algorithm in our method is to attempt to fit an ideal cylinder into the center of the β-barrel region within the density map. The main axis of the cylinder (the axis between the centers of the two circles at the end) should align with the axis of the β-barrel. Additionally, the entire cylinder should fit within the empty region within the β-barrel and avoid contact with the β-barrel walls (Fig. 3).

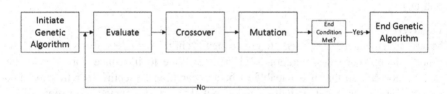

Fig. 3.   General flowchart describing the genetic algorithm process

The individual used in our genetic algorithm method is an ideal cylinder. Each ideal cylinder is defined by a 1 by 7 vector containing seven parameters, $[x_1, y_1, z_1, x_2, y_2, z_2, r]$. These parameters represent the two points at the center of the two circles at the two ends of the cylinder along with radius of the two circles. Each cylinder is assumed to be hollow to match the β-barrel. Each cylinder has a fitness score that describes how likely a cylinder is fit to the center region of a β-barrel. The lower the fitness score, the more likely the cylinder is fit correctly to a β-barrel. The initial population of ideal cylinders is randomly generated. For each randomly generated cylinder, two random 3D points are selected within the confines of the density map and a random radius, between 3 Å and 10 Å, is chosen. For all the cryo-EM density maps tested, a population size of 200 ideal cylinders was used. For all density maps, the genetic algorithm was run for 100 generations.

Fitness Score

The fitness score describes how likely an ideal cylinder candidate is fit to a β-barrel region. The fitness score, F, is given by:

$$F = MSE + BP \tag{1}$$

where MSE is the mean-squared error and BP is a penalty for voxels located inside the cylinder. The mean-squared error, MSE, is calculated by:

$$MSE = \frac{\sum_i^N (d_i)^2}{N} \tag{2}$$

where N is the number of voxels in the density map remaining after preprocessing and d_i is the shortest distance between the i-th voxel in the density map and the surface of the cylinder. The BP is a penalty for when voxels are located inside the candidate cylinder. In a β-barrel, there should not be any voxels/density located inside of it and this BP attempts to account for this. The BP is calculated by:

$$BP = \sum_j^M (d_j)^3 \tag{3}$$

where M is the number of voxels located inside the volume of the cylinder and d_j is the shortest distance between the j-th internal voxel and the surface of the cylinder. The BP is not averaged as we wanted it to make more weight compared to MSE when calculating F.

Crossover

The crossover function is used to create new "child" cylinders from the most fit cylinders in the current generation [22]. This is done to introduce variation to the genetic algorithm so that it is capable of better searching the entire search space. For each generation, the most fit half of the population are designated as parents. Each parent is paired up in order of their fitness (i.e. most fit with the second most fit) and two child cylinders are created from each pair. The first child cylinder will randomly select one cylinder end-point from each parent and the second child will select the two unpicked ones. The radius for both child cylinders consists of the average radius between the two parent cylinders. The parent and child cylinders are then combined to form the next generation for the genetic algorithm.

Mutation

The mutation function is used to introduce new "genes" to the population, so that the search can avoid becoming trapped within local minima/maxima. During crossover, every child cylinder has a chance of undergoing mutation. When mutating, one of the three parameters defining the cylinder will be changed by a mutation factor. The mutation factor is determined using a non-uniform mutation strategy, where the mutation factor starts off large in early generations of the genetic algorithm and decreases as generations pass by [23].

2.3   Postprocessing

The postprocessing step in our method is used to discover the true shape of the β-barrel. As the shape of β-barrels is rarely regular, it is difficult to fit any single regular shape to all possible β-barrels. Our method avoids this by using ray tracing to trace the true shape of the β-barrel. Ray tracing is a computer graphics technique that attempts to imitate how light rays work [24]. We leverage ray tracing by shooting rays around the fitted cylinder to find the voxels that make up the walls of the β-barrel. The fitted cylinder is split up into evenly divided circular slices along the axis of the cylinder. Within each slice, rays are shot from the axis of the cylinder every 1° towards the surface of the cylinder (Fig. 4).

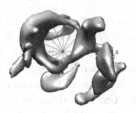

Fig. 4. Ray Tracing for a single slice of the fitted cylinder. The red dot is a point on the axis of the fitted cylinder. The orange lines represent the rays being casted from the axis. (Color figure online)

For each ray, all voxels within 1.0 Å from the casted ray are added to a set. If there are no voxels in the set, it is assumed that the ray was casted into empty space. Otherwise, the scalar projection onto the casted ray is calculated for every voxel in the set. Any voxels with a negative scalar projection are removed. The voxel with the smallest scalar projection is determined to have "intersected" the ray. The ratio of intersecting rays over the total number of rays casted is calculated for every cylinder slice. Starting from the two ends, slices that do not achieve a ratio of at least 0.6 are removed until a slice with at least a 0.6 ratio is reached. If all the slices were removed, then our method suggests that there is no β-barrel. This step is done to remove non β-barrel voxels located above and below the β-barrel. Lastly, as some β-barrels have gaps in its walls, rays may be casted through these gaps and intersect with the surrounding non β-barrel voxels. To remove these voxels, voxels that are at least 1.5 σ from the mean distance between all the voxels in the β-barrel voxel set and the axis of the fitted cylinder are removed.

2.4   Accuracy

To quantify the accuracy of our method, we calculated the sensitivity and specificity based on the α-carbons (Cα) of each amino acid detected. A Cα is determined to be detected if there are detected voxels 2.0 Å away from it. Sensitivity indicates the percentage of β-barrel Cα correctly detected (true positives) [25]. Sensitivity is given by:

$$\text{Sensitivity} \; = \; \# \; \beta\text{-barrel C}\alpha \text{ detected/total } \# \; \beta\text{-barrel C}\alpha \qquad (4)$$

Specificity indicates the percentage of non β-barrel Cα correctly detected (true negatives) [25]. Specificity is given by:

$$\text{Specificity} \; = \; \# \text{ non } \beta\text{-barrel C}\alpha \text{ detected/total } \# \text{ non } \beta\text{-barrel C}\alpha \qquad (5)$$

3 Results

3.1 Simulated Density Maps

Ten simulated density maps were tested using our method. Each density map was generated using the program *pdb2mrc* from EMAN [26] at a resolution of 9 Å and a sampling of 1 Å per pixel. These ten proteins were all obtained from the CATH database (http://www.cathdb.info/) under "Beta Barrel" architecture section.

In Fig. 5, we show an example of one of the simulated density map tested, protein 1AJZ chain A. All the β-barrel Cα were detected on the 1AJZ protein using our method. However, a specificity of only 80.0% was obtained. In Fig. 5C, we can see the non β-barrel voxels detected above the β-barrel. This is because the fitted cylinder extends above the β-barrel and our method includes some of these non β-barrel voxels near the cylinder in the final voxel set. Although some of these outlier voxels are removed, the voxels right above and below the β-barrel seem to surround the fitted cylinder enough to manage at least a 0.6 ratio of ray hits to avoid being removed by our postprocessing method.

In Table 1, we describe the results of this method on the ten simulated density maps. The average sensitivity among the ten simulated density maps is 97.7% which

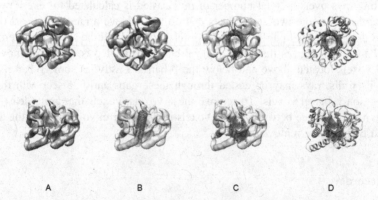

A   B   C   D

Fig. 5. β-barrel detection from simulated density map. (A) Simulated density map of 1AJZ chain A at 9 Å resolution. (B) The fitted cylinder (red) (C) the detected β-barrel surface (red) (D) the detected β-barrel surface (red) superimposed over the true PDB structure. The top view (first row) and a side view (second row) are shown in (A), (B), (C), and (D) (Color figure online)

Table 1. Accuracy of β-barrel detection on simulated density maps

PDB ID[a]	TL[b]	Barrel[c]	TP[d]	FP[e]	Sens.[f]	Spec..[g]
1AJZ_A	282	37	37	49	1.000	0.800
1AL7_A	350	34	33	56	0.971	0.823
1JB3_A	127	46	45	10	0.978	0.877
1NNX_A	93	45	45	10	1.000	0.792
1TIM_A	247	50	50	52	1.000	0.736
4HIK_A	136	45	43	23	0.956	0.747
1Y0Y_A	335	35	35	100	1.000	0.667
3GP6_A	166	93	89	25	0.957	0.657
3ULJ_A	96	55	54	7	0.982	0.829
2DYI_A	162	41	38	23	0.927	0.810
		Average			0.977	0.774

[a]PDB_chain
[b]Total number of Cα in the protein chain
[c]Total number of Cα on the β-barrel
[d]True positive Cα detected
[e]False positive Cα detected
[f]Sensitivity of β-barrel detection, see Eq. 4
[g]Specificity of β-barrel detection, see Eq. 5

suggests that our method is very good at finding the general shape of the β-barrel. However, the specificity is only 77.4%. This is due to all the non β-barrel (loop and turn) Cα that are located very close to the two ends of β-which is very challenging to detect and remove from the final set of voxels for the β-barrel.

3.2 Experimental Density Maps

Six experimental density maps, obtained from EMDB [4], were tested. All six had corresponding PDB files that gave us the true protein structure. In Fig. 6, we show the

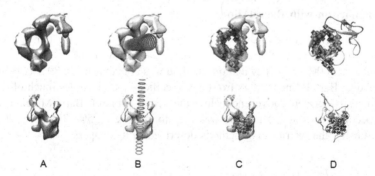

Fig. 6. β-barrel detection from experimental density map. (A) Experimental density map of 4CSU chain K. (B) The fitted cylinder (red) (C) The detected β-barrel surface (red) (D) The detected β-barrel surface (in red) superimposed over the true PDB structure. The top view (first row) and a side view (second row) are shown in (A), (B), (C), and (D) (Color figure online)

results of one of the experimental cryo-EM density maps, EMD 2605 aligned with protein 4CSU chain K. For this density map, a sensitivity of 89.3% (25 of 28 β-barrel Cα detected) and a specificity of 92.5% were obtained. The β-barrel α-carbons that were not detected were ones that were at the top end of individual β-strands and were positioned just far enough away from the barrel wall so that it fell outside our 2.0 Å threshold. The detected non β-barrel Cα were located right above the actual β-barrel where it is challenging to detect and remove as outliers.

In Table 2, we can see the results of this method on the six experimental density maps. The average sensitivity among the six experimental density maps is 82.8%, which is significantly lower than the sensitivity for the simulated maps. Due to the noisy and incomplete nature of experimental cryo-EM data, this is not surprising and reflects the difficulty of automatic β-barrel detection from experimental density maps.

Table 2. Accuracy of β-barrel detection on experimental density maps

EMDB_PDB_ID (Res)[a]	TL[b]	Barrel[c]	TP[d]	FP[e]	Sens.[f]	Spec.[g]
1657_2WWQ_W (5.8 Å)	94	30	23	17	0.767	0.734
1780_3IZ5_M (5.5 Å)	140	28	22	24	0.786	0.786
1849_3IZU_L (8.25 Å)	123	31	24	1	0.774	0.989
1849_3IZU_W (8.25 Å)	94	40	34	9	0.850	0.833
2605_4CSU_K (5.5 Å)	121	28	25	7	0.893	0.925
6396_5A9Z_AL (6.4 Å)	122	30	27	25	0.900	0.728
		Average			0.828	0.833

[a]EMDB_PDB_chain (resolution)
[b]Total number of Cα in the protein chain
[c]Total number of Cα on the β-barrel
[d]True positive Cα detected
[e]False positive Cα detected
[f]Sensitivity of β-barrel detection, see Eq. 4
[g]Specificity of β-barrel detection, see Eq. 5

3.3   Comparison with BarrelMiner

In Fig. 8, we show a comparison between the results produced using our method and the results produced using BarrelMiner on density map EMD 2605. BarrelMiner was able to achieve 100% sensitivity and our method was able to achieve 89.3%. However, as seen above, BarrelMiner suffers from low specificity as it includes much of the non β-barrel density located above and below the actual β-barrel. BarrelMiner achieved 66.7% specificity, while our method was able to achieve 92.5%. These results were calculated using the metrics and methods described in this paper.

3.4   Density Maps with No β-Barrels

To further test our method, two protein chains that are known to not contain any β-barrels were tested. These two were tested to determine if our method could predict

Fig. 7. Comparison of BarrelMiner and our method using 4CSU chain K. (A) Detected β-barrel surface using our method (red) superimposed over PDB. (B) BarrelMiner β-barrel surface (yellow) superimposed over PDB. (C) β-barrel surfaces (yellow) using BarrelMiner and our method (red) superimposed over PDB. (Color figure online)

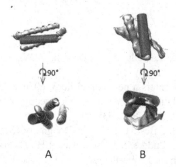

Fig. 8. β-barrel detection for simulated density maps containing no β-barrels. (A) Density map and fitted cylinder for 1COS chain A. (B) Density map and fitted cylinder for 4R80 chain A.

the presence of β-barrels. In Fig. 7, you can see the results for the two non β-barrel protein chains tested, PDB 1COS chain A and 4R80 chain A. For both protein chains, a cylinder was fitted as best as possible using the genetic algorithm, but neither cylinder had slices able to achieve the ray hit ratio of at least 0.6 and our method correctly determined that no β-barrel exists.

4   Conclusion and Future Work

In this paper, we describe a new approach that attempts to automatically and accurately detect and extract the β-barrel region from cryo-EM density maps. It utilizes a genetic algorithm to fit an ideal cylinder to the center of the β-barrel region before applying ray tracing identify the voxels that make up the β-barrel region of the density map. Our approach has been tested on both experimental and simulated cryo-EM density maps. The results from these tests have proven that our proposed approach is capable of automatically identifying the β-barrel from cryo-EM density maps. However, the accuracy was significantly lower when applied to experimental density maps. This was due to the noise and incompleteness inherent in experimental cryo-EM density maps. Further work needs to be done on improving this method to enable better detection of β-barrels from less than ideal density maps. Additionally, as the ray tracing portion consumed the bulk of the computation time, optimizations to speed up the ray tracing algorithm should be looked into in future work.

Our method was coded in C++ and all tests were performed on a desktop machine with an Intel i7-4790k @ 4.0 GHz processor and 16 GB memory. The execution time needed to obtain the results varied between five seconds and two minutes depending on the number of voxels contained in the input density maps.

Acknowledgement. This work was supported by the Graduate Research Award from the Computing and Software Systems division of University of Washington Bothell and the startup fund 74-0525.

References

1. Adrian, M., Dubochet, J., Lepault, J., McDowall, A.W.: Cryo-electron microscopy of viruses. Nature **308**, 32–36 (1984)
2. Kühlbrandt, W.: Cryo-EM enters a new era. eLife **3**, e03678 (2014)
3. Zhou, Z.H.: Atomic resolution cryo electron microscopy of macromolecular complexes. Adv. Protein Chem. Struct. Biol. **82**, 1–35 (2011)
4. Lawson, C.L., Baker, M.L., Best, C., Bi, C., Dougherty, M., Feng, P., van Ginkel, G., Devkota, B., Lagerstedt, I., Ludtke, S.J., Newman, R.H., Oldfield, T.J., Rees, I., Sahni, G., Sala, R., Velankar, S., Warren, J., Westbrook, J.D., Henrick, K., Kleywegt, G.J., Berman, H. M., Chiu, W.: EMDataBank.org: unified data resource for CryoEM. Nucleic Acids Res. **39**, D456–D464 (2011)
5. Al Nasr, K., Sun, W., He, J.: Structure prediction for the helical skeletons detected from the low resolution protein density map. BMC Bioinf. **11**, S44 (2010)
6. Baker, M.L., Ju, T., Chiu, W.: Identification of secondary structure elements in intermediate-resolution density maps. Struct. Lond. Engl. **1993**(15), 7–19 (2007)
7. Si, D., Ji, S., Nasr, K.A., He, J.: A machine learning approach for the identification of protein secondary structure elements from electron cryo-microscopy density maps. Biopolymers **97**, 698–708 (2012)
8. Si, D., He, J.: Combining image processing and modeling to generate traces of beta-strands from cryo-EM density images of beta-barrels. In: 2014 36th Annual International Conference of the IEEE Engineering in Medicine and Biology Society, pp. 3941–3944 (2014)
9. Rusu, M., Wriggers, W.: Evolutionary bidirectional expansion for the tracing of alpha helices in cryo-electron microscopy reconstructions. J. Struct. Biol. **177**, 410–419 (2012)
10. Jiang, W., Baker, M.L., Ludtke, S.J., Chiu, W.: Bridging the information gap: computational tools for intermediate resolution structure interpretation. J. Mol. Biol. **308**, 1033–1044 (2001)
11. Si, D., He, J.: Tracing beta strands using StrandTwister from cryo-EM density maps at medium resolutions. Struct. Lond. Engl. **1993**(22), 1665–1676 (2014)
12. Li, R., Si, D., Zeng, T., Ji, S., He, J.: Deep convolutional neural networks for detecting secondary structures in protein density maps from cryo-electron microscopy. In: 2016 IEEE International Conference on Bioinformatics and Biomedicine (BIBM), pp. 41–46 (2016)

13. Dal Palù, A., He, J., Pontelli, E., Lu, Y.: Identification of alpha-helices from low resolution protein density maps. In: Computational Systems Bioinformatics Conference, pp. 89–98 (2006)
14. Si, D., He, J.: Modeling beta-traces for beta-barrels from cryo-EM density maps. Biomed. Res. Int. **2017**, 1793213 (2017)
15. Si, D., He, J.: Beta-sheet detection and representation from medium resolution cryo-EM density maps. In: Proceedings of the International Conference on Bioinformatics, Computational Biology and Biomedical Informatics, pp. 764:764–764:770. ACM (2013)
16. Si, D.: Automatic detection of beta-barrel from medium resolution Cryo-Em density maps. In: Proceedings of the 7th ACM International Conference on Bioinformatics, Computational Biology, and Health Informatics, pp. 156–164. ACM, New York (2016)
17. Wimley, W.C.: The versatile β-barrel membrane protein. Curr. Opin. Struct. Biol. **13**, 404–411 (2003)
18. Flower, D.R.: The lipocalin protein family: structure and function. Biochem. J. **318**, 1–14 (1996)
19. Pettersen, E.F., Goddard, T.D., Huang, C.C., Couch, G.S., Greenblatt, D.M., Meng, E.C., Ferrin, T.E.: UCSF Chimera–a visualization system for exploratory research and analysis. J. Comput. Chem. **25**, 1605–1612 (2004)
20. Baker, M.L., Baker, M.R., Hryc, C.F., Ju, T., Chiu, W.: Gorgon and pathwalking: macromolecular modeling tools for subnanometer resolution density maps. Biopolymers **97**, 655–668 (2012)
21. Fraser, A.S., Fraser, A.S.: Simulation of genetic systems by automatic digital computers I. Introduction. Aust. J. Biol. Sci. **10**, 484–491 (1957)
22. Holland, J.H.: Adaptation in Natural and Artificial Systems: An Introductory Analysis with Applications to Biology, Control and Artificial Intelligence. MIT Press, Cambridge (1992)
23. Zhao, X., Gao, X.-S., Hu, Z.-C.: Evolutionary programming based on non-uniform mutation. Appl. Math. Comput. **192**, 1–11 (2007)
24. Goldstein, R.A., Nagel, R.: 3-D visual simulation. Trans. Soc. Comput. Simul. **16**, 25–31 (1971)
25. Altman, D.G., Bland, J.M.: Statistics notes: diagnostic tests 1: sensitivity and specificity. BMJ **308**, 1552 (1994)
26. Ludtke, S.J., Baldwin, P.R., Chiu, W.: EMAN: semiautomated software for high-resolution single-particle reconstructions. J. Struct. Biol. **128**, 82–97 (1999)

Mining K-mers of Various Lengths in Biological Sequences

Jingsong Zhang[1], Jianmei Guo[2], Xiaoqing Yu[3], Xiangtian Yu[1], Weifeng Guo[4], Tao Zeng[1(✉)], and Luonan Chen[1,5(✉)]

[1] Institute of Biochemistry and Cell Biology,
Shanghai Institutes for Biological Sciences,
Chinese Academy of Sciences, Shanghai 200031, China
jasun@dmbio.info, graceyu1985@163.com, {zengtao,lnchen}@sibs.ac.cn
[2] Department of Computer Science and Engineering,
East China University of Science and Technology, Shanghai 200237, China
gjm@ecust.edu.cn
[3] Department of Applied Mathematics, Shanghai Institute of Technology,
Shanghai 201418, China
xqyu@sit.edu.cn
[4] School of Automation, Northwestern Polytechnical University, Xi'an 710072, China
shaonianweifeng@126.com
[5] Collaborative Research Center for Innovative Mathematical Modelling,
Institute of Industrial Science, University of Tokyo, Tokyo 153-8505, Japan

Abstract. Counting the occurrence frequency of each k-mer in a biological sequence is an important step in many bioinformatics applications. However, most k-mer counting algorithms rely on a given k to produce single-length k-mers, which is inefficient for sequence analysis for different k. Moreover, existing k-mer counters focus more on DNA sequences and less on protein ones. In practice, the analysis of k-mers in protein sequences can provide substantial biological insights in structure, function and evolution. To this end, an efficient algorithm, called VLmer (Various Length k-mer mining), is proposed to mine k-mers of various lengths termed *vl-mers* via inverted-index technique, which is orders of magnitude faster than the conventional forward-index method. Moreover, to the best of our knowledge, VLmer is the first able to mine k-mers of various lengths in both DNA and protein sequences.

Keywords: Sequential pattern mining · K-mer counting · K-mers of various lengths · Biological sequence analysis

1 Introduction

K-mer counting, which identifies frequent contiguous subsequences of length-k in a sequence database, is a fundamental data-mining problem with broad applications, including genome assembly [1], error correction of sequencing reads [2],

---

J. Zhang and J. Guo—Contributed equally to this work.

© Springer International Publishing AG 2017
Z. Cai et al. (Eds.): ISBRA 2017, LNBI 10330, pp. 186–195, 2017.
DOI: 10.1007/978-3-319-59575-7_17

protein-protein interaction prediction [3], finding mutations [4], sequence classification [5], sequence alignment [6]. Many previous studies contributed to efficient k-mer counting, such as Tallymer [7], Jellyfish [8], BFCounter [9], KMC [10], DSK [11], KAnalyze [12], KMC 2 [13] and KCMBT [14]. In recent years, k-mer counting is gaining momentum and has become an active topic in sequence analysis community.

The k-mer counting algorithms developed so far have good performance. Unfortunately, almost all of the previous algorithms rely on a fixed k to split the given sequence(s) and output all possible contiguous subsequences along with their frequencies. This leads to the inefficiency during the sequence analysis for various values of k, since a user usually have to perform the k-mer counting algorithm many times. For example, as shown in [15,16], at least 6 sizes (17-mer, 21-mer, 25-mer, 41-mer, 55-mer and 77-mer) are selected to count the k-mers. In addition, Kurta et al. [7] used k-mers ranging from 10 to 500 to annotate the plant genomes, which needs to manually run the k-mer counter 491 times. Currently, there is no method that is able to automatically generate k-mers of various lengths.

The frequent k-mers in protein sequences are often the conserved composition patterns reflecting structural and functional features [17]. Miranda et al. [18] deeply analyzed the sequence specificity of pentatricopeptide repeat (PPR) proteins by enriched k-mers. However, previous work on k-mer counting focuses more on DNA sequences and less on protein ones, which hinders the identification of conserved regions in protein sequences. Consequently, a major challenge is how to design an effective algorithm to ensure that the k-mer counter outputs all k-mers of various lengths and meanwhile such counter is suitable for both DNA and protein sequences.

K-mer counting has been studied extensively, yet still efficient implementations take several hours or even days on large sequence databases. The inverted index, in computer science, is an important data structure that stores a mapping from content to its locations in a database file. Compared to the forward index structure, the inverted index restructures the representative format of files and contributes to a high efficiency of information retrieval, especially on large databases. Inspired by the well applications of inverted index [19], this technique can be explored to alleviate the efficiency problem encountered by previous k-mer counting methods.

In this paper, we propose VLmer that has four steps to efficiently mine k-mers of various lengths (i.e., vl-mers) in both DNA and protein sequences. In the first step, the initial biological sequences are transformed to structure like "(vl-mer, positions)" by the inverted index technique. In the second step, VLmer uses a pattern-growth scheme to generate the candidates of frequent vl-mers. In the third step, a few pruning techniques are explored to prune the unpromising vl-mers. In our experiments, we compare VLmer to ConSpan, which is a modified version of both CCSpan [20] and ConSgen [21] algorithms that are most related to our work, in terms of mining efficiency.

2 Methods

2.1 Preliminaries

The term "k-mer" typically refers to the fixed-length contiguous subsequences. Unfortunately, it's quite difficult to pre-specify a proper length k to count k-mers which have the promise to provide biological insights. We introduce an alternative term "vl-mer" to extend the traditional definition of k-mer as follows.

Definition 1 (vl-mer). *Given two sequences $S_1 = a_1 a_2 \cdots a_i$ and $S_2 = b_1 b_2 \cdots b_j$, where $|S_1| \leq |S_2|$, S_1 is a vl-**mer** of S_2 if S_1 is a contiguous subsequence (i.e., substring) of S_2.*

Based on this definition, the term vl-mer refers to all the possible contiguous subsequences (i.e., all the k-mers of various lengths) of a given sequence. Note that, this extended definition of k-mer uses non-fixed-length instead of fixed-length adopted in traditional definition to constraint the subsequences.

For a vl-mer s, it can be represented by a tuple (s, idx), where idx is the index (i.e., position) of s in sequence database D. For example, assume the database D consists of only a sequence $S = CCTCCCGCCTCA$, the tuple representation of vl-mer $s = CCTC$ is set $\{(CCTC, 0), (CCTC, 7)\}$.

Definition 2 (frequent vl-mer). *Given a minimum support threshold σ, a vl-mer s is frequent in sequence database D if $Sup_D(s) \geq \sigma$.*

For simplicity, we use notation vl-mer to denote *frequent vl-mer* if not explicitly stated.

We are now ready to formulate our problem. Given a sequence database D and a support threshold σ, discover a complete set of vl-mers such that each of which is frequent.

2.2 Inverted Projection

As we discussed earlier, the inverted index data structure is preferable to the forward index one in terms of the speed of the query in information retrieval domain, which motivates us to utilize such inverted index technique to transform the input sequence for vl-mer mining. For clarity, we call inverted index data structure *inverted projection* and revise the notion of it in sequence analysis community as follows.

Definition 3 (inverted projection). *Given a single sequence S as the input sequence database D, suppose all distinct characters of S be a set $I = \{i_1, i_2, \cdots, i_m\}$. The inverted projection of S is a set R expressed in tuples, denoted as $(f, <f\_indexes>)$ such that (1) each tuple consists of a character f ($f \in I$) and its position index(es) appearing in S, and (2) there is a one-to-one correspondence between the f of tuple $(f, <f\_indexes>)$ in R and the element $i_k \in I$, where $1 \leq k \leq m$.*

Note that, unlike the forward index data structure, the inverted projection uses a set of $(f, <f\_indexes>)$ pairs to equivalently represent the input sequence. The element $<f\_indexes>$ is a list consisting of one or more non-negative integers, each of which corresponds to a position number of vl-mers f in the original sequence. Table 1 shows the inverted projection of the sequence $S = CCTCCCGCCTCAGTTCGCGCCGCGCCTCGGCTTGGAACGC$. The shift of input sequence significantly reduces the number of scans of the sequence, so as to minimize the computation cost during the mining process.

Table 1. An inverted projection of sequence S

Character	Index list
C	0, 1, 3, 4, 5, 7, 8, 10, 15, 17, 19, 20, 22, 24, 25, 27, 30, 37, 39
T	2, 9, 13, 14, 26, 31, 32
G	6, 12, 16, 18, 21, 23, 28, 29, 33, 34, 38
A	11, 35, 36

2.3 Candidate Generation of vl-mers

Most k-mer counting algorithms, due to the forward index data structure, need to split all single-length contiguous subsequences as k-mers and compute their occurrence frequencies, rendering the counting operation to be costly. To the best of our knowledge, instead of the forward index, an alternative but clever data structure is the inverted projection. Therefore, we propose a pattern-growth approach based on the inverted projection to generate candidates so as to efficiently mine vl-mers. Based on this approach, each length-k candidate is easily produced by the original sequence, frequent $(k-1)$-mers and their indexes, rather than using a n-gram model [22,23] to enumerate all the possible snippets of input sequences.

2.4 Pruning Techniques

Upon generating a new candidate by the pattern-growth approach based on the inverted projection, we want to immediately check whether or not it is a distinct yet frequent vl-mer. We study some properties of frequent vl-mers in this subsection, which underpin the design of pruning scenarios.

Definition 4 (max-suffix). *Given two sequences $s_1 = a_1 a_2 \cdots a_i$ and $s_2 = b_1 b_2 \cdots b_j$, s_1 is a max-suffix of s_2, denoted as $s_1 \sqsubset_{suf} s_2$, if (1) $|s_1| \geqslant 1$, $|s_2| - |s_1| = 1$; and (2) $a_1 = b_2, a_2 = b_3, \cdots, a_i = b_j$.*

Theorem 1. *Given a sequence s ($|s| \geqslant 2$), suppose s is frequent in sequence database D, then the max-suffix of s is frequent in D.*

Proof (PROOF). Letting F be a set consisting of all subsequences of s, then each element of F, i.e., $\forall s' \in F$ is frequent by virtue of Theorem 1 in [21]. Letting the max-suffix of s be s_{suf}, then s_{suf} satisfies $s_{suf} \in F$. Thus, s_{suf} is frequent. The theorem holds immediately.

Lemma 1. *Given a sequence s ($|s| \geqslant 2$) and a sequence database D, if there exists no frequent max-suffix of s, i.e., $Sup_D(s_{suf}) < \sigma$ holds, then s can be safely pruned.*

Proof (PROOF). The proof of the above lemma is obvious according to Theorem 1, and thus it is omitted here.

By exploring some properties of vl-mers above, three effective pruning techniques, repeated candidate pruning, max-suffix pruning, and support pruning, are introduced to prune the unpromising vl-mers.

2.5   VLmer Algorithm

For delineating VLmer, we first introduce two data structures. First, the inverted index data structure, namely $(f, <f\_indexes>)$ is employed for storing the temporary output vl-mers, where f is a vl-mer itself and $f\_indexes$ represent its indexes in original sequence database. The size of f's indexes indicates the frequency of f. Second, the triple data structure $(f, f.count, B)$ from [21] formalizes the final output vl-mers, where f is also a vl-mer, $f.count$ is the actual support of f, and the last attribute variable "B" takes on the value "Y" by default. The vl-mers of F can be organized into a set $\{\{F_1\}, \{F_2\}, \cdots, \{F_i\}\}$ consisting of i different partitions, each of which is a subset of vl-mers.

Algorithm 1. VLmer(S, σ)

Input: sequence S, support threshold σ

$\quad F_{k-1} \leftarrow \emptyset;$ // initialize F_{k-1} to store $(k-1)$-mers

$\quad F_k \leftarrow \emptyset;$ // initialize F_k to store k-mers

$\quad F_{sta} \leftarrow \emptyset;$ // initialize F_{sta} to store all vl-mers

$\quad F \leftarrow \emptyset;$ // initialize F to store all patterns

1: $F_{k-1} \leftarrow$ 1-mer-gen(S, σ); // inverted projecting

2: $F_{sta} \leftarrow$ sta-gen(F_{k-1});   // standard 1-mers

3: $F \leftarrow F_{sta}$;   // add 1-mers

4: **while** $F_{k-1}.count > 0$ **do**

5:     $F_k \leftarrow$ k-mer-gen(S, F_{k-1}, σ); //generate k-mers

6:     **if** $F_k.count > 0$ **then**

7:         $F_{sta} \leftarrow$ sta-gen(F_k);   // standard k-mers

8:         $F \leftarrow \cup_k F_{sta}$;   // add standardized k-mers

9:     **end if**

10:     $F_{k-1} \leftarrow F_k$;

11: **end while**

Output: F;

Algorithm 1 sketches VLmer that performs the frequent vl-mer mining. As shown, two parameters include a sequence S as the input database and a support threshold σ. Global variable F_{k-1} and F_k store the frequent $(k-1)$-mers and k-mers respectively. Each pattern of both F_{sta} and F is represented by the inverted index structure $(f, <f\_indexes>)$, while F_{sta} and F store vl-mers with the triple data structure $(f, f.count, B)$. During the mining process, F_{sta} is the currently generated single-length vl-mers, and F saves all output vl-mers. Function 1-mer-gen() first produces an inverted projection by projecting the original input sequence, and then generates the frequent 1-mers conforming to the $(f, <f\_indexes>)$ structure (line 1). Such 1-mers are transformed into the standard triple structure $(f, f.count, B)$ (line 2), and then delivered to set F (line 3). Those generated 1-mers are viewed as frequent $(k-1)$-mers to feed function k-mer-gen() for checking the longer k-mers (2-mers). For each non-empty set F_{k-1}, i.e., $F_{k-1}.count > 0$ (line 4), k-mer-gen() produces a set of length-specified frequent k-mers based on S, σ, and the generated $(k-1)$-mers (line 5). Function k-mer-gen() continues its scan until the output set F_k is empty. When a non-empty F_k is generated, each k-mer of them is transformed into the above triple data structure and such a set F_k as a whole is added into set F (line 8). The output of VLmer is such a pattern set $F = \{(f, f.count, B)|f.count \geq \sigma\}$. The main ideas of the above functions are detailed in Subsects. 2.2 to 2.4, and thus we do not recount them here.

3   Results

3.1   Datasets

In our experiments, we used both DNA and protein sequences to study the performance of the VLmer algorithm.

3.2   Effectiveness Study

Unlike previous k-mer counting algorithms, we can conveniently obtain all k-mers of various lengths (i.e., all vl-mers) by performing VLmer algorithm only once. Figure 1 depicts the characteristics of the vl-mers on DNA sequence AL607040. Figure 1(a) shows the distribution of vl-mers against their length for support thresholds (σs) varying from 2 to 6. Note that the smaller σ we choose, the more vl-mers will be generated, which is consistent with previous k-mer counting or sequential pattern mining. From Fig. 1(a), we can see that the number of vl-mers equals or is close to 4^l when the vl-mer-length $l \leq 5$ for all test values of σ. These vl-mers are almost the exhaustive enumeration of all possible combinations of the four base-pares. Intuitively, such vl-mers may be impossible to reveal any biological significance and can be discarded during some sequence analysis, such as sequence classification and TFBS identification. The peak values of the number of vl-mers mainly locate at length 6 ($\sigma \geq 3$) and length 7 ($\sigma = 2$), while they are far less 4^6 and 4^7 respectively. These frequent vl-mers with enough lengths may reflect some biological significance that has been demonstrated [24].

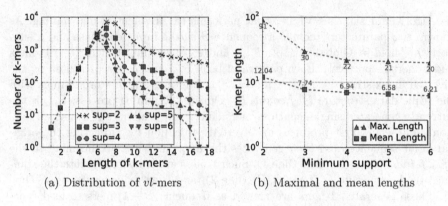

(a) Distribution of vl-mers   (b) Maximal and mean lengths

Fig. 1. vl-mer analysis on DNA sequence AL607040.

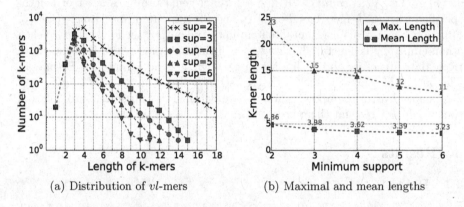

(a) Distribution of vl-mers   (b) Maximal and mean lengths

Fig. 2. vl-mer analysis on protein sequence XP_011987916.

The maximal and mean lengths of vl-mers generated by our algorithm are also presented for varied support thresholds. From Fig. 1(b), the maximal length of mined vl-mers increases from 20 to 91, with the support thresholds lowering from 6 to 2. An interesting point is that when we set the support threshold as $\sigma = 2$, the maximal length of vl-mers reaches 91. These long frequent snippets are often the conserved regions. Figure 1(b) also shows the mean length of vl-mers at each σ value, from which one can see that most mean lengths are consistent with their corresponding σ positions of the curve-vertexes.

We also use protein sequence XP_011987916 as the test dataset, to report the characters of mined vl-mers. Figure 2(a) shows the distribution of vl-mer with varied support thresholds. Except for the snippets of $\sigma = 2$, all protein vl-mers are relatively short at 16 and below, while the lengths of DNA vl-mers as shown in Fig. 1(a) are far greater than 18. It is easy to see that most protein vl-mers fall in the spectrum of 3-mers to 6-mers while DNA vl-mers at the range of 6-mers to 10-mers. Like the trend in Fig. 1(b), when the support threshold tends towards a low value, for example, at $\sigma = 2$ as shown in Fig. 2(b), the length of protein vl-mers increases dramatically.

3.3 Efficiency Study

We assessed VLmer efficiency in both runtime and memory usage for the two real datasets in terms of the support threshold. We compare our approach with Con-Span, which is a modified version of both CCSpan [20] and ConSgen [21] algorithms that are most related to our work. Figure 3(a) presents the running time of the two algorithms at different support thresholds on dataset AL607040. The execution time of ConSpan increases from 52.02 to 356.85 s, while VLmer only takes from 1.26 to 11.15 s. At a low support ($\sigma = 2$), VLmer can be 32 times faster than ConSpan. When we raise the σ, for example, at $\sigma = 6$, our algorithm obtain a better performance that reaches 41 times compared to ConSpan.

Figure 3(b) shows the comparison of the memory consumption between the two algorithms on DNA dataset AL607040 shared with the above experiments. In most cases, VLmer and ConSpan have very similar performance in memory usage, while our algorithm requires a smaller memory space in comparison with ConSpan when the σ value is lowered to 2.

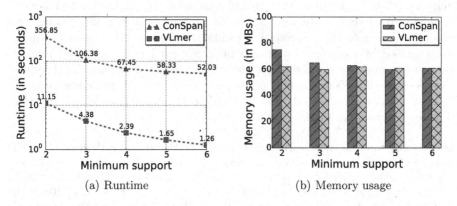

(a) Runtime   (b) Memory usage

Fig. 3. Runtime and memory usage comparison on DNA sequence AL607040.

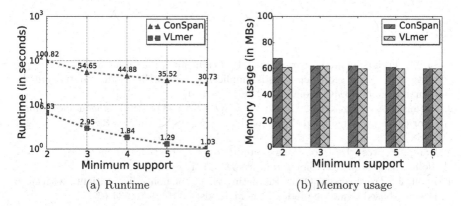

(a) Runtime   (b) Memory usage

Fig. 4. Runtime and memory usage comparison on protein sequence XP_011987916.

We also compare the running time and memory consumption between VLmer and ConSpan on protein sequence XP_011987916. In Fig. 4(a), the execution time of the two algorithms is illustrated. One can see that, the time consumption of ConSpan ranges from 52.02 ($\sigma = 6$) to 356.85 ($\sigma = 2$) s, while VLmer only from 1.03 to 6.63 s. Obviously, VLmer is significantly faster than ConSpan. As shown in Fig. 4(b), the two algorithms occupy a similar memory space.

From the above efficiency study, we conclude that VLmer has better overall performance for both DNA and protein sequences compared to ConSpan.

4   Conclusion

In this paper, we introduced the problem of mining k-mers of various lengths, i.e., vl-mers, in biological sequences. We presented a novel algorithm, VLmer, which efficiently mines all distinct vl-mers. VLmer first utilizes the inverted index technique to project the original sequences. Then, a pattern-growth approach is adopted to generate potential vl-mers, each of which accurately records their occurrence positions in the original sequences. Three pruning techniques, i.e., repeated candidate pruning, max-suffix pruning, and support pruning, are explored to remove the unpromising candidate vl-mers. All possible vl-mers are generated by running VLmer only once. We used both DNA and protein sequences to evaluate the the performance of VLmer. Our experimental results demonstrated that VLmer is able to analyze both DNA and protein sequences. In the future, we plan to study how to push gap constraint into VLmer in order to mine conserved patterns.

Acknowledgements. This work was supported by the Strategic Priority Research Program of the Chinese Academy of Sciences (No. XDB13040700); the National Natural Science Foundation of China (Nos. 91439103; 91529303; 61602460; 31200987); Shanghai Municipal Natural Science Foundation (Nos. 17ZR1406900; 2016M601660); the China Postdoctoral Science Foundation (Nos. 2016M600338; 2016M601660); and the JSPS KAKENHI Grant (No. 15H05707).

References

1. Li, W., Freudenberg, J., Miramontes, P.: Diminishing return for increased mappability with longer sequencing reads: implications of the k-mer distributions in the human genome. BMC Bioinform. **15**(1), 2 (2014)
2. Bremges, A., Singer, E., Woyke, T., Sczyrba, A.: MeCorS: metagenome-enabled error correction of single cell sequencing reads. Bioinformatics **32**(14), 2199–2201 (2016)
3. Hamp, T., Rost, B.: Evolutionary profiles improve protein-protein interaction prediction from sequence. Bioinformatics **31**(12), 1945–1950 (2015)
4. Zhou, J., Troyanskaya, O.G.: Predicting effects of noncoding variants with deep learning-based sequence model. Nat. Methods **12**(10), 931–934 (2015)
5. Kim, D., Song, L., Breitwieser, F.P., Salzberg, S.L.: Centrifuge: rapid and sensitive classification of metagenomic sequences. Genome Res. **26**(12), 1721–1729 (2016)

6. Horwege, S., Lindner, S., Boden, M., Hatje, K., Kollmar, M., Leimeister, C.-A., Morgenstern, B.: Spaced words and KMACS: fast alignment-free sequence comparison based on inexact word matches. Nucleic Acids Res. **42**, W1–W7 (2014)
7. Kurtz, S., Narechania, A., Stein, J.C., Ware, D.: A new method to compute k-mer frequencies and its application to annotate large repetitive plant genomes. BMC Genom. **9**(1), 517 (2008)
8. Marçais, G., Kingsford, C.: A fast, lock-free approach for efficient parallel counting of occurrences of k-mers. Bioinformatics **27**(6), 764–770 (2011)
9. Melsted, P., Pritchard, J.K.: Efficient counting of k-mers in DNA sequences using a bloom filter. BMC Bioinform. **12**(1), 1 (2011)
10. Deorowicz, S., Debudaj-Grabysz, A., Grabowski, S.: Disk-based k-mer counting on a PC. BMC Bioinform. **14**(1), 1 (2013)
11. Rizk, G., Lavenier, D., Chikhi, R.: DSK: k-mer counting with very low memory usage. Bioinformatics **29**(5), 652–653 (2013)
12. Audano, P., Vannberg, F.: Kanalyze: a fast versatile pipelined k-mer toolkit. Bioinformatics **30**(14), 2070–2072 (2014)
13. Deorowicz, S., Kokot, M., Grabowski, S., Debudaj-Grabysz, A.: KMC 2: fast and resource-frugal k-mer counting. Bioinformatics **31**(10), 1569–1576 (2015)
14. Mamun, A.-A., Pal, S., Rajasekaran, S.: KCMBT: a k-mer Counter based on Multiple Burst Trees. Bioinformatics **32**(18), 2783–2790 (2016)
15. Li, R., Zhu, H., Ruan, J., Qian, W., Fang, X., Shi, Z., Li, Y., Li, S., Shan, G., Kristiansen, K., et al.: De novo assembly of human genomes with massively parallel short read sequencing. Genome Res. **20**(2), 265–272 (2010)
16. Shariat, B., Movahedi, N.S., Chitsaz, H., Boucher, C.: HyDA-Vista: towards optimal guided selection of k-mer size for sequence assembly. BMC Genom. **15**(10), S9 (2014)
17. Degnan, P.H., Ochman, H., Moran, N.A.: Sequence conservation and functional constraint on intergenic spacers in reduced genomes of the obligate symbiont buchnera. PLoS Genet. **7**(9), e1002252 (2011)
18. Miranda, R.G., Rojas, M., Montgomery, M.P., Gribbin, K.P., Barkan, A.: RNA binding specificity landscape of the pentatricopeptide repeat protein PPR10. RNA **23**(4), 586–599 (2017)
19. Zhang, R., Xue, R., Yu, T., Liu, L.: Dynamic and efficient private keyword search over inverted index-based encrypted data. ACM Trans. Internet Technol. (TOIT) **16**(3), 21 (2016)
20. Zhang, J., Wang, Y., Yang, D.: CCSpan: mining closed contiguous sequential patterns. Knowl.-Based Syst. **89**, 1–13 (2015)
21. Zhang, J., Wang, Y., Zhang, C., Shi, Y.: Mining contiguous sequential generators in biological sequences. IEEE/ACM Trans. Comput. Biol. Bioinf. **13**(5), 855–867 (2016)
22. Zhang, J., Wang, Y., Wei, H.: An interaction framework of service-oriented ontology learning. In: Proceedings of the 21st ACM International Conference on Information and Knowledge Management, pp. 2303–2306. ACM (2012)
23. Zhang, J., Wang, Y., Yang, D.: Automatic learning common definitional patterns from multi-domain Wikipedia pages. In: 2014 IEEE International Conference on Data Mining Workshop (ICDMW), pp. 251–258. IEEE (2014)
24. Leung, K.-S., Wong, K.-C., Chan, T.-M., Wong, M.-H., Lee, K.-H., Lau, C.-K., Tsui, S.K.: Discovering protein-DNA binding sequence patterns using association rule mining. Nucleic Acids Res. **38**(19), 6324–6337 (2010)

Coestimation of Gene Trees and Reconciliations Under a Duplication-Loss-Coalescence Model

Bo Zhang[1] and Yi-Chieh Wu[2(✉)]

[1] Department of Mathematics, Harvey Mudd College, Claremont, CA, USA
bzhang@hmc.edu
[2] Department of Computer Science, Harvey Mudd College, Claremont, CA, USA
yjw@cs.hmc.edu

Abstract. Accurate gene tree-species tree reconciliation is fundamental to understanding evolutionary processes across species. However, within eukaryotes, the most popular algorithms consider only a restricted set of evolutionary events, typically modeling only duplications and losses or only coalescences. Recent work has unified duplications, losses, and coalescences through an intermediate locus tree; however, the associated reconciliation algorithms assume that the gene tree is known and do not account for gene tree reconstruction error. Here, we demonstrate that independent reconstruction of the gene tree followed by reconciliation substantially degrades accuracy compared to using the true gene tree. To address this challenge, we present DLC-Coestimation, a Bayesian method that simultaneously reconstructs the gene tree and reconciles it with the species tree. We have applied our method on two clades of flies and fungi and demonstrate that it outperforms existing approaches in ortholog, duplication, and loss inference. This work demonstrates the utility of coestimation methods for inferences under joint phylogenetic and population genomic models.

Keywords: Phylogenetics · Reconciliation · Coalescence · Incomplete lineage sorting · Gene duplication and loss

1 Introduction

Phylogenetic tree reconciliation is fundamental to understanding how genes have evolved within and between species. For a *gene family*, or a set of genes with detectable common ancestry, the reconciliation problem takes as input two trees: a *gene tree* that depicts the evolutionary relationships among genes within the gene family, and a *species tree* that depicts the evolutionary relationships among a set of species. We can think of a gene tree as evolving "inside" a species tree, with the *reconciliation* between a gene tree and a species tree explaining this nesting and postulating evolutionary events to account for any observed incongruence.

Electronic supplementary material The online version of this chapter (doi:10.1007/978-3-319-59575-7_18) contains supplementary material, which is available to authorized users.

© Springer International Publishing AG 2017
Z. Cai et al. (Eds.): ISBRA 2017, LNBI 10330, pp. 196–210, 2017.
DOI: 10.1007/978-3-319-59575-7_18

For eukaryotic species, the two most popular reconciliation methods allow for only gene duplication and loss (which we refer to as *duplication-loss models*; Fig. 1A; [1–9]) or only coalescence (which we refer to as *coalescent models*; Fig. 1B; [10–18]). While duplication-loss models can address paralogous families, they assume that incomplete lineage sorting, in which polymorphisms survive several rapid speciations then eventually fix or go extinct in a pattern incongruent with the species tree, is negligible. In contrast, while coalescent models can address such population-related effects, they assume only orthologous genes. Thus, each class of models provides only a partial view of gene family evolution.

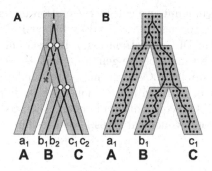

Fig. 1. Different views of gene trees and species trees. (A) In the duplication-loss model, incongruence between the gene tree (black) and species tree (blue) indicates the presence of gene duplications (yellow star) and gene losses (red ×). (B) In a multispecies coalescent model, incongruence between the gene tree and species tree indicates the presence of incomplete lineage sorting (ILS). (Color figure online) [This figure and caption are adapted with permission from Wu et al. [19] and Rasmussen and Kellis [20].]

Recently, Rasmussen and Kellis [20] presented a generative model, DLCoal, for studying duplications, losses, and coalescence and how they interact with one another. In addition to the gene tree and species tree, this model postulated an intermediate *locus tree* that describes how new loci are created and destroyed. Given an input gene tree and species tree, the corresponding algorithm DLCoal-Recon estimates the maximum *a posteriori* reconciliation. More recently, Wu et al. [19] introduced the *labeled coalescent tree* (LCT), which describes the species tree, locus tree, gene tree, and the reconciliations between them in a single structure. The associated reconciliation algorithm DLCpar infers a most parsimonious LCT, that is, one that minimizes the total cost of inferred duplications, losses, and deep coalescence. Both DLCoalRecon and DLCpar showed improved accuracy for inferring evolutionary events compared to duplication-loss methods. And by allowing for paralogs, they were more applicable than coalescent methods.

However, both DLCoalRecon and DLCpar assume that the gene tree and species tree are known and do not account for reconstruction error. In practice, both trees must be estimated, and any reconstruction error is propagated into the reconciliation problem. Gene tree reconstruction is particularly susceptible to error as, unlike species tree reconstruction methods, it cannot benefit from the use of well-behaved gene families or multigene phylogeny methods [21,22]. For duplication-loss models, gene tree reconstruction error has been shown to decrease the accuracy of inferred events [7], motivating several approaches for mitigating error. One class of methods relies on Bayesian inference to simultaneously reconstruct the gene tree topology and its reconciliation with the species tree [3,9]. Another class of methods relies on gene tree error correction; that is, they consider local rearrangements of an initial gene tree to find an error-corrected gene tree with minimum reconciliation cost [4,23,24]. But so far, gene tree reconstruction error has not been considered under a duplication-loss-coalescence model.

To address this shortcoming, we consider evolution under this unified model in which we assume the species tree is fixed and known but allow for errors in the gene tree. We present DLC-Coestimation, a Bayesian method that seeks the maximum *a posteriori* estimate of the gene tree and its reconciliation with the species tree. Finally, we apply DLC-Coestimation to simulated data set of 12 flies and biological data set of 16 fungi and demonstrate its improved performance compared to existing DLC-reconciliation methods. The DLC-Coestimation software is freely available for download at https://www.cs.hmc.edu/~yjw/software/dlc-coestimation.

2   Methods

2.1   Unified Model of Gene Family Evolution

In this section, we review the DLCoal model that unifies the duplication-loss and coalescence models for gene family evolution (Fig. 2; [20]). We then extend it to incorporate sequence evolution. The DLCoal model makes the following assumptions:

1. Any incongruence between the gene tree and species tree topologies can be explained through duplication, loss, and coalescence. Each duplication creates a unique new locus that is unlinked with the original locus, allowing coalescence within the original and new loci to occur independently, and there is no gene conversion between duplicated loci.
2. Duplication and loss events do not fix differently in descendant species; that is, they do not undergo hemiplasy [25]. Equivalently, all duplications and losses either always go extinct or fix in all descendant lineages, allowing us to separate the duplication-loss process from the coalescent process.
3. Each extant species is represented by a single haploid sample; that is, within each gene family, multiple genes from the same extant species are sampled from multiple loci in a single individual as opposed to being sampled from the same locus across multiple individuals.

Assumption 1 is applicable to evolution within eukaryotic species, and assumption 2 was shown to affect only a small number of gene trees in simulation with biologically realistic parameters [20]. We are currently investigating a relaxation of assumption 3 in a separate work.

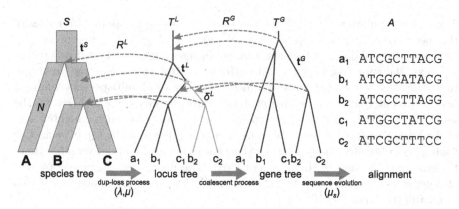

Fig. 2. Generative process. Given a species tree S with known topology and divergence times, a top-down duplication-loss process generates a locus tree T^L. The locus tree contains duplication nodes (star) and daughter nodes δ^L. From the locus tree, a bottom-up coalescent process generates a gene tree T^G. Mappings between the trees represented by R^G and R^L indicate how one tree "fits inside" the other. From the gene tree, sequences evolve to generate an alignment A. [Parts of this figure and caption are adapted with permission from Rasmussen and Kellis [20].]

In addition to the usual gene tree and species tree, the DLCoal model introduces a third kind of tree, the locus tree. Whereas the gene tree represents how gene lineages evolve over time, the locus tree represents how loci are created and destroyed. In brief, the locus tree evolves within the species tree according to the duplication-loss model, and the gene tree evolves within the locus tree according to a modified multispecies coalescent model known as the multilocus coalescent model. We now describe the technical details of the generative process that relates these three trees. Parts of the next three paragraphs are reproduced verbatim, with permission, from Rasmussen and Kellis [20].

We start with a species tree with topology S and branch lengths $\mathbf{t}^S$ expressed in units of time (generations). The topology S is rooted, full, and binary with a set $V(S)$ of nodes and a set $E(S)$ of directed branches (u, v). For node $v \in V(S)$, we let $p(v)$ denote its parent and $e(v)$ denote the branch $(p(v), v)$. We assume that the effective population sizes N are given, and we let $N(v)$ represent the constant population size for branch $e(v)$.

The locus tree is generated by a top-down birth-death process within the species tree [3,9,20,26,27]. We assume a constant rate of gene duplication λ and gene loss μ expressed in events/gene/generation. The locus tree has topology T^L and branch lengths $\mathbf{t}^L$ expressed in units of time (generations). The birth-death process also generates an associated reconciliation R^L that maps each node in T^L to a node (in the case of speciation) or a branch (in the case of duplication) in the species tree S. For simplicity, we often consider R^L as a node-to-node mapping in which, for $v \in V(T^L)$ and $u \in V(S)$, $R^L(v) = e(u)$ is equivalent to $R^L(v) = u$. For each duplication node, one of the children is randomly selected to evolve in the new locus; the set of such *daughter* nodes is denoted δ^L. We define the population sizes N^L for the locus tree using the population sizes of the species tree, that is, for $v \in V(T^L)$, $N^L(v) = N(R^L(v))$. As a postprocessing step, we prune all doomed lineages, that is, lineages with no extant descendants.

Next, the gene tree is generated bottom-up within the locus tree according to a *multilocus coalescent process*, which is similar to a multispecies coalescent process except that the gene tree evolves within the locus tree rather than the species tree, and further, complete coalescence is required at each daughter edge of the locus tree (that is, only one gene lineage is present at the top of each edge leading to a daughter node). The gene tree has topology T^G and divergence times $\mathbf{t}^G$ expressed in units of time (generations). The multilocus coalescent process also generates an associated reconciliation R^G that maps each node in T^G to a branch in the locus tree T^L.

Finally, in addition to the three-tree model, we introduce a fourth object, the alignment data A, which evolves along the gene tree according to a substitution model (e.g. JC [28], HKY [29], GTR [30]). For simplicity, we consider only a substitution rate of μ_s in substitutions/site/generation here though our model allows for more complexity.

2.2   Coestimation of Gene Trees and Reconciliations

Along with this unified model, Rasmussen and Kellis [20] developed the reconciliation algorithm DLCoalRecon that, given a gene tree topology T^G, a species tree topology S, and model parameters $\theta = (\mathbf{t}^S, N, \lambda, \mu)$, infers the maximum *a posteriori* reconciliation between the gene tree and species tree as captured by the three-tree reconciliation structure $\mathbb{R} = (T^L, R^G, R^L, \delta^L)$. DLCoalRecon assumed that the gene tree topology was previously inferred from a sequence alignment A using existing phylogenetic methods.

In contrast, we now describe our DLC-Coestimation algorithm for simultaneously estimating the gene tree topology and reconciliation. Our algorithm takes as input a sequence alignment A, a species tree topology S, and model parameters $\theta = (\mathbf{t}^S, N, \lambda, \mu, \mu_s)$. Our goal is to infer the maximum *a posteriori* gene tree topology T^G and reconciliation $\mathbb{R} = (T^L, R^G, R^L, \delta^L)$:

$$\hat{T^G}, \hat{\mathbb{R}} = \underset{T^G, \mathbb{R}}{\operatorname{argmax}} P(T^G, \mathbb{R}|A, S, \theta) \tag{1}$$

As a reminder, T^L is the locus tree topology, R^G the reconciliation between the gene tree and the locus tree, R^L the reconciliation between the locus tree and the species tree, and δ^L the set of daughter nodes.

Because the alignment A is given, maximizing the above posterior probability is equivalent to maximizing the joint probability $P(T^G, \mathbb{R}, A|S, \theta) = P(T^G, T^L, R^G, R^L, \delta^L, A|S, \theta)$. Next, we introduce the gene tree and locus tree branch lengths $\mathbf{t}^G$ and $\mathbf{t}^L$ and take into account conditional independencies to separate the variables for the locus tree, gene tree, and alignment (Supplemental Sect. S1):

$$P(T^G, \mathbb{R}, A|S, \theta) = \iint P(T^L, \mathbf{t}^L, R^L, \delta^L|S, \theta^S)$$
$$\times P(T^G, R^G, \mathbf{t}^G|T^L, \mathbf{t}^L, \delta^L, \mathbf{N}^L) \qquad (2)$$
$$\times P(A|T^G, \mathbf{t}^G, \underline{\mu}_s) \, d\mathbf{t}^G \, d\mathbf{t}^L$$

That is, we approach our optimization problem using Bayesian inference, in which we decompose the joint probability into a prior and a likelihood. Our model prior is further factored into two terms: the reconciled locus tree prior and the reconciled gene tree prior. Thus, each term of (2) corresponds to one component of the generative evolutionary process presented in Sect. 2.1:

- $P(T^L, \mathbf{t}^L, R^L, \delta^L|S, \theta^S)$, where $\theta^S = (\mathbf{t}^S, N, \lambda, \mu)$, captures the duplication-loss process that generates the locus tree from the species tree. We decompose this probability using the factorization of Rasmussen and Kellis [20] into

$$P(\delta^L|T^L, R^L, S) \times P(T^L, R^L|S, \theta^S) \times P(\mathbf{t}^L|T^L, R^L, S, \theta^S). \qquad (3)$$

The first term $P(\delta^L|T^L, R^L, S)$ captures the fact that there are two ways to choose a daughter node for each duplication in the locus tree. The second term $P(T^L, R^L|S, \theta^S)$ captures the process of generating a reconciled locus tree topology from the species tree, and the third term $P(\mathbf{t}^L|T^L, R^L, S, \theta^S)$ captures the distribution of locus tree branch lengths. Each of these terms has been previously derived (see [3,9,20,31,32]).

- $P(T^G, R^G, \mathbf{t}^G|T^L, \mathbf{t}^L, \delta^L, \mathbf{N}^L)$ captures the multilocus coalescent process that generates the gene tree from the locus tree. We decompose this into

$$P(T^G, R^G|T^L, \mathbf{t}^L, \delta^L, \mathbf{N}^L) \times P(\mathbf{t}^G|T^G, R^G, T^L, \mathbf{t}^L, \mathbf{N}^L). \qquad (4)$$

Similar to above, the first term $P(T^G, R^G|T^L, \mathbf{t}^L, \delta^L, \mathbf{N}^L)$ captures the process of generating a reconciled gene tree topology from the locus tree and has been previously derived [20]. The second term $P(\mathbf{t}^G|T^G, R^G, T^L, \mathbf{t}^L, \mathbf{N}^L)$ captures the distribution of gene tree branch lengths and is derived in Supplemental Sect. S2.

- $P(A|T^G, \mathbf{t}^G, \mu_s)$ captures the process that generates the sequence alignment from the gene tree. This probability is the likelihood under a specific substitution model and can be computed efficiently using Felsenstein's pruning algorithm [33]. We compute this term using the Phylogenetic Likelihood Library [34].

2.3 Efficient Implementation

Putting the above components together, we factor (2) into

$$P(T^G, \mathbb{R}, A|S, \theta) = \iint P(\delta^L|T^L, R^L, S) \times P(T^L, R^L|S, \theta^S)$$
$$\times P(T^G, R^G|T^L, \mathbf{t}^L, R^L, \delta^L, \mathbf{N}^L) \times P(A|T^G, \mathbf{t}^G, \mu_s)$$
$$\times P(\mathbf{t}^G|T^G, R^G, T^L, \mathbf{t}^L, \mathbf{N}^L)$$
$$\times P(\mathbf{t}^L|T^L, R^L, S, \theta^S) \, d\mathbf{t}^G \, d\mathbf{t}^L.$$
$$(5)$$

We perform the integration using the Monte Carlo method by sampling from $P(\mathbf{t}^G|T^G, R^G, T^L, \mathbf{t}^L, \mathbf{N}^L)$ and $P(\mathbf{t}^L|T^L, R^L, S, \theta^S)$. We sample over $\mathbf{t}^G$ as described in Supplemental Sect. S2, and and we sample over $\mathbf{t}^L$ as described in Arvestad et al. [3] and Rasmussen and Kellis [9].

Using (5), we can compute the probability of any proposed gene tree topology T^G and reconciliation $\mathbb{R} = (T^L, R^G, R^L, \delta^L)$. To estimate the maximum *a posteriori* T^G and $\mathbb{R}$, we heuristically search over the space of possible solutions using an iterative hill-climbing approach. We initialize our search similarly to DLCoalRecon, with an initial gene tree topology T^G obtained using any existing phylogenetic method and a reconciliation $\mathbb{R}$ that has locus tree topology T^L congruent with the gene tree T^G, mappings R^G and R^L that are Last Common Ancestor (LCA) mappings [2], and, if needed, randomly chosen daughter nodes δ^L. Next, we iteratively improve the locus tree and gene tree components. That is, we fix the gene tree components and optimize for the locus tree components; then, we fix the locus tree components and optimize for the gene tree components. We repeat this process for a user-specified number of iterations, and our algorithm outputs the proposed gene tree and reconciliation with the highest posterior probability.

When optimizing for the locus tree components, we fix the gene tree topology T^G and search for the reconciliation $\mathbb{R}$ that maximizes $P(T^G, \mathbb{R}, A|S, \theta)$. Since T^G is fixed, this problem reduces to the three-tree model of Rasmussen and Kellis [20], in which we maximize $P(\mathbb{R}|T^G, S, \theta)$, and can be solved using the associated DLCoalRecon algorithm.

When optimizing for the gene tree components, we fix the locus tree topology T^L, the locus tree-species tree reconciliation R^L, and daughter nodes δ^L and search for the gene tree topology T^G and the gene tree-locus tree reconciliation R^G that maximize the joint probability

$$
\iint P(T^G, R^G|T^L, \mathbf{t}^L, R^L, \delta^L, \mathbf{N}^L) \times P(A|T^G, \mathbf{t}^G, \mu_s)
$$
$$
\times P(\mathbf{t}^G|T^G, R^G, T^L, \mathbf{t}^L, \mathbf{N}^L) \times P(\mathbf{t}^L|T^L, R^L, S, \theta^S)\, d\mathbf{t}^G\, d\mathbf{t}^L.
\tag{6}
$$

By comparing (6) with (5), we see that $P(\delta^L|T^L, R^L, S)$ and $P(T^L, R^L|S, \theta^S)$ are not needed because we have fixed T^L, R^L, and δ^L. To maximize this probability, we search over the space of possible T^G and R^G, again using a hill-climbing approach. For each proposal, we either propose a new gene tree topology T^G using subtree pruning and regrafting (SPR) or propose a new reconciliation R^G by rearranging the mapping [35]. Finally, we use the standard approach of log probabilities to prevent underflow in our calculations and additionally allow for a regularization parameter β (default $\beta = 0.01$) to weight the relative contributions of the prior and likelihood. Specifically, when optimizing (6), we evaluate the expected value of $\log(P(T^G, R^G|T^L, \mathbf{t}^L, R^L, \delta^L, \mathbf{N}^L)) + \beta \log(P(A|T^G, \mathbf{t}^G, \mu_s))$ using the Monte Carlo method over $\mathbf{t}^G$ and $\mathbf{t}^L$.

3 Results

3.1 Simulated Data Set of 12 Flies

We applied our algorithm to a simulated 12 *Drosophila* data set that has been previously used to evaluate DLC-reconciliation algorithms [19,20]. This data set used the species tree (Supplemental Fig. S1A) of the *Drosophila* 12 Genomes Consortium [36] with estimated divergence times [37], gene duplication and loss rates of 0.0012 events/gene/million years [38], a generation time of 10 generations/yr [39,40], and effective population sizes of 1–100 million individuals. While *Drosophila melanogaster* is estimated to have an effective population size of $\sim$1.15 million [41], the data set includes a wide range of population sizes to induce various levels of incongruence. For each population size, 500 gene trees were simulated.

To introduce gene tree error, for each gene tree, we simulated alignments of 1000 nucleotides under a HKY model [29] with a substitution rate of 5×10^{-9} substitutions/site/generation [42,43] and using seq-gen [44]. We then reconstructed gene trees using RAxML [45] and TreeFix [24] and either reconciled gene trees with the species tree using DLCoalRecon [20] and DLCpar [19], or using DLC-Coestimation with the different reconstructed trees as initial estimates (Fig. 3).

We find that reconciling reconstructed gene trees instead of true gene trees substantially degrades performance. For example, for an effective population size of 25 million, our 500 simulated gene trees contain 232 duplications, 216 losses, and $33, 182$ pairs of orthologous genes. DLCoalRecon applied to true (simulated) gene trees yields similar numbers of events and orthologs at high sensitivity and precision, with 242 duplications (90.5% sensitivity, 86.8% precision), 216 losses (98.6%, 98.6%), $33, 285$ ortholog pairs (99.7%, 99.4%), and 96.0% locus tree topological accuracy. In contrast, DLCoalRecon applied to reconstructed (TreeFix) gene trees yields decreased metrics across every dimension, with 239 duplications (69.0%, 66.9%), 339 losses (98.1%, 62.5%), $32, 298$ ortholog pairs (96.8%, 99.4%), and 90.8% topological accuracy. Though impressively, DLCoalRecon is able to achieve these metrics despite poor gene tree topological accuracy (8.8%), and performance across these metrics does not decrease substantially even as gene tree reconstruction degrades with population size. While DLCpar outperforms DLCoalRecon for both true and reconstructed gene trees, the performance gap between reconciliations on true and reconstructed gene trees remains.

By simultaneously reconstructing and reconciling gene trees, DLC-Coestimation improves over existing DLC-reconciliation methods. For example, for the same data set above, DLC-Coestimation slightly increases gene tree topological accuracy (10.2%). More impressively, across other metrics, its performance exceeds DLCoalRecon and either exceeds or is comparable to DLCpar, with 237 duplications (77.6%, 75.9%), 231 losses (98.1%, 91.8%), $33, 029$ ortholog pairs (98.8%, 99.3%), and 93.2% locus tree topological accuracy.

While DLC-Coestimation outperforms DLCoalRecon for every population size, it underperforms DLCpar for small populations and outperforms DLCpar

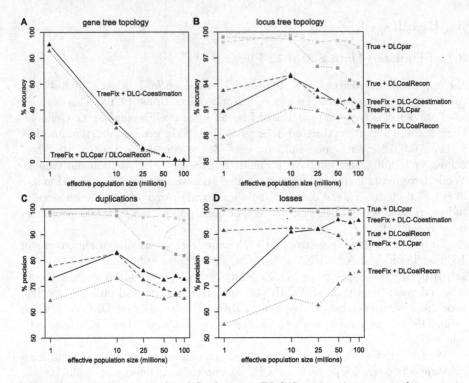

Fig. 3. Evaluation on a simulated fly data set. DLC-Coestimation was used to reconstruct and reconcile gene trees given simulated alignments and initialized with reconstructed (TreeFix) gene trees. DLCpar and DLCoalRecon were also used to reconcile reconstructed gene trees. DLC-Coestimation improves over DLCoalRecon in both (A, B) the accuracy of reconstructed gene tree and locus tree topologies and (C, D) the precision of inferred duplications and losses. DLC-Coestimation performance against DLCpar varies with population size. For comparison, DLCpar and DLCoalRecon were also used to reconcile simulated (true) gene trees. Note that, to highlight differences between programs, y-axes for these plots may not start at 0. Additional results can be found in Supplemental Fig. S2. [Simulated data sets and True+DLCoalRecon results are from Rasmussen and Kellis [20]. True+DLCpar results are from Wu et al. [19].]

for large populations. As the population size increases with a constant generation time, ILS rate increases. So our finding suggests that DLC-Coestimation is better able to handle data sets with low phylogenetic signal, a type of data set that will become increasingly prevalent as we sequence denser clades. DLCpar relies on a parsimony framework, so its performance may increase with different event costs. However, choosing appropriate event costs remains a challenge.

For increased ILS rates, DLC-Coestimation is more robust than DLCoal-Recon to gene tree reconstruction errors (Supplemental Fig. S2). For example, for the same data set above, DLCoalRecon applied to RAxML trees yields 574 duplications (25.4%, 10.3%), 3425 losses (65.7%, 4.1%), 19,407 ortholog pairs (57.9%, 99.0%), and 22.0% locus tree topological accuracy. (DLCpar is NP-hard [46], so we did not apply it to RAxML trees as the amount of incongruence between RAxML gene trees and the known species tree makes the algorithm too inefficient.) In contrast, DLC-Coestimation initialized with the same gene trees yields 232 duplications (67.7%, 67.7%), 254 losses (96.3%, 81.9%), 32,798 ortholog pairs (97.8%, 99.0%), and 90.2% topological accuracy. That is, despite starting from a worse initial gene tree estimate, DLC-Coestimation performs comparably to DLCoalRecon applied to better TreeFix gene trees.

DLC-Coestimation errors could be attributed either to a limit in the power of our model to identify the correct reconciliation or to limitations in our present implementation of a heuristic search strategy. However, while we find that increasing the number of searches increases performance for a population of 25 million, it has minimal effect on performance for a population of 1 million. This finding suggests that better search heuristics could lead to performance increases in some cases but also that our model may not always be able to identify the correct reconciliation.

3.2   Biological Data Set of 16 Fungi

We also assessed the performance of DLC-Coestimation on a biological data set of 5351 gene trees across 16 fungal genomes (Supplemental Fig. S1B; [47]). This data set has been used extensively by ourselves and others to evaluate several phylogenetic algorithms [9,19,20,24,48].

For this comparison, we ran DLC-Coestimation using parameters previously estimated for DLCoalRecon and a substitution rate of 3.3×10^{-10} substitutions/site/generation [49]. Additionally, as we have previously found that DLC-reconciliation methods outperform non-ILS-aware methods, we focus here on comparing DLC-Coestimation to DLCpar and DLCoalRecon. As the truth is not known for real data, we used several informative metrics to assess the quality of our inferences (Table 1, Supplemental Table S1).

Table 1. Evaluation on a real fungal data set.

Program[a]	% orths[b]	# orths[c]	# dups[c]	# losses[c]	DCS[d]
DLC-coestimation	99.1	583,943	4,375	4,992	0.944
DLCpar	99.1	590,113	4,535	5,535	0.899
DLCoalRecon	99.0	583,490	4,472	5,378	0.927

[a]DLC-reconciliation methods were applied to reconstructed TreeFix trees. Additional results can be found in Supplemental Table S1. [DLCpar and DLCoalRecon results are from Wu et al. [19].]
[b]Percentage of 183,374 syntenic orthologs recovered.
[c]Number of pairwise orthologs, duplications, and losses inferred across all gene trees.
[d]Average duplication consistency score. Scores range from 0 to 1, with a higher score indicating more consistent duplications.

Our first metric assesses the ability to recover syntenic orthologs (one-to-one homologs with conserved gene order that are highly likely to be orthologous).

We find that all DLC-reconciliation programs recover a similar percentage of syntenic orthologs (99.0–99.1%) and infer similar number of orthologs (0.1–1.1% difference).

Our second metric evaluates the total number of inferred duplications and losses. We find that DLC-Coestimation infers substantially fewer duplications (2.2–3.6% difference) and losses (7.8–10.8%) than DLCpar and DLCoalRecon, suggesting that DLC-Coestimation is better able to remove spurious duplication and loss events that result from ILS.

Our third metric considers the duplication consistency score [50], which measures the plausibility of inferred duplications. For each duplication node, this score computes the percentage of species overlap in the two child subtrees; the assumption is that erroneous duplications are often followed by compensating losses and therefore yield a low score. We find that DLC-Coestimation slightly outperforms DLCpar and DLCoalRecon, with a higher average score and a consistently higher score distribution (Supplemental Fig. S3).

4   Discussion

In this work, we have presented a new method DLC-Coestimation for simultaneously reconstructing and reconciling gene trees under a duplication-loss-coalescent-model. Our analysis shows that DLC-Coestimation yields improved inferences compared to applying existing DLC-reconciliation methods on gene trees reconstructed with popular and top-performing methods.

We envision several possible future improvements to DLC-Coestimation. One limitation of our approach is that it currently performs substantially slower than independent reconstruction and reconciliation (Supplemental Table S2). Therefore, we might reasonably question whether the increased accuracy of DLC-Coestimation is worth the additional computational effort. However, we note that because DLCpar relies on an exhaustive search over the space of reconciliations, it is NP-hard and in particular does not scale well as gene tree-species tree incongruence increases. In contrast, though DLC-Coestimation relies on a heuristic search, its hill-climbing approach is guaranteed to complete, and our experiments demonstrate that the search often finds an accurate solution in practice. Additionally, we have not implemented many optimizations so far; for example, although we use optimized libraries, computing the sequence likelihood is orders of magnitude slower than computing the gene tree and locus tree topology prior. As our heuristic search makes local rearrangements to the gene tree, locus tree, and reconciliations between these and the species tree, we should be able to reuse many of our computations between proposals.

Furthermore, we have yet to investigate the effect of our regularization hyperparameter that trades-off sequence likelihood with the topology prior. While we believe that hyperparameter tuning could further improve performance, one challenge is that properly selecting a hyperparameter would require several manually-curated gene trees for validation. Alternatively, we are currently investigating whether a hyperparameter exists that works well across a range of data sets.

In our study, we have also made several assumptions, for example, that model parameters can be estimated accurately by other methods. While in most cases these existing methods [15,51] would suffice, another research direction would be to simultaneously optimize model parameters along with the gene tree and reconciliation. There has also been recent work on jointly inferring species trees and gene trees [52] and gene trees and sequence alignments [53] under the duplication-loss-only model, indicating that incorporating further coestimation may be possible under the duplication-loss-coalescence model.

Acknowledgments. We thank Matthew D. Rasmussen, Ran Libeskind-Hadas, and Mark Huber for helpful comments, feedback, and discussions. This work was supported by funds from the Department of Computer Science and the Dean of Faculty of Harvey Mudd College.

References

1. Goodman, M., Czelusniak, J., Moore, G.W., Romero-Herrera, A.E., Matsuda, G.: Fitting the gene lineage into its species lineage, a parsimony strategy illustrated by cladograms constructed from globin sequences. Syst. Zool. **28**(2), 132–163 (1979)
2. Page, R.D.M.: Maps between trees and cladistic analysis of historical associations among genes, organisms, and areas. Syst. Biol. **43**(1), 58–77 (1994)
3. Arvestad, L., Berglund, A.-C., Lagergren, J., Sennblad, B.: Gene tree reconstruction and orthology analysis based on an integrated model for duplications and sequence evolution. In: Proceedings of the Eighth Annual International Conference on Research in Computational Molecular Biology, RECOMB 2004, pp. 326–335. ACM, New York (2004)
4. Durand, D., Hallórsson, B.V., Vernot, B.: A hybrid micro-macroevolutionary approach to gene tree reconstruction. J. Comput. Biol. **13**(2), 320–335 (2006)
5. Górecki, P., Tiuryn, J.: DLS-trees: a model of evolutionary scenarios. Theoret. Comput. Sci. **359**(1–3), 378–399 (2006)
6. Li, H., Coghlan, A., Ruan, J., Coin, L.J., H'erich'e, J.-K., Osmotherly, L., Li, R., Liu, T., Zhang, Z., Bolund, L., Wong, G.K.-S., Zheng, W., Dehal, P., Wang, J., Durbin, R.: TreeFam: a curated database of phylogenetic trees of animal gene families. Nucleic Acids Res. **34**, 572–580 (2006)
7. Hahn, M.: Bias in phylogenetic tree reconciliation methods: implications for vertebrate genome evolution. Genome Biol. **8**(7), 141 (2007)
8. Rasmussen, M.D., Kellis, M.: Accurate gene-tree reconstruction by learning gene- and species-specific substitution rates across multiple complete genomes. Genome Res. **17**(12), 1932–1942 (2007)
9. Rasmussen, M.D., Kellis, M.: A Bayesian approach for fast and accurate gene tree reconstruction. Mol. Biol. Evol. **28**(1), 273–290 (2011)
10. Kingman, J.F.C.: The coalescent. Stoch. Proc. Appl. **13**(3), 235–248 (1982)
11. Pamilo, P., Nei, M.: Relationships between gene trees and species trees. Mol. Biol. Evol. **5**(5), 568–583 (1988)
12. Takahata, N.: Gene genealogy in three related populations: consistency probability between gene and population trees. Genetics **122**(4), 957–966 (1989)
13. Maddison, W.P.: Gene trees in species trees. Syst. Biol. **46**(3), 523–536 (1997)
14. Rosenberg, N.A.: The probability of topological concordance of gene trees and species trees. Theor. Popul. Biol. **61**(2), 225–247 (2002)

15. Rannala, B., Yang, Z.: Bayes estimation of species divergence times and ancestral population sizes using DNA sequences from multiple loci. Genetics **164**(4), 1645–1656 (2003)
16. Degnan, J.H., Rosenberg, N.A.: Gene tree discordance, phylogenetic inference and the multispecies coalescent. Trends Ecol. Evol. **24**(6), 332–340 (2009)
17. Wakeley, J.: Coalescent Theory: An Introduction. Roberts & Company Publishers, Greenwood Village (2009)
18. Heled, J., Drummond, A.J.: Bayesian inference of species trees from multilocus data. Mol. Biol. Evol. **27**(3), 570–580 (2010)
19. Wu, Y.-C., Rasmussen, M.D., Bansal, M.S., Kellis, M.: Most parsimonious reconciliation in the presence of gene duplication, loss, and deep coalescence using labeled coalescent trees. Genome Res. **24**(3), 475–486 (2014)
20. Rasmussen, M.D., Kellis, M.: Unified modeling of gene duplication, loss, and coalescence using a locus tree. Genome Res. **22**, 755–765 (2012)
21. Delsuc, F., Brinkmann, H., Philippe, H.: Phylogenomics and the reconstruction of the tree of life. Nat. Rev. Genet. **6**(5), 361–375 (2005)
22. Burleigh, J.G., Bansal, M.S., Eulenstein, O., Hartmann, S., Wehe, A., Vision, T.J.: Genome-scale phylogenetics: inferring the plant tree of life from 18,896 gene trees. Syst. Biol. **60**(2), 117–125 (2011)
23. Górecki, P., Eulenstein, O.: A linear time algorithm for error-corrected reconciliation of unrooted gene trees. In: Chen, J., Wang, J., Zelikovsky, A. (eds.) ISBRA 2011. LNCS, vol. 6674, pp. 148–159. Springer, Heidelberg (2011). doi:10.1007/978-3-642-21260-4_17
24. Wu, Y.-C., Rasmussen, M.D., Bansal, M.S., Kellis, M.: TreeFix: statistically informed gene tree error correction using species trees. Syst. Biol. **62**(1), 110–120 (2013)
25. Avise, J.C., Robinson, T.J.: Hemiplasy: a new term in the lexicon of phylogenetics. Syst. Biol. **57**(3), 503–507 (2008)
26. Dubb, L.: A likelihood model of gene family evolution. Ph.D. thesis, University of Washington, Seattle (2005)
27. Åkerborg, Ö., Sennblad, B., Arvestad, L., Lagergren, J.: Simultaneous Bayesian gene tree reconstruction and reconciliation analysis. Proc. Natl. Acad. Sci. U.S.A. **106**(14), 5714–5719 (2009)
28. Jukes, T.H., Cantor, C.R.: Evolution of protein molecules. In: Munro, M.N. (ed.) Mammalian Protein Metabolism, vol. III, pp. 21–132. Academic Press, New York (1969)
29. Hasegawa, M., Kishino, H., Yano, T.-A.: Dating of the human-ape splitting by a molecular clock of mitochondrial DNA. J. Mol. Evol. **22**(2), 160–174 (1985)
30. Tavaré, S.: Some probabilistic and statistical problems in the analysis of DNA sequences. Lect. Math. Life Sci. **17**, 57–86 (1986)
31. Arvestad, L., Berglund, A.-C., Lagergren, J., Sennblad, B.: Bayesian gene/species tree reconciliation and orthology analysis using MCMC. Bioinformatics **19**(Suppl. 1), 7–15 (2003)
32. Arvestad, L., Lagergren, J., Sennblad, B.: The gene evolution model and computing its associated probabilities. J. ACM **56**(2), 1–44 (2009)
33. Felsenstein, J.: Inferring Phylogenies, 2nd edn. Sinauer Associates, Sunderland (2003)
34. Flouri, T., Izquierdo-Carrasco, F., Darriba, D., Aberer, A.J., Nguyen, L.-T., Minh, B.Q., Von Haeseler, A., Stamatakis, A.: The phylogenetic likelihood library. Syst. Biol. **64**(2), 356–362 (2015)

35. Doyon, J.-P., Chauve, C., Hamel, S.: An efficient method for exploring the space of gene tree/species tree reconciliations in a probabilistic framework. IEEE/ACM Trans. Comput. Biol. Bioinform. **9**(1), 26–39 (2012)

36. *Drosophila* 12 Genomes Consortium: Evolution of genes and genomes on the *Drosophila* phylogeny. Nature **450**(7167), 203–218 (2007)

37. Tamura, K., Subramanian, S., Kumar, S.: Temporal patterns of fruit fly (*Drosophila*) evolution revealed by mutation clocks. Mol. Biol. Evol. **21**(1), 36–44 (2004)

38. Hahn, M.W., Han, M.V., Han, S.-G.: Gene family evolution across 12 *Drosophila* genomes. PLoS Genet. **3**(11), 197 (2007)

39. Sawyer, S.A., Hartl, D.L.: Population genetics of polymorphism and divergence. Genetics **132**(4), 1161–1176 (1992)

40. Pollard, D.A., Iyer, V.N., Moses, A.M., Eisen, M.B.: Widespread discordance of gene trees with species tree in *Drosophila*: evidence for incomplete lineage sorting. PLoS Genet. **2**(10), 173 (2006)

41. Charlesworth, B.: Fundamental concepts in genetics: effective population size and patterns of molecular evolution and variation. Nat. Rev. Genet. **10**, 195–205 (2009)

42. Kimura, M.: Evolutionary rate at the molecular level. Nature **217**(5129), 624–26 (1968)

43. Haag-Liautard, C., Dorris, M., Maside, X., Macaskill, S., Halligan, D.L., Charlesworth, B., Keightley, P.D.: Direct estimation of per nucleotide and genomic deleterious mutation rates in *Drosophila*. Nature **445**(7123), 82–85 (2007)

44. Rambaut, A., Grassly, N.C.: Seq-Gen: an application for the Monte Carlo simulation of DNA sequence evolution along phylogenetic trees. Comput. Appl. Biosci. **13**, 235–238 (1997)

45. Stamatakis, A.: RAxML-VI-HPC: maximum likelihood-based phylogenetic analyses with thousands of taxa and mixed models. Bioinformatics **22**(21), 2688–2690 (2006)

46. Bork, D., Cheng, R., Wang, J., Sung, J., Libeskind-Hadas, R.: On the computational complexity of the maximum parsimony reconciliation problem in the duplication-loss-coalescence model. Algorithm Mol. Biol. **12**(6) (2017). https://almob.biomedcentral.com/articles/10.1186/s13015-017-0098-8

47. Butler, G., Rasmussen, M.D., Lin, M.F., Santos, M.A.S., Sakthikumar, S., Munro, C.A., Rheinbay, E., Grabherr, M., Forche, A., Reedy, J.L., Agrafioti, I., Arnaud, M.B., Bates, S., Brown, A.J.P., Brunke, S., Costanzo, M.C., Fitzpatrick, D.A., de Groot, P.W.J., Harris, D., Hoyer, L.L., Hube, B., Klis, F.M., Kodira, C., Lennard, N., Logue, M.E., Martin, R., Neiman, A.M., Nikolaou, E., Quail, M.A., Quinn, J., Santos, M.C., Schmitzberger, F.F., Sherlock, G., Shah, P., Silverstein, K.A.T., Skrzypek, M.S., Soll, D., Staggs, R., Stansfield, I., Stumpf, M.P.H., Sudbery, P.E., Srikantha, T., Zeng, Q., Berman, J., Berriman, M., Heitman, J., Gow, N.A.R., Lorenz, M.C., Birren, B.W., Kellis, M., Cuomo, C.A.: Evolution of pathogenicity and sexual reproduction in eight Candida genomes. Nature **459**(7247), 657–662 (2009)

48. Wapinski, I., Pfeffer, A., Friedman, N., Regev, A.: Natural history and evolutionary principles of gene duplication in fungi. Nature **449**(7158), 54–61 (2007)

49. Lynch, M., Sung, W., Morris, K., Coffey, N., Landry, C.R., Dopman, E.B., Dickinson, W.J., Okamoto, K., Kulkarni, S., Hartl, D.L., Thomas, W.K.: A genome-wide view of the spectrum of spontaneous mutations in yeast. Proc. Natl. Acad. Sci. U.S.A. **105**(27), 9272–9277 (2008)

50. Vilella, A.J., Severin, J., Ureta-Vidal, A., Heng, L., Durbin, R., Birney, E.: EnsemblCompara GeneTrees: complete, duplication-aware phylogenetic trees in vertebrates. Genome Res. **19**(2), 327–335 (2009)
51. Hahn, M.W., De Bie, T., Stajich, J.E., Nguyen, C., Cristianini, N.: Estimating the tempo and mode of gene family evolution from comparative genomic data. Genome Res. **15**(8), 1153–1160 (2005)
52. Boussau, B., Szöllősi, G.J., Duret, L., Gouy, M., Tannier, E., Daubin, V.: Genome-scale coestimation of species and gene trees. Genome Res. **23**(2), 323–330 (2013)
53. Liu, K., Raghavan, S., Nelesen, S., Linder, C.R., Warnow, T.: Rapid and accurate large-scale coestimation of sequence alignments and phylogenetic trees. Science **324**(5934), 1561–1564 (2009)

A Median Solver and Phylogenetic Inference Based on DCJ Sorting

Ruofan Xia[1,2], Jun Zhou[2], Lingxi Zhou[2], Bing Feng[2], and Jijun Tang[1,2,3]([✉])

[1] School of Computer Science and Technology, Tianjin University, Tianjin, China
jtang@cse.sc.edu
[2] University of South Carolina, Columbia, SC 29205, USA
[3] Institute of Computational Biology, Tianjin University, Tianjin, China

Abstract. Genome rearrangement is known as one of the main evolutionary mechanisms on the genomic level. Phylogenetic analysis based on rearrangement played a crucial role in biological research in the past decades, especially with the increasing availability of fully sequenced genomes. In general, phylogenetic analysis tries to solve two problems: Small Parsimony Problem (SPP) and Big Parsimony Problem (BPP). Maximum parsimony is a popular approach for SPP and BPP which relies on iteratively solving a NP hard problem, the median problem. As a result, current median solvers and phylogenetic inference methods based on the median problem all face serious problems on scalability and cannot be applied to datasets with large and distant genomes.

In this paper, we propose a new median solver for gene order data that combines double-cut-and-join sorting with the Simulated Annealing algorithm (SAMedian). Based on the median solver, we built a new phylogenetic inference method to solve both SPP and BPP problems. Our experimental results show that the new median solver presents an excellent performance on simulated datasets and the phylogenetic inference tool built based on the new median solver has a better performance than other existing methods.

Keywords: Simulated annealing · Phylogenetic inference · Median problem · Small phylogeny problem · Big phylogeny problem

1 Introduction

A genome is used to represent the complete set of DNA (genes) in an organism. Different features and characteristics from genes have been used to reconstruct phylogenetic trees and ancestral genomes, including gene sequence, copy number [1] and rearrangement events [2–5]. The most common rearrangement events include reversal, fission, fusion, transposition, and translocation. Sankoff and Blanchette [6] proposed the first algorithm to reconstruct phylogeny from genome rearrangement events. Since then, genome rearrangement analysis is widely used by biologists, mathematicians, and computer scientists. Various methods [7] have been developed to reconstruct phylogenetic trees and ancestral genomes from gene order, including parsimony-based methods such as

© Springer International Publishing AG 2017
Z. Cai et al. (Eds.): ISBRA 2017, LNBI 10330, pp. 211–222, 2017.
DOI: 10.1007/978-3-319-59575-7_19

GRAPPA [8] and GASTS [9], as well as likelihood-based methods such as MLGO [10]. The core of most existing methods is to solve the median problem, which is defined as given three genomes, find the median genome (ancestor) that minimizes the sum of distances from the median to the three given genomes. Yancopoulos *et al.* [11] proposed a simplified model which uses the universal double-cut-and-join (DCJ) operation to account for all rearrangement events and the median problem can be seen as DCJ median problem. Later, several methods are proposed to solve the DCJ median problem. Among these parsimony-based methods, the ASMedian [12] tool outperforms all others. ASMedian iteratively searches Adequate Subgraphs and decomposes the median problem into smaller sub-problems. This method dramatically reduces the solution space and is very efficient when the genomes are closely related. However, it becomes quite slow when the genomes are distant. Given the number of genes as N and the average number of events is r, ASMedian becomes extremely time consuming and the accuracy rate drops significantly when the ratio r/N is over 0.5.

In this paper, we propose a method using simulated annealing algorithm (SAMedian) to solve the median problem based on DCJ-sorting between two genomes. We build a new phylogenetic inference method (SA_GRAPPA) by introducing our median solver into GRAPPA. Experimental results show that our phylogenetic reconstruction method produces a more accurate result than existing tools especially when the input data are large and distant.

2   Background

2.1   Genome Rearrangements

Given a set of n genes $\{1, 2, \cdots, n\}$, a genome can be represented by these genes following an order. To state the strandedness of genes, each gene is assigned with an orientation that is either positive, written i, or negative, written $-i$. Two genes i and j are said to be *adjacent* in genome G if i is immediately followed by j, or, equivalently, $-j$ is immediately followed by $-i$.

Define the head of a gene i by i^h and its tail by i^t. We refer $+i$ as an indication of direction from head to tail $(i^h \rightarrow i^t)$ and otherwise $-i$ as $(i^t \rightarrow i^h)$. There are a total of four scenarios for two consecutive genes a and b in forming an *adjacency*: $\{a^t, b^t\}$, $\{a^h, b^t\}$, $\{a^t, b^h\}$, and $\{a^h, b^h\}$. If gene c is at the first or last place of a linear chromosome, then we have a corresponding singleton set, $\{c^t\}$ or $\{c^h\}$, called a *telomere*.

Assign G as a genome with signed ordering $\{g_1, g_2, \cdots, g_n\}$, an *inversion* between indices i and j $(i \leq j)$ of produces a new genome with linear ordering

$$g_1, g_2, \cdots, g_{i-1}, -g_j, -g_{j-1}, \cdots, -g_i, g_{j+1}, \cdots, g_n$$

There are additional operations for multi-chromosomal genomes, such as *translocation* (one end segment in one chromosome is exchanged with one end segment in the other chromosome), *fission* (one chromosome splits and becomes two), and *fusion* (two chromosomes combine to become one).

Genome graph is consisted of vertices and edges to represent a genome. The vertices are the telomeres and adjacencies while the edges are the connection between gene tail and head. Figure 1 gives a detailed example.

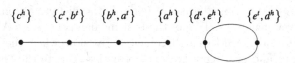

$$\{c^h\} \quad \{c^t, b^t\} \quad \{b^h, a^t\} \quad \{a^h\} \; \{d^t, e^h\} \quad \{e^t, d^h\}$$

Fig. 1. Genome graph for $\{\{c^h\}, \{c^t, b^t\}, \{b^h, a^t\}, \{a^h\}, \{d^t, e^h\}, \{e^t, d^h\}\}$

2.2 Adjacency Graph and DCJ Distance

The DCJ operation has been widely used because of its mathematical simplicity and robustness in practice. The DCJ operation acts on two vertices u and v of a graph by cutting two vertices and rejoining four ends in a new way. There are three ways for the DCJ operation:

- If both $u = \{p, q\}$ and $v = \{r, s\}$ are internal vertices, they could be replaced by the two vertices $\{p, r\}$ and $\{q, s\}$ or by two vertices $\{p, s\}$ and $\{r, q\}$.
- If $u = \{p, q\}$ is internal and $v = \{r\}$ is external, they could be replaced by $\{p, r\}$ and $\{q\}$ or by $\{q, r\}$ and $\{p\}$.
- If both $u = \{q\}$ and $v = \{r\}$ are external, they could be replaced by $\{q, r\}$. An inverse case, a single internal vertex $\{q, r\}$ it also can be replaced by two external vertices $\{q\}$ and $v = \{r\}$.

Lemma 1. *Applying a single DCJ operation changes the number of circular or linear components by at most one.*

Given two genomes A and B, the DCJ sorting is to find the shortest sequence of DCJ operations that transform A into B. The length of such sequence is called the DCJ distance between A and B, denoted by $d_{DCJ}(A, B)$.

The adjacency graph $AG(A, B)$ is a bipartite multi-graph whose set of vertices are the adjacencies and telomeres of A and B. For each $u \in A$ and $v \in B$ there are $|u \cap v|$ edges between u and v. Let A and B be the two genomes defined on the same set of N genes, which we also call equal content, then we have

$$d_{DCJ}(A, B) = N - (C + I/2)$$

where C is the number of cycles and I is the number of odd paths in $AG(A, B)$. The application of a single DCJ operation changes the number of odd paths in the adjacency graph by $-2, 0, 2$. and the number of the circle in adjacency graph by $-1, 0, 1$.

2.3   The Median Problem

Given three genomes (leaves) G_1, G_2, G_3 and a genome M, the median score is defined as $d(G_1, M) + d(G_2, M) + d(G_3, M)$, where $d(G_i, M)$ represents the DCJ distance from G_i to M. The DCJ Median Problem is to find the median genome which has the minimum median score (sum of the distances from the median to the three given genomes). Two of the best median solvers are ASMedian [12] and GAMedian [13]). ASMedian becomes really slow when the genomes are large and distant and also tends to severely underestimate the true number of evolutionary events. GAMedian combines genetic algorithm (GA) with genomic sorting to solve the DCJ median problem in a limited time and space. Since the GA method needs to generate a large population during each generation, it is too slow to converge for distant genomes, despite its great accuracy.

2.4   Simulated Annealing

The primitive idea of SA comes from Metropolis *et al.* [14]. He proposed the algorithm to simulate the cooling of material in a heat bath, which is known as annealing. If we heat a solid up to a melting point and then cool it, the cooling rate would determine the structural properties of the solid. Metropolis's algorithm simulates the cooling process by gradually lowering the temperature of the system until it converges to a steady state. In 1982, Kirkpatrick *et al.* [15,16] applied Metropolis's algorithm to solve the optimization problems. Finding an optimal solution for certain optimization problems could be an incredibly difficult task for the reason that when a problem gets sufficiently large we need to search through an enormous number of possible solutions to find the optimal one. Simulated annealing works greatly in searching for feasible solutions and converges to an optimal solution. It is now viewed as a generic probabilistic metaheuristic for the global optimization problem. Applying the Simulated Annealing algorithm to solve the DCJ median problem needs to overcome some major obstacles: obtaining the initial state and the neighbor state, selecting the best-fit approach of cooling schedule, inducing an acceptance function that the system can avoid falling into local optimal.

3   Methods

In this section, we present our SA-based algorithm for the median problem. Our algorithm design contains four phases. (1) we start our SA system with an initial state and temperature: the initial state is generated by DCJ sorting and the temperature will be cooled by Exponential Multiplicative Cooling method. (2) we use two different settings to develop the neighbor of the current state, one is by a certain number of random DCJ operations while the other is by DCJ sorting. (3) we check the new neighbor with the acceptance function: if this neighboring state is better than the current, we accept it directly; otherwise, the acceptance probability is associated with the current temperature and the difference between these two states. (4) the system repeats step one to step three iteratively until it meets the termination condition.

3.1   SA Median

Initialization. Given three (leaf) genomes, for any pair of the given genomes G_i and G_j, the median genome might be at the sorting path from G_i to G_j. Based on this idea, we design the initial stage to sort each of the three original genomes towards the other two with a random number of steps, which generates six candidate genomes. The state (candidate median genome) for the current generation is randomly picked from the six genomes and is used as the input median for the next generation.

Neighbors of a State. The neighbors of a genome are produced by altering the current genome in a certain way. We developed two different approaches to find neighboring genomes. The simplest way is to randomly apply a certain number of DCJ operations on the current genome (naïve approach), which is very unlikely to converge as the search space is very large (there are $2^n n!$ possible genomes for n genes). The other more complex approach is to apply DCJ sortings (sorting approach) to better direct the search, an approach successfully used in the GAMedian.

This approach works as follows: from the second generation, as the current median genome G is given, it will generate three candidate genomes by sorting m steps from G to the three original leaf genomes; we randomly pick one from these three candidates as the potential input median for the next generation.

We then compare the potential median to the current median based on their median scores to accept or reject the new genome, using the reliable acceptance criteria defined as follows.

Acceptance Function. First, we check if the neighboring state is a better choice which has lower median score than the current state. If it is better, we accept it unconditionally. Otherwise, we need to consider two factors: how bad is the neighboring state and how high is the current temperature. We employ the standard acceptance formula so that our algorithm which is more likely to accept worse neighbor state at high temperatures.

$$Acceptance = \begin{cases} \exp{-\Delta E / T} & \text{if } \Delta E \geq 0 \\ 1 & \text{if } \Delta E < 0 \end{cases} \quad (1)$$

where the ΔE is the difference from the energy of the neighbor to that of the current state. T is the temperature of the current generation and exp is the exponential. The principle is that the possibility to accept will depend on the value of T and ΔE in the exponential function.

Initial Temperature and Cooling Scheme. The initial temperature and cooling schedule play critical roles in SA algorithms. Based on our experimental observations, the results greatly depend on the values of temperature T in each generation, while T depends on the initial temperature T_0 and the cooling schedule α.

The procedures we use to pick a reasonable estimate value of T_0 are as follows: Given P_0 and average $\Delta Cost$, the equation to compute T_0 is

$$P_0 = \exp(-\Delta E)/T_0.$$

At the first several states, we want to accept worse candidates as much as possible. We set up the initial acceptance percentage as P_0, and estimate the $\Delta Cost$ from experiment result, then we can obtain T_0 by formula $(\ln P_0)/(-\Delta Cost)$.

For the cooling schedule, there are multiple different cooling approaches for different specific problems. After our experiments, we select the approach of Exponential Multiplicative Cooling, which is proposed by Kirkpatrick *et al.* [17]. T_0 is the initial temperature, T_n is the temperature after n iterations, and α is the cooling rate.

$$T_n = T_0 \cdot \alpha^n \qquad (0.8 \leq \alpha \leq 0.9)$$

The maximum number of iterations for our SAMedian solver was set as G but it could be terminated early if it reached the perfect median score. The detailed description of our algorithm is shown in Algorithm 1.

Input: three genomes as leaf genomes
Output: *bestS* as a genome which have the smallest DCJ sum distance to the three leaf genomes.
Initialization: S_0: one genome which is one DCJ sorting distance from a leaf genome, T_0: initialized temperature, $G = Max_{gen}$ as left over cycle number, α: cooling rate, *bestS* = S_0, current temperature T, current solution *currentS* = S_0.
While: $G > 0$
　　generate new genome *newS* by DCJ sorting from *currentS*
　　$\delta Cost = (newS - CurrentS)$;
　　If $\delta Cost < 0$ **Then**;
　　　　currentS = *newS*;
　　　　If $\delta newS < bestS$ **Then**;
　　　　　　bestS = *newS*;
　　Else if *(Random(0,1)* $< \exp -\Delta E/T$) **Then**
　　　　currentS − *newS*;
　　$T = \alpha T$;
　　$G = G - 1$;
Return *bestS*;

Algorithm 1: Simulated Annealing algorithm

GRAPPA is one of the parsimony-based methods to infer ancestral gene orders and phylogenies simultaneously. It searches the tree space and scores potentially good trees to find the best tree. To obtain the score of a tree, it iteratively solves each median problem defined on an internal node until there is no improvement. Currently, Caprara's [18] reversal median solver and the DCJ median solver (ASMedian) are included in GRAPPA. We replace the current median solvers in GRAPPA with the new median solver to build our own phylogenetic inference and ancestral genome reconstruction method (SA_GRAPPA).

4   Experimental Results

We use simulated datasets to evaluate accuracy and efficiency of our tools which is widely used to assess the quality of phylogenetic methods. Our model tree simulation follows Lin's et al. [19] birth-death model. Following the model tree, We first initialize a permutation of n genes as root. From the root permutation, we generate the rest internal and leaf genomes by conducting r random double-cut-and-join (DCJ) events along corresponding branches. r is an average branch length (event number) for each dataset, and we used diameter (d) to represent the ratio r/n. We use m to represent the total number of genomes generated. For each parameter setting, we run 20 trials to get the average result.

4.1   Comparison with ASMedian and GAMedian

To show the performance of our median solver SAMedian, we set the simulation data generation parameters leaf nodes number as 3, n as 200 and d ranges from 0.1 to 1 for our simulation data. To evaluate the accuracy of our sorting-based approach, we compare our method with ASMedian and GAMedian, and the result is presented in Fig. 2.

Our result shows that the computation time of ASMedian increases dramatically as r increases. Since GAMedian has to maintain a large genome pool to obtain the optimal solution, the time usage is the longest among all the methods. On the other hand, our SA method keeps at a consistent speed, even when r becomes quite high. Table 1 shows the comparison of time usage. Meanwhile, the accuracy of median scores is very close to that obtained by ASMedian and GAMedian. Figure 2 lists the median score comparison result (lower is better). Since ASMedian applies the parsimony approach, its median score is optimal in each case. GAMedian obtains a similar result after an excessive amount of time. Our method returns a score very close to that of ASMedian and GAMedian for each dataset, most cases are the same. Because our method is a meta-heuristic, it is capable of solving more complicated datasets than ASMedian could.

SAMedian has a great improvement over speed compared with GAMedian. The running time of GAMedian is determined by the time it spends in each generation. For the number of genes (n), if n is 200, the running time in each generation costs about 2.5 s; as the maximum number of generation is set at 500, therefore the total amount of running time is over 1000 s. Meanwhile, we find out that if n is larger than 1000 and diameter r is over 0.6, it needs more than 1200 generations to obtain the optimal, and each generation costs more than 60 s–as a result, the total running time is over 20 h. Even though the GAMedian presents an excellent performance on the median problem, it costs too much time, especially when the gene number is large.

On the other hand, SAMedian is much faster than GAMedian: it only takes 0.2 s to solve one median problem with 200 genes, and takes 3 s with 1000 genes. Therefore, the SAMedian solver is a better solution to explore phylogeny reconstruction and ancestral inference problem, which requires iteratively solving many instances of the median problems.

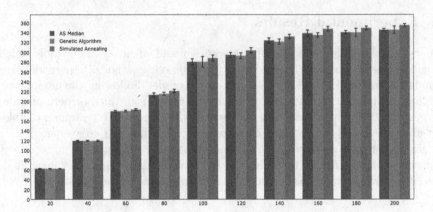

Fig. 2. Comparison of median scores between ASMedian and GAMedian on genomes with 200 genes. The x-axis is the expected distance from a leaf to the median, diameter is ranged from 0.1 to 1. The y-axis is the median score for the resulted median.

Table 1. Comparison of time usage among our Median method, ASMedian and GAMedian. Each genome has 200 genes. (second)

r/n	0.1	0.2	0.3	0.4	0.5	0.6	0.7	0.8	0.9	1.0
ASMedian	1.2	1.3	1.1	260.0	610.5	613.6	620.3	670.0	660.4	675.3
GAMedian	1100	1178	1187	1146	1175	1151	1114	1101	1201	1298
SAMedian	**0.20**	**0.22**	**0.20**	**0.21**	**0.23**	**0.20**	**0.20**	**0.22**	**0.24**	**0.25**

We evaluate SAMedian with the other two by calculating how similar the inferred median genome and the true genome are, using two measurements: how far away the inferred median are from the true, and how accurate the inferred median is in term of genomic structure. Figure 3 shows the average DCJ distance from the inferred median to the true ancestor. Our method generates the median genomes closer to the true scenario, which is comparable to ASMedian. Our method has slightly longer branches than that of the GAMedian.

The accuracy of the genomic structure of the median genome can be measured by comparing the adjacencies presented in both the inferred median and the true ancestor. Suppose the set of adjacencies in the inferred median genome is A and the set of adjacencies in the true ancestor is B. The *accuracy of adjacency* is defined as the proportion of the adjacencies in both A and B to all the adjacencies either in A or B, as the expression $|A \cap B|/|A \cup B|$. Therefore, based on the adjacency Fig. 4, we could obtain a similar result as ASMedian and our method outperforms ASMedian when diameter goes bigger ($r \geq 80$). The result from SAMedian is slightly worse than the GAMedian.

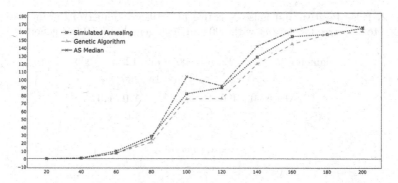

Fig. 3. Distance between the inferring median genome and true ancestor under different event number. The results for ASMedian, GAMedian and our SAMedian are shown in red, green and blue. X-axis represents the event number, the y-axis is the distance. (Color figure online)

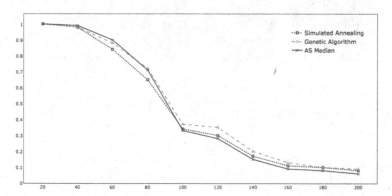

Fig. 4. Adjacency accuracy of the inferred median genome to true ancestor under different number of events. The results for ASMedian, GAMedian and our SAMedian are shown in red, green and blue, respectively. X-axis represents the number of events, the y-axis is the accuracy of adjacency. (Color figure online)

4.2   Phylogeny Reconstruction and Ancestor Inference

To show the ability of our method for phylogeny and ancestral genome reconstruction, we compare our result with the powerful tool, GASTS, by using simulation data. The parameter setting for our simulation data generation is m as 12, n as 500 or 1000 while d is 1, 2, 3 or 4 correspondingly.

GASTS is a tool to find the most parsimony tree from gene-order data. Both methods are able to infer accurate phylogenies and ancestral genomes by comparing to true scenarios. We also compare our method with Intermediate Genomes [20] method, which uses the concept of intermediate genomes, arising in optimal pairwise rearrangement scenarios, to reconstruct the ancestral gene orders by reading a given phylogeny (i.e. solves the SPP problem).

Table 2. False Positive and False Negative bipartition number of the inferred tree topology to true tree for dataset with 500 and 1000 genes with 12 leaf genomes.

diameter	m = 12, n = 500				m = 12, n = 1000			
	1n	2n	3n	4n	1n	2n	3n	4n
SA_GRAPPA	0	0.6	**2.1**	**2.4**	0	**1.0**	**1.6**	**1.8**
GASTS	0	0.6	3.6	6.9	0	1.6	2.4	6.3

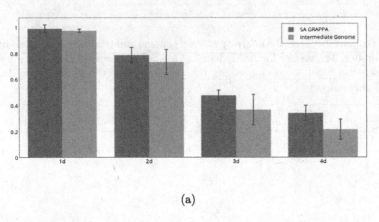

(a)

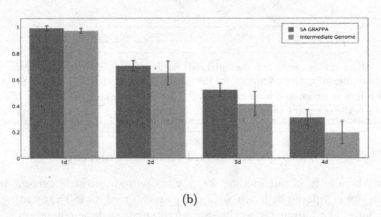

(b)

Fig. 5. Adjacency accuracy of the inferred internal genome to true ancestor for 500 (top) and 1000 (bottom) genes with 12 leaf genomes.

For big phylogeny problem (BPP), we compare the inferred tree topology with the true scenario as shown in the Table 2 by getting false positive and negative rate. Here we can see our SA_GRAPPA is able to infer tree topologies closer to the true tree than GASTS on both 500 and 1000 genes dataset.

For small phylogeny problem (SPP), we compare the adjacency accuracy of the inferred internal genome to true nodes as shown in Fig. 5, which shows that our SA_GRAPPA outperforms the current the Intermediate Genome method and obtains much more correct adjacencies on both 500 and 1000 genes dataset.

5   Conclusions

In this paper, we introduce a DCJ sorting based Simulated Annealing algorithm to solve the well-known three-genome median problem. Our median solver, SAMedian, presents a great potential in approximating the optimal solution for the three-genome problem. DCJ sorting is essential for our SA median method for the reason that SimpleSA fails to converge. We can see that our SA median solver is much more efficient than ASMedian and GAMedian, especially when the input has a big event and/or gene number. The median inferred from our method approximates better to true scenario than ASMedian and worse than GAMedian. Since ASMedian tends to underestimate evolutionary distance, the result from ASMedian is likely to have a lower median score but far from the true ancestor. Although the GAMedian frequently gives the best result, it is quite limited by its speed and scalability. Meanwhile, our method presents an excellent performance on phylogeny reconstruction, better than other existing reconstruction methods, such as Intermediate Genome and GASTs.

Although our method shows a great performance in our experiment, several adapted changes are needed in our future work. First, to extend our work to unequal content by considering insertion, deletion, and duplication. As distance estimation under unequal content has been considered by earlier work [21–23], our method is easy to extend to handle unequal content. Second, on the implementation level, we can apply parallel programming to speed up our application.

References

1. Zhou, J., Lin, Y., Rajan, V., Hoskins, W., Feng, B., Tang, J.: Analysis of gene copy number changes in tumor phylogenetics. Algorithms Mol. Biol. **11**(1), 26 (2016)
2. Feng, B., Zhou, l., Tang, J.: Ancestral genome reconstruction on whole genome level. Curr. Genomics (2017)
3. Zhou, J., Lin, Y., Hoskins, W., Tang, J.: An iterative approach for phylogenetic analysis of tumor progression using FISH copy number. In: Harrison, R., Li, Y., Măndoiu, I. (eds.) ISBRA 2015. LNCS, vol. 9096, pp. 402–412. Springer, Cham (2015). doi:10.1007/978-3-319-19048-8_34
4. Cai, J., Liu, X., Vanneste, K., Proost, S., Tsai, W.-C., Liu, K.-W., Chen, L.-J., He, Y., Xu, Q., Bian, C., et al.: The genome sequence of the orchid phalaenopsis equestris. Nat. Genet. **47**(1), 65–72 (2015)
5. Zhou, J., Lin, Y., Rajan, V., Hoskins, W., Tang, J.: Maximum parsimony analysis of gene copy number changes. In: Pop, M., Touzet, H. (eds.) WABI 2015. LNCS, vol. 9289, pp. 108–120. Springer, Heidelberg (2015). doi:10.1007/978-3-662-48221-6_8
6. Sankoff, D., Blanchette, M.: Multiple genome rearrangement and breakpoint phylogeny. J. Comput. Biol. **5**(3), 555–570 (1998)

7. Zhou, J., Hu, F., Hoskins, W., Tang, J.: Assessing ancestral genome reconstruction methods by resampling. In: 2014 IEEE International Conference on Bioinformatics and Biomedicine (BIBM), pp. 25–31. IEEE (2014)

8. Moret, B.M.E., Siepel, A.C., Tang, J., Liu, T.: Inversion medians outperform breakpoint medians in phylogeny reconstruction from gene-order data. In: Guigó, R., Gusfield, D. (eds.) WABI 2002. LNCS, vol. 2452, pp. 521–536. Springer, Heidelberg (2002). doi:10.1007/3-540-45784-4_40

9. Xu, A.W., Moret, B.M.E.: GASTS: parsimony scoring under rearrangements. In: Przytycka, T.M., Sagot, M.-F. (eds.) WABI 2011. LNCS, vol. 6833, pp. 351–363. Springer, Heidelberg (2011). doi:10.1007/978-3-642-23038-7_29

10. Hu, F., Zhou, J., Zhou, L., Tang, J.: Probabilistic reconstruction of ancestral gene orders with insertions and deletions. IEEE/ACM Trans. Comput. Biol. Bioinform. $11(4)$, 667–672 (2014)

11. Yancopoulos, S., Attie, O., Friedberg, R.: Efficient sorting of genomic permutations by translocation, inversion and block interchange. Bioinformatics $21(16)$, 3340–3346 (2005)

12. Xu, A.W., Sankoff, D.: Decompositions of multiple breakpoint graphs and rapid exact solutions to the median problem. In: Crandall, K.A., Lagergren, J. (eds.) WABI 2008. LNCS, vol. 5251, pp. 25–37. Springer, Heidelberg (2008). doi:10.1007/978-3-540-87361-7_3

13. Gao, N., Yang, N., Tang, J.: Ancestral genome inference using a genetic algorithm approach. PLoS ONE $8(5)$, 62156 (2013)

14. Metropolis, N., Rosenbluth, A.W., Rosenbluth, M.N., Teller, A.H., Teller, E.: Equation of state calculations by fast computing machines. J. Chem. Phys. $21(6)$, 1087–1092 (1953)

15. Kirkpatrick, S.: Optimization by simulated annealing: quantitative studies. J. Stat. Phys. $34(5–6)$, 975–986 (1984)

16. Černý, V.: Thermodynamical approach to the traveling salesman problem: an efficient simulation algorithm. J. Optim. Theory Appl. $45(1)$, 41–51 (1985)

17. Kirkpatrick, S., Gelatt, C.D., Vecchi, M.P., et al.: Optimization by simulated annealing. Science $220(4598)$, 671–680 (1983)

18. Caprara, A.: On the practical solution of the reversal median problem. In: Gascuel, O., Moret, B.M.E. (eds.) WABI 2001. LNCS, vol. 2149, pp. 238–251. Springer, Heidelberg (2001). doi:10.1007/3-540-44696-6_19

19. Lin, Y., Rajan, V., Moret, B.M.: Fast and accurate phylogenetic reconstruction from high-resolution whole-genome data and a novel robustness estimator. J. Comput. Biol. $18(9)$, 1131–1139 (2011)

20. Feijão, P.: Reconstruction of ancestral gene orders using intermediate genomes. BMC Bioinform. $16(14)$, 3 (2015)

21. Braga, M.D.V., Willing, E., Stoye, J.: Genomic distance with DCJ and indels. In: Moulton, V., Singh, M. (eds.) WABI 2010. LNCS, vol. 6293, pp. 90–101. Springer, Heidelberg (2010). doi:10.1007/978-3-642-15294-8_8

22. Shao, M., Moret, B.M.E.: On computing breakpoint distances for genomes with duplicate genes. In: Singh, M. (ed.) RECOMB 2016. LNCS, vol. 9649, pp. 189–203. Springer, Cham (2016). doi:10.1007/978-3-319-31957-5_14

23. Hu, F., Lin, Y., Tang, J.: MLGO: phylogeny reconstruction and ancestral inference from gene-order data. BMC Bioinform. $15(1)$, 354 (2014)

Addressing the Threats of Inference Attacks on Traits and Genotypes from Individual Genomic Data

Zaobo He[1], Yingshu Li[1(✉)], Ji Li[1], Jiguo Yu[2], Hong Gao[3], and Jinbao Wang[4]

[1] Department of Computer Science, Georgia State University, Atlanta, GA, USA
zhe4@student.gsu.edu, yili@gsu.edu
[2] Qufu Normal University, Rizhao, Shandong, China
[3] School of Computer Science and Technology, Harbin Institute of Technology,
Harbin, Heilongjiang, China
[4] The Academy of Fundamental and Interdisciplinary Sciences,
Harbin Institute of Technology, Harbin, Heilongjiang, China

Abstract. The decreasing cost of DNA-sequencing empowers high availability of genetic-oriented services, which further promote growing number of genomes and traits of individuals being accessible online. Notoriously, these data are sensitive and may further lead to more sensitive data leakage. In this paper, we formulate the trait and genotype inference problem and develop an efficient inference method based on factor graph and belief propagation. An adversary then can infer the potential traits and genotypes of the victims whose portions of data are observed, depending on trait/SNP associations available from GWAS catalog. To protect against such inference attacks, we detail privacy and utility metrics then propose a genomic data-sanitization method that can effectively tradeoff genomic data openness and privacy.

Keywords: SNP/trait associations · Belief propagation · Factor graph · Data-sanitization

1 Introduction

Rapidly growth of technology in DNA sequencing had been offering significant genetic-oriented services, from genetic diagnosis to specifical genomic medicine to the test of genetic compatibilities. For example, a DNA sequencing platform was recently authorized by Food and Drug Administration to expand the use of genomes in genetic medicine [8]. Furthermore, Genome-wide association studies (GWAS) are making efforts to uncover the associations between genomes (*i.e.*, Single Nucleotide Polymorphism (SNP)) and human traits (like diseases). As a consequence, geneticists need to collect large scale of human genomes and traits to boost the progress of research and services. Recent years have witnessed the massive growth in genomes, shared by individuals in order for genetic services, which offered great potential. Furthermore, a large body of genetic trait-related websites (*e.g.*, patientslikeme.com [1]), genomic data sharing platforms

© Springer International Publishing AG 2017
Z. Cai et al. (Eds.): ISBRA 2017, LNBI 10330, pp. 223–233, 2017.
DOI: 10.1007/978-3-319-59575-7_20

(*e.g.*, OpenSNP [2]), and social networks, in which individuals release their traits and genomes, are emerged to enable significant amount of valuable data to be available.

Although the vast individual data hold significant benefit, they also raise stringent privacy concerns. Genomes are associated with sensitive SNPs in which some ones are closely related to diseases. Once such SNPs are identified, the owner would be placed into discrimination risk (from insurance company or job market) [6]. For instance, GWAS catalog reported massive diseases that are associated with a set of SNPs [5]. Furthermore, individual genomes also encode complex correlations with their relatives' DNA sequences. Consequently, an individual who discloses its genomes without any relatives' consent, will also place their relatives into risk. For example, the publishing of Henrietta Lacks's genomes sparks controversy regarding the potential reveal of private information about the relatives [3]. Therefore, in order for realizing the full benefit of genomes, effective tradeoff between data openness and privacy is necessary.

Unfortunately, massive publicly available auxiliary information further aggravates this threat. For example, case-control studies in GWAS report SNPs and associated traits, risk allele frequency and corresponding statistics to GWAS catalog [5]. Associated such auxiliary information with the huge amount of individual genomes available online, an adversary can launch significant inference attacks to infer the traits and genotypes of individuals and their relatives. As a consequence, some individuals determine to just release portions of genomic data and traits. Releasing a portion of both cannot, however, completely protect against inference attacks for the unreleased portions. Unreleased parts of traits and genotypes could be reconstructed with the help of released portions and massive auxiliary information. For example, James Watson shared his full DNA sequence expect for Apolipoprotein E, known as main squeeze for the predictor of Alzheimer's disease. However, James Watson becomes an unsuccessful example that reminds the correlations among SNPs (*i.e.*, *linkage disequilibrium*) could be employed to infer such sensitive SNPs [23].

In this paper, we explore how to launch inference attacks to infer target traits and genotypes with known SNPs of individuals and publicly available statistics, namely, GWAS catalog [5]. Then, we propose a data-sanitization method that can effectively sanitize known SNPs to protect against such inference attacks, while do not much degrade the benefit of shared genomic data. To explore how to infer target traits and genotypes by adversaries, we construct an effective reconstruction method by representing the SNPs, traits and SNP/trait associations on a probability graphical model and operate belief propagation for inference. Previous reconstruction methods generally have high computational complexity which grows with the scale of SNPs and individuals. Considering the SNPs are in the order of tens of millions, which prevents the existing methods from obtaining precise inference results. Our work does consider the magnitude property of SNPs and empower the inference method on target traits and genotypes in linear complexity. To protect against such inference attacks, we formalize the genomic privacy and utility metrics of individuals and develop a data-sanitization method to realize privacy/utility tradeoff. Compared with previous contributions, our data-sanitization method can optimally balance the genomic data privacy and openness.

2   Preliminaries

This section briefly introduces two concepts used latterly, namely, SNPs and GWAS catalog.

2.1   Single Nucleotide Polymorphism

In human beings, 99.9% of their NDA are same, the remaining 0.1% makes one individual unique. SNP is the common DNA variation making the genetic variation significant. SNPs carry significant information about an individual, such as disease predisposition, phenotype change.

There are two nucleotides on a SNP locus: (1) a major allele (if allele frequency is larger than 50%), and (2) a minor allele (if allele frequency is less than 50%). Both major allele (represented by B) and minor allele (represented by b) take values from nucleotides (A, T, G, C). On a SNP locus, one allele is inherited from mother and the other one is inherited from father. Thus, one SNP locus' content could be: BB (both nucleotides are major alleles), Bb (one major allele and one minor allele) or bb (both nucleotides are minor alleles).

2.2   GWAS Catalog

GWAS catalog is a series of statistics regarding case-control studies under GWAS. Case-control studies are performed by analyzing the genotypes between: case group (participants with traits) and control group (participants without traits). After conducting statistical test over the SNPs of these two groups, two alleles in an SNP locus can be identified: one risk allele and one non-risk allele. Risk allele is the one that is more frequently carried by individuals in case group compared with in control group. Furthermore, for an arbitrary allele, the ratio of its frequency in case group and that in control group is also reported as *odds ratio*. Then, SNPs that are associated with traits and significant statistical indicators such as Risk Allele Frequency (RAF) (the frequency of individuals carrying such risk allele), non-Risk Allele Frequency (nRAF), odds ratio, *etc*, are reported to GWAS catalog.

3   Problem Formulation

3.1   Genomic Data Model

For an arbitrary individual, the set of SNPs are defined as S, $|S| = n$. s_i is defined to be the content of SNP i ($i \in S$), $s_i \in \{0, 1, 2\}$ (for simplicity, we denote the genotype BB, Bb or bb as 0, 1, 2, respectively). Some SNPs of an individual are known by adversary (some individuals or their relatives share full or part of their SNPs for obtaining services or helping genetic research) while others do not. We denote the set of publicly available SNPs as S_K, while unknown SNPs as S_U.

As auxiliary information obtained from GWAS catalog, the set of potential traits are represented by T. t_j is defined to be the trait j ($j \in T$) of an individual. For each trait t_j, there are a set of associated SNPs in GWAS catalog. For each s_i associated with t_j, the risk allele r_i^j of s_i can be extracted. Furthermore, the odds ratio of r_i^j, O_i^j and the RAF in control group $f_i^{j^o}$ can be extracted. Although the RAF in case group $f_i^{j^a}$ is not given directly, it can be easily determined by $f_i^{j^o}$ and O_i^j [26]. Similarly, the known traits shared by individuals are denoted as T_K and the unknown ones are denoted as T_U.

3.2   Adversary Model

The objective of an adversary is to infer target traits and genotypes in X_U, $X_U = T_U \cup S_U$ of an individual. A powerful adversary is assumed to launch inference attacks with extensive available knowledge: (i) the known SNPs from individuals who share part or full their SNPs (*i.e.*, S_K), (ii) the known traits shared by individuals (*i.e.*, T_K), (iii) the GWAS catalog which contains the interdependent information among traits and SNPs and statistical information (*i.e.*, $\mathcal{C}(T, s_i, r_i^j, O_i^j, f_i^{j^o})$).

4   Inference Attack

We formulate the inference attack as calculating the Marginal Probability Distribution (MPD) of target SNPs in S_U or target traits in T_U, given the known SNPs S_K, known trait set T_K, statistical information from GWAS catalog $\mathcal{C}$. Then, the joint probability distribution of all unknown variables X_U, $X_U = T_U \cup S_U$, is $p(X_U|S_K, T_K, \mathcal{C})$, conditioned on the available knowledge. Then, the MPD of an unknown variable $x_i \in X_U$ can be derived from.

$$p(x_i|S_K, T_K, \mathcal{C}) = \sum_{X_U \setminus x_i} p(X_U|S_K, T_K, \mathcal{C}) \tag{1}$$

where $X_U \setminus x_i$ is to sum out over all variables in X_U except x_i.

Unfortunately, the scale of summing terms presents exponential increase with the growth of the number of SNPs and traits. Considering individual genomes includes tens of million of SNPs, calculating MPD of an SNP directly is unfeasible. Hence, we consider factorizing the joint probability distribution into a set of local functions, and each function describes the dependency relationships among SNPs and traits, by taking a subset of SNPs and traits as variables.

By operating belief propagation over factor graph, the joint probability distribution can be factorized and then the MPD can be calculated with linear complexity. We construct a factor graph which includes two types of *variable* nodes: (i) *SNP variable node*: taking each SNP as a variable node, and (ii) *trait variable node*: taking each trait as a variable node. And one type of factor node: representing the association between SNPs and traits. The factor graph is constructed in following way:

- Trait variable node t_j and SNP variable node s_i connect to factor node f_{ji} if s_i is one of the SNPs associated with trait t_j.

By operating belief propagation over factor graph, the global conditional distribution $p(X_U|S_K, T_K, \mathcal{C})$ is factorized into several local functions and each function takes a subset of SNPs and traits as variables:

$$p(X_U|S_K, T_K, \mathcal{C}) = \frac{1}{Z} \prod_{i \in S} \prod_{j \in T} f_{ji}(s_i, t_j, \mathcal{C}) \qquad (2)$$

where Z is a constant normalization factor.

A factor graph with 2 traits $T = \{t_1, t_2\}$, 3 SNPs $S = \{s_1, s_2, s_3\}$ for an individual, is shown in Fig. 1. As shown in Fig. 1, for trait t_1 and t_2, the associated SNP sets are $\{s_1\}$ and $\{s_1, s_2, s_3\}$, respectively.

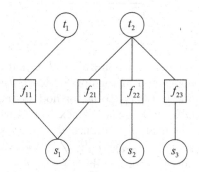

Fig. 1. A factor graph with 2 traits $T = \{t_1, t_2\}$ and 3 SNPs $S = \{s_1, s_2, s_3\}$.

Given a factor graph structure, we next need to specify the probability dependency between factor node and variable node. We first treat the prevalence rate of each trait $p(t_j)$ as prior knowledge, which can be acquired from internet or public statistics (such as CDC [4]). Then, given each associated trait, the conditional probability of s_i is necessary to be determined. With this goal, we first turn to figure out conditional probability of allele, given each associated trait. The conditional probability of each allele given associated trait can be specified by RAF and nRAF, as shown in Table 1 for SNP s_i with one of neighbor factor nodes t_j.

Based on the conditional allele probability, the next step is to calculate conditional genotype probability. With allele r_i^j and ρ_i^j, the genotype of s_i for trait t_j can be one of the following: $r_i^j r_i^j$, $r_i^j \rho_i^j$ and $\rho_i^j \rho_i^j$. Therefore, the genotype

Table 1. Conditional probability of risk allele r_i^j and non-risk allele ρ_i^j, given one of neighbor factor nodes t_j of s_i

	t_j	$\bar{t}_j$
r_i^j	$f_i^{j^a}$	$f_i^{j^o}$
ρ_i^j	$1 - f_i^{j^a}$	$1 - f_i^{j^o}$

frequency can be easily obtained by transforming Table 1, as shown in Table 2. Similarly, the probability of trait conditional on one of associated SNPs can be easily derived from Table 2 based on Bayesian posterior probability.

Table 2. Genotype probability of $r_i^j r_i^j$, $r_i^j \rho_i^j$ and $\rho_i^j \rho_i^j$, given one of s_i' neighbor factor nodes t_j

	t_j	$\bar{t}_j$
$r_i^j r_i^j$	$\sqrt{f_i^{j^a}}$	$\sqrt{f_i^{j^o}}$
$r_i^j \rho_i^j$	$f_i^{j^a}(1 - f_i^{j^a})$	$f_i^{j^a}(1 - f_i^{j^o})$
$\rho_i^j \rho_i^j$	$\sqrt{1 - f_i^{j^a}}$	$\sqrt{1 - f_i^{j^o}}$

Belief propagation is a massage-passing algorithm that iteratively passes messages between variable nodes and factor nodes. Hence, we define the massages passing from variable node (s_i or t_j) to factor node as μ, the massage passing from factor node to variable node as λ. We next take variable nodes t_2 and s_1, factor node f_{21} in Fig. 1 as example to illustrate the massage-passing between variable and factor nodes. The massage $\mu_{v \to f}^{(n)}(s_1^{(n)})$ passing from variable node s_1 to factor node f_{21} denotes the probability of $s_1 = \kappa, (\kappa = 0, 1, 2)$ in n-th iteration. The message $\lambda_{f \to v}^{(n)}(s_1^{(n)})$ passing from factor node f_{21} to variable node s_1 denotes the probability of $s_1 - \kappa, (\kappa = 0, 1, 2)$ in n th iteration, given trait/SNP associations.

A variable node v sends massage to neighbor factor node f by multiplying all messages from neighbor factor nodes except f. Taking the factor graph in Fig. 1 as an example, the message from s_1 to f_{21} (denoted as $s \to f$) is:

$$\mu_{s \to f}^{(n)}(s_1^{(n)}) = \frac{1}{Z} \times \prod_{f^* \in N(s_1) \backslash f_{21}} \lambda_{f^* \to s}^{(n-1)}(s_1^{(n-1)}) \tag{3}$$

where $N(s_1) \backslash f_{21}$ is all neighbor factor nodes of s_1 except f_{21} (in Fig. 1, $N(s_1) \backslash f_{21} = \{f_{11}\}$).

The message from t_2 to f_{21} can be formulated similarly.

Then, according to the principle of belief propagation, factor node f sends message to neighbor variable node v by multiplying all massages from f' neighbors except v, and then multiplying the obtained product with the factor, and then summing out all the neighbor variable nodes of f except v. The massage from f_{21} to variable node s_1 (denoted as $f \to s$) is

$$\lambda_{f \to s}^{(n)}(s_1^{(n)}) = \sum_{t_2} f_{21}(s_1, t_2) \prod_{v^* \in N(f_{21}) \setminus s_1} \mu_{v^* \to f}^{(n)}(v^*) \qquad (4)$$

Note that $f_{21}(s_1, t_2) \propto p(s_1 | t_2)$ and it can be obtained from Table 2.

The message from f_{21} to t_2 can be formulated similarly.

The massage-passing iteration starts with passing massage by variable nodes. We now specify the boundary conditions in the above iterations. At the first iteration (*i.e.*, $n = 1$), for any SNP variable node $s_i \in S_U$, since no massages are sent from s_i' neighbor factor nodes, $\mu_{s_i \to f}^{(1)}(s_i^{(1)}) = 1$ for each potential values of s_i. On the other hand, for any SNP variable node $s_i \in S_K$ and $s_i = \kappa$, $\mu_{s \to f}^{(1)}(s_i^{(1)} = \kappa) = 1$ and $\mu_{s \to f}^{(1)}(x_j^{(1)} = \kappa') = 0$ for other potential SNP values, where $\kappa' \in \{\{0, 1, 2\} \setminus \kappa\}$. The massages for trait variable node t_j are set with same ways. Until all unknown variables are converged (or the passing massages are converged), the iterations can be stopped.

Finally, the MPD of each unknown variable in T_U is obtained by multiplying all massages to each variable.

5 Genome Privacy-Utility Tradeoff

In this section, our objective is to propose a data-sanitization method that can optimize the tradeoff between genome privacy and utility. As a consequence, releasing the sanitized genomic data can protect against inference attacks on target traits and genotypes, while do not much degrade the data benefit. We first introduce privacy and utility metrics to measure the genome privacy and utility due to sanitization method executed. Then, a data-sanitization method to sanitize genotypes is developed that can optimally balance data openness and privacy.

5.1 Privacy and Utility Metrics

For any SNP sharer, it is expected that adversary cannot effectively infer his/her target traits and genotypes. Here, privacy is measured by the ambiguity of inference results; namely, the larger ambiguity of inference results is, the larger is the uncertainty of adversary. The ambiguity of inference results can be quantified by the entropy of $p(x_i | S_K, T_K, C)$:

$$H_i = \frac{-\sum_{x_i} p(x_i | S_K, T_K, C) \log p(x_i | S_K, T_K, C)}{\log(3)} \qquad (5)$$

where x_i is either target SNP ($x_i \in \{0, 1, 2\}$) or trait ($x_i \in \{0, 1\}$). The larger the entropy is, the larger is the ambiguity of $p(x_i | S_K, T_K, C)$.

Then, a parameter δ is introduced to bound H_i as privacy metric:

Definition 5.1 δ-privacy. *The released SNPs satisfy δ-privacy if $H_i \geq \delta$ for each SNP s_i.*

For data utility, it is expected that as many actual SNPs are released as possible, while guaranteeing δ-privacy.

Definition 5.2 Utility. *The utility of a set of SNPs is measures as the expected number of released SNPs.*

5.2   Data-Sanitization Method

To protect against inference attacks on x_i, we propose a data-sanitization method based on sanitizing the neighbor SNPs of x_i. For an arbitrary SNP s_j, if there exists a path from s_j to x_i in factor graph, s_j is one of the neighbor SNPs of x_i. For example, s_3 is one of the neighbor SNPs of t_1, because there exists a path from s_i to t_1: $s_3 \rightarrow t_2 \rightarrow s_1 \rightarrow t_1$.

Our objective is to find a subset of neighbor SNPs of each x_i so that sanitizing them can maximize data utility while guaranteeing privacy constraint. For this purpose, the concept of *vulnerable neighbor SNP* is introduced:

Definition 5.3 Vulnerable neighbor SNP. *The vulnerable neighbor SNP of x_i is a neighbor SNP of x_i, whose sanitizing will decrease the prediction accuracy on x_i.*

Considering obfuscated SNPs (replace actual SNP content with another one) brings uncontrollable results when making genetic analysis, we sanitize SNPs by taking *removing* method. The privacy of x_i upon removing its vulnerable neighbor SNP x_k is $H_i(N_i - x_k)$, where N_i is the neighbor SNPs of x_i.

With Definition 5.3, the problem of realizing privacy/utility tradeoff can be stated as identifying a subset of vulnerable neighbor SNPs of each x_i to sanitize, who are responsible for maximizing data utility of released SNPs while guaranteeing privacy constraints of each SNP and trait.

To solve the above problem, we first prove the ambiguity of inference results, *i.e.*, Eq. (5) has *monotonicity* and *submodularity* property, when the increasing number of SNPs are sanitized. Monotonicity property means that if we sanitize more SNPs, we can only improve privacy.

Theorem 5.1 Monotonicity. *The privacy function of an arbitrary variable $x_i \in X_U$, $H_i : N_i \rightarrow \mathbb{R}^*$ is monotonically nondecreasing, i.e., $H_i(N_i \cup s_k) \leq H_i(N_i)$, where $s_k \in N_i$ and N_i is vulnerable neighbor SNPs of x_i.*

Theorem 5.2 Submodularity. *The privacy function of an arbitrary variable $x_i \in X_U$, $H_i : N_i \rightarrow \mathbb{R}^*$ has submodularity property, i.e., $H_i(U_i \cup s_k) - H_i(U_i) \leq H_i(V_i \cup s_k) - H_i(V_i)$, where $U_i \subseteq V_i \subseteq N_i$, $s_k \in N_i$, and U_i, V_i are the set of vulnerable neighbor SNPs of x_i.*

Theorems 5.1 and 5.2 shows that the problem of finding a SNP sanitization method is transformed to the minimization of submodular, nondecreasing, nonnegative function with constraints that is knapsack-like. Then, we can utilize the greedy algorithm proposed in [25] to solve this problem.

6   Related Works

Several algorithms have been proposed for inference attacks on genotypes, haplotype, or phenotypes (e.g., disease) based on probability graphical models. Bayesian networks are generally utilized in order for mapping the association between phenotypes and genes [11,26,28], or the association between disease genes and genetic maps [10,21]. Factor graph are also proved to be an effective model in mapping the association among genetic relations, linkage disequilibrium and genes, in the context of kin privacy [18]. [10,21] are proposed to infer target genotypes when phenotypes are given. Based on the constructed factor graph, [18] aims to infer the genotypes given genomes released by relatives and LD values among SNPs. Another work based on Markov chain Monte Carlo (MCMC) sampling to infer genotype given genotypes with large scale of phenotypic and lifestyle knowledge if individuals [24]. Moreover, genotype imputation [16] is also an significant technique utilized by geneticists to infer unknown SNPs based on known genotype data. [22] reviews the statistical techniques for imputing genotypes and describes the factors which are correlated to imputation performance. None of these works proposes an effective privacy preserving method to address privacy. Recent works have proved that individual privacy can be easily breached by inference attacks and significant countermeasures have also been proposed for location privacy, social networks [7,13,15], or mobile networks [14,27,29].

For preserving genome privacy, several contributions have proved that data anonymization is inefficient to preserve genomic privacy [9,12,19]. [19] proves that an individual's genotype can be de-anonymized since it is linked to auxiliary information such as phenotypic traits, by which individuals can be re-identified. [12] demonstrated that surnames could be inferred based on individual's genomes by investigating short tandem repeats on the Y chromosome. Moreover, encryption and differential privacy is popularly used to promoting genome openness while preserving privacy [9,17,20].

In contrast with the previous works, in this work, we not only propose an efficient inference model for inferring individual's traits given released traits, SNPs, and publicly available GWAS catalog. Furthermore, we tradeoff data openness (utility) and genome privacy by introducing the utility and privacy metrics and proposing an effective data-sanitization method, which can maximize data openness while preserving genome privacy.

7   Conclusions

In this work, we have proposed an inference method for predicting the traits and genotypes of individuals, relying on portions of observable genomic data. Meanwhile, we have proposed privacy and utility metrics to quantify privacy and utility, based on which a data-sanitization method has been proposed to realize privacy/utility tradeoff. The proposed inference method can launch inference attacks with linear complexity relying on factor graph and belief propagation, considering massive SNPs and traits. To protect against such inference attacks, we quality trait and genotype privacy based on adversary uncertainty and prediction error, and sanitize neighbor SNPs of traits and genotypes to tradeoff genomic privacy and data openness.

Acknowledgments. This work is partly supported by the National Science Foundation (NSF) of China under grant 61632010, 61602129.

References

1. https://www.patientslikeme.com/
2. https://opensnp.org/
3. http://www.nytimes.com/2013/03/24/opinion/sunday/the-immortal-life-of-henrietta-lacks-the-sequel.html?pagewanted=all
4. https://www.cdc.gov/nchs/fastats/hypertension.htm
5. The NHGRI-EBI catalog of published genome-wide association studies. https://www.ebi.ac.uk/gwas/docs/about
6. Ayday, E., Cristofaro, E.D., Hubaux, J., Tsudik, G.: The chills and thrills of whole genome sequencing (2013). CoRR abs/1306.1264
7. Cai, Z., He, Z., Guan, X., Li, Y.: Collective data-sanitization for preventing sensitive information inference attacks in social networks. IEEE Trans. Dependable Secur. Comput. **PP**(99), 1 (2016)
8. Collins, F.S., Hamburg, M.A.: First FDA authorization for next-generation sequencer. New Engl. J. Med. **369**(25), 2369–2371 (2013)
9. Erlich, Y., Narayanan, A.: Routes for breaching and protecting genetic privacy. Nat. Rev. Genet. **15**(6), 409–421 (2014)
10. Fishelson, M., Geiger, D.: Exact genetic linkage computations for general pedigrees. Bioinformatics **18**, S189 (2002)
11. Guo, X., Zhang, J., Cai, Z., Du, D.-Z., Pan, Y.: DAM: a Bayesian method for detecting genome-wide associations on multiple diseases. In: Harrison, R., Li, Y., Măndoiu, I. (eds.) ISBRA 2015. LNCS, vol. 9096, pp. 96–107. Springer, Cham (2015). doi:10.1007/978-3-319-19048-8_9
12. Gymrek, M., McGuire, A.L., Golan, D., Halperin, E., Erlich, Y.: Identifying personal genomes by surname inference. Science **339**(6117), 321–324 (2013)
13. Han, M., Li, J., Cai, Z., Han, Q.: Privacy reserved influence maximization in GPS-enabled cyber-physical and online social networks. In: 2016 IEEE International Conferences on Social Computing and Networking (SocialCom), pp. 284–292. IEEE (2016)
14. He, Z., Cai, Z., Han, Q., Tong, W., Sun, L., Li, Y.: An energy efficient privacy-preserving content sharing scheme in mobile social networks. Pers. Ubiquit. Comput. **20**(5), 833–846 (2016)
15. He, Z., Cai, Z., Sun, Y., Li, Y., Cheng, X.: Customized privacy preserving for inherent data and latent data. Pers. Ubiquit. Comput. **21**(1), 43–54 (2017)
16. Howie, B., Fuchsberger, C., Stephens, M., Marchini, J., Abecasis, G.R.: Fast and accurate genotype imputation in genome-wide association studies through pre-phasing. Nat. Genet. **44**(8), 955–959 (2012)
17. Humbert, M., Ayday, E., Hubaux, J.P., Telenti, A.: Reconciling utility with privacy in genomics. In: Proceedings of the 13th Workshop on Privacy in the Electronic Society, WPES 2014, pp. 11–20. ACM (2014)

18. Humbert, M., Ayday, E., Hubaux, J.P., Telenti, A.: Addressing the concerns of the lacks family: quantification of kin genomic privacy. In: Proceedings of the 2013 ACM SIGSAC Conference on Computer and Communications Security, pp. 1141–1152. ACM (2013)
19. Humbert, M., Huguenin, K., Hugonot, J., Ayday, E., Hubaux, J.P.: De-anonymizing genomic databases using phenotypic traits. Proc. Priv. Enhanc. Technol. **2015**(2), 99–114 (2015)
20. Johnson, A., Shmatikov, V.: Privacy-preserving data exploration in genome-wide association studies. In: Proceedings of the 19th ACM SIGKDD International Conference on Knowledge Discovery and Data Mining, KDD 2013, pp. 1079–1087. ACM, New York (2013)
21. Lauritzen, S.L., Sheehan, N.A.: Graphical models for genetic analyses. Stat. Sci. **18**, 489–514 (2003)
22. Marchini, J., Howie, B.: Genotype imputation for genome-wide association studies. Nat. Rev. Genet. **11**(7), 499–511 (2010)
23. Nyholt, D.R., Yu, C.-E., Visscher, P.M.: On Jim Watson's APOE status: genetic information is hard to hide. Eur. J. Hum. Genet. **17**(2), 147–149 (2009)
24. O'Connell, J., Sharp, K., Shrine, N., Wain, L., Hall, I., Tobin, M., Zagury, J.F., Delaneau, O., Marchini, J.: Haplotype estimation for biobank-scale data sets. Technical report, Nature Publishing Group (2016)
25. Sviridenko, M.: A note on maximizing a submodular set function subject to a knapsack constraint. Oper. Res. Lett. **32**(1), 41–43 (2004)
26. Wang, Y., Wu, X., Shi, X.: Using aggregate human genome data for individual identification. In: 2013 IEEE International Conference on Bioinformatics and Biomedicine, pp. 410–415, December 2013
27. Zhang, L., Cai, Z., Wang, X.: Fakemask: a novel privacy preserving approach for smartphones. IEEE Trans. Netw. Serv. Manag. **13**(2), 335–348 (2016)
28. Zhang, L., Pan, Q., Wu, X., Shi, X.: Building Bayesian networks from GWAS statistics based on independence of causal influence. In: 2016 IEEE International Conference on Bioinformatics and Biomedicine (BIBM), pp. 529–532, December 2016
29. Zheng, X., Cai, Z., Li, J., Gao, H.: Location-privacy-aware review publication mechanism for local business service systems. In: The 36th Annual IEEE International Conference on Computer Communications (INFOCOM) (2017)

Phylogenetic Tree Reconciliation: Mean Values for Fixed Gene Trees

Paweł Górecki[1]([⊠]), Alexey Markin[2], Agnieszka Mykowiecka[1],
Jarosław Paszek[1], and Oliver Eulenstein[2]

[1] Faculty of Mathematics, Informatics and Mechanics, University of Warsaw,
Warsaw, Poland
{gorecki,agnieszka.mykowiecka,j.paszek}@mimuw.edu.pl
[2] Department of Computer Science, Iowa State University, Ames, USA
{amarkin,oeulenst}@iastate.edu

Abstract. Phylogenetic tree reconciliation is a widely used approach
for analyzing the inconsistencies between the evolutionary histories of
genes, and the species through which they have evolved. An important
aspect of tree reconciliation are the cost functions involved that are the
minimum number of evolutionary events explaining such inconsistencies.
Mean values for these functions are fundamental when analyzing tree
reconciliations. Here we describe mean value formulas when a history of
genes is fixed for the cost functions for the events gene duplication, gene
loss and gene duplication-loss, under the uniform model of species trees.
We show that these formulas can be efficiently computed, and finally
analyze the mean values using empirical and simulated data.

Keywords: Tree reconciliation · Duplication-loss model · Deep coalescence · Speciation · Gene duplication · Gene loss · Bijectively labelled
tree · Uniform model of trees · Mean value

1 Introduction

Phylogenetic tree reconciliation is a powerful tool for analyzing the inconsistencies between the evolutionary histories of genes, and the species through which
they have evolved. Through algorithmic advances in tree reconciliation such analyzes have become common practice in various biological research areas, such
as molecular biology and microbiology [21]. For example tree reconciliation is
used to illuminate the dynamics of gene family evolution in terms of complex
evolutionary processes [5,20]. Reconciling trees is also one of the most reliable
approaches for identifying truly orthologous genes [1,2], which is a fundamental
task in understanding the evolution of genetic function [19].

Tree reconciliation is a process that takes two trees as input, a *gene tree* that
is the evolutionary history of genes, and a *species tree* that is the evolutionary
history of the species hosting the genes. It seeks an embedding of the gene tree
into the species tree (i.e., the evolution of the gene tree along the branches of

© Springer International Publishing AG 2017
Z. Cai et al. (Eds.): ISBRA 2017, LNBI 10330, pp. 234–245, 2017.
DOI: 10.1007/978-3-319-59575-7_21

the species tree) that explains possible inconsistencies between the two trees by inferring the minimum number of evolutionary events, such as gene duplication, gene loss, the combination of gene duplication-loss, and deep coalescence.

An important aspect of tree reconciliation is its associated cost that is the (minimum) number of evolutionary events inferred by the process. This, for example, allows the comparative analysis of gene trees in the context of their corresponding species trees [25, 26], which is a standard approach for synthesizing large-scale species trees from collections of discordant gene trees [3, 6].

The widespread usage of tree reconciliation in practice has led to a growing interest in analyzing reconciliation cost functions. This includes analyzing the *diameters* of such functions that are the maximum costs when one or both tree topologies are given [11–14]. More recently, the mean values of reconciliation cost functions have been studied when either a gene tree or a species tree is given. The *mean value for a gene tree* for a reconciliation cost function is the mean of the costs between the gene tree and all of its corresponding species trees. The *mean value of a species tree* is defined similarly. These mean values have been studied under two classic probability models for phylogenetic trees that are the uniform model and the Yule-Harding model [18, 24, 28].

Here we study the mean values for a gene tree under the uniform distribution for the reconciliation functions for each of the events, gene duplication and loss, gene duplication, and gene loss.

Previous Work. The pioneering work of Goodman et al. [9] introduced the approach for reconciling a gene tree with a corresponding species tree, where both of these trees are rooted and full binary. This approach is embedding the gene tree into the species tree using a *mapping* that relates every gene in the gene tree to its *host species* that is the most recent species that could have contained the gene. Consequently, the mapping is relating every leaf-gene of the gene tree to the species from which it has been sampled. When restricted to the leaf-genes, the mapping is referred to as *leaf-labeling*. Based on this mapping the evolutionary events, gene duplication, gene loss, and the combination of gene duplication and subsequent loss (in short, duplication-loss) are identified. A gene is a *gene duplication* when it has a child with the same host species, and a *gene loss* is accounted for by a maximum subtree in the species tree that has no host species (i.e., no mapping from the gene tree). While other embeddings are possible [15] the mapping describes the most parsimonious embedding in terms of the number of gene duplication and loss events [4, 7, 15]. The reconciliation cost function associated with each of these events counts the number of their occurrences in terms of gene duplications, gene losses, and gene duplications plus losses, and are termed *duplication*, *loss*, and *duplication-loss* cost functions respectively. The deep coalescence cost function, introduced by Maddison [22], is also based on the reconciliation approach. Edges in the species tree may have embedded edges from the gene tree, which are called *lineages*. The *deep coalescence cost function* counts for every edge in the species tree the number of lineages minus one, which are thought to be caused by deep coalescence events. From the mathematical point of view, the gene loss cost function is a linear combination of gene duplication and deep coalescence cost functions [16,31], and therefore, any property derived for these two functions can naturally be translated into gene loss and gene duplication-loss cost functions. All of the described reconciliation functions have been defined for general leaf-labelings and for bijective leaf-labelings.

The focus of this work are the mean values of the described reconciliation cost functions for bijective leaf-labelings under the uniform distribution of phylogenetic trees. Mean value formulas have been described for a given species tree for the deep coalescence cost function [29]. More recently such formulas have also been described for the gene duplication, gene loss, and gene duplication-loss cost functions [17]. For the computation time to obtain these mean values let n be the size of the given species tree. The mean values for a given species tree under the uniform model can be computed in $O(n)$ time for the deep coalescense cost function, and in time $O(n^3)$ for the gene duplication, gene loss, and gene duplication-loss cost functions [17]. Mean value formulas for a given gene tree have only been described for the deep coalescence cost function [29], and this value is computable in $O(n)$ time, where n is the size of the given gene tree.

Our Contributions. In this article we develop the formulas to compute the mean values for the reconciliation cost using gene duplication and loss, gene duplication, and gene loss events when the gene tree is given under a uniform distribution for the species trees. We show that these formulas can be computed in time $O(n^3)$ for a given gene tree of size n. Finally, we conducted comparative studies for fixed gene and fixed species tree means for our reconciliation costs and performed an analysis of an empirical dataset consisting of thousands of gene family trees.

2   Basic Definitions

We follow the basic definitions and notation from [16,31]. Let X be a non-empty set of n species (taxa). The set of all full binary and rooted trees whose leaves are bijectively labeled by the species in X is denoted by $R(X)$. Trees in $R(X)$ are denoted by using the standard nested parenthesis notation. Given a tree $T \in R(X)$, we denote its node and edge sets by V_T and E_T respectively. The root of T is denoted by $\mathsf{root}(T)$ and the parent of a non-root node v is denoted by $\mathsf{par}(v)$. We denote the least common ancestor of nodes $v, w \in V_T$ in tree T by $\mathsf{lca}_T(v, w)$. A *cluster* (or also called clade) of a node $v \in V_T$ is the set of all leaf labels of the subtree of T rooted at v.

In phylogenetic tree reconciliation a gene tree is embedded into its corresponding species tree. In this work we assume that both types of trees have the same bijective labelling of leaves. Therefore, we assume that every gene tree and every species tree is an element of $R(X)$. For a (gene) tree $G \in R(X)$ and a (species) tree $S \in R(X)$ *the least common ancestor mapping between G and S,* or *lca-mapping*, $\mathsf{M}\colon V_G \to V_S$, is defined as $\mathsf{M}(g) = s$ if g and s are leaves with the same label, and $\mathsf{M}(g) = \mathsf{lca}_S(\mathsf{M}(g'), \mathsf{M}(g''))$ if g has two children g' and g''. An internal node g is called a *duplication*, or an *S-duplication*, if $\mathsf{M}(g) = \mathsf{M}(a)$

for a child a of g. Every internal non-duplication node is called a *speciation*. The duplication cost, denoted by $D(G, S)$, is the total number of S-duplications in G [25]. The deep coalescence cost function [22,23,31] can be expressed by $DC(G, S) := \sum_{g \in V_G \setminus \{\text{root}(G)\}} (\| M(g), M(\text{par}(g)) \| - 1)$, where $\|a, b\|$ is the number of edges on the simple path connecting nodes $a, b \in S_V$. The reader is referred to [29] for alternative definitions of DC. Finally, we can provide formulas for the loss and duplication-loss cost functions [31]: $L(G, S) := 2 D(G, S) + DC(G, S)$ and $DL(G, S) := D(G, S) + L(G, S)$. For a more detailed introduction to the model please refer to [15,22,25].

3 Results

In the uniform model of binary trees an equal probability is assigned to each possible leaf labeled binary tree with n leaves. In this model rooted trees can be generated by uniform and random insertions of one edge to any edge including the rooting edge at each step. For example, given a rooted tree $(a, (b, c))$, the following five four-labelled trees can be created by inserting a new edge with a leaf d: $(((a, d), b), c)$, $((a, (b, d)), c)$, $(((a, b), d), c)$, $((a, b), (c, d))$, and $(((a, b), c), d)$.

We analyse the mean of the duplication cost in the uniform model of rooted leaf-labeled trees. Let $R(X)$ denote the set of all bijectively labeled rooted trees over a non-empty set X. Then, the mean of duplication cost for a fixed gene tree $G \in R(X)$ under a probabilistic model of species trees is:

$$\overline{D}_u(G) = \sum_{S \in R(X)} \mathbb{P}(S) \, D(G, S). \tag{1}$$

Recall that size of $R(X)$ is $b(n) = (2n - 3)!!$, where $k!!$ is the double factorial, i.e., $k!! = k \cdot (k - 2)!!$ and $0!! = (-1)!! = 1$. Hence, in the uniform model for every tree $T \in R(X)$ has probability $\mathbb{P}(T) = \frac{1}{b(n)}$.

Now we introduce a notion of a (rooted) split. Every non-leaf node $v \in V_T$, induces a *split* $A|B$, where A and B are the clusters of children of v. The set of all splits in T is denoted by $\text{Spl}(T)$. As an example, $\text{Spl}(((a, b), (c, d))) = \{\{\{a, b\}, \{c, d\}\}, \{\{a\}, \{b\}\}, \{\{c\}, \{d\}\}\}$, which we describe by using the simplified split notation: $\{ab|cd, a|b, c|d\}$.

For a split $A|B$ induced by a node v from a fixed gene tree $G \in R(X)$, by $\xi_n^{\text{Dup}}(A, B)$ we denote the number of species trees S from $R(X)$ such that v is an S-duplication node. Similarly, we define $\xi_n^{\text{Spec}}(A, B)$ for speciation nodes.

Lemma 1. *For a gene tree G with n leaves,*

$$\sum_{A|B \in \text{Spl}(G)} \xi_n^{\text{Dup}}(A, B) + \xi_n^{\text{Spec}}(A, B) = b(n) \cdot (n - 1).$$

Now, the mean (1) is equivalent to

$$\overline{D}_u(G) = \frac{1}{b(n)} \sum_{A|B \in \text{Spl}(G)} \xi_n^{\text{Dup}}(A, B) = n - 1 - \frac{1}{b(n)} \sum_{A|B \in \text{Spl}(G)} \xi_n^{\text{Spec}}(A, B). \tag{2}$$

Similarly to [17], it is more convenient to count directly the number of speciation nodes rather then duplications.

Lemma 2. *For a species tree G with n leaves and a split $A|B$ present in G*

$$\xi_n^{\mathsf{Spec}}(A,B) = \sum_{i=0}^{m}\sum_{j=0}^{m-i}\binom{m}{i}\binom{m-i}{j}b(|A|+i)b(|B|+j)b(m-i-j+1).$$

where $m = n - |A| - |B|$.

Proof. Let $v \in G$ has the split $A|B$. A species tree S that induces a speciation node v mapped into a node s from S can be constructed as follows. Let z be an element not in X. Let A' and B' be two disjoint supersets of A and B, respectively. Then, a species tree $S \in R(X)$ such that s has split $A'|B'$ can be constructed by replacing the leaf z in a tree $R((X \setminus (A' \cup B')) \cup \{z\})$ by a tree (S_A, S_B) such that $S_A \in R(A')$ and $S_B \in R(B')$. Then, v is a speciation node mapped to the root of (S_A, S_B) in S. On the other hand note that every S such that v from G is a speciation node mapped to a node in S, is inferred exactly once in the above procedure. $\qquad\square$

Now, we can state the main result that follows from Lemma 2 and Eq. 2.

Theorem 1 (Fixed gene tree mean of D under the uniform model). *For a given gene tree G with n leaves*

$$\overline{D}_u(G) = n - 1 - \frac{1}{b(n)}\sum_{\substack{A|B\in\mathsf{Spl}(G)\\m=n-|A|-|B|}}\sum_{i=0}^{m}\sum_{j=0}^{m-i}\binom{m}{i}\binom{m-i}{j}b(|A|+i)b(|B|+j)b(m-i-j+1).$$

To obtain the mean formula for DL cost we recall the result from [29] (see Corollary 13) on the deep coalescence cost. For a gene tree G with n leaves:

$$\overline{DC}_u(G) = -(2n-1) + 2n\frac{(2n-2)!!}{b(n)} - \frac{(2n-2)!!}{b(n)}\sum_{v\in V_G}\frac{(2|C_v|-3)!!}{(2|C_v|-2)!!},$$

where C_v denotes the cluster of a node v.

Finally, we have the result for DL and L (see also similar results for fixed species tree from [17]).

Theorem 2 (Fixed gene tree mean of DL and L). *For a gene tree G we have $\overline{DL}_u(G) = 3 \cdot \overline{D}_u(G) + \overline{DC}_u(G)$ and $\overline{L}_u(G) = 2 \cdot \overline{D}_u(G) + \overline{DC}_u(G)$.*

Proof. It follows from the definition of gene loss and duplication-loss functions and the properties of mean values. $\qquad\square$

Given the mean formulas for DC and D it is now straightforward to obtain the exact formulas for the means of DL and D. We omit these details for brevity. See an example of mean values depicted in Fig. 1.

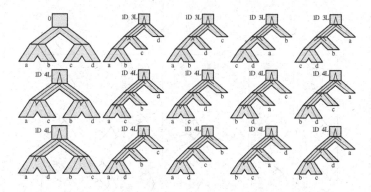

Fig. 1. Embeddings (scenarios) of $G = ((a, b), (c, d))$ into every species four-leaf species tree [15]. Each scenario is summarized with two numbers denoting the number of gene duplications (D) and the number of gene losses (L). We have 14 gene duplications, 31 speciation nodes and 52 gene losses in total. In this example, $\overline{D}_u(G) = 14/15$, $\overline{L}(G) = 52/15$ and $\overline{DL}_u(G) = 66/15$.

Computing the mean of deep coalescence for a fixed gene tree can be completed in $O(n)$ steps under assumption that double factorials are memorized and the required size of clusters is stored with the nodes of the standard pointer-like implementation of trees. For the mean of the remaining cost functions, however, we need two additional loops. Therefore, the time complexity of computing $\overline{D}_u(G)$, $\overline{L}_u(G)$ and $\overline{DL}_u(G)$ is $O(n^3)$.

4   Experimental Evaluation

4.1   Mean Values for Tree Shapes

Here we analyze the mean values of our analyzed reconciliation cost functions for all tree shapes with $3, 4, \ldots 9$ leaves ordered by their Furnas rank [8], which are depicted in Table 1. We observe that tree shapes with the same number of splits induce the same mean values (e.g., the two red colored tree shapes) which follows directly from the mean value formulas for deep coalescence and duplication cost functions. This property also holds for the mean values when a species tree is fixed [17]. Note, while in [17] the mean value of the duplication cost function for a fixed species tree was conjectured to grow monotonically with the Furnas rank, this is not the case for the corresponding mean values when a gene tree is fixed as indicated in Table 1. Moreover, we can observe that the mean value of the duplication cost function is maximum for caterpillar trees while it is minimum for the most balanced once.

Moreover, we compared the mean values for fixed species tree shapes from [17] with their corresponding values when the gene tree is fixed. Therefore, we computed the mean values for all gene tree shapes with up to 20 leaves, e.g., for $n = 20$ there are 293547 trees. Figure 2 depicts two diagrams which represent the means for a fixed species tree shapes [17] and the corresponding means for a

Table 1. *Mean values for all gene tree shapes with* $n \in \{3, 4, \ldots, 9\}$ *leaves.* The shapes are shown ordered by their Furnas rank [8]. The table is patterned after [17,29]. The two red shapes for $n = 9$ have the same number of splits, which implies equal values of the corresponding mean values.

$n \leq 6$	0.67 2.00 2.67 0.67	1.47 4.93 6.40 2.00	0.93 3.4[illegible] 4.40 1.[illegible]	2.32 8.59 10.91 3.94	1.79 7.07 8.86 3.49	1.64 6.53 8.17 3.26	3.21 12.87 16.08 6.44	2.68 11.30 13.97 5.94	2.53 10.74 13.26 5.68	2.45 10.43 12.88 5.52	1.92 8.85 10.77 5.02	2.32 9.92 12.24 5.27
$n = 7$	4.12 17.71 21.83 9.46	3.59 16.09 19.67 8.91	3.44 15.[illegible]0 18.94 8.[illegible]3	3.36 15.18 18.54 8.46	2.83 13.56 16.38 7.90	3.23 14.65 17.88 8.18	3.31 14.97 18.28 8.34	2.78 13.34 16.12 7.78	2.63 12.76 15.39 7.51	3.13 14.15 17.27 7.89	2.60 12.52 15.12 7.33	
$n = 8$	5.04 23.05 28.10 12.97	4.51 21.39 25.90 12.37	4.36 20.[illegible]9 25.15 12[illegible]07	4.28 20.45 24.73 11.88	3.75 18.79 22.54 11.29	4.16 19.90 24.05 11.59	4.24 20.23 24.46 11.75	3.70 18.56 22.27 11.16	3.55 17.96 21.51 10.86	4.05 19.37 23.42 11.27	3.52 17.71 21.23 10.67	4.21 20.07 24.27 11.66
	3.67 18.40 22.08 11.06	3.52 17.80 21.32 10.76	3.44 17.[illegible]6 20.90 10[illegible]57	2.91 15.80 18.71 9.98	3.32 16.91 20.23 10.28	3.99 19.02 23.00 11.04	3.45 17.35 20.81 10.44	3.30 16.75 20.05 10.14	3.93 18.72 22.65 10.85	3.40 17.05 20.45 10.26	2.87 15.39 18.25 9.66	
$n = 9$	5.98 28.88 34.86 16.92	5.44 27.17 32.62 16.29	5.29 26.[illegible]5 31.84 15[illegible]97	5.22 26.20 31.42 15.77	4.68 24.50 29.18 15.13	5.09 25.63 30.72 15.45	5.17 25.97 31.14 15.63	4.64 24.27 28.90 14.99	4.48 23.64 28.13 14.67	4.99 25.08 30.07 15.11	4.45 23.38 27.83 14.47	5.14 25.80 30.94 15.52
	4.61 24.10 28.70 14.89	4.45 23.48 27.93 14.57	4.38 23.[illegible]2 27.50 14[illegible]37	3.84 21.42 25.27 13.73	4.25 22.55 26.80 14.05	4.92 24.71 29.63 14.87	4.39 23.01 27.40 14.23	4.24 22.38 26.62 13.91	4.87 24.40 29.26 14.67	4.33 22.70 27.03 14.03	3.80 20.99 24.79 13.40	5.12 25.68 30.79 15.44
	4.58 23.97 28.56 14.81	4.43 23.35 27.78 14.49	4.35 23.[illegible]0 27.35 14[illegible]29	3.82 21.29 25.12 13.65	4.23 22.43 26.65 13.97	4.31 22.77 27.07 14.15	3.78 21.06 24.84 13.51	3.62 20.44 24.06 13.19	4.12 21.88 26.00 13.63	3.59 20.18 23.77 13.00	4.88 24.44 29.32 14.68	4.34 22.73 27.08 14.04
	4.19 22.11 26.30 13.73	4.12 21.76 25.88 13.53	3.58 20.[illegible]6 23.64 12[illegible]89	3.99 21.19 25.18 13.21	4.79 23.92 28.71 14.34	*4.26 22.22* *26.48 13.71*	4.10 21.60 25.70 13.39	*4.26 22.22* *26.48 13.71*	3.72 20.52 24.24 13.07	3.57 19.89 23.46 12.75	Key: $\overline{D}_u$ $\overline{L}_u$ $\overline{DL}_u$ $\overline{DC}_u$	

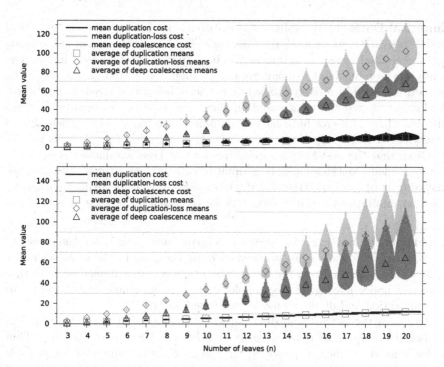

Fig. 2. *Top:* frequency diagram of mean values of duplication, duplication-loss and deep coalescence costs for all fixed gene tree shapes for $n = 3, 4, \ldots 20$ under the uniform model of species trees. For each n, mean values for every cost were grouped into bins of size 0.01. The width of each bin is proportional to $\log_2 K$, where K is the number of gene tree shapes having the mean value in this bin. *Bottom:* the same type of diagram for means of fixed species tree taken from [17]. (Color figure online)

fixed gene tree shapes, respectively. While we are expecting that the blue ovoids and the red ovoids will increasingly overlap with an increasing number of taxa, we observe that this occurs earlier (i.e., for smaller sizes of taxa) for species tree shapes. For the duplication cost function we observe a broader range of means in the upper diagram, while the range for the other cost functions appears to be broader for the species tree shapes. In combination with our previous observations from Table 1, we conclude that the properties of the duplication cost function differs significantly when comparing the two types of fixed tree means.

4.2 Empirical Study

In this section we study the distribution of mean values for the duplication and duplication-loss cost functions for gene trees obtained from a baseline empirical dataset. Additionally, we evaluate how the duplication and duplication-loss costs compare to the respective mean values.

Empirical Dataset. To evaluate the distributions of mean values on empirical phylogenetic datasets we analyzed the classic *TreeFam ver.9* dataset [27] consisting of gene family trees of 109 mostly animal species (with 71 taxa in the gene family trees on average). Among around 15 thousand rooted gene trees in the dataset, the 4070 bijectively labeled and strictly bifurcated trees were selected. We further filtered the trees based on their size; that is, we removed all trees with less than 10 leaves in order to eliminate otherwise arising outliers due to insufficient tree size.

Given that the best-known species tree for the TreeFam dataset is not completely refined (contains many large multifurcations), we estimated the species tree using a popular supertree tool, *duptree2* [30]; the tool approximates a species tree that minimizes the duplication cost for the given set of gene trees.

Experimental Setting. In order to compare mean values for gene trees of different sizes and topologies we need to bring them up to the same scale. We achieve this by normalizing the mean values by respective diameters. Note that diameters under fixed gene tree topologies can be computed exactly, both for the duplication and duplication-loss cost functions [10, 13].

To assess the mean value distributions for trees taken from the empirical dataset, we compare them to *complete* distributions for trees of fixed size. That is, for a fixed number of leaves, t, we compute mean values for all possible tree topologies with t leaves. This is repeated for $t = 10, 12, 14$, and 16. Apart from serving as a complete distribution reference, these data also allows us to empirically observe how the mean-value distributions progress with the increase of taxa.

Results and Discussion. Figure 3 illustrates that the mean values under the duplication cost function for the TreeFam gene trees are concentrated around the value 0.9. That is, the mean values are very close to respective cost diameters, which implies that for all the trees under consideration, most of possible species trees have a very high (close to the maximum) duplication cost. It also suggests that the proximity of a duplication cost (normalized by the diameter) to 0 indicates a high confidence in the species tree.

Further, the complete distributions of duplication means for all possible tree topologies over varying taxa size are shown on Fig. 4 (left hand side) closely resemble the distribution on empirical datasets. The figure also demonstrates that the duplication-mean distribution does not seem to change much with the increase of taxa.

The empirical distribution for the duplication-loss means on Fig. 3 (left-hand side, red histogram) is rather spread on the interval from approximately 0.25 to 0.7 with multiple picks. Figure 4 (right hand side) additionally shows that duplication-loss mean values (normalized by the respective diameters) gradually decrease with increasing taxa number. Given that the TreeFam dataset contains trees of varying size, the shifts in mean values for gene trees of larger size, explain the wide range of duplication-loss means on Fig. 4.

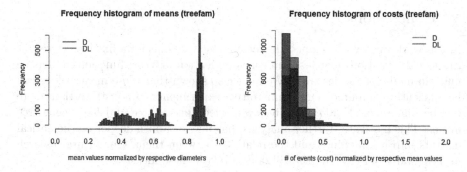

Fig. 3. Comparison of (i) mean values normalized by diameters and (ii) costs normalized by mean values for the duplication (D) and the duplication-loss (DL) cost functions (TreeFam dataset). (Color figure online)

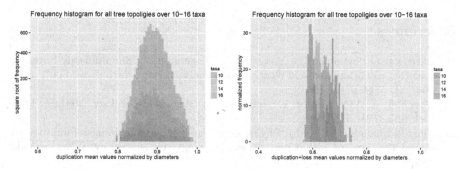

Fig. 4. *Left*: distribution of *duplication*-mean values normalized by respective diameters. The frequencies of the histogram were scaled by a square root to achieve a more comprehensive visualization. *Right*: distribution of duplication-loss-mean values normalized by respective diameters. Distributions are shown for all possible tree topologies over 10, 12, 14, and 16 taxa respectively.

Further, the mean values play an important role in the normalization of reconciliation costs, since it allows us to relate reconciliation costs that are otherwise significantly affected by topologies of the gene trees. The histogram on the right-hand side of Fig. 3 shows duplication and duplication-loss costs normalized by respective mean values. While the majority of trees are concentrated below the value 0.5 (i.e., the cost is significantly smaller than the respective mean), there are some outliers for which the cost is close to the mean or even exceeds it. Such trees can be thought of as not strongly correlating with the corresponding species tree (or even correlating negatively), and they can represent gene families of interest for a researcher. Alternatively, when the reconciliation cost between a gene tree and a species tree exceeds the mean value, it might indicate possible errors in the gene tree.

5   Conclusions

In this work we have developed the mean value formulas for a fixed gene tree for the gene duplication, gene loss and gene duplication-loss cost functions under the uniform model of species trees. We have also shown that these mean values can be efficiently computed. Our comparative experiments demonstrate that there can be fundamental differences between fixed species tree and fixed gene tree means. This motivates further analyzes that may establish deeper mathematical insights into mean values and the relations between them. Our future research in mean values of tree shapes will dovetail with these ideas.

Acknowledgements. This material is based upon work supported by the grants of the National Science Foundation under Grant No. 1617626 and the NCN #2015/19/B/ST6/00726.

References

1. Akerborg, O., Sennblad, B., Arvestad, L., Lagergren, J.: Simultaneous Bayesian gene tree reconstruction and reconciliation analysis. Proc. Natl. Acad. Sci. U.S.A. **106**(14), 5714–5719 (2009)
2. Altenhoff, A.M., Dessimoz, C.: Inferring orthology and paralogy. Methods Mol. Biol. **855**, 259–279 (2012)
3. Bininda-Emonds, O.R. (ed.): Phylogenetic Supertrees: Combining Information to Reveal the Tree of Life. Computational Biology, vol. 4. Springer, Netherlands (2004)
4. Bonizzoni, P., Della Vedova, G., Dondi, R.: Reconciling a gene tree to a species tree under the duplication cost model. Theor. Comput. Sci. **347**(1–2), 36–53 (2005)
5. David, L.A., Alm, E.J.: Rapid evolutionary innovation during an Archaean genetic expansion. Nature **469**(7328), 93–96 (2011)
6. Eulenstein, O., Huzurbazar, S., Liberles, D.: Reconciling phylogenetic trees. In: Evolution After Gene Duplication, pp. 185–206. Wiley (2010)
7. Eulenstein, O.: Vorhersage von Genduplikationen und deren Entwicklung in der Evolution. Ph.D. thesis, Rheinische Friedrich-Wilhelms-Universität Bonn, Bonn, Germany (1008)
8. Furnas, G.W.: The generation of random, binary unordered trees. J. Classif. **1**(1), 187–233 (1984)
9. Goodman, M., et al.: Fitting the gene lineage into its species lineage, a parsimony strategy illustrated by cladograms constructed from globin sequences. Syst. Zool. **28**(2), 132–163 (1979)
10. Górecki, P., Eulenstein, O.: Bijective diameters of gene tree parsimony costs (2017, submitted)
11. Górecki, P., Eulenstein, O.: Deep coalescence reconciliation with unrooted gene trees: linear time algorithms. In: Gudmundsson, J., Mestre, J., Viglas, T. (eds.) COCOON 2012. LNCS, vol. 7434, pp. 531–542. Springer, Heidelberg (2012). doi:10.1007/978-3-642-32241-9_45
12. Górecki, P., Eulenstein, O.: Maximizing deep coalescence cost. IEEE-ACM Trans. Comput. Biol. Bioinform. **11**(1), 231–242 (2014)
13. Górecki, P., Eulenstein, O.: Gene tree diameter for deep coalescence. IEEE-ACM Trans. Comput. Biol. Bioinform. **12**(1), 155–165 (2015)

14. Górecki, P., Paszek, J., Eulenstein, O.: Unconstrained gene tree diameters for deep coalescence. In: Proceedings of the 5th ACM Conference on Bioinformatics, Computational Biology, and Health Informatics, BCB 2014, pp. 114–121. ACM, New York (2014)

15. Górecki, P., Tiuryn, J.: DLS-trees: a model of evolutionary scenarios. Theor. Comput. Sci. **359**(1–3), 378–399 (2006)

16. Górecki, P., Eulenstein, O., Tiuryn, J.: Unrooted tree reconciliation: a unified approach. IEEE-ACM Trans. Comput. Biol. Bioinform. **10**(2), 522–536 (2013)

17. Górecki, P., Paszek, J., Mykowiecka, A.: Mean values of gene duplication and loss cost functions. In: Bourgeois, A., Skums, P., Wan, X., Zelikovsky, A. (eds.) ISBRA 2016. LNCS, vol. 9683, pp. 189–199. Springer, Cham (2016). doi:10.1007/978-3-319-38782-6_16

18. Harding, E.F.: The probabilities of rooted tree-shapes generated by random bifurcation. Adv. Appl. Probab. **3**(1), 44–77 (1971)

19. Ihara, K., Umemura, T., Katagiri, I., Kitajima-Ihara, T., Sugiyama, Y., Kimura, Y., Mukohata, Y.: Evolution of the archaeal rhodopsins: evolution rate changes by gene duplication and functional differentiation. J. Mol. Biol. **285**(1), 163–174 (1999)

20. Kamneva, O.K., Knight, S.J., Liberles, D.A., Ward, N.L.: Analysis of genome content evolution in PVC bacterial super-phylum: assessment of candidate genes associated with cellular organization and lifestyle. Genome Biol. Evol. **4**(12), 1375–1390 (2012)

21. Kamneva, O.K., Ward, N.L.: Reconciliation approaches to determining HGT, duplications, and losses in gene trees. In: Goodfellow, M., Chun, J., Sutcliffe, I.C. (eds.) New Approaches to Prokaryotic Systematics, Methods in Microbiology, chap. 9, vol. 41, pp. 183–199. Academic Press (2014)

22. Maddison, W.P.: Gene trees in species trees. Syst. Biol. **46**, 523–536 (1997)

23. Maddison, W.P., Knowles, L.L.: Inferring phylogeny despite incomplete lineage sorting. Syst. Biol. **55**(1), 21–30 (2006)

24. McKenzie, A., Steel, M.: Distributions of cherries for two models of trees. Math. Biosci. **164**(1), 81–92 (2000)

25. Page, R.: From gene to organismal phylogeny: reconciled trees and the gene tree/species tree problem. Mol. Phylogenet. Evol. **7**(2), 231–240 (1997)

26. Page, R.D.M.: Maps between trees and cladistic analysis of historical associations among genes, organisms, and areas. Syst. Biol. **43**(1), 58–77 (1994)

27. Ruan, J., et al.: TreeFam: 2008 update. Nucleic Acids Res. **36**, D735–D740 (2008)

28. Steel, M.A., Penny, D.: Distributions of tree comparison metrics – some new results. Syst. Biol. **42**(2), 126–141 (1993)

29. Than, C.V., Rosenberg, N.A.: Mean deep coalescence cost under exchangeable probability distributions. Discret. Appl. Math. **174**, 11–26 (2014)

30. Wehe, A., Burleigh, J.G.: Scaling the gene duplication problem towards the tree of life: accelerating the rSPR heuristic search (2010)

31. Zhang, L.: From gene trees to species trees II: species tree inference by minimizing deep coalescence events. IEEE-ACM Trans. Comput. Biol. Bioinform. **8**, 1685–1691 (2011)

Computer Assisted Segmentation Tool:
A Machine Learning Based Image Segmenting
Tool for TrakEM2

Augustus N. Tropea[1](✉), Janey L. Valerio[1], Michael J. Camerino[2],
Josh Hix[2], Emmalee Pecor[2], Peter G. Fuerst[3], and S. Seth Long[1]

[1] Department of Natural Sciences and Mathematics,
Lewis-Clark State College, Lewiston, ID 83501, USA
antropea@lcmail.lcsc.edu
[2] North Idaho College, Coeur d'Alene, ID 83814, USA
[3] Department of Biological Sciences,
University of Idaho, Moscow, ID 83844, USA

Abstract. The recent availability of serial block face scanning electron microscopy has permitted researchers to reconstruct cells and neurons by manually identifying and coloring objects. This technique was instrumental in work such as uncovering the anatomical basis for direction selectivity of vision [1]. Unfortunately, reconstruction involves an expenditure of time which can be expensive or prohibitive. We have developed the Computer Assisted Segmentation Tool (CAST), which produces results that appear similar to manual segmentation with reduced personnel time requirements. Results are shown for serial block face electron micrograph (SBEM) images of Mus musculus retinal axons; however, CAST is capable of operation on other image types. CAST is available under an open source license in a modified version of the TrakEM2 plugin for the popular Fiji image analysis suite. Usage and installation instructions can be found at http://isoptera.lcsc.edu/segmentation_tool/.

Keywords: Machine learning · Image segmentation · Neural network · Fast marching algorithm · Fiji · Track EM2 · Trainable Weka Segmentation · Neuron

1 Introduction

Since the invention of the electron microscope in 1926, Electron Microscopy (EM) technology has advanced to the point that a scanning electron microscope can be combined with a microtome to create Sequential Block-Face Scanning Electron Microscopy (SBEM or SBFSEM). Using SBEM, a stack of high-resolution images can be created, with sufficient detail to reconstruct neural structures [1]. Reconstructed neural structures can be used to study conditions affecting neurons. For example, Briggman et al. used 3D reconstruction to discover how direction selectivity in the retina is mediated by the anatomical arrangement of synapses [1]. Reconstruction of 3D structures in SBEM imagery has been performed in a manual manner by outlining the structure of interest in each image of a stack. Once outlines are complete, the structure can be visualized in 3D [2].

© Springer International Publishing AG 2017
Z. Cai et al. (Eds.): ISBRA 2017, LNBI 10330, pp. 246–257, 2017.
DOI: 10.1007/978-3-319-59575-7_22

Since available person-hours are always in high demand, some attempt must be made to maximize the quality and quantity of manual image segmentation produced on a fixed budget. For example, touch screen technology can be used to allow coloring of structures using a stylus instead of a mouse [3]. As another example, automated detection of neural boundaries has been pursued to improve outlining efficiency [4]. This latter strategy, of automated boundary detection, has been tested in several works [4–9]. As part of our work understanding how retinal neural circuits are organized, we propose the Computer Assisted Segmentation Tool (CAST) to fill the need for a readily-available segmentation tool. Some tools such as Raveler [10], Ilastik [11], Rhoana [12], and Knossos [13] are available for segmentation of EM imagery. While each one of these tools does some kind of image segmentation, none of the tools mentioned solves the targeted neuron segmentation problem, as described previously. CAST is designed to aid in studying neural connectivity in a targeted area, so the segmentation of an entire image is unnecessary when trying to identify one or two specific neurons. Eyewire is an interactive online game which allows users to assist in mapping neural connectivity through the process of semi-automatically outlining neurons. CAST is designed to do the same thing, but locally with user specific data sets. CAST is also designed to be a more general outlining tool, offering semi-automatic outlining for more than just neurons. A Fast Marching tool has been integrated into the Fiji Image Analysis suite; however, it has not proven suitable in some cases (see Results section). Integrating CAST into the TrakEM2 Fiji plugin allows a hybrid workflow involving both manual and semi-automatic segmentation. Availability of both tools enables manual segmentation of any part of any structure that cannot be segmented automatically. Using this hybrid workflow, CAST performance that is imperfect yet largely correct can be utilized for improved efficiency. Integration with TrakEM2 also allows semi-automatic segmentation within partially-segmented data sets to preserve and build upon existing work. Finally, integrating CAST into TrakEM2 avoids the need to re-train manual segmentation personnel to use different software.

In our previous work we performed reconstruction of rod spherules from SBEM imagery [14]. Since reconstruction of the *Mus musculus* retina is ongoing, we have focused development of CAST on segmentation of *Mus musculus* retinal neurons, specifically retinal bipolar cells. Results are presented on *Mus musculus* axons, and comparison is made with the previous fast marching approach. To assess the accuracy of the tool, an in-depth comparison is made to manual outlines.

Denk and Horstmann [15] have explained the importance of visualizing and reconstructing 3D structures in electron microscope images, and CAST is intended to reduce the time cost of this reconstruction and visualization. The following sub sections discuss the different platforms this tool was built for, and previous work done in the field of image segmentation.

1.1 Implementation

Fiji is an open source software suite for image processing. "Fiji uses modern software engineering practices to combine powerful software libraries with a broad range of scripting languages to enable rapid prototyping of image processing algorithms" [16]. Fiji is the base software package that many other plugins are built to interact with.

The popular neural circuit reconstruction program, TrakEM2, is a plugin written for Fiji that allows the user to reconstruct, measure, and visualize neural structures in 3D. The 3D morphological reconstructions are created by outlining structures of interest throughout the sequence of EM images. TrakEM2 then creates a 3D representation of the structure using the outlines. The CAST tool has been built directly into TrakEM2, and is accessible through the toolbar just like the freehand tool. This makes it possible for someone who is previously familiar with the TrakEM2 program to start using this tool without any kind of specialized training.

1.2 Materials

EM images used in this work were acquired by briefly perfusing mice with saline. The retinas were then removed and fixed in a cacodylate buffer. Retinas were embedded for EM and imaged by Renovo Inc. For a more detailed description see the description that was previously stated in [14].

2 Algorithms

CAST uses Multi-Layer Perceptron (MLP), a type of artificial neural network (ANN) that is often applied to classification problems [17]. An ANN consists of nodes called neurons, and learns based on how the information is passed through the network. Information comes into a MLP through the input layer, is passed through a series of hidden layers, and produces a classification based on which neurons are activated.

The training set used for CAST was created using Trainable Weka Segmentation in Fiji [18]. Each pixel that is outlined in Trainable Weka Segmentation is used as an example to train the neural network. The retina training set is comprised of 5,891 pixel examples. The accuracy of the training set was tested using Weka Explorer, which calculated the accuracy of correctly classifying pixels to be 96.4013% [19]. Trainable Weka Segmentation provides many different filters to choose from for feature extraction. To best optimize the results of the tool the following filters were used: Gaussian blur, Hessian, Membrane projections, Sobel filter, difference of Gaussians, and derivatives. These are all of the default filters with the exception of the derivatives filter. Gaussian blur performs n individual convolutions with Gaussian kernels using the normal n variations of σ. Sobel filter calculates the gradient at each pixel and is commonly used for boundary detection. This is helpful because there is a change in gradients between the borders and inside area of retinal neurons. Hessian creates a matrix at each pixel and generates features based on the different matrix operations. Difference of Gaussians calculates two Gaussian blur images from the original image and subtracts one from the other. Membrane projections use six different kernels to find the sum, mean, standard deviation, median, maximum, and minimum of the pixels in the image. The derivatives filter was added to the default settings because the contrast between pixels is a key feature. With these filters selected ninety-seven features/attributes were produced. For more information on the filters refer to the Fiji website [20]. The network used consisted of 97 input nodes, 1 hidden layer with 49 nodes, and 2 output nodes. To determine the number of nodes used in the hidden layer the following formula was used, $\lfloor \frac{attributes + classes}{2} \rfloor$.

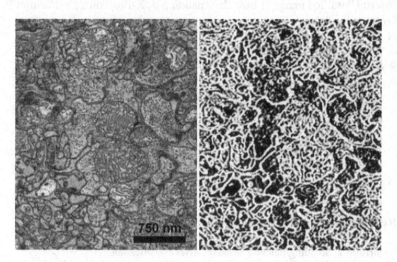

Fig. 1. ANN image processing. Left: Electron micrograph of the retina. **Right**: The ANN converts the image into one in which continuous features are consistently colored.

The network was trained using ten-fold cross validation. The learning method used was the standard method of backpropagation (gradient decent). An image is given to the network as input, and the network produces a binary image based on its classifications. This can be seen above in Fig. 1. The pixels the network classifies as a boundary are white, and the pixels it classifies as not a boundary are black.

The Expand Area algorithm was used to explore outward from a single point.

Input: Point p inside the structure, Stopping function T_i
Output: Set of points inside the structure
$P \leftarrow [p]$
$C \leftarrow Neighborsof(p)$
foreach $c \in C$
   If $c \notin P \land Intensity(c) \geq T_i$
     Add c to P
     Add $Neighborsof(c)$ to C
Return P

Algorithm 1. Expand Area

The stopping function in Expand Area is an intensity threshold function [21]. Expand Area would stop if the difference of intensities is greater than a predefined threshold value. However, the intensity threshold function proved to be insufficient at finding boundaries, because a single misclassified pixel in a boundary can result in a leak. Algorithm 2 prevents leaks of this nature by checking for a difference over a number of pixels.

Input: Classified image I, Boundary padding d, X-Position x, Y-Position y
Output: Boolean
$M \leftarrow 0$
$m \leftarrow 255$
$R \leftarrow False$
foreach $\{i \in \mathbb{Z} | -d \le i \le d\}$
  foreach $\{j \in \mathbb{Z} | -d \le j \le d\}$
    If $M < I(x + i, y + j)$
      $M \leftarrow I(x + i, y + j)$
    If $m > I(x + i, y + j)$
      $m \leftarrow I(x + i, y + j)$
If $\frac{(M-m)}{\left(\frac{M+m}{2}+1\right)} > 1$
  $R \leftarrow True$
Return R

Algorithm 2. Stopping function based on ANN classification

Using Algorithm 2, leaks are reduced; however, detecting boundaries based on intensity gradient is sensitive to imaging parameters, cell type, etc. In some cases, a thick membrane poorly stained may not present an adequate gradient to trigger the stopping condition. Therefore, as a further refinement, an ANN was used to detect boundaries and generate a binary image showing boundary and non-boundary. Expand Area operates using the binary image produced by the neural network. When operating on a binary image, m and M in Algorithm 2 are either 0 or 1.

Each pixel, assuming the pixel is not on the edge of an image, has eight neighbors or eight possible directions of travel. Expand Area works by exploring each of these neighbors. Once the algorithm runs into a boundary, a white pixel, it stops exploring in that direction. This process is then shifted to a neighbor and repeated until there are no neighbors left to explore, or it has reached a predefined stopping point. The process is initiated with a single point supplied by the user via the mouse.

3 Results

3.1 Comparison with Fast Marching Method

In order to develop CAST we utilized an artificial neural network (ANN) and tested its ability against fast marching. The fast marching algorithm has previously been used to solve boundary value problems. Fast marching works similarly to Dijkstra's algorithm, using a travel time function [22]. The algorithm starts at one or more points. All the neighbors of the starting points are initially labeled as far away or unvisited. Each neighbor is then explored, and considered using the travel time function or "the propagation of the interface is done via the construction of the arrival time function" [22]. Once all of the neighbors have been evaluated and when all fronts propagating in opposite directions have met, fast-marching segmentation is finished [22]. The applications of the fast marching method range from fluid interactions to noise reduction and image segmentation. The fast marching method and its applications to image segmentation have been integrated into TrakEM2. Fast marching is used in TrakEM2 to provide semi-automatic segmentation of structures. This method does well finding boundaries when the contrast between the inner area and the boundary is high. However, fast marching does have trouble with structures that have a lower contrast between the boundary and inner area. The fast marching algorithm also has the problem of leaking into adjacent structures. This happens if there is a gap in the boundary. Even if there is a perceived one-pixel gap in the boundary, fast marching will leak into the adjacent structure. Once fast marching leaks into the adjacent structure, it will continue until it finds another boundary or it has reached its predefined stopping bounds. Figure 2 shows a graphical representation of the leak problem.

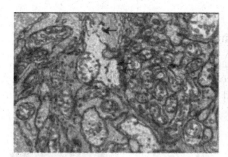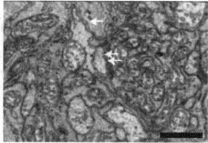

750 nm

Fig. 2. Fast Marching vs. ANN. Left: Fast marching results in mistaken identification of continuous areas (red arrows). **Right**: The ANN is better able to predict continuous areas with less leakage (green arrows; note absent leakage). (Color figure online)

3.2 Image Segmentation Using Neural Networks

Classifiers such as perceptrons, the basic component of neural networks, have been found useful for identifying membranes in EM images [6]. There have been many different approaches to image segmentation, including graphs and Random Forest classifiers [8], but most of these involve an ANN. One kind of neural network that has been widely used for image segmentation is the Convolutional Neural Network (CNN). The CNN can compute the probability of a pixel being a boundary based on the pixels around it [6]. The Sliding Window Network (SW-net) works by taking a sample of pixels around a center pixel, classifying the center pixel, and then moving over by one pixel [7]. The product of these methods can be a classification of the input image. The accuracy of this classification is dependent on the training data, the features that were selected, and the method of classification that was used. This method of image segmentation takes a variable amount of time to produce accurate classifications.

ANNs require significant computing power, which can be an important factor if the data set is large. However, once the initial training process is complete far less computing power is needed to classify images. Use of ANNs has become common in image processing due to generally high performance, and our results do not differ in that we also find classification by ANN to provide superior performance.

Fast marching works well with sharp contrast and solid boundaries. However, it does not perform well with low contrasts and fragmented boundaries, as demonstrated (Fig. 2A and B). This is due to the nature of the fast marching algorithm, its reliance on the travel time function, and the fact that fast marching only considers part of the selected area. This makes fast marching prone to leaks and missing low contrast boundaries. The artificial neural network examines each pixel in the selected area. This allows the neural network to make an accurate classification despite low contrasts and fragmented boundaries. While leaks do occasionally occur using this method, the total number of leaks is much smaller relative to the fast marching method. Fast marching also tends to miss areas that the neural network would not. Outlines made using the neural network are subjectively much more accurate than those made using fast marching.

3.3   Comparing CAST to Manual Outlines

To determine the accuracy of the outlines made by CAST compared to those made by a person, the following process was used: two different structures from the same image stack were selected, structure A (blue) and structure B (yellow) as shown below in Fig. 3.

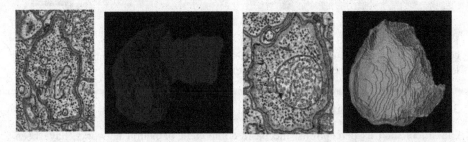

Fig. 3. Left: Sample outline of structure A with 3D visualization. **Right:** Sample outline of structure B with 3D visualization. (Color figure online)

To obtain a sample for comparison, structure A was outlined by three different people (Outliners 1-3), and Structure B was outlined by four different people (Outliners 4-7). Both structures were then outlined using CAST. A direct comparison of pixel to pixel would not show accuracy, but would instead show the different methods by which structures were outlined. For example, some of the people who outlined the structures followed the outside or went on top of the membranes, whereas CAST outlined the inner area of a structure and stopped at the inside of a membrane. So, to compare the outlines a method was created where the outside edges of each outline were compared. By comparing the outside edges of the outlines, or more specifically the distance between them, an accurate comparison between the manual outlines and those made by the ANN could be made. The reason for choosing the outside edges is because that is where the 3D reconstruction would start.

This method of comparison is done on an image by image basis, but is averaged over the entire stack. The process starts by filtering out everything except for the outlines which turns the original images into binary images. Once there is nothing in the images but the outlines, the points on the outside edges of each outline are added to a list. Now that there are two lists of edge points, the distance is taken from an edge point in the first list to an edge point at the same position in the second list. There is a specific order in which the pixels are inserted into the list which means that a pixel at the same position in the other list is the closest comparable point. After averaging all of the distances between the edge points, an average slice distance is calculated. This distance represents how far away the two outlines are from each other on average for that slice. Then, an overall distance is made by averaging all of the slice distances over the entire stack, which represents how far away the two outlines are from each other on average throughout the entire stack. This entire process can be seen below in Fig. 4.

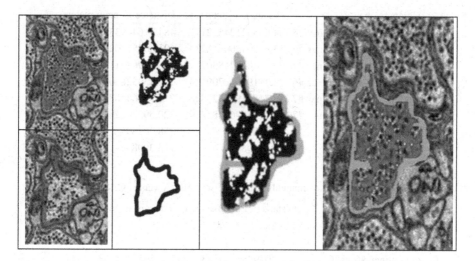

Fig. 4. A visualization of how two different outlines are compared.

The CAST outlines were compared to manual outlines, and the manual outlines were compared to one another. These outlines were all made using the same image set. This shows the level of consistency of manual outlines, and provides a basis for comparison with CAST. The comparison for structure A is first shown for manual outlines (See Table 1) and then for the comparison of manual outlines to CAST outlines (See Table 2). The same thing was done with the second structure, structure B. The manual outlines compared with manual outlines can be seen in Table 3 and the comparison between the manual outlines and CAST outlines can be seen in Table 4.

Table 1. Comparison of manual outlines for structure A.

Comparison	Distance in pixels	Distance in nano meters
Outliner #1 vs. Outliner #2	76.76753277844435	575.7564958383326
Outliner #1 vs. Outliner #3	73.02362443244613	547.677183243346
Outliner #2 vs. Outliner #3	102.0939481056508	765.704610792381
Average	83.9617	629.71276
Standard deviation	15.81417	118.60628

Table 2. Comparison of manual outlines of structure A versus CAST outlines.

Comparison	Distance in pixels	Distance in nano meters
Outliner #1 vs. CAST	95.30304500232596	714.7728375174447
Outliner #2 vs. CAST	97.11740386593159	728.3805289944869
Outliner #3 vs. CAST	111.86170045630398	838.9627534222799
Average	101.42738	760.70537
Standard deviation	9.08181	68.11355

Table 3. Comparison of manual outlines for structure B.

Comparison	Distance in pixels	Distance in nano meters
Outliner #4 vs. Outliner #5	64.57552124532171	484.3164093399128
Outliner #4 vs. Outliner #6	53.47594360350234	401.06957702626755
Outliner #4 vs. Outliner #7	63.33462374830779	475.00967811230845
Outliner #5 vs. Outliner #6	33.71191018709654	252.83932640322408
Outliner #5 vs. Outliner #7	28.71861934427826	215.38964508208696
Outliner #6 vs. Outliner #7	42.745693634818196	320.5927022611365
Average	47.76039	358.20289
Standard deviation	15.11998	113.39985

Table 4. Comparison of manual outlines of structure A versus CAST outlines.

Comparison	Distance in pixels	Distance in nano meters
Outliner #4 vs. CAST	79.95801682574454	599.685126193084
Outliner #5 vs. CAST	47.629824959831716	357.2236871987379
Outliner #6 vs. CAST	49.84039086063373	373.802931454753
Outliner #7 vs. CAST	51.11915791771337	383.3936843828503
Average	57.13685	428.52636
Standard deviation	15.28223	114.61676

As can be seen from the data above, the manual outlines and CAST outlines were marginally close when comparing averages and standard deviation. Structure A had more vesicles close to the border of the membrane which caused CAST to slightly leak. This explains why the averages and standard deviation of structure A were further apart than with structure B. Even though the averages may be off by approximately 10 pixels, the pixels are very small which makes it an insignificant number. This shows that the outlines made by CAST are still comparable to manual outlines.

In this study, we produced an application, CAST, to speed up manual reconstruction of images gathered from multiple different media. Future work on the project will continue with the same objective, streamlining the process of neural circuit reconstruction. Simplifying this process can be achieved through further automation. Currently the tool presents a method for semi-automatically outlining objects in two dimensions. This two-dimensional limitation results from Expand Area's reliance on a neural network. Expand Area works in three dimensions [21], but the current neural network only works in two dimensions. At this time the role of the user is to move through an image stack manually and identify the structure of interest from one slice to the next. Since there are upwards of six hundred images in a given stack, this task by itself is extremely time consuming.

4   Conclusions and Future Work

The original purpose of this tool was to increase the productivity of researchers working on neural circuit reconstructions. In an attempt to improve upon previous work and increase accessibility, the tool was integrated into the Fiji/TrakEM2 platform. Utilizing an artificial neural network, the tool can accurately recognize and fill a structure. This tool is an improvement over the existing fast marching tool due to its accuracy. This improved accuracy comes from the integration of machine learning. Even though this tool was developed for EM images of retinal neurons, the tool can still be used on other image data.

Future work will focus on automating this task. The main problem in attempting to automate this process, is identifying the same structure from one slice to the next. There are two reasons this problem is such a challenge. First, the structure of interest will not look exactly the same from one slice to another. Since each slice is at a different depth, each slice can be expected to have a slightly different look than the slices above or below. This results from the voxels in an image stack looking more like vertical rectangles than cubes. Second, the structure of interest is not always in the same place from one slice to another even if the stack is aligned. This could be due to inconsistency in the image data, or the structure could be in such a position that it appears to move from slice to slice. A possible solution could be reached through expanding upon previous work that involved recursive training of a neural network to classify boundaries on three-dimensional images [4]. However, structures moving from slice to slice could cause a problem. If part of the structure of interest shifted out of the area being examined by the neural network somewhere in the stack, the results produced by the network would be off.

One possible solution would be to move the region being examined by the neural network on a slice-to-slice basis. The shift would be based on where the structure of interest is located on the current slice. If the movement of a structure can be predicted from one slice to another, the neural network could follow the structure through the stack and make an accurate classification. Following a structure through the stack could be accomplished by following the structures trajectory and drift, where trajectory refers to the direction of the structure, and drift refers to the distance that a structure moves from one slice to another. Using trajectory and drift as a guide, an educated guess can be made on where to look for the structure of interest in the next slice. The ultimate goal of all this being the ability to outline an entire neuron in three dimensions with only one click.

Acknowledgments. We would like to thank the students of the Fall 2016 CS492 Bioinformatics class at Lewis-Clark State College for providing the manual outlines referenced in this study. This research was supported by the INBRE program, NIH Grant No. P20 GM103408 (National Institute of General Medical Sciences).

References

1. Briggman, K., Denk, W.: Towards neural circuit reconstruction with volume electron microscopy techniques. Curr. Opin. Neurobiol. **16**, 562–570 (2006)
2. Cardona, A., Saalfeld, S., Schindelin, J., Arganda-Carreras, I., Preibisch, S., Longair, M., Tomancak, P., Hartenstein, V., Douglas, R.: TrakEM2 software for neural circuit reconstruction. PLoS One **7**, e38011 (2012)
3. Platt, J.: U.S. Patent No. 6,380,929. U.S. Patent and Trademark Office, Washington, DC (2002)
4. Lee, K., Zlateski, A., Vishwanathan, A., Seung, H.: Recursive training of 2D-3D convolutional networks for neuronal boundary detection. arxiv preprint arXiv:1508.04843 (2015)
5. Jurrus, E., Watanabe, S., Giuly, R., Paiva, A., Ellisman, M., Jorgensen, E., Tasdizen, T.: Semi-automated neuron boundary detection and nonbranching process segmentation in electron microscopy images. Neuroinformatics **11**, 5–29 (2012)
6. Ciresan, D., Giusti, A., Gambardella, L.M., Schmidhuber, J.: Deep neural networks segment neuronal membranes in electron microscopy images. In: Advances in Neural Information Processing Systems, pp. 2843–2851 (2012)
7. Tschopp, F.: Efficient convolutional neural networks for pixelwise classification on heterogeneous hardware systems. arxiv preprint arXiv:1509.03371 (2015)
8. Tajoddin, B.: Semi-automatic segmentation for serial section electron microscopy images (2012)
9. Chalfoun, J., Majurski, M., Dima, A., Stuelten, C., Peskin, A., Brady, M.: FogBank: a single cell segmentation across multiple cell lines and image modalities. BMC Bioinform. **15** (2014)
10. Olbris, D., Winston, P., Chklovskii, D.: Raveler—a software for editing large segmented electron microscopy datasets
11. Sommer, C., Straehle, C., Koethe, U., Hamprecht, F.: Ilastik: interactive learning and segmentation toolkit. In: 2011 IEEE International Symposium Biomedical Imaging: From Nano to Macro, pp. 230–233 (2011)
12. Knowles-Barley, S., Kaynig, V., Jones, T., Wilson, A., Morgan, J., Lee, D., Pfister, H.: RhoanaNet pipeline: dense automatic neural annotation. arxiv preprint arXiv:1611.06973 (2016)
13. Kornfeld, J., Svara, F., Nguyen, M.-T., Pfeiler, N., Pronkin, M., Shatz, O., Spaar, S., Alex, S., Valerio, J.: knossos-project/knossos. https://github.com/knossos-project/knossos

14. Li, S., Mitchell, J., Briggs, D., Young, J., Long, S., Fuerst, P.: Morphological diversity of the rod spherule: a study of serially reconstructed electron micrographs. PLoS One **11**, e0150024 (2016)
15. Denk, W., Horstmann, H.: Serial block-face scanning electron microscopy to reconstruct three-dimensional tissue nanostructure. PLoS Biol. **2**, e329 (2004)
16. Schindelin, J., Arganda-Carreras, I., Frise, E., Kaynig, V., Longair, M., Pietzsch, T., Preibisch, S., Rueden, C., Saalfeld, S., Schmid, B., Tinevez, J., White, D., Hartenstein, V., Eliceiri, K., Tomancak, P., Cardona, A.: Fiji: an open-source platform for biological-image analysis. Nat. Methods **9**, 676–682 (2012)
17. Ruck, D., Rogers, S., Kabrinsky, M.: Feature selection using a multilayer perceptron. J. Neural Netw. Comput. **2**, 40–48 (1990)
18. Arganda-Carreras, I., Kaynig, V., Schindelin, J., Cardona, A., Seung, H.: Trainable weka segmentation: a machine learning tool for microscopy image segmentation (2014)
19. Hall, M., Frank, E., Holmes, G., Pfahringer, B., Reutemann, P., Witten, I.: The WEKA data mining software. ACM SIGKDD Explor. Newsl. **11**, 10 (2009)
20. Arganda-Carreras, I., Cardona, A., Kaynig, V., Rueden, C., Schindelin, J.: Trainable Weka Segmentation – ImageJ (2016). http://imagej.net/Trainable_Weka_Segmentation
21. Long, S., Holder, L.: Graph-based shape analysis for MRI classification. Int. J. Knowl. Discov. Bioinform. **2**, 19–33 (2011)
22. Cardinal, M., Meunier, J., Soulez, G., Maurice, R., Therasse, E., Cloutier, G.: Intravascular ultrasound image segmentation: a three-dimensional fast-marching method based on gray level distributions. IEEE Trans. Med. Imaging **25**, 590–601 (2006)

Accelerating Electron Tomography Reconstruction Algorithm ICON Using the Intel Xeon Phi Coprocessor on Tianhe-2 Supercomputer

Zihao Wang[1,2], Yu Chen[1,2], Jingrong Zhang[1,2], Lun Li[1,3], Xiaohua Wan[1],
Zhiyong Liu[1(✉)], Fei Sun[2,4,5(✉)], and Fa Zhang[1(✉)]

[1] Key Laboratory of Intelligent Information Processing,
Institute of Computing Technology, Chinese Academy of Sciences,
Beijing 100190, China
{zyliu,zhangfa}@ict.ac.cn
[2] University of Chinese Academy of Sciences, Beijing, China
feisun@ibp.ac.cn
[3] School of Mathematical Sciences, University of Chinese Academy of Sciences,
Beijing, China
[4] National Key Laboratory of Biomacromolecules,
CAS Center for Excellence in Biomacromolecules, Institute of Biophysics,
Chinese Academy of Sciences, Beijing 100101, China
[5] Center for Biological Imaging, Institute of Biophysics,
Chinese Academy of Sciences, Beijing 100101, China

Abstract. Electron tomography (ET) is an important method for studying three-dimensional cell ultrastructure. Combining with a subvolume averaging approach, ET provides new possibilities for investigating in situ macromolecular complexes in sub-nanometer resolution. Because of the limited sampling angles, ET reconstruction usually suffers from the 'missing wedge' problem. With a validation procedure, Iterative Compressed-sensing Optimized NUFFT reconstruction (ICON) demonstrates its power in the restoration of validated missing information for low SNR biological ET dataset. However, the huge computational demand has become a bottleneck for the application of ICON. In this work, we developed the strategies of parallelization for NUFFT and ICON, and then implemented them on a Xeon Phi 31SP coprocessor to generate the parallel program ICON-MIC. We also proposed a hybrid task allocation strategy and extended ICON-MIC on multiple Xeon Phi cards on Tianhe-2 supercomputer to generate program ICON-MULT-MIC. With high accuracy, ICON-MIC has a significant acceleration compared to the CPU version, up to 13.3x, and ICON-MULT-MIC has good weak and strong scalability efficiency on Tianhe-2 supercomputer.

Keywords: Electron tomography · ICON · Parallel NUFFT · Hybrid task allocation strategy · Tianhe-2 supercomputer · MIC acceleration

---

Y. Chen—Contributes equally to this work.

© Springer International Publishing AG 2017
Z. Cai et al. (Eds.): ISBRA 2017, LNBI 10330, pp. 258–269, 2017.
DOI: 10.1007/978-3-319-59575-7_23

1   Introduction

Electron tomography (ET) is an important method for studying three-dimensional cell ultrastructure [1,2]. Combining with a sub-volume averaging approach [3], ET provides new possibilities for investigating in situ structures and conformational dynamics of macromolecular complexes in sub-nanometer resolution [4]. However, because of the physical restriction of the sample stage and the specificity of the biological samples, the sampling angles are usually limited within $-70°$ to $70°$ leading to missing information, also called 'missing wedge' problem [5]. Thus, traditional ET methods, such as WBP [6], SIRT [7], INFR [8] usually suffer from the 'missing wedge' artifacts, which severely weaken the further biological interpretation [9].

Recent years, the topic of solving 'missing wedge' problem in ET has been widely discussed and many algorithms have been proposed. FIRT [10] and DART [11] apply prior constrains, including density smoothness, density non-negativity, etc., to the reconstructed tomogram to compensate the 'missing wedge' problem. Compressed sensing electron tomography tried to solve the reconstruction problem as an underdetermined problem based on a theoretical framework called 'compressed sensing' (CS) [12] and demonstrated certain success for the data with a high signal to noise ratio (SNR) (e.g., material science data) [13–15]. To cope with the low SNR case (e.g., biological cryo-ET data), Deng et al. [16] proposed ICON by combining CS and non-uniform fast Fourier transform (NUFFT) together. ICON not only can restore the missing information but also can measure the fidelity of the information restoration using a validation procedure.

Although ICON has demonstrated its power in restoring validated missing information for low SNR biological ET dataset, the huge computational demand becomes a bottleneck for its wide application. As the high performance computing platforms becoming more and more popular, many ET reconstruction algorithms have been ported to the heterogeneous system containing acceleration units (graphics processing units (GPU) [17], many integrated core (MIC) [18], field-programmable gate array (FPGA), and so on). Moreover, these ET reconstruction algorithms will be more efficient if they are implemented on super-computers such as Tianhe-2 [19] which is one of world's Top5 supercomputer.

In this work, we developed the strategies of parallelization for ICON and implemented them on a Xeon Phi 31SP coprocessor to generate the parallel program ICON-MIC. In this step, to achieve high acceleration, we developed parallel versions of NUFFT and adjoint NUFFT on MIC. And then we proposed a hybrid task allocation strategy (TLLB) and extended ICON-MIC on multiple Xeon Phi cards on Tianhe-2 supercomputer to generate program ICON-MULT-MIC. Experimental results show that ICON-MIC has high accuracy and exhibit significant acceleration factors compared to the CPU version ICON (ICON-CPU) and ICON-MULT-MIC has good weak and strong scalability efficiency. Both ICON-MIC and ICON-MULT-MIC are developed into software packages which can be downloaded from our homepage: http://ear.ict.ac.cn.

2   Related Work

2.1   Iterative Compressed-Sensing Optimized NUFFT Reconstruction (ICON)

ICON is an iterative reconstruction algorithm based on the theoretical framework of 'compressed sensing' and the complete workflow of ICON can be divided into 4 steps: **'Pre-processing'**, **'Gray value adjustment'**, **'Reconstruction and pseudo-missing-validation'** and **'Verification filtering'** [16]. A series of tests show that **'Reconstruction and pseudo-missing-validation'** accounts for at least 95% of the execution time of ICON. Thus, the major task for accelerating ICON is paralleling this step effectively on MIC. The parallelization of **'reconstruction'** and **'pseudo-missing-validation'** are similar and only **'reconstruction'** will be discussed in this paper. The major steps of ICON **'reconstruction'** can be briefly described as followed.

Step 1. Fidelity preservation step using steepest descent method [20].

$$r = A^h W A x^k - A^h W f \tag{1}$$

$$\alpha = \frac{r^T r}{r^T A^h W A r} \tag{2}$$

$$y^{k+1} = x^k + \alpha * r \tag{3}$$

where x^k is the two dimensional (2D) reconstructed image of the kth iteration. A is the projection operation, we here define A as a nonuniform Fourier sampling matrix, which performs Fourier transform on the non-integer grid points. A^h stands for the conjugate transpose of A. W follows INFR's description [8] and contains the weights that account for the non-uniform sampling in the Fourier space (similar to the ramp filtering in WBP). f is the Fourier transform of acquired projections. r is the residual. α is the coefficient used to control the step of updating. y^{k+1} is the intermediate updating result of the $(k+1)th$ iteration.

Step 2. Prior sparsity restriction step

$$x^{k+1} = H(y^{k+1}) = \begin{cases} 0, & if \ \ y^{k+1} < 0 \\ y^{k+1}, & if \ \ y^{k+1} \geq 0 \end{cases} \tag{4}$$

where y^{k+1} is the intermediate updating result of the $(k+1)th$ iteration. $H(\cdot)$ is a logic function. x^{k+1} is the 2D reconstructed slice of the $(k+1)th$ iteration.

We classified the operations of these two steps into three types: a. Element-wise operations of matrices; b. The summation of a matrix; c. The NUFFT and the adjoint NUFFT. For each type of operation, a strategy for parallelization is proposed in Sect. 3.

2.2   Non-uniform Fast Fourier Transform (NUFFT)

First, we give a brief description of NUFFT. Given the Fourier coefficients $\hat{f}_k \in \mathbb{C}, k \in I_N$ and $I_N = \{k = (k_t)_{t=0,\ldots,d-1} \in \mathbb{Z}^d : -\frac{N_t}{2} \le k_t < \frac{N_t}{2}, t = 0, \ldots, d - 1\}$ as input, NUFFT tries to evaluate the following trigonometric polynomial efficiently at the reciprocal points $x_t \in [-\frac{1}{2}, \frac{1}{2}), t = 0, \ldots, M - 1$:

$$f_j = f(x_j) = \sum_{k \in I_N} \hat{f}_k e^{-2\pi i k x_j}, j = 0, \ldots, M - 1 \tag{5}$$

Correspondingly, the adjoint NUFFT tries to evaluate Eq. (6) at the frequencies k:

$$\hat{h}_k = \sum_{j=0}^{M-1} f_j e^{2\pi i k x_j} \tag{6}$$

NFFT3.0 [21] is a successful and widely used open source C library for NUFFT and adjoint NUFFT. However, to our knowledge, no corresponding library on MIC is available yet. Thus, we parallelled the NUFFT and the adjoint NUFFT based on the algorithms described in NFFT3.0 and the algorithm of 2D NUFFT is displayed in Algorithm 1 for deep analysis.

Algorithm 1: NUFFT

Input: $M, N = \{N_1, N_2\}, \sigma = \{\sigma_1, \sigma_2\}, m, x_j \in [-\frac{1}{2}, \frac{1}{2})^2, j = 0, \ldots, M - 1, \hat{f}_k$
    $\in \mathbb{C}, k \in I_N n = \sigma N = \{n_1, n_2\} = \{\sigma_1 N_1, \sigma_2 N_2\}$

1: For $k \in I_N$ compute
   $\hat{g}_k = \dfrac{\hat{f}_k}{|I_n| c_k(\tilde{\varphi})}$
   $c_k(\tilde{\varphi}) = \hat{\varphi}(k_1)\, \hat{\varphi}(k_2)$

2: For $l \in I_n$ compute by 2-variate FFT
   $g_l = \sum\limits_{k \in I_N} \hat{g}_k e^{-2\pi i k(n^{-1} \odot l)}$

3: For j=0, ...,M-1 compute
   $f_j = \sum\limits_{l \in I_{n,m}(x_j)} g_l \tilde{\psi}(x_j - n^{-1} \odot l)$
   $I_{n,m}(x_j) = \{l \in I_n : n \odot x_j - m1 \le l \le n \odot x_j + m1\}$
   $\tilde{\psi}(x) = \varphi(x_1)\varphi(x_2)$

$\varphi(x)$ and $\hat{\varphi}(k)$ are the window functions. In this work, the (dilated) Gaussian window functions (Eqs. (7) and (8)) are used.

$$\varphi(x) = (\pi b)^{-\frac{1}{2}} e^{-\frac{(nx)^2}{b}} (b = \frac{2\sigma}{2\sigma - 1}\frac{m}{\pi}) \tag{7}$$

$$\hat{\varphi}(k) = \frac{1}{n} e^{-b\left(\frac{\pi k}{n}\right)^2} \tag{8}$$

where x is a component of the reciprocal points x. k is a component of the frequencies k. σ is a component of the oversampling factors σ with $\sigma > 1$. n is one component of $n = \sigma N$. $m \in \mathbb{N}$ and $m \ll n$. In this work, $\sigma = 2$ and $m = 6$.

The operations of NUFFT can be classified into three types: a. Element-wise operations of matrices; b. Two dimensional fast Fourier transform (FFT); c. Calculation of window functions.

3   Acceleration of ICON Using MIC Coprocessors

3.1   Parallel Element Wise Matrix Operations

In ET reconstruction, the size of a matrix is usually lager than the number of processing units on MIC, so we need to select a proper number of processing units (threads) to balance the control and computing resources. According to our experiments on the Xeon Phi 31SP card, if the thread number is close or equal to 228 (57 cores with each core having 4 hardware threads), all threads joining in the computation will cause performance degradation. So we use 200 threads when parallelizing the element wise operations and we divide a matrix into 200 parts and assign each part to one thread for calculation. Experiments show that the matrices constituted large arrays in ICON has high allocation cost using 4 KB pages. In order to reduce the allocation cost, we use 2 MB pages in offload mode which also reduces the TLB misses and page faults. We take advantage of the 512-bit vector processing unit (VPU) [22] on each core which means 16 single-precision or 8 double-precision operations can be executed at one time in order to achieve a high computational throughput for element-wise matrix operations.

3.2   NUFFT and Adjoint NUFFT Parallelization

As mentioned in Sect. 2.2, there are three classes of operations in NUFFT and adjoint NUFFT. The strategy for the parallelization of element-wise matrix operations in NUFFT and adjoint NUFFT is the same as the strategy described in Sect. 3.1. To achieve a high performance of FFT, we take advantage of the FFT library named Intel MKL FFT [23]. Thus, we use Intel's FFT interface with making memory alignment and changing the layout of multi-dimensional data in the coprocessor memory to achieve high efficiency. We make memory alignment and change the layout of multi-dimensional data in the coprocessor memory [24] to achieve high efficiency.

For calculation of window functions, since ICON is an iterative algorithm, NUFFT and adjoint NUFFT will be repeated many times. To cut down the time of calculation and memory transfer, we use data persistence technology showed in Fig. 1 by pre-computing the window functions, and storing them in device memory. We used a resin embedded ET dataset (see Sect. 4.1 for detail) to test the performance of parallel NUFFTs on MIC comparing to NFFT3.0. NFFT3.0 ran on one core (thread) of an Intel® Xeon™ CPU E5-2620 v2 @ 2.1 GHz (6 cores per CPU), parallel NUFFTs ran on a Xeon Phi 31SP coprocessor of Tianhe-2. The test datasets include the image sizes of 512 * 512, 1k * 1k, 2k * 2k, 4k * 4k. Experiments results show that parallel NUFFTs are 10 times faster than the library NFFT3.0 in Fig. 2.

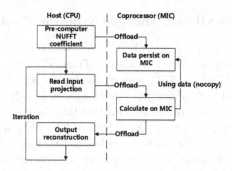

Fig. 1. NFFT Pre-computing using data persistence

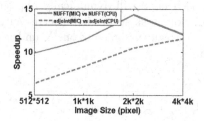

Fig. 2. The speedups of parallel NUFFTs compared to NFFT3.0.

3.3 Efficient Summation of a Matrix

Commonly, CPU program will execute the summation using one single thread. However, for MIC, the computational capability of one thread is too weak to sum up a whole matrix in a reasonable time.

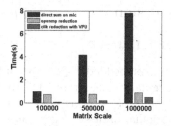

Fig. 3. Summation of a matrix

OpenMP reduction clause is usually used to avoid executing the summation on one core of MIC. However, it cannot take advantage of the 512-bit vector processing unit on MIC. We use array notation which is part of Intel Cilk Plus [25] to help the compiler with vectorization in order to achieve an efficient utilization of all available processing resources. We compared these three summation strategies mentioned above in Fig. 3. Using Intel Cilk Plus reduction with VPU consumes the least running time, and thus it is used in ICON-MIC.

3.4   Extend ICON-MIC on Multiple Xeon Phi Cards on Tianhe-2

To further satisfy the huge amount of computational requirements, we extended the ICON-MIC on multiple Xeon Phi cards on Tianhe-2 to generate ICON-MULT-MIC. To make ICON-MULT-MIC compatible for the architecture of Tianhe-2, we proposed a hybrid task allocation model named Two-Level Load Balancing (TLLB) by taking advantage of Message Passing Interface (MPI). TTLB combines the static allocation (for the level on Xeon Phi cards) with dynamic allocation (for the level on CPU nodes) and it can be described as Fig. 4.

In ET, the reconstruction of a 3D volume can be divided into a series of similar tasks. Since each node on Tianhe-2 has three Xeon Phi cards, we separate all tasks into a series of task subsets and each task subset contains three tasks. During reconstruction, each node will dynamically request for one task subset at one time after the previous task subset is finished. Within one node, each of those three tasks will be statically assigned to one Xeon Phi card and ICON-MIC is used to process them.

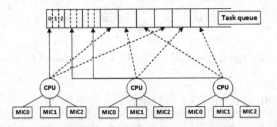

Fig. 4. TLLB for ICON on Tihanhe-2

4   Results and Discussion

4.1   Resin Embedded ET Dataset

We test ICON-MIC on a resin embedded ET dataset of MDCK cell section. The tilt angles of the dataset originally range from $-68°$ to $+68°$ with $1°$ increment. In order to verify ICONs' ability of restoring miss information, we extract every two projections from the original dataset to generate a new tilt series with $2°$ increment for the following experiments. The tilt series are aligned using atom-align [26]. The original image size is 4k * 4k with a pixel size of 0.72 nm. We also binned the tilt series with factors of 2, 4, 8 to generate datasets of 2k * 2k, 1k * 1k and 512 * 512, respectively.

4.2   Reconstruction Precision

Firstly, we evaluate the numerical accuracy of ICON-MIC using the root-mean-square relative error (RMSRE) ε, as Eq. (9).

$$\varepsilon = \sqrt{\frac{\sum_{i=1}^{N*N} \left(\frac{P_i - C_i}{C_i}\right)^2}{N*N}} \qquad (9)$$

where $N*N$ is the size of one slice; C is the slice reconstructed by ICON-CPU; C_i is the value of the ith pixel in C; P is the slice reconstructed by ICON-MIC; P_i is the value of the ith pixel in P; ε is the root-mean-square relative error. All reconstructed slices are first normalized into $(0,1]$ using Eq. (10).

$$I_{norm} = \frac{I - minI}{maxI - minI} + c \qquad (10)$$

where I_{norm} is the normalized slice; I is the originally reconstructed slice; $minI$ is the minimum value of I; $maxI$ is the maximum value of I; c is a small constant to avoid 0 in I_{norm}, in this work, $c = 10^{-7}$. The RMSREs of ICON-MIC increase slowly with the image size, but they are in the range of $(10^{-6}, 10^{-5})$ yielding a reasonable numerical accuracy as showed in Fig. 5.

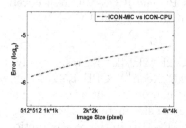

Fig. 5. The RMSREs of ICON-MIC

We further investigate the reconstruction accuracy by the pseudo-missing-validation procedure [21]. Here, the $-0.29°$ tilt (the minimum tilt) projection was excluded as the omit-projection ('ground truth'), see Fig. 6(a). We re-project the reconstructed tomograms at $-0.29°$. The re-projections of ICONs (Fig. 6(c, d)) are identical with each other and the NCCs between each other are all 1. The re-projections of ICONs are clearer in detailed structures and more similar to the 'ground truth', compared to WBP (Fig. 6(b)). Such visual assessments are further verified quantitatively by comparing the FRC curves between the re-projections and the 'ground truth'. The FRCs of ICONs coincide with each other, and they are better than that of WBP (Fig. 6(e)). The coincident FRCs of ICONs further demonstrate the accuracy of ICON-MIC from the perspective of restoring miss information.

4.3   Speed Up

We evaluate the acceleration of ICON-MIC by comparing the running time of reconstructing one slice under 200 iterations. We reconstruct the datasets with

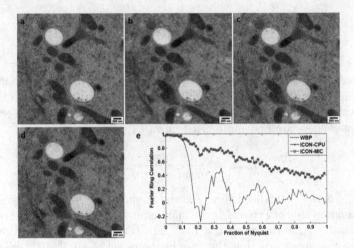

Fig. 6. Evaluate ICON-MIC by the pseudo-missing-validation procedure. (a) The omit-projection ('Ground truth'); (b–d) the re-projections of the omit-tomograms reconstructed by WBP, ICON-CPU and ICON-MIC respectively; (e) the pseudo-missing-validation FRCs of WBP, ICON-CPU and ICON-MIC.

sizes of 512 * 512, 1k * 1k, 2k * 2k, 4k * 4k, respectively. ICON-CPU run on one core (thread) of an Intel® Xeon™ CPU E5-2620 v2 @ 2.1 GHz, ICON-MIC run on a Xeon Phi 31SP coprocessor of Tianhe-2. The acceleration of ICON-MIC improves when the slice size increases (Fig. 7 and Table 1). The maximum speedups are 13.3x for ICON-MIC in the reconstruction of a 4k * 4k slice. With the efficient acceleration, the reconstruction time of one 4k * 4k slice is reduced from hours to minutes.

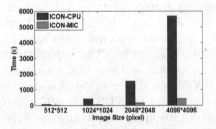

Fig. 7. The comparison of time-consuming of ICON-CPU and ICON-MIC

4.4   ICON-MULT-MIC on Tianhe-2 Supercomputer

We tested the weak scalability and the strong scalability of ICON-MULT-MIC on Tianhe-2 supercomputer. In weak scalability test, we fix the number of tasks assigned to one processor, and in the strong scalability test, we fix the total

Table 1. The speedups of ICON-MIC compared to ICON-CPU

Image size	$512 * 512$	$1024 * 1024$	$2048 * 2048$	$4096 * 4096$
ICON-MIC	5.2x	9.4x	10.9x	13.3x

number of tasks in all nodes; and then we valuate how the executing time varies with the number of processors.

Firstly, we tested the weak scalability with image sizes of $1024 * 1024$ and $2048 * 2048$. We gradually increase the number of Xeon Phi cards from 3 to 48. As the node gradually increases, the total number of image being processed also increases. The executing time showed in Fig. 8(a) increases from 875 s to 900 s, which indicated that the developed parallelization strategy is good for weak scalability.

Secondly, we tested the strong scalability. We fix the test image size to be $1024 * 1024$ and the total number of images to be 48 which equal to the biggest number of Xeon Phi cards. We only increase the number of Xeon Phi cards from 3 to 48. From the Fig. 8(b), we can observe that the parallel efficiency decreases to 92% when using 12 Xeon Phi cards and further to 83% when using 48 Xeon Phi cards. The observed degradation of the strong scalability efficiency is acceptable.

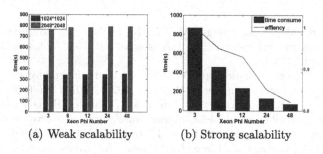

    (a) Weak scalability   (b) Strong scalability

Fig. 8. Scalability results on Tianhe-2

5 Conclusion

In the present work, we analyze the iterative framework of ICON and classify the operations of ICON's major steps into three types. Accordingly, we design the strategies of parallelization for ICON and implement them on single MIC card to generate parallel program ICON-MIC. We also develop a parallel version of NUFFT and adjoint NUFFT on MIC. To satisfy the huge amount of computation requirements, we proposed the hybrid task allocation strategy (TLLB) and expanded the ICON-MIC on multiple Xeon Phi cards to generate ICON-MULT-MIC.

We test ICON-MIC on a resin embedded ET dataset of MDCK cell section. The RMSREs of ICON-MIC are about 10e-6 yielding an reasonable numerical accuracy. The high reconstruction accuracy demonstrates that ICON-MIC have the same ability of restoring miss information as ICON-CPU. Experimental results also show ICON-MIC have a good acceleration, 13.3x for ICON-MIC in the reconstruction of one 4k * 4k slice and ICON-MULT-MIC has good weak and strong scalability efficiency. Experimental results indicate that ICON-MULT-MIC can use the heterogeneous computational resources of Tianhe-2 supercomputer effectively.

Acknowledgments. This research is supported by the NSFC projects Grant Nos. U1611263, U1611261, 61232001, 61472397, 61502455, 61672493 and Special Program for Applied Research on Super Computation of the NSFC-Guangdong Joint Fund (the second phase), the Strategic Priority Research Program of Chinese Academy of Sciences (Grant No. XDB08030202), the National Basic Research Program (973 Program) of Ministry of Science and Technology of China (2014CB910700). The authors would like to thank Prof. Wanzhong He (NIBS, Beijing) for providing the resin embedded ET dataset. All the intensive computations were performed on Tianhe-2 supercomputer at the National Supercomputer Center in Guangzhou (NSCC-GZ), China and Center for Biological Imaging, Institute of Biophysics, Chinese Academy of Sciences (http://cbi.ibp.ac.cn).

References

1. Fridman, K., Mader, A., Zwerger, M., Elia, N., Medalia, O.: Advances in tomography: probing the molecular architecture of cells. Nat. Rev. Mol. Cell Biol. **13**(11), 736–742 (2012)
2. Lučić, V., Rigort, A., Baumeister, W.: Cryo-electron tomography: the challenge of doing structural biology in situ. J. Cell Biol. **202**(3), 407–419 (2013)
3. Castaño-Díez, D., Kudryashev, M., Arheit, M., Stahlberg, H.: Dynamo: a flexible, user-friendly development tool for subtomogram averaging of cryo-em data in high-performance computing environments. J. Struct. Biol. **178**(2), 139–151 (2012)
4. Bharat, T.A., Russo, C.J., Löwe, J., Passmore, L.A., Scheres, S.H.: Advances in single-particle electron cryomicroscopy structure determination applied to subtomogram averaging. Structure **23**(9), 1743–1753 (2015)
5. Penczek, P., Marko, M., Buttle, K., Frank, J.: Double-tilt electron tomography. Ultramicroscopy **60**(3), 393–410 (1995)
6. Radermacher, M.: Weighted back-projection methods. In: Frank, J. (ed.) Electron Tomography, pp. 245–273. Springer, New York (2007)
7. Gilbert, P.: Iterative methods for the three-dimensional reconstruction of an object from projections. J. Theor. Biol. **36**(1), 105–117 (1972)
8. Chen, Y., Förster, F.: Iterative reconstruction of cryo-electron tomograms using nonuniform fast Fourier transforms. J. Struct. Biol. **185**(3), 309–316 (2014)
9. Lučić, V., Förster, F., Baumeister, W.: Structural studies by electron tomography: from cells to molecules. Annu. Rev. Biochem. **74**, 833–865 (2005)
10. Chen, Y., Zhang, Y., Zhang, K., Deng, Y., Wang, S., Zhang, F., Sun, F.: FIRT: filtered iterative reconstruction technique with information restoration. J. Struct. Biol. **195**(1), 49–61 (2016)

11. Batenburg, K.J., Sijbers, J.: DART: a practical reconstruction algorithm for discrete tomography. IEEE Trans. Image Process. **20**(9), 2542–2553 (2011)
12. Donoho, D.L.: Compressed sensing. IEEE Trans. Inf. Theory **52**(4), 1289–1306 (2006)
13. Goris, B., Van den Broek, W., Batenburg, K., Mezerji, H.H., Bals, S.: Electron tomography based on a total variation minimization reconstruction technique. Ultramicroscopy **113**, 120–130 (2012)
14. Leary, R., Saghi, Z., Midgley, P.A., Holland, D.J.: Compressed sensing electron tomography. Ultramicroscopy **131**, 70–91 (2013)
15. Saghi, Z., Divitini, G., Winter, B., Leary, R., Spiecker, E., Ducati, C., Midgley, P.A.: Compressed sensing electron tomography of needle-shaped biological specimens-potential for improved reconstruction fidelity with reduced dose. Ultramicroscopy **160**, 230–238 (2016)
16. Deng, Y., Chen, Y., Zhang, Y., Wang, S., Zhang, F., Sun, F.: ICON: 3D reconstruction with missing-informationrestoration in biological electron tomography. J. Struct. Biol. **195**(1), 100–112 (2016)
17. Palenstijn, W., Batenburg, K., Sijbers, J.: Performance improvements for iterative electron tomography reconstruction using graphics processing units (GPUs). J. Struct. Biol. **176**(2), 250–253 (2011)
18. Dahmen, T., Marsalek, L., Marniok, N., Turoňová, B., Bogachev, S., Trampert, P., Nickels, S., Slusallek, P.: The ettention software package. Ultramicroscopy **161**, 110–118 (2016)
19. Liao, X., Xiao, L., Yang, C., Lu, Y.: MilkyWay-2 supercomputer: system and application. Front. Comput. Sci. **8**(3), 345–356 (2014)
20. Goldstein, A.A.: On steepest descent. J. Soc. Ind. Appl. Math. Ser. A: Control **3**(1), 147–151 (1965)
21. Keiner, J., Kunis, S., Potts, D.: Using NFFT 3—a software library for various nonequispaced fast Fourier transforms. ACM Trans. Math. Softw. (TOMS) **36**(4), 19 (2009)
22. Duran, A., Klemm, M.: The intel® many integrated core architecture. In: 2012 International Conference on High Performance Computing and Simulation (HPCS), pp. 365–366. IEEE (2012)
23. Wang, E., Zhang, Q., Shen, B., Zhang, G., Lu, X., Wu, Q., Wang, Y.: High-Performance Computing on the Intel® Xeon Phi^{TM}, vol. 5, p. 2. Springer, Heidelberg (2014)
24. Asai, R., Vladimirov, A.: Intel Cilk Plus for complex parallel algorithms: Enormous Fast Fourier Transforms (EFFT) library. Parallel Comput. **48**, 125–142 (2015)
25. Robison, A.D.: Cilk Plus: language support for thread and vector parallelism. Talk at HP-CAST **18**, 25 (2012)
26. Han, R., Zhang, F., Wan, X., Fernández, J.J., Sun, F., Liu, Z.: A marker-free automatic alignment method based on scale-invariant features. J. Struct. Biol. **186**(1), 167–180 (2014)

Modeling the Molecular Distance Geometry Problem Using Dihedral Angles

Michael Souza[1(✉)], Carlile Lavor[2], and Rafael Alves[2]

[1] Federal University of Ceará (DEMA-UFC), Fortaleza, CE 60440-900, Brazil
michael@ufc.br
[2] University of Campinas (IMECC-UNICAMP), Campinas, SP 13081-970, Brazil
clavor@ime.unicamp.br, rafaelsoalves@uol.com.br

Abstract. An alternative formulation based on dihedral angles to the molecular distance geometry problem with imprecise distance data is presented. This formulation considers the additional hypothesis of a particular ordering such that all distances $||x_i - x_j|| = d_{ij}$, $|i - j| < 3$, are known. Considering that bond length and angles are given a priori in a protein backbone, there is always at least one of such ordering in instances involving real protein data. This hypothesis reduces by 2/3 the number of variables of the problem and allows us to calculate the derivatives of the standard Cartesian coordinates representation with respect to the dihedral angles. Numerical experiments illustrate the correctness and viability of the proposed formulation.

Keywords: Distance geometry · Modeling · Dihedral angles · Optimization

1 Introduction

In its more general formulation, the molecular distance geometry problem (MDGP) with imprecise distances consists of determining a configuration $x = (x_1, \ldots, x_n) \in \mathbb{R}^{3n}$ satisfying the inequalities

$$l_{ij} \leq ||x_i - x_j|| \leq u_{ij}, \forall (i,j) \in E \subset \{1, \ldots, n\} \times \{1, \ldots, n\}, \qquad (1)$$

where l_{ij} and u_{ij} are given positive real numbers [7].

Out of necessity, mathematical models are based on simplifications of reality. The greater the abstraction, the more extensive the scope of results. However, sometimes, additional hypotheses do not restrict the number of practical applications. The present work explores one of these cases, where we consider the additional hypothesis of existence of a particular order $\{k_1, \ldots, k_n\}$ such that all distances $||x_{k_i} - x_{k_j}||$ for $|k_i - k_j| < 3$ are known. As we shall see in Sect. 2, this additional hypothesis does not exclude real instances, since there is always at least one such order in instances involving real protein data [2].

In Sect. 3, we will show that the ordering hypothesis allows to reformulate the MDGP in terms of dihedral angles instead of Cartesian coordinates. With this, we reduced by 2/3 the number of variables of the original problem.

© Springer International Publishing AG 2017
Z. Cai et al. (Eds.): ISBRA 2017, LNBI 10330, pp. 270–278, 2017.
DOI: 10.1007/978-3-319-59575-7_24

A natural approach to solve the MDGP is to turn it into an optimization problem and apply a standard optimization algorithm or some specialized heuristic [6]. The standard optimization formulation defined by Crippen et al. [3] is the global optimization problem

$$(P) \quad \min_{x} \left\{ f(x) \equiv \sum_{(i,j)\in E} p_{ij}(x_i - x_j) \right\}, \tag{2}$$

where the penalty function $p_{ij} : \mathbb{R}^3 \to \mathbb{R}$ is given by

$$p_{ij}(x) = \max(l_{ij} - ||x||, 0) + \max(||x|| - u_{ij}, 0). \tag{3}$$

It is easy to see that $f(x) \geq 0$ and $f(x) = 0$ if, and only if, x is a solution of the problem.

In general, the algorithms for the optimization formulation of the MDGP make use of derivatives [8, 11]. With this motivation, we present the explicit form of the derivatives of the parametrization with respect to the dihedral angles in Sect. 4.

In Sect. 5, we present some numerical experiments validating the formulation and its derivatives.

2   Biological Motivation

One of the most important applications of the MDGP is the determination of protein structures using data from nuclear magnetic resonance (NMR) spectroscopy. As is well known, there is a direct connection between the three-dimensional structure of proteins and the functions they perform. Proteins are composed of amino acids and, although there are more than 500 amino acids, only 22 variations of them occur naturally [1,5]. Therefore, each naturally synthesized protein can be represented as a *string*, where each character represents one of 22 different amino acids. Unfortunately, there is no sufficiently precise and robust method to obtain the geometry and therefore the function of a protein from the knowledge of the sequence of the amino acid residues that comprise it. Instead, current experimental techniques analyze samples from each individual protein [10].

Experimental techniques for determining protein geometries differ by precision (resolution) and by the environments in which they can be applied. Despite the relatively low resolution, NMR is one of the most applied techniques in determining protein structures (see Table 1). One of the appeals of NMR is the possibility of applying it in aqueous media normally found in living organisms. This advantage is important because it allows analyzing dynamic aspects as the ones related to the temperature variation [4]. NMR does not directly provide the positions of the atoms. Instead, it gives the distances between pairs of hydrogen atoms whose distances are less than 5–6 Å [10].

Not all interatomic distances need to be estimated by resonance. In fact, some interatomic distances are typical and do not vary with the specific geometry of

Table 1. Number of protein structures in Protein Data Bank (PDB) obtained by different experimental techniques (From: PDB - Accessed in 01/30/2017).

Method	#Proteins
X-Rray	105,656
NMR	10,257
Electron microscopy	993
Hybrid	97
Other	181
Total	117,184

the protein, since they depend only on the type of chemical bonding and the elements involved [10]. Thus, the interatomic distances can be divided into two groups. The first formed by approximately exact distances

$$||x_i - x_j|| = d_{ij} \tag{4}$$

which do not rely on conformation. And the second group,

$$l_{ij} \leq ||x_i - x_j|| \leq u_{ij}, \tag{5}$$

formed by imprecise distances (inequalities) estimated experimentally.

In the original formulation of the MDGP, there is no differentiation between these two groups. However, considering that bond lengths and bond angles are fixed [10], we can define an atomic ordering $\{k_1, \ldots, k_n\}$ such that

$$||x_{k_i} - x_{k_j}|| = d_{k_i k_j} \in \mathbb{R}, \quad \forall (k_i, k_j) \text{ such that } |k_i - k_j| < 3. \tag{6}$$

In other words, there is always an enumeration that guarantees the existence of exact distances connecting a given atom to its two immediate successors. For simplicity, we will consider without loss of generality that the ordering $\{1, \ldots, n\}$ has this property.

3   Parametrization via Dihedral Angles

Despite simplicity, this hypothesis of ordering has very sophisticated implications. One is that, from the knowledge of the distances d_{ij}, $|i - j| < 3$, we can uniquely determine, except for translations and rotations, triangle T_k involving the triple $(k, k+1, k+2)$ of consecutive atoms. This implication may be extended to an arbitrary number of atoms. For example, for each quadruple of consecutive atoms, say, $(k, k+1, k+2, k+3)$, we can associate a structure formed by the triangles T_k and T_k+1 having the edge $(k+1, k+2)$ in common (see Fig. 1). In general, a sequence $\{1, 2, \ldots, n\}$ will determine an ordered structure $T = \{T_1, \ldots, T_{n-2}\}$ formed by $n - 2$ triangles, where each triangle T_k shares the edge $(k, k+1)$ with its predecessor T_{k-1} and the edge $(k+1, k+2)$ with its successor T_{k+1}. Note that the only possible relative movement between two triangles that share an edge is a rotation around this same edge. Thus, the degrees of freedom or flexibility of the structure T are given by the dihedral angles ω_k between the successive triangles T_k and T_{k+1}. In other words, the protein backbone structure can be characterized exclusively by the dihedral angles ω_k.

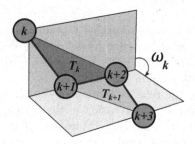

Fig. 1. Each triple of consecutive atoms defines a single triangle except for translations and rotations. Each pair of consecutive triangles with a common edge defines a dihedral angle ω_k.

If we want to modify only the relative distance between the triangles T_k and T_{k+1}, we simply rotate all the triangles T_j from $j > k$ around the edge $(k + 1, k + 2)$ since, in this way, the only pair of neighboring triangles that will exhibit relative motion will be $(T_k, T_k + 1)$. Note that rotating the triangles T_j for $j > k$ around the edge $(k + 1, k + 2)$ means simply rotating the vertices j for $j > (k + 2)$ around this same edge. That is, we can explore the entire space of possible configurations by using rotations around the edges shared by neighboring triangles.

We conclude that the ordering hypothesis reduces the MDGP to the problem of finding a configuration $T = \{T_1, T_2, \ldots, T_{n-2}\}$ of triangles with vertices $x_j \in \mathbb{R}^3$ that satisfy the inequality constraints of Eq. (5), since the equality equations are automatically satisfied by the definition of triangles T_k. Since the configurations T are characterized exclusively by the dihedral angles ω_k, the solution x of the MDGP is determined by $\omega = (\omega_1, \ldots, \omega_{n-3})$. In other words, we get a new problem, called MDGP$_\omega$, in the variable $x(\omega)$.

Definition 1 (MDGP$_\omega$). Determine $\omega \in \mathbb{R}^{n-3}$ (vector of dihedral angles) such that

$$l_{ij} \leq ||x_i(\omega) - x_j(\omega)|| \leq u_{ij}, \forall (i, j) \in E \subset \{1, \ldots, n\} \times \{1, \ldots, n\}, \quad (7)$$

where l_{ij} and u_{ij} are given positive real numbers.

The MDGP$_\omega$ formulation has some advantages. The first one is the reduction in the number of variables. Instead of $3n$ real coordinates, we use $n - 3$ dihedral angles. Additionally, in real instances there may be restrictions for the dihedral angles formed by certain chemical bonds which would be much more complicated to characterize in terms of the Cartesian coordinates. However, these advantages are obtained at the cost of increasing the complexity of the representation and, consequently, the increase of the time required to evaluate the model function and its derivatives.

4   Calculating Rotations

We can explicitly write $x(\omega)$ using rotations. Obtaining an explicit representation of $x(\omega)$ will be fundamental in order to define the objective function of the optimization formulation of the MDGP$_\omega$ and to calculate the derivatives required by the most efficient optimization algorithms.

As discussed earlier, the relative motion of the triangle T_{k+1} with respect to the triangle T_k is given by the rotation of the vertex x_{k+3} around the edge $(k + 1, k + 2)$ shared by T_k and T_{k+1}. Moreover, if we want to modify only the angle ω_k, then we have to apply the same rotation at all vertices x_j for $j > (k + 2)$.

Since rotations play an important role in argumentation, it will be useful to adopt a synthetic representation for them. Therefore, define $R(\theta, p, u, v)$ as being the result of the rotation of point $p \in \mathbb{R}^3$ by the angle θ around the axis given by the line containing points $u, v \in \mathbb{R}^3$. The analytic expression for the point $y = R(\theta, p, u, v) \in \mathbb{R}^3$ is given by

$$
y = (1 - \cos(\theta)) \begin{bmatrix} z_1(w_2^2 + w_3^2) - w_1(\kappa - z_1 w_1) \\ z_2(w_1^2 + w_3^2) - w_2(\kappa - z_2 w_2) \\ z_3(w_1^2 + w_2^2) - w_3(\kappa - z_3 w_3) \end{bmatrix}
$$
$$
+ \sin(\theta) \begin{bmatrix} -z_3 w_2 + z_2 w_3 - w_3 p_2 + w_2 p_3 \\ z_3 w_1 - z_1 w_3 + w_3 p_1 - w_1 p_3 \\ -z_2 w_1 + z_1 w_2 - w_2 p_2 + w_1 p_2 \end{bmatrix} + \cos(\theta) p, \tag{8}
$$

where $w = (u - v)/\|u - v\|$ is the normalized direction and $\kappa = w \cdot (u - p)$ [9].

The idea is to represent $x(\omega)$ using successive rotations. For this, we need an initial configuration $x^0 = (x_1^0, \ldots, x_n^0) \in \mathbb{R}^{3n}$ that satisfies the constraints of Eq. (4) and has dihedral angles $\omega_k = 0$. Finally, starting from the reference configuration x^0, we will construct a sequence $x^1(\omega), \ldots, x^{n-3}(\omega) \in \mathbb{R}^{3n}$ and we will assign $x(\omega) = x^{n-2}(\omega)$.

The iterative step in constructing the sequence $\{x^i\}_{i=1}^{n-2}$ is given by

$$
x_j^i = x_j^i(\omega) = \begin{cases} x_j^{i-1}(\omega), & \text{if } j < (i+3) \\ R(\omega_i, x_j^{i-1}(\omega), x_{i+1}^{i-1}(\omega), x_{i+2}^{i-1}(\omega)), & \text{otherwise,} \end{cases} \tag{9}
$$

where $x_j^i \in \mathbb{R}^3$ represents the position occupied by the j-th vertex after the application of the rotation associated with ω_i. By definition, the configuration x^1 is obtained from x^0 applying the rotation by angle ω_1 and axis (x_2^0, x_3^0) on all triangles T_k for $k > 1$. With this, all triangles but T_1 are rotated and, moreover, the dihedral angle between T_1 and T_2 will be ω_1. Note that this rotation modifies (updates) all coordinates x_j^0 for $j \geq 4$. In the second step, the configuration x^2 is obtained from x^1, but now rotating the triangles T_k for $k > 2$. In this way, only the dihedral angle between T_2 and T_3 is modified and, even more, its value becomes exactly ω_2. The process is repeated successively until all the dihedral angles have been fixed ($i = n - 2$). It is important to note that this procedure allows to arbitrarily define each of the dihedral angles.

The derivatives $\sigma_{jk}^i = \partial x_j^i(\omega)/\partial \omega_k$ can be calculated directly from Eq. (9). Note that the rotation of angle ω_k only affects the index points $j > (k + 2)$ of the configurations x^i for $i \geq k$. Therefore, using the chain rule, we get

$$\sigma_{jk}^i = \frac{\partial x_j^i(\omega)}{\partial \omega_k} = \nabla^t R(\omega_i, x_j^{i-1}, x_{i+1}^{i-1}, x_{i+2}^{i-1})(\delta_{ik}, \sigma_{jk}^{i-1}, \sigma_{i+1,k}^{i-1}, \sigma_{i+2,k}^{i-1}), \qquad (10)$$

if $j > (k + 3)$ with $i \geq k$, and $\sigma_{jk}^i = 0$ otherwise.

Although laborious, the derivative (Jacobian) of the rotation R can be calculated from Eq. (8). We can thus apply any of the MDGP optimization methods in the MDGP_ω formulation.

5   Numerical Experiments

In this section we will present some computational results validating the MDGP_ω formulation. Our objective is to illustrate the validity of the formulas presented in the previous section and the applicability of the proposal. We will consider the formulation

$$(P_\omega) \quad \min_\omega f(x(\omega)), \qquad (11)$$

derived from Eq. (3), where $x(\omega) = (x_1(\omega), \ldots, x_n(\omega)) \in \mathbb{R}^{3n}$ is given by Eq. (9) and the control variables are the dihedral angles $\omega = (\omega_1, \ldots, \omega_{n-3}) \in \mathbb{R}^{n-3}$.

The functions $\|x\|$ and $\max\{\alpha, 0\}$ with $\alpha \in \mathbb{R}$ on the formulation (P_ω) are not differentiable. Thus, to verify the validity of the derivatives of $x(\omega)$, we will use the hyperbolic approximations

$$\theta_\tau(y) = \sqrt{\tau^2 + \sum_{k=1}^{3} y_k^2} \quad \text{and} \quad \phi_\tau(\alpha) = (\alpha + \sqrt{\alpha^2 + \tau^2})/2, \qquad (12)$$

where the parameter τ controls the approximation quality [11,12]. In fact, we have that $\theta_\tau(y) = \|y\|$ and $\phi_\tau(\alpha) = \max\{y, 0\}$, when $\tau \to 0$ (see Fig. 2).

With these substitutions, we obtain a differentiable and arbitrarily close formulation $(P_{\tau,\omega})$ to the problem (P_ω). In addition to differentiability, the θ_τ and ϕ_τ functions reduce the number of local minima by convexifying the parcels of the objective function of the problem $(P_{\tau,\omega})$. However, the convexification is achieved with values τ relatively high causing the smoothed problem $(P_{\tau,\omega})$ to be very different from the original problem. To remedy this situation, the paper [11] proposes a heuristic, called SPH, where a sequence of problems $(P_{\tau_i,\omega})$ is

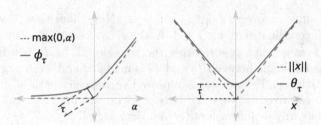

Fig. 2. Hyperbolic smoothing versions of $\max(\alpha, 0)$ and $||x||$. The parameter τ controls the approximation degree.

solve for $\tau_0 > \tau_1 > \ldots > \tau_m$ with $\tau_m \to 0$. The solution of the problem $(P_{\tau_i, \omega})$ is taken as the starting point of the next problem $(P_{\tau_{i+1}, \omega})$. Thus, as experimentally exemplified in that paper, trajectories that converge to less interesting local minimizers are generally avoided.

The SPH heuristic and the ideas expressed in Eqs. (8)–(10) were implemented in Matlab. We called SPH_ω the application of the SPH routine on the modified problem (P_ω).

The developed codes were tested on 100 instances of 8 randomly generated points. In each of these instances, we consider all equality distances $d_{ij} = ||x_i - x_j||$ for $|i - j| < 3$ and set the limits $l_{ij} = (1 - \epsilon)d_{ij}$ and $u_{ij} = (1 + \epsilon)d_{ij}$ with $\epsilon = 0.1$ for the distances $d_{ij} < 5$ with $|i - j| \geq 3$.

In our computations results, a set of coordinates $x(\omega) \in \mathbb{R}^{3n}$ solves the MDGP if

$$(1 - \mu)l_{ij} \leq ||x_i(\omega) - x_j(\omega)|| \leq (1 + \mu)u_{ij}, \quad \forall (i, j) \in E \tag{13}$$

for tolerance parameter $\mu = 0.1$.

Table 2 presents the results of the experiments comparing the performance of the SPH_ω heuristic with Quasi-Newton (QN) algorithm used by the Matlab local minimization routine **fminunc**. We considered the same randomly generated initial points in both alternatives. As we can see, the SPH_ω proposal is effective in obtaining solutions in 90% of the instances against 67% of the QN algorithm. This results corroborate the hypothesis of correctness of both the formulation and the derivatives of MDGP$_\omega$ model.

Table 2. Computational results.

	ns	nf	Time (sec)
QN	67	328.18	0.35
SPH$_\omega$	90	101.56	0.98

However, these promising results are attenuated by the time required to solve the problem (P_ω). In fact, the time of an evaluation of the function $f(x(\omega))$ was on average 10x higher than that of the function $f(x)$. Despite the use of a interpreted language environment as Matlab, this performance is mainly due to the recursive calls used in the implementation of the derivatives with respect to the dihedral angles.

6 Conclusions and Future Works

We presented a new formulation for the molecular distance geometry problem based on dihedral angles. Instead of the traditional approach that makes use of the Cartesian coordinates, this formulation uses an specific ordering to reduce the problem to the dihedral angles defined by each quadruple of consecutive atoms. This formulation reduces by 2/3 the number of variables of the original formulation.

We also presented the calculations of the derivatives of the Cartesian coordinates as a function of the dihedral angles. Finally, we illustrate through numerical experiments that our proposal can be used in conjunction with heuristics based on Cartesian coordinates.

The current version of the code makes intensive use of recursive calls which demands high computational time. We plan to explore opportunities for parallelism and scalability in future work. Note that there is no mandatory reason to use x_{i-1} and x_{i-2} as the axis of $x_i(\omega)$ rotation. In fact, any given pair (x_j, x_k) related to already fixed points could be used to define the rotations of x_i such that $||x_i - x_j||$ and $||x_i - x_k||$ are known. In this case, each of these rotations could be applied in parallel once they are independent. Another advantage of this improvement would be the reduction of recursive calls because x_i would not necessarily depend on x_{i-1}.

The instances and source codes used in this work can be obtained from the GitHub repository https://goo.gl/BnrMNZ.

Acknowledgments. The authors are grateful for the support of the Brazilian research agencies CNPq, CAPES, FAPESP, the Federal University of Ceará and the University of Campinas.

References

1. Atkins, J.F., Gesteland, R.: The 22nd amino acid. (Perspectives: biochemistry). Science **296**(5572), 1409–1411 (2002)
2. Cassioli, A., Günlük, O., Lavor, C., Liberti, L.: Discretization vertex orders in distance geometry. Discret. Appl. Math. **197**, 27–41 (2015). doi:10.1016/j.dam.2014.08.035
3. Crippen, G.M., Havel, T.F., et al.: Distance Geometry and Molecular Conformation, vol. 74. Research Studies Press, Taunton (1988)
4. Dyson, H.J., Wright, P.E.: Insights into protein folding from NMR. Ann. Rev. Phys. Chem. **47**(1), 369–395 (1996). doi:10.1146/annurev.physchem.47.1.369
5. Hao, B., Gong, W., Ferguson, T.K., James, C.M., Krzycki, J.A., Chan, M.K.: A new UAG-encoded residue in the structure of a methanogen methyltransferase. Science **296**(5572), 1462–1466 (2002). doi:10.1126/science.1069556

6. Liberti, L., Lavor, C., Mucherino, A., Maculan, N.: Molecular distance geometry methods: from continuous to discrete. Int. Trans. Oper. Res. **18**(1), 33–51 (2011). doi:10.1111/j.1475-3995.2009.00757.x

7. Liberti, L., Lavor, C., Maculan, N., Mucherino, A.: Euclidean distance geometry and applications. SIAM Rev. **56**(1), 3–69 (2014). doi:10.1137/120875909

8. Moré, J.J., Wu, Z.: Distance geometry optimization for protein structures. J. Glob. Optim. **15**(3), 219–234 (1999). doi:10.1023/a:1008380219900

9. Murray, G.: Rotation about an arbitrary axis in 3 dimensions (2017). http://inside.mines.edu/fs_home/gmurray/ArbitraryAxisRotation/

10. Schlick, T.: Molecular Modeling and Simulation: An Interdisciplinary Guide, vol. 21. Springer, New York (2010). doi:10.1007/978-1-4419-6351-2

11. Souza, M., Xavier, A.E., Lavor, C., Maculan, N.: Hyperbolic smoothing and penalty techniques applied to molecular structure determination. Oper. Res. Lett. **39**(6), 461–465 (2011). doi:10.1016/j.orl.2011.07.007

12. Souza, M., Lavor, C., Muritiba, A., Maculan, N.: Solving the molecular distance geometry problem with inaccurate distance data. BMC Bioinform. **14**(Suppl 9), S7 (2013). doi:10.1186/1471-2105-14-s9-s7

What's Hot and What's Not? - Exploring Trends in Bioinformatics Literature Using Topic Modeling and Keyword Analysis

Alexander Hahn, Somya D. Mohanty[✉], and Prashanti Manda

Department of Computer Science, University of North Carolina,
Greensboro, NC 27455, USA
mohanty.somya@uncg.edu

Abstract. Scientists exploring a new area of research are interested to know the "hot" topics in that area in order to make informed choices. With exponential growth in scientific literature, identifying such trends manually is not easy. Topic modeling has emerged as an effective approach to analyze large volumes of text. While this approach has been applied on literature in other scientific areas, there has been no formal analysis of bioinformatics literature.

Here, we conduct keyword and topic model-based analysis on bioinformatics literature starting from 1998 to 2016. We identify top keywords and topics per year and explore temporal popularity trends of those keywords/areas. Network analysis was conducted to identify clusters of sub-areas/topics in bioinformatics. We found that *"big-data"*, *"next generation sequencing"*, and *"cancer"* all experienced exponential increase in popularity over the years. On the other hand, interest in drug discovery has plateaued after the early 2000s.

Keywords: Bioinformatics · Scientific literature · Data mining · Topic modeling · Text analysis · Temporal mining

1 Introduction

Scientific literature holds a rich record of the ever-changing landscape of thought and observations in a wide variety of domains. Within a particular domain, researchers are increasingly interested in exploring scientific literature to gain insights on how research develops and evolves over time [24]. For instance, this kind of analytical data-driven insight can benefit researchers as they delve into new areas by providing knowledge of current popular topics and how the focus on different topics has shifted through time [1,24]. While the advent of digital publishing and open access science have led to greater access to scientific content, the sheer volume has made it very difficult for researchers to analyze literature at a high level and identify temporal trends in the evolution of research areas. [24]. This problem is particularly relevant in the thriving field of bioinformatics

© The Author(s) 2017
Z. Cai et al. (Eds.): ISBRA 2017, LNBI 10330, pp. 279–290, 2017.
DOI: 10.1007/978-3-319-59575-7_25

that encompasses several sub-areas garnering interest from biologists, computer scientists, and mathematicians.

Several approaches have been developed for analyzing text to identify semantic content, the most notable being topic modeling. Topic modeling is a text mining technique that identifies the hidden thematic/latent structure in collections of documents thereby allowing us to efficiently summarize large volumes of text [6]. Topic modeling algorithms take documents in a corpus and identify 'salient words grouping them to form 'topics'. Each document in a corpus is represented as a probabilistic mixture of topics while each topic consists of a mixture of words. In this manner, topic modeling algorithms discover patterns in textual data via topic generation and use those topics to connect documents with similar content [1]. This approach of analyzing text has been used in disparate domains such as social sciences, business analytics, and computer science.

While there are several topic modeling algorithms [6,10,11], Latent Dirichlet Allocation (LDA) [6] is one of the most widely used approaches and has been shown to be effective at finding distinct topics from a corpus [7,24]. In LDA, the topic distribution is assumed to have a Dirichlet prior unlike other algorithms such as LSA [10] and pLSA [11].

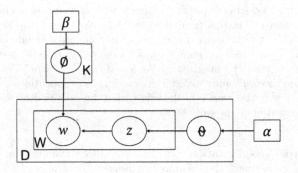

Fig. 1. LDA model representation for W words over D documents with K topics [6]. The two boxes represent replicates with the outer box representing documents and the inner box representing topics and words within a document.

LDA is a generative statistical model that models each of D documents in a corpus as a mixture of K topics where each topic corresponds to a multinomial distribution of W words [6] (Fig. 1). Other parameters in the model are defined as follows:

- α: Dirichlet prior on the topic distributions of each document
- β: Dirichlet prior on the word distributions of each word,
- θ_d: Topic distribution for document d,
- φ_k: Word distribution for topic k,
- z_{ij}: Topic for the ith word in document j, and
- w_{ij}: A particular word.

One of the input parameters of the LDA algorithm is the number of topics (K) to be identified from the corpus. Several studies have developed approaches to determine the optimal number of topics [4,22]. While there are likelihood based measures that help determine the right number of topics, these measures cannot be used alone to find the best model [3].

Here, we present our work on analyzing decades of Bioinformatics scientific literature to identify broad research themes and how those themes evolve across time. The goal of this work is to provide an exploration of different research areas within bioinformatics, identify "hot" areas and show how these areas interact with one another. We conduct a two-pronged analysis to achieve this goal. First, we analyze keywords and their popularity in each year to understand trends in popular research. A network of top keywords is built to identify clusters within these popular areas to observe interactions. Next, we apply topic modeling on abstracts to identify salient research themes at greater detail than keywords. These themes are complementary to themes identified from keywords. A network of topics is created to show how these research themes overlap and interact with each other. We explore temporal analysis of 10 curated topics to identify how research topics trend over time.

2   Related Work

Several studies have demonstrated the use of topic modeling to analyze scientific literature. Paul and Girju conducted analysis of literature in Computational Linguistics, and Education [17]. Their work shows how topics change over time in each field and how topics across fields are related. Similarly, Bolelli and Gilesb analyzed publications to identify research topics in computer science, influential authors, and trends related to those topics [7]. In a recent study, Kane et al. used topic models to compare the development of research on crops such as wheat, rice, sorghum, etc. [13]. Results from the topic models revealed interesting trends on how research on perennial crops was advancing and that is different from the progress on individual crops.

Much closer to our work is Altena et al.'s study on understanding the term big data from a text analysis of bio-medical literature [3]. While there are similarities in the literature corpus and techniques being applied, Altena et al.'s work differs from this study in that they restrict their study to big data literature in the bio-medical field while we analyze all areas of bioinformatics literature. In addition, we aim to search for over-arching patterns and trends in bioinformatics rather than focusing on one particular concept such as big data. Lastly, Suominen et al. performed topic modeling using LDA on scientific literature from Web of Science to compare how latent topics identified by LDA correlate with human assigned keyword categorization. The only use of topic models relating to bioinformatics to the best of our knowledge has been to answer specific research questions such as cluster analysis on medical, biological genotyping data [23] and toxicogenomics data analysis [15]. There is a notable lack of topic modeling based text analysis aimed at the wide corpus of bioinformatics literature to identify salient research topics and their evolution over time. Our work here aims to fill this gap.

3   Methods

3.1   Data Collection - Creating the Corpus

Scientific literature for this study was obtained by searching the Scopus database (https://www.scopus.com/) using the search term "bioinformatics". Scopus adds relevant index terms selected from controlled vocabularies to all publications (https://www.elsevier.com/__data/assets/pdf_file/0007/69451/scopus_content_coverage_guide.pdf). In addition to author keywords and titles, these index terms are used for searching. A list of publications matching the search term were retrieved using the Scopus API. Next, Scopus was queried to retrieve additional data such as authors, keywords, abstract, year of publication, and other metadata corresponding to the publication. For each publication, a document was created by concatenating the corresponding title, keywords, and abstract. A corpus of scientific literature was created by putting together documents corresponding to each publication.

3.2   Keyword-Based Analysis

Publications were analyzed based on their keywords in the following ways:

1. The number of publications per year was examined to identify any significant trends in research output across years.
2. The number of unique keywords observed in each publication year was extracted to explore correlations with the distribution of publication output.
3. A list of 25 keywords selected from the top keywords per year was curated and temporal analysis of their popularity across years was conducted. The popularity of a keyword computed using its occurrence frequency across documents per year was normalized to the [0, 1] range using min-max scaling. This analysis identified research areas experiencing upward spikes and rise in popularity and those experiencing decline.
4. A network of the top 25 keywords per year was built to explore relationships, inter-connectivity, and to identify clusters among these keywords. The network arranges the set of keywords into clusters and identifies intra- and inter-cluster interactions. Keywords in the network are weighted based on the prominence of their association with different publications. The larger the proportion of publications a keyword is associated with, the larger the keyword appears on the network. The network was constructed using Gephi (https://gephi.org/), an open source tool for network building and analysis. Clusters/communities in the keyword network are detected and optimized using the Louvain method [9]. After initial clusters are formed, the modularity optimization component further optimizes the clusters.

3.3   Topic Model Based Analysis

Latent Dirichlet Allocation was applied on the literature corpus - a collection of documents, one corresponding to each publication. 6 topic models were created using K (number of topics) in the range of 25 to 150 at increments of 25. These topic models were evaluated through manual examination. For each model, 20 top words per topic were examined to assess scientific coherence of the words as a set, overlap in topic words across topics, and human understandability. The selected model was used for all subsequent analyses.

After model selection, publications were analyzed based on their topics in the following ways:

1. The top 10 salient words relevant to 10 curated topics in the model were extracted and reported. This report provides a descriptive view of the topic model and verifies if topics identified by the model match natural human perception of the sub-areas of research within biology/bioinformatics.
2. A topic similarity network of all topics was built to identify topic clusters and their interplay. This allows for the identification of exciting clusters of research areas within bioinformatics.

4 Results

Searching for the term "bioinformatics" on the Scopus database resulted in 85,106 publications between the years of 1998 and 2016. When grouped by year, we see an upward trend in the number of publications per year (Fig. 2) except for years 2012 and 2013. Surprisingly, there appears to be a noticeable drop in publications in those two years.

4.1 Keyword-Based Analysis

We found 100,754 unique keywords across the 85,106 publications spanning across 18 years with an average of about 3 keywords per publication. The trend in the distribution of unique keywords in publications per year (Fig. 2) is very similar to the distribution of yearly publication numbers.

Temporal Keyword Trends. We manually curated 25 interesting keywords from top keywords in each year. Figure 3 shows the popularity trends of these 25 curated keywords. *"big data"*, *"proteomics"*, *"rna seq"*, *"cancer"*, *"next generation sequencing"*, and *"transcriptomics"* are among the areas that exhibit an increasing presence in publications over the last decade. It is interesting to see the emergence of big data applications within bioinformatics around 2010 accompanied by an exponential increase in relevant publications. *"rna seq"*, or rna sequencing, is another area that emerged during the later parts of the past decade and has emerged as a very popular research area. Unsurprisingly, the trend of *"next generation sequencing"* is similar to *"rna seq"*. Overall, *"next generation sequencing techniques"*, *"cancer informatics"*, *"biomarkers"*, *"metabolomics"*, *"mirna"*, *"machine learning"*, and *"big data"* are promising areas of research based on these trends. The emphasis on *"cancer"*, *"biomarkers"*, and *"big data"* indicate that health informatics is a sought after specialization. However, surprisingly, the same positive trend is not observed in the area of *"drug discovery"* which has plateaued over time. *"functional genomics"*, *"ontologies"*, and *"neural networks"* show mixed trends.

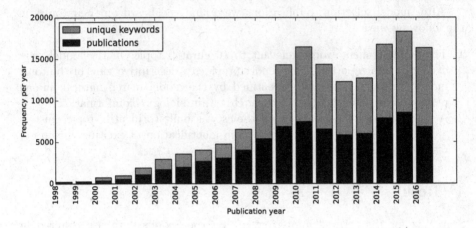

Fig. 2. Distribution of publications and unique keywords per year

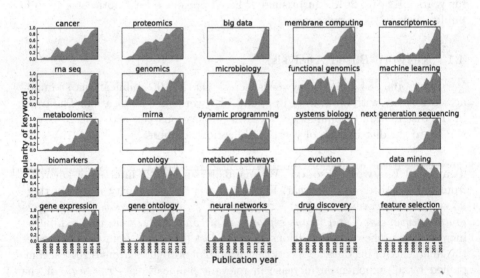

Fig. 3. Temporal trends of popularity of keywords over time.

Keyword Network. The network built using the top 25 keywords per year comprises 6 clusters shown in blue, pink, purple, green, brown, and grey (Fig. 4). It is evident that the blue cluster is central to the network with substantial overlap with other clusters. For lack of space, we only show the central blue and the green cluster in greater detail (Figs. 5 and 6). The blue cluster (Fig. 5) is largely focused on health informatics - in particular the study of different types of cancer such as *"colorectal"*, *"prostate"*, *"breast"*, etc. The cluster accurately identifies that microarray and gene expression analyses have been significant contributors to the study of cancer in the past decades [8,12,18]. It also hints at more recent approaches to cancer analytics which include using *"gene ontology"*, *"text mining"*, and machine learning approaches such as *"clustering"*, etc. [14,20].

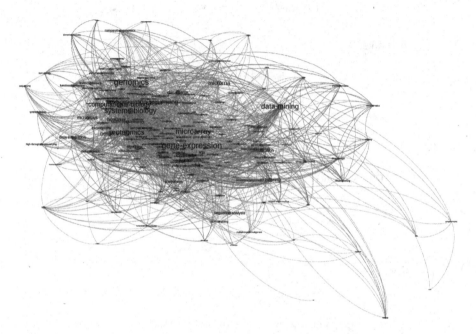

Fig. 4. Network of the top 25 keywords per year from 1998–2016 (Color figure online)

The green cluster (Fig. 6) focuses largely on sequence analysis and alignment using algorithms and techniques from graph theory. The green cluster contains certain nodes that are a bit distant from the rest of the cluster. These words include *"MPI"*, *"hadoop"*, *"mapreduce"*, *"cuda"*, and *"membrane"*, *"cloud computing"*. Interestingly, all these words pertain to big-data approaches that have recently come into play to analyze high throughput data from next generation sequencing approaches [19,21]. As sequencing data becomes more and more complex and voluminous, we can expect these words to become more central to this cluster over time.

The brown cluster focuses on computational techniques such as data mining, machine learning, feature selection for drug design and discovery, protein-structure prediction, pattern recognition, structural bioinformatics, etc. Moving

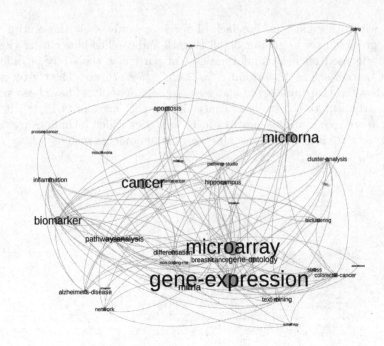

Fig. 5. Cancer informatics cluster (Color figure online)

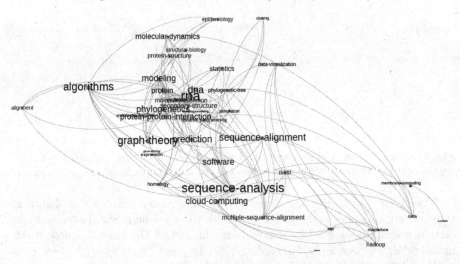

Fig. 6. Sequence analysis cluster (Color figure online)

on to the pink cluster, we see *"data integration"*, *"database"*, *"semantic web"*, and ontologies being used for the study of phenotypes, evolution, and phylogenies. This cluster points to the increasing applications of ontologies and data integration for the study of evolutionary phenotypes [16]. The grey cluster is largely related to proteomics, systems biology, functional genomics, analysis of microrna etc. The purple cluster is related to next generation sequencing, gene expression analyses, genomics, transcriptome, and genetics.

5 Topic Model-Based Analysis

Six topic models were created using number of topics $(K) \in$ [25, 50, 75, 100, 125, 150]. After careful manual evaluation of the topics and their top words, the model built using 50 topics was selected based on topic coherence and human understanding. Increasing the number of topics would make each individual topic more specific and might increase overlap between topics. Decreasing the number of topics, would result in more high-level abstract topics. A snapshot of the 50 topic model is shown in Table 1 by illustrating salient words in 10 curated topics.

Table 1. Salient words of the 10 selected topics

Topic 15	Patients, cancer, early, treatment, biomarkers, molecular, gene, expression studies, diagnosis
Topic 21	Cancer, gene, expression, mirna, association, studies, tumor, microarray, disease, cells
Topic 1	parallel, sequence, alignment, algorithm, performance, rna, gpu, implementation, memory, speedup
Topic 17	Medical, human, imaging, techniques, segmentation, algorithm, features, detection, information, gene
Topic 22	Cell, rna, transcription, infection, viruses, host, molecular, systems, dna, replication
Topic 14	large, species, phylogenetic, tree, sequence, gene, network, evolutionary, algorithms, performance
Topic 36	Biological networks, understanding, complex, functional, pathways, metabolic, processes, protein, microarray
Topic 37	Proteomics, peptides, mass, spectrometry, genome, clinical, variants, identification, genome, sequence
Topic 34	Snps, genetic, methods, sequencing, variants, association, single, haplotype, gwas, algorithm
Topic 9	Biological, database, web framework, scientific, workflows knowledge, management, cloud, computational

Topic Similarity Network. Next, we built a topic similarity network of all topics. In this network, nodes indicate topics represented by topic number and edges represent similarity between topics computed using the complement of Hellinger distance [5] between the probability distributions of two topics. The topic similarity network reveals four clusters (shown in blue, green, purple, and brown) (Fig. 7).

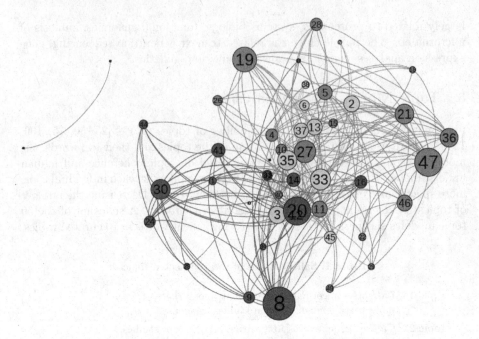

Fig. 7. A network of the 50 topic model (Color figure online)

Topics in the purple cluster correspond mainly to health informatics, clinical informatics, specifically focusing on cancer informatics. The topics in this cluster are characterized by words such as *"drug discovery"*, *"tumor"*, *"mirna studies"*, *"gene expression"*, *"association studies"*, *"target cells"*, *"differentially expressed"*, *"phage"*, *"genetic variants"*, *"biomarkers"*, *"early treatment"*, *"clinical diagnosis"*, etc. The topics hint at ontologies, pathways, networks, text mining, and association studies as some of the computational tools used in this area of research.

Sequence alignment, sequence similarity, and other related applications are seen prominently in the green cluster. Other top areas in this cluster include phylogenetic trees, evolutionary algorithms, protein structure and prediction, protein interactions, and distributed computing. The blue cluster represents research in proteomics, genome sequencing, annotation, and assembly tools. Other areas represented in this cluster include metabolomics, protein structures, mass spectrometry, community software, and genome databases. Interestingly, the brown cluster which contains only two topics representing studies on water quality and treatment, is an outlier to the other clusters. Prominent words in these two topics include ph level, removal, water quality, nanoparticles, adsorption, iron, and concentration indicating work on water treatment advances using adsorption [2]. It is not surprising that these topics have little similarity with the other areas.

Overall, these topics indicate research areas such as health and cancer informatics, proteomics, genome annotation and assembly, sequence alignment, and the computational techniques used in each of these areas.

6   Conclusion

In this study, we conducted scientific literature analysis of bioinformatics publications from 1998 to 2016 using keyword and topic modeling based analysis. We discovered research areas within bioinformatics that are experiencing a rise in popularity and those witnessing waning interest. The trends show that there is increasing research in cancer informatics and that cancer research has shifted towards using big data techniques in recent years. The presence of big data techniques can also be seen in other areas such as sequence alignment and genome annotation. Machine learning, feature selection, network analysis, ontologies, data mining, distributed computing, parallel computing, hadoop, web applications, and community databases are some of the prominent computational techniques seen in bioinformatics.

References

1. Alghamdi, R., Alfalqi, K.: A survey of topic modeling in text mining. Int. J. Adv. Comput. Sci. Appl. (IJACSA) **6**(1) (2015)
2. Ali, I., Gupta, V.: Advances in water treatment by adsorption technology. Nat. Protoc. **1**(6), 2661–2667 (2006)
3. Altena, A.J., Moerland, P.D., Zwinderman, A.H., Olabarriaga, S.D.: Understanding big data themes from scientific biomedical literature through topic modeling. J. Big Data **3**(1), 23 (2016)
4. Arun, R., Suresh, V., Veni Madhavan, C.E., Narasimha Murthy, M.N.: On finding the natural number of topics with latent dirichlet allocation: some observations. In: Zaki, M.J., Yu, J.X., Ravindran, B., Pudi, V. (eds.) PAKDD 2010. LNCS, vol. 6118, pp. 391–402. Springer, Heidelberg (2010). doi:10.1007/978-3-642-13657-3_43
5. Beran, R.: Minimum hellinger distance estimates for parametric models. Ann. Stat. **5**, 445–463 (1977)
6. Blei, D.M., Ng, A.Y., Jordan, M.I.: Latent Dirichlet allocation. J. Mach. Learn. Res. **3**(Jan), 993–1022 (2003)
7. Bolellia, L., Gilesb, S.: What is trendy? generative models for topic detection in scientific literature
8. Cheang, M.C., van de Rijn, M., Nielsen, T.O.: Gene expression profiling of breast cancer. Annu. Rev. Pathmechdis. Mech. Dis. **3**, 67–97 (2008)
9. De Meo, P., Ferrara, E., Fiumara, G., Provetti, A.: Generalized louvain method for community detection in large networks. In: 2011 11th International Conference on Intelligent Systems Design and Applications (ISDA), pp. 88–93. IEEE (2011)
10. Deerwester, S., Dumais, S.T., Furnas, G.W., Landauer, T.K., Harshman, R.: Indexing by latent semantic analysis. J. Am. soc. Inf. Sci. **41**(6), 391 (1990)
11. Hofmann, T.: Probabilistic latent semantic indexing. In: Proceedings of the 22nd annual international ACM SIGIR conference on Research and development in information retrieval, pp. 50–57. ACM (1999)
12. Hoopes, L.: Genetic diagnosis: DNA microarrays and cancer. Nat. Educ. **1**(1), 3 (2008)
13. Kane, D.A., Rogé, P., Snapp, S.S.: A systematic review of perennial staple crops literature using topic modeling and bibliometric analysis. PloS One **11**(5), e0155788 (2016)

14. Kourou, K., Exarchos, T.P., Exarchos, K.P., Karamouzis, M.V., Fotiadis, D.I.: Machine learning applications in cancer prognosis and prediction. Comput. Struct. Biotechnol. J. **13**, 8–17 (2015)

15. Lee, M., Liu, Z., Huang, R., Tong, W.: Application of dynamic topic models to toxicogenomics data. BMC Bioinf. **17**(13), 368 (2016)

16. Manda, P., Balhoff, J.P., Lapp, H., Mabee, P., Vision, T.J.: Using the phenoscape knowledgebase to relate genetic perturbations to phenotypic evolution. Genesis **53**(8), 561–571 (2015)

17. Paul, M.J., Girju, R.: Topic modeling of research fields: an interdisciplinary perspective. In: RANLP, pp. 337–342 (2009)

18. Perez-Diez, A., Morgun, A., Shulzhenko, N.: Microarrays for cancer diagnosis and classification. In: Mocellin, S. (ed.) Microarray Technology and Cancer Gene Profiling, pp. 74–85. Springer, New York (2007)

19. Vijayakumar, S., Bhargavi, A., Praseeda, U., Ahamed, S.A.: Optimizing sequence alignment in cloud using hadoop and MPP database. In: 2012 IEEE 5th International Conference on Cloud Computing (CLOUD), pp. 819–827. IEEE (2012)

20. Wu, T.J., Schriml, L.M., Chen, Q.R., Colbert, M., Crichton, D.J., Finney, R., Hu, Y., Kibbe, W.A., Kincaid, H., Meerzaman, D., et al.: Generating a focused view of disease ontology cancer terms for pan-cancer data integration and analysis. Database **2015**, bav032 (2015)

21. Xue, Q., Xie, J., Shu, J., Zhang, H., Dai, D., Wu, X., Zhang, W.: A parallel algorithm for dna sequences alignment based on MPI. In: 2014 International Conference on Information Science, Electronics and Electrical Engineering (ISEEE), vol. 2, pp. 786–789. IEEE (2014)

22. Zhao, W., Chen, J.J., Perkins, R., Liu, Z., Ge, W., Ding, Y., Zou, W.: A heuristic approach to determine an appropriate number of topics in topic modeling. BMC Bioinf. **16**(13), S8 (2015)

23. Zhao, W., Zou, W., Chen, J.J.: Topic modeling for cluster analysis of large biological and medical datasets. BMC Bioinf. **15**(11), S11 (2014)

24. Zhou, H.K., Yu, H.M., Roland, H.: Topic discovery and evolution in scientific literature based on content and citations. Frontiers **1** (2016)

Structure Modeling and Molecular Docking Studies of Schizophrenia Candidate Genes, Synapsins 2 (*SYN2*) and Trace Amino Acid Receptor (*TAAR6*)

Naureen Aslam Khattak[1], Sheikh Arslan Sehgal[2],
Yongsheng Bai[1(✉)], and Youping Deng[3,4(✉)]

[1] Department of Biology and The Center for Genomic Advocacy,
Indiana State University, 600 Chestnut Street, Terre Haute, IN 47809, USA
naslam@sycamores.indstate.edu,
Yongsheng.Bai@indstate.edu
[2] University of Chinese Academy of Sciences, Beijing, China
arslan.msbi4@iiu.edu.pk
[3] Department of Complementary and Integrative Medicine,
University of Hawaii John A. Burns School of Medicine,
Honolulu, HI 96813, USA
dengy@hawaii.edu
[4] National Center of Colorectal Disease,
Nanjing Municipal Hospital of Chinese Medicine, The Third Affiliated Hospital,
Nanjing University of Chinese Medicine, Nanjing 210001, China

Abstract. Schizophrenia (SZ) is a severe manifesting psychiatric neural disorder with abnormal behavior, disorganized speech and figment of the imagination. The Synapsin II (SYN2) and Trace Amine Associated Receptor (TAAR6) genes has direct association with SZ. In the current study, the 3-dimensional structure of SYN2 and TAAR6 protein is proposed and the protein-protein docking analysis was applied to explore the binding interactions of the candidate proteins. The comparative modeling was performed with the suitable template (Q86VA8 for SYN2 and H0YF79 for TAAR6) which represents the query coverage (71%, 87%), sequence identity (67%, 34%) and the e-value (0.0, 1e-43) respectively. The structure quality of the predicted model of SYN2 and TAAR6 presents 90.7%, and 96.5% residues in the favored region of Ramachandran plot analysis respectively, suggests the good quality models construction. The phylogenetic analysis suggests that the TAAR6 sequence is conserved in chimpanzee and gorilla (>80% homology) whereas the SYN2 is closely related with macaque. The protein docking analysis of SYN2 shows five ionic interactions with Lys-256, Lys-539, Arg-475, Gln-536 and Gln-529 with His-121, Glu-467, Glu-472, Arg-458 and Asp-477 of CAPON. The TAAR6 have two interactions of Glu-33 and Gly-171 with Arg-85 and Lys-52 of the PPP3CC. Current computational study may play a significant role to recruit, analyze and cure the mysteries of schizophrenia neurodegenerative disorder.

Keywords: Bioinformatics · Computational biology · TAAR6 · SYN2 · Protein modeling · Molecular docking · Phylogenetic · Schizophrenia

© Springer International Publishing AG 2017
Z. Cai et al. (Eds.): ISBRA 2017, LNBI 10330, pp. 291–301, 2017.
DOI: 10.1007/978-3-319-59575-7_26

1 Introduction

Schizophrenia is a chronic neurological disorder with serious social and personal impact [1]. The severe symptoms of SZ may also include affective flattening, avolition, alogia. The SZ symptoms usually starts affecting in the late teens to early twenties of patient life and occur in 1% of total population [2]. The meta-data analysis suggests that the risk of SZ in males are 40% higher as compare to the females [3]. Several candidate genes are considered for the SZ susceptibility [4]. Positional cloning, linkage analysis and candidate gene approaches are successfully used to explore the disease-causing genes [5].

Pearlson and Foley [6] suggests that copy number variant (CNVs) and single nucleotide polymorphism (SNPs) within the population or mutations like deletion/insertion can alter cellular/multiple or single processes may leads to the occurrence of schizophrenia disease. It takes a lot of efforts and time to collect large samples for the SZ patients, which is one of the factor that causes a delay in publishing the SZ data on disease genotype and phenotype. The phenotypic observation of the patients with SZ shows heterogeneity which are responsible for the major obstacle in the research. The small number of candidate genes which potentially controls the homogeneous phenotypes of SZ may facilitate to dissect the disease pathogenicity [7]. The gene linkages and meta-analysis of genome scan [8] suggests highly susceptible candidate genes for schizophrenia are present on chromosomes 1q, 3p, 5q, 6p, 8p, 11q, 14p, 20q and 22q [9–11].

The trace amine receptor family 6 (TAAR6) was first reported as a schizophrenia susceptible gene and belongs to the trace amine receptor family. The TAAR has super family of G-protein-coupled receptors and have a core interest in depression and schizophrenia. These are endogenous amine and chemically similar to the classic biogenic amines like serotonin, dopamine, histamine and norepinephrine and are widely expressed in the brain cells [12].

The synapsin2 (SYN2) is a strong candidate gene for the SZ and mapped on chromosome 3p2. This gene plays a major role in the neurotransmission regulation, neural plasticity and the synaptogenesis [13, 14]. Three synapsin genes of human (SYN1, SYN2 and SYN3) have been reported [15]. These phosphoproteins are central regulators of transmitter release vesicle fusions [16]. In current study, 3D structures and protein-protein docking analysis of SYN2 and TAAR6 were performed to elucidates the connection of SYN2 and TAAR6 candidate proteins with SZ.

2 Materials and Methods

The methodology used in the current study is described below.

2.1 Comparative Modeling

The amino acid sequence of SYN2 and TAAR6 proteins (582aa and 318aa) were used for the comparative modeling analysis. The sequences in FASTA format were retrieved from UniProt Knowledgebase [17] with accession number Q86VA8 and H0YF79 respectively. The canonical sequences of SYN2 and TAAR6 were subjected to PSI_ BLAST [18] search by using Protein Data Bank (PDB) [19]. The protein structure prediction software MODELLER [20] was used to predict the candidate proteins 3D structure. The reliability of predicted 3D model was further evaluated by Rampage [21] and ERRAT [22]. Rampage generated Ramachandran plot and ERRAT evaluated the quality factor of the predicted structures.

2.2 Phylogenetic Analysis

The molecular evolutionary genetic algorithm (MEGA 5) [23] was applied on SYN2 and TAAR6 candidate proteins to investigate their ancestral relationship. Distance based approach were utilized by exploring the Neighbor-Joining method. The bootstrap value was subject to 1000 replication to construct the phylogenetic tree.

2.3 Protein-Protein Docking

For protein- protein interactions, the STRING [24] server was used to find the direct (physical) and indirect (functional) relationship of SYN2 and TAAR6. The PatchDock [25] and Gramm-X server [26] were employed to demonstrate the protein-protein docking for both candidate proteins. Post docking analysis were performed by PyMol [27]. The methodology followed in current study is represented in Fig. 1 and the bioinformatics techniques used in this study are mentioned in Table 1.

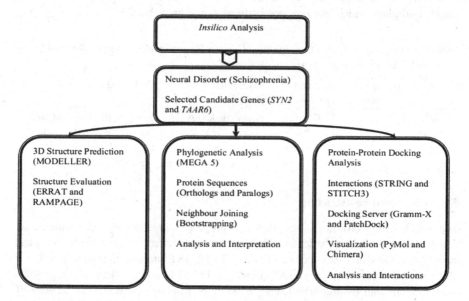

Fig. 1. The flowchart represents the overall strategy used in the current study

Table 1. Bioinformatics analysis tool used in the current study

Tools/databases	Output/function
BioGPS	Expression profiling
UniProt	Amino acid sequence retrieval
MODELLER	3D Structure prediction
Chimera	Visualization, Superimposition, Interaction
ERRAT	3D Structure evaluation
ENSEMBL	Phylogenetic Sequences Retrieval
MEGA	Phylogenetic Analysis
STRING	Protein-Protein Physical Interactions
GRAMM-X	Protein-Protein Docking
PyMol	Binding interactions of docked Protein-protein complexes

3   Results and Discussions

The consideration was given to the literature survey to explore the most likely candidate genes in schizophrenia disorder. Human protein reference database (HPRD) [28] was used to extract the information based on molecular function, biological process and cellular location of both candidate proteins. SYN2 shows catalytic activity and ATP binding site as a molecular function and it biological processes include synaptic transmission and Neurotransmitter secretion. On the other hand, TAAR6 is responsible for G-protein couple receptor activity, therefore play an active role in G-protein coupled receptor signaling pathway. Both SYN2 and TAAR6 are present in synaptic vesicle and plasma membrane respectively as shown in Table 2.

Table 2. Cytogenetic location from Ensembl Genome Browser and gene expression profiling of SYN2 and TAAR6 candidate genes involved in SZ.

Gene	Genomic location	Start and End BP	Molecular function	Biological Function
TAAR6	Chr 3	12045862–12232907	Catalytic and ATP binding activity	Synaptic transmission and Neurotransmitter secretions
SYN2	Chr 6	132891461–132892498	G-protein coupled receptor activity	G-protein coupled receptor signaling pathways

3.1   Comparative Modeling

The MODELER 9v10, homology modeling program was implemented for the candidate proteins model construction. The five templates were chosen based on query coverage, e-value and sequence identity by PSI-BLAST analysis as shown in Table 3. The selected template for the *SYN2* (PDB ID, 1PK8) has 71% query coverage, 67% identity score, and E-value 0.0. The PDB ID of 2VT4 shows 34% identity score, 87% query coverage and E-value 1e-87 for *TAAR6*. Table 4 represents the appropriate template selection for *SYN* and *TAAR6*. A total of 50 predicted models were generated

Table 3. The selective five templates of SYN and TAAR 2 candidate proteins

Gene	Accession ID	Total score	Query coverage	E-value	Max-identity
SYN2	1PK8	569	71%	0.0	67%
	1I7L	659	53%	0.0	97%
	1AUX	536	52%	0.0	77%
	1AUV	508	52%	2e−177	74%
	2P0A	507	58%	1e−176	68%
TAAR6	2VT4	154	87%	1e−43	34%
	2Y00	151	88%	3e−42	34%
	2R4S	150	87%	1e−41	31%
	2R4R	149	88%	2e−41	31%
	3KJ6	149	88%	2e−44	31%

Table 4. The suitable template selection for SYN and TAAR2 candidate proteins for the comparative modeling analysis

Protein	Accession number	Template	Amino acids	Query coverage	E-value	Identity
SYN2	Q86VA8	1PK8	582 aa	71%	0.0	67%
TAAR6	H0YF79	2VT4	318 aa	87%	1e−43	34%

by MODELLER. The 10 predicted models were selected by using the object function value (DOPE). The best 3D models with lowest DOPE value was picked as the final predicted protein model and subjected to the model assessment.

The predicted models were superimposed to observe the structure relevance. The Figs. 2 and 3 illustrates the predicted model structures, superimposition and protein structure evaluation of SYN and TAAR. Ramachandran plot and ERRAT was used to analyzed the overall quality and the protein reliability of the predicted models. Ramachandran plot showed the distribution of amino acids in favored, allowed and outlier regions. The higher residues (>90%) present in favored regions represents the good quality model.

3.2 Phylogenetic Analysis

The paralogs of *TAAR6* and *SYN2* were retrieved from ENSEMBL and confirmed from the UCSC Genome database. The three paralogs of *TAAR6* (*TAAR2, TAAR5* and *TAAR6*) and *SYN* (*SYN1, SYN2,* and *SYN3*) were used as an input to predict the phylogenetic history. The ciona first mutate and produced three clusters and each cluster represents a gene topology. Rodents, teleosts and birds are at their specific positions in clusters in bifurcations. The SYN of human has conserved sequence with primates with indels. The ciona first mutate and formed *SYN3* then *SYN2* and *SYN1* were bifurcate. *SYN3* showed two main clusters of teleosts and mammals. Human is closely related with macaque. *SYN2* gene is only predicted in rodents, tetrapods and primates and represents in tree at their concerning position. *SYN1* also showed same results like *SYN2* (Fig. 4).

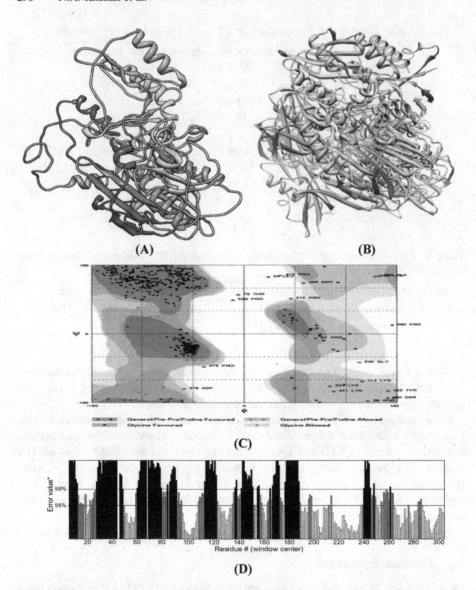

Fig. 2. (A) The 3D structure of SYN2 (template 1PK8) with 71% query coverage (B) The superimposition of SYN2 (blue colour) and 1PK8 (brown colour) (C) Rampage and ERRAT (D) evaluation structure server showed 90.7%, residues lied in favoured region, 5.7% in the allowed region with 3.6% in outlier region (Color figure online)

3.3   Protein-Protein Docking

The protein functional interactors of SYN2 and TAAR were extracted from the STRING database. The C-Terminal PDZ Domain Ligand of Neuronal Nitric Oxide

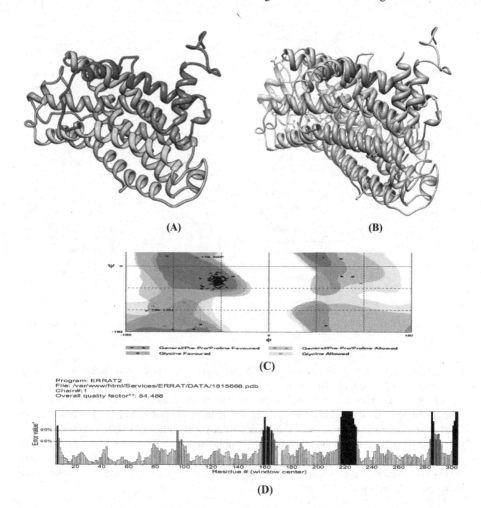

Fig. 3. (A) The 3D structure of TAAR6 generated by 2VT4 template with 87% query coverage. (B) Superimposition of TAAR6 with 2VT4. Blue colour representing template and brown color shown predicted model. (C) Ramachandran plot showed 96.5% residues in favoured region, 2.5% in allowed region and 0.9% in outlier region for TAAR6 candidate gene structure analysis (D) TAAR6 predicted structure showed 84.488% quality factor by ERRAT evaluation tool. (Color figure online)

Synthase (CAPON) shows high interacting score of 0.945 with SYN2 and therefore selected for SYN II protein docking analysis.

The TAAR has a score of 0.802 with the Protein Phosphatase 3 Catalytic Subunit Gamma (PPP3CC) for protein docking. The 3D structure of CAPON protein was not reported in PDB therefore, it was constructed by the MODELLER program. The 3D structure of PPP3CC for TAAR2 was retrieved from Swiss Model accession number 1M63A.

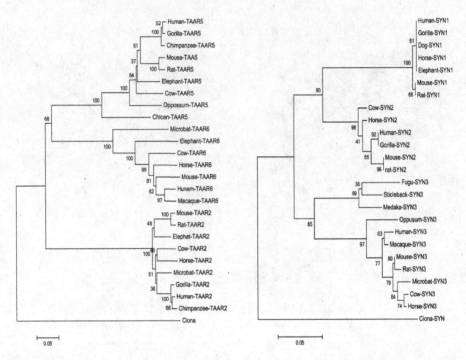

Fig. 4. Evolutionary history of (A) *TAAR6* and (B) *SYN* gene was inferred by using Neighbour-joining method. Numbers represents on branches is bootstrap values of 1000 replications. Uncorrected p-distance parameter was used. Scale bar represents amino acid substitution per site. Phylogenetic analysis of *TAAR* indicates that human is closely related to Gorilla, Macaque and Chimpanzee. The human from gorilla bifurcates 100 mya. Teleosts and rodents are closely related among their respective organisms. The SYN gene of human is closely related to vertebrates and rodents. Organisms with feathers (Chicken and Zebra finch) are closely related to each other. Tetrapods are present in their respective branch in species tree. Teleosts are shown at close branches

Both SYN2 and TAAR6 shows ionic interaction with their functionally interactive proteins, CAPON and PPP3CA respectively. The SYN2 and CAPON have five ionic interactions. Oxygen atom of lys256 and Gln536 amino acids of SYN2 interacted with the nitrogen atoms of His121 and Arg458 with bond distance 2.0 and 3.2 respectively. Lys539, Arg475, Gln539 interacted with Glu467, Glu472, Asp477 with bond distance 2.6, 2.5, 2.9 respectively. The TAAR6 and PPPCC3 proteins have two ionic interactions. Oxygen atoms of Glu33 and Gly171 amino acids of TAAR6 interacted with the nitrogen atoms of Arg85 and lys52 of ligand protein (PPP3CC) with bond distance 2.2 and 2.2 respectively. The PyMol visualization of SYN with CAPON and TAAR6 with PPP3CC is shown in Fig. 5 and Table 5.

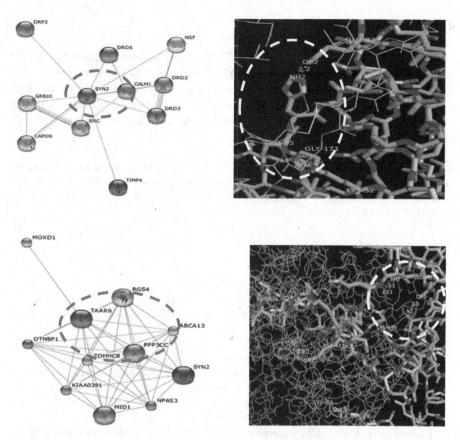

Fig. 5. (A) The *SYN2* functional protein partner generated by STRING database (B) The protein-protein docking of *SYN2* with *CAPON* (C) *TAAR2* protein interaction network and (D) shows the *PPP3CC* interaction with *TAAR2*. Red circles indicate the functional amino acid with partner protein. Brown circle shows the selective functional partner for each candidate protein (Color figure online)

Table 5. The active binding residues analysis for SYN2 and TAAR with CAPON and PPP3CC

Receptor	Interacting protein	Functional amino acid	Bond distance	Interactions type
SYN2	CAPON	Lys-256/O → His-121/NE2	2.0	Ionic Bonding
		Lys-539/N2 → Glu-467/OE1	2.6	Ionic Bonding
		Arg-475/NH1 → Glu-472/OE2	2.5	Ionic Bonding
		Gln-536/OE1 → Arg-458/NH2	3.2	Ionic Bonding
		Gln-529/N → Asp-477/OD2	2.9	Ionic Bonding
TAAR6	PPP3CC	Glu-33/OE2 → Arg-85/NH2	2.2	Ionic Bonding
		Gly-171/O → Lys-52/NH2	2.2	Ionic Bonding

4   Conclusions

The predicted 3D structures of SYN and TAAR2 may assist to understands the 3D conformational changes of the protein SYNS2 and TAAR6 in schizophrenia. The successfully identified protein-protein interactions with key binding residues may play a potential role to examine the diseases pathogenicity. Site-directed mutagenesis can further help to investigates the in-vitro impact of identified key residues on wild type and mutant protein.

Acknowledgements. We thank Indiana State University for providing support to complete the manuscript. This work was supported by the Indiana State University start-up funds to YB. The work was also supported by the NIH Grant 5P30GM114737, the NIH Grant P20GM103466, and the NIH Grant U54 MD007584.

References

1. Austin, J.: Schizophrenia an update and review. J. Genet. Couns. **14**, 329–340 (2005)
2. Gottesman II, S.: Schizophrenia the Epigenetic Puzzle. Cambridge University Press, Cambridge (1984)
3. Aleman, A., Kahn, R.S., Selten, J.P.: Sex differences in the risk of schizophrenia: evidence from meta-analysis. Arch. Gen. Psychiatr. **60**, 565–571 (2003)
4. Craddock, N., O'Donovan, M.C., Owen, M.J.: Genes for schizophrenia and bipolar disorder. Implications for psychiatric nosology. Schizophr. Bull. **32**, 9–16 (2006)
5. Karayiorgou, M., Gogos, J.A.: Schizophrenia genetics: uncovering positional candidate genes. Eur. J. Hum. Genet. **14**, 512–519 (2006)
6. Pearlson, G.D., Folley, B.S.: Schizophrenia, psychiatric genetics, and Darwinian psychiatry: an evolutionary framework. Schizophr. Bull. **34**, 722–733 (2008)
7. Boks, M.P., Leask, S., Vermunt, J.K., Kahn, R.S.: The structure of psychosis revisited: the role of mood symptoms. Schizophr. Res. **93**, 178–185 (2007)
8. Liu, H., Heath, S.C., Sobin, C., Roos, J.L., Galke, B.L., Blundell, M.L.: Genetic variation at the 22q11 PRODH2/DGCR6 locus presents an unusual pattern and increases susceptibility to schizophrenia. Proc. Natl. Acad. Sci. **99**, 3717–3722 (2002)
9. Harrison, P.J., Owen, M.J.: Genes for schizophrenia Recent findings and their pathophysiological implications. Lancet **361**, 417–419 (2003)
10. O'Donovan, M.C., Williams, N.M., Owen, M.J.: Recent advances in the genetics of schizophrenia. Hum. Psychiatry **8**, 217–224 (2003)
11. O'Donovan, M.C., Craddock, N., Norton, N., Williams, H., Peirce, T., Moskvina, V., Nikolov, I., Hamshere, M., Carroll, L., Georgieva, L.: Identification of loci associated with schizophrenia by genome-wide association and follow-up. Nat. Genet. **40**, 1053–1055 (2008)
12. Duan, J., Martinez, M., Sanders, A.R., Hou, C., Saitou, N., Kitano, T., Mowry, B.J., Crowe, R.R., Silverman, J.M., Levinson, D.F., Gejman, P.V.: Polymorphisms in the trace amine receptor 4 (TRAR4) gene on chromosome 6q23.2 are associated with susceptibility to schizophrenia. Am. J. Hum. Genet. **75**, 624–638 (2004)
13. Hilfiker, S., Pieribone, V.A., Czernik, A.J., Kao, H.T., Augustine, G.J., Green-gard, P.: Synapsins as regulators of neurotransmitter release. Phil. Trans. R Soc. Lond. B **354**, 269–279 (1999)

14. Kao, H.T., Porton, B., Hilfiker, S., Stefani, G., Pieribone, V.A., DeSalle, R.: Molecular evolution of the synapsin gene family. J. Exp. Zool. **285**, 360–377 (1999)
15. Tokumaru, H., Umayahara, K., Pellegrini, L.L., Ishizuka, T., Saisu, H., Betz, H.: SNARE complex oligomerization by synaphin/complexin is essential for synaptic vesicle exocytosis. Cell **104**, 421–432 (2001)
16. Karlin, S., Chen, C., Gentles, A.J., Cleary, M.: Associations between a human disease genes and overlapping gene groups and multiple amino acid runs. Proc. Nat. Acad. Sci. **99**, 17008–17013 (2002)
17. Apweiler, R., Bairoch, A., Wu, C.H.: UniProt: the Universal Protein knowledgebase. Nucleic Acids Res. **32**(Database issue), D115–D119 (2004). doi:10.1093/nar/gkh131
18. Altschul, S.F., Madden, T.L., Schäffer, A.A., et al.: Gapped BLAST and PSI-BLAST: a new generation of protein database search programs. Nucleic Acids Res. **25**, 3389–3402 (1997)
19. Berman, H.M., Westbroo, J.K., Feng, Z., Gilliland, G., Bhat, T.N., Weissig, H., Shindyalov, I.N., Bourne, P.E.: The protein data bank. Nucleic Acids Res. **28**, 235–242 (2000). www.rcsb.org
20. Eswar, N., Eramian, D., Webb, B., Shen, M.Y., Sali, A.: Protein Structure modeling with MODELLER. Methods Mol. Biol. **426**, 145–159 (2008)
21. Laskowski, R.A., MacArthur, M.W., Moss, D.S., Thornton, J.M.: PROCHECK- a program to check the stereochemical quality of protein structures. J. Appl. Cryst. **26**, 283–291 (1993)
22. Colovos, C., Yeates, T.O.: Verification of protein structures: patterns of nonbonded atomic interactions. Protein Sci. **2**, 1511–1519 (1993)
23. Koichiro, T., Glen, S., Daniel, P., Alan, F., Sudhir, K.: MEGA6: molecular evolutionary genetics analysis version 6.0. Mol. Biol. Evol. **30**, 2725–2729 (2013)
24. Szklarczyk, D., Franceschini, A., Kuhn, M., Simonovic, M., Roth, A., Minguez, P., et al.: The STRING database in 2011: functional interaction networks of proteins globally integrated and scored. Nucleic Acids Res. **39**, 561–568 (2011)
25. Schneidman-Duhovny, D., Inbar, Y., Nussinov, R., Wolfson, H.J.: PatchDock and SymmDock: servers for rigid and symmetric docking. Nucleic Acids Res. **33**, W363–W367 (2005)
26. Tovchigrechko, A., Vakser, I.A.: GRAMM-X public web server for protein-protein docking. Nucleic Acids Res. **34**, W310–W314 (2006)
27. The PyMOL Molecular Graphics System, Version 1.8 Schrödinger, LLC
28. Prasad, K.S.T., Goel, R., Kandasamy, K., Keerthikumar, S., et al.: Human protein reference database update. Nucleic Acids Res. **37**, D767–D772 (2009)

Accurate Prediction of Haplotype Inference Errors by Feature Extraction

Rogério S. Rosa and Katia S. Guimarães$^{(\boxtimes)}$

Informatics Center, UFPE, Recife, Brazil
{rsr,katiag}@cin.ufpe.br

Abstract. An important problem in Bioinformatics is Haplotype Inference (HI), that consists of computationally inferring haplotype sequences from genotype data. Haplotype data is highly informative for illness propensity detection, but it is much costly and time consuming to acquire; that gives the HI Problem an overwhelming relevance. In this paper, we formally demonstrate that specific genomic data features can be very strong indicators of error propensity in each one of four well-known HI methods studied. We apply Statistical analyses to explore the relevance of biologically meaningful properties extracted from the genotype sequences, and develop models to predict the accuracy expected in the haplotype inference results, for different methods and error metrics. The quality and the stability of our models are demonstrated by statistical evidence. One of our estimated models presents nearly perfect accuracy for all four methods studied. Our results provide useful insights to help develop more effective HI methods.

Keywords: Haplotype inference · Genotype data · Statistical analyses · Sequences

1 Introduction

Haplotype information is valuable in understanding species evolution, as well as in association studies that try to correlate the propensity to certain diseases with patterns inherited through the haploid cells [1]. Since capturing haplotypes directly from experiments is both difficult and expensive [2], it is highly desirable to determine haplotypes from genotypes (available on large scale at low cost) through computational approaches.

A genotype (haplotype) $g(h)$ is a sequence over the alphabet $\{0, 1, 2\}$ ($\{0, 1\}$), each position of which is called a *site*. It can be computationally represented by a vector of symbols $\{0, 1, 2\}$ ($\{0, 1\}$), where symbol 2 represents an heterozygous site (meaning that the nucleotides at this site, in the corresponding haplotypes, are different), while 0 and 1 represent homozygous sites (the nucleotides in the corresponding haplotypes are the same).

R.S. Rosa—This work was developed with financial support from Brazilian sponsoring agency CAPES, which the authors gratefully acknowledge.

© Springer International Publishing AG 2017
Z. Cai et al. (Eds.): ISBRA 2017, LNBI 10330, pp. 302–313, 2017.
DOI: 10.1007/978-3-319-59575-7_27

Inferring haplotype from genotypic data can be formally described as follows. A matrix $H_{(2n)m}$ with n individuals and m single nucleotide polymorphisms (SNPs) contains $2n$ haplotypes. Each haplotype pair $(h_{(2i-1)}, h_{(2i)})$, $1 \leq i \leq n$, generates genotype g with m SNPs. Then a genotype vector g, with m sites, can be explained (resolved) by two haplotype vectors h_1 and h_2, where sites $h_1(i)$ and $h_2(i)$, $1 \leq i \leq m$, follow the rule given by Eq. 1.

$$\begin{cases} h_1(i) = h_2(i) = g(i), \text{ if } g(i) \in \{0,1\}; \\ h_1(i) = 1 - h_2(i), \quad \text{ if } g(i)=2. \end{cases} \tag{1}$$

Several computational methods have been proposed for Haplotype Inference (HI), such as ShapeIt [3], Impute2 [4], Beagle [5], Haplorec [6], fastPHASE [7], PTG [8], among others. These approaches usually consider one of the main biological models, called Parsimony Principle. Computationally speaking, in the Parsimony model, the HI Problem is NP-Hard, which means that, to this day, there is no algorithm that can solve this problem in a exact fashion within reasonable time. Therefore, one must resort to heuristics that can offer results with acceptable accuracy within viable amount of time.

In a previous study, Rosa and Guimarães [9] showed that, although different algorithms for HI may present similar error scores, most of those errors occur in different loci along the genotype sequence. Identifying regions of the genotype where each method has a higher propensity to make mistakes could help improve existing methods, and could also shad some light on unfavorable scenarios, eventually leading to an ensemble approach, based on biological features of the sequences. In this paper, we seek to reveal features of the data that could be associated with inference errors in the different methods. We assess, based on different error metrics, the inference errors of four HI methods. Methods Haplorec and fastPHASE were selected for their widespread use. Method Beagle was included because it is a more recent and very competitive tool. Method PTG was also included because it is a strongly parsimony principle-based tool.

Linkage Disequilibrium and Heterozygosity are important genomic features that have long been associated to haplotyping (e.g. [10]). In this work we present an extensive analysis of the behavior of those methods in regions with different LD and HTZ values. Different computational choices for estimating the LD of a given SNP are assessed. The influence of shorter and longer stretches of upstream and downstream neighboring SNPs in each method's error is investigated, and Multiple Linear Regression (MLR) models are developed to predict the error in each haplotype inference method, based on several genomic features. Statistical tests and residual analysis are used to validate those models. We also considered methods ShapeIt [3] and Impute2 [4], but they have not been included in this study because they require additional input, bringing the need of specific variables and also adding a clear bias to the models. A comparative analysis of the models developed unveils a number of interesting features.

2 Predicting Errors in HI Methods

Since it is well known that haplotyping is strongly related to different genomic features, studying possible associations between features of the genomic data and the occurrence of errors in haplotype inference can help understand the behavior of the current HI methods and perhaps lead in the design of better techniques, more resilient to the impact of those features.

When a certain genomic sequence is shared by a large number of elements of a given population, it is said that the sequence is conserved, because it passes from generation to generation without suffering many changes. From a computational point of view, it means that there is a higher similarity among the sequences considered. Based on the Parsimony Principle, HI methods search for the smallest number of haplotype sequences that explain a given set of genotypes (a parsimonious solution), hence the conservation level of the dataset is a most valuable feature.

Another feature that is central to this problem is the Linkage Disequilibrium (LD), which measures the correlation level of a given SNP (a column in the genotypes matrix) and its neighbors (neighboring columns). We scrutinize the influence of SNPs to the left and/or right, at different ranges.

We also assess the influence of several classic genomic features, including allele frequency, endogamic level, and Hardy-Weinberg equilibrium. We make an effort to translate those features into more straightforward indicators, based on the presence and distribution of symbols 2 (bases to be inferred by the HI methods) in the dataset, in order to optimize the computational time.

The pipeline illustrating the several steps involved in the analysis is presented in Fig. 1-A. Initially, several features are extracted from the dataset (regressor variables and HI solutions by the different methods). Then, for each combination of HI method and error metric, a subset of regressor variables (features) with stronger prediction power is selected. The MLR models are estimated and validated, generating the prediction performance statistics.

Feature extraction is the crucial basis for the regression analysis performed, since the features extracted through that process will be used to predict the propension of each HI method to generate wrong answers. Figure 1-B shows the blow-up for the feature extraction and construction of the data bases used in the actual analysis process. The genotype datasets collected from the HapMap Project [11] were processed by four HI methods: fastPHASE, Haplorec, Beagle, and PTG, creating an HI Solutions Base.

The errors of the Haplotype Inference, as well as the genotype features were computed considering two distinct views of the HI Solutions Base: (1) A set of individuals (solution matrix lines), which we call Individuals Base, and (2) A set of SNPs (solution matrix columns), which we call SNPs Base.

For the SNPs Base the features collected from the genotype datasets are:

1. Heterozygosity (HTZ), given by the number of symbols 2 in a given SNP column;
2. Linkage Disequilibrium (LD), given by the correlation described in Sect. 3.

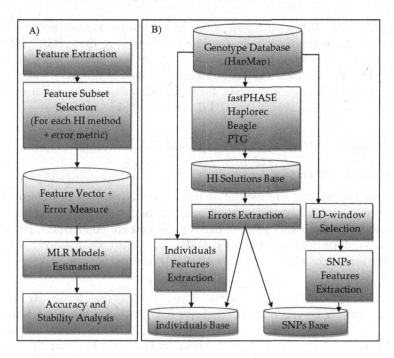

Fig. 1. (A) Pipeline of prediction analysis, and (B) blow-up of feature extraction process.

The LD of a given SNP is computed based on the correlation of that SNP with its neighbors, considered in different numbers. Location of neighboring SNPs relative to a given SNP (that is, upstream (or preceding) versus downstream (or following) SNPs, since that is a genomically important detail. Another variation considered was exclusion of non-informative SNPs. (For more details on that, see Sect. 3). All these filtering processes and parameters choices were done in order to capture the LD impact more effectively.

For the Individuals Base the features collected from the genotype dataset that capture the information of ambiguity were the following:

1. Number of symbols 0 (NS0);
2. Number of symbols 1 (NS1);
3. Heterozygosity (HTZ), similar to SNPs Base;
4. Density of Heterozygosity (DHZ), represented by the number of neighbors with symbol 2 paired with symbols 2 in each individual;
5. Number of blocks of symbols 2 (NB2), where a block is a sequence of identical symbols in an individual;
6. Average length of blocks of symbols 2 (LB2); and
7. Conservation level (CSV).

Extensive experiments were done in order to assess the impact of such features to the occurrence of inference errors. The inference error metrics (response variables for the MLR models) are different for the Individuals and the SNPs bases. In the Individuals base, we consider two measures, Error Rate (ER) and Switch Error (SE), while for the SNPs Base, we use Switch Distance (SD). A more detailed technical explanation can be found in the next section. The results of the MLR models designed for prediction of the errors in HI methods fastPHASE, Haplorec, Beagle, and PTG are presented in Sect. 4.

3   Methods and Experiments Design

3.1   Feature Extraction

Genotype Database: The dataset used in our experiments were collected from different populations in the HapMap Base. It is comprised of sets of haplotypes/ genotypes from Chromosome 20 of five different ethnic populations. Each segment of genotype collected has 1000 SNPs. The Chromosome positions from where the sequences were taken were randomly selected.

Errors Extraction: The error metrics more frequently used for the HI Problem are Error Rate (ER) [12] and Switch Error (SE) [13]. For the Error Rate, the correct haplotypes are aligned to the inferred ones, the number of mismatches is computed, and the Error Rate is given by the ratio between the number of mismatches and the total number of sites in the dataset. The Swith Error is calculated as $(N-1-SD)/(N-1)$, where N denotes the number of heterozigous loci, and SD (Switch Distance) denotes the minimum number of block exchanges required between the two inferred haplotypes, in order to make them identical to the original ones. Since SE is applied on genotype fragments, and here we are analyzing specific SNPs, we use the intermediate metric SD.

LD-Window Selection: This step is necessary for SNPs Base only. Given a genotype matrix H_n with n SNPs, and a window of width w, the correlation score for a SNP k, $LD(k, w)$, is the average correlation between SNP k and its w neighbors, it is computed for left ($LD(k, w) = \frac{1}{w} \sum_{i=1}^{w} cor(H_k, H_{k-i})$), right ($LD(k, w) = \frac{1}{w} \sum_{i=1}^{w} cor(H_k, H_{k+i})$) and left-right window ($LD(k, w) = \frac{1}{2w} \sum_{i=1}^{w} cor(H_k, H_{k-i}) + cor(H_k, H_{k+i})$).

For each neighbor SNP H_{k-i} (H_{k+i}) of correlation (Pearson's coefficient) $cor(H_k, H_{k-i})$ ($cor(H_k, H_{k+i})$) when the Exact Fisher's Test, with 95% of confidence, indicates no evidence of dependency between H_k and H_{k-i} (H_k and H_{k+i}), then SNP H_{k-i} (H_{k+i}) is discarded.

The procedures for identifying the more informative SNP window were organized as follows:

1. The LD patterns in each dataset was computed considering different window widths, including or not, SNPs without correlation evidence, as described above. For analysis the LD values were partitioned into four ranges: $[0, 0.25]$, for low LD; $(0.25, 0.5]$, for mid-low LD; $(0.5, 0.75)$, for mid-high LD, and $[0.75, 1.0]$, for high LD, thus creating LD mosaics.

2. For each population in the dataset, HTZ patterns were computed, considering the mosaics constructed in Step (1) above;

3. For each genotype dataset resolved by each of the four Haplotype Inference methods considered (fastPHASE, Haplorec, Beagle, and PTG), two error measures were computed (Switch Error/Switch Distance and Error Rate), and the error average of the four methods were taken, for each dataset separately;

4. For each LD mosaic, for each of the four LD ranges previously defined, two measures are taken: (1) $h = \frac{HTZ_{local}}{HTZ_{total}}$, where HTZ_{local} denote HTZ computed only for SNPs of one LD range, and HTZ_{total} is HTZ computed for the full dataset considered, (2) $error = \frac{SD_{local}}{SD_{total}}$, where SD_{local} denote SD computed only for SNPs of one LD range, and SD_{total} is SD computed for the full dataset considered, and (3) A quality indicator given by the ratio error/h.

SNPs Features Extraction: For the SNPs Base the features collected from the genotype datasets were: HTZ and LD, given by the correlation as described previously in this Section, obtained by mosaic W10_L_WITH_FISHER. SNPs base contains 6 attributes: 4 results for SD and the regressors HTZ and LD.

Individuals Features Extraction: For the Individuals Base the features collected from the genotype dataset that capture the information of ambiguity. Features NS0 and NS1 represent the number of homozygous sites in a given Individual. HTZ is given by the ambiguity level of the individual, i.e., the number of symbols 2 that must be resolved by the haplotyping method. DHZ is indicated by the number of symbols 2 that are next to another 2 on the same SNP. NB2 represents the number of contiguous sequences of two or more symbols 2 in the individual. LB2 is the average length of the sequences represented in NB2. For a fragment of genotype F with m SNPs of an individual, CSV is represented by probability p given by $p(F) = p(F(1)) \prod_{i=2}^{m} p(F(i)|F(i-1))$.

3.2   Feature Selection, Validation and Performance

A MLR model is composed by p regressor variables x, p regression coefficients β, a response variable y, and a noise ε ($y = \varepsilon + \beta_1 x_1 + ... + \beta_p x_p$), where p is the number of regressors. In our models, for the SNPs Base, $p = 2$, and for the Individuals Base, $p = 7$.

Twenty four MLR models were estimated, eight for the Individuals Base (four methods with two error metrics each), and sixteen for the SNPs Base (four methods with four LD levels each). The Step-Wise method [14] was applied to select among the explanatory variables (feature attributes) a suitable model for each one of the response variables (error attributes).

The training and test schemes were done with the following steps. (1) 15% of the samples for each population was randomly selected for testing, and the remaining samples were used for training; (2) For each answer variable, a MLR model was estimated using the training dataset; (3) The models constructed in

Step (2) were used to predict the errors in the test data; (4) Steps (1) through (3) were repeated 120 times, in order to generate a set of predicted samples large enough to execute t-tests with confidence greater than 95%; (5) Hypothesis tests were done with paired samples.

4   Results

4.1   Selecting the More Informative LD Window

Genotype datasets present different LD patterns along the SNPs sequence [6,7]. Estimation of those patterns is the core of methods Haplorec and fastPHASE, which apply Markov models to create LD mosaics. While Haplorec uses that technique to estimate windows with different LDs, excluding non-informative SNPs from the inference process, fastPHASE uses it to create groups of individuals and SNPs with similar LD patterns, in order to solve those genotypes in separate sets.

A given SNP may present different correlation levels (high, low, non-existing) with different sets of neighboring SNPs. With that concern in mind, we seek to assess what would be a more informative or appropriate window width for neighbors, and also to analyze the impact of using, within a given window, SNPs without statistical evidence of correlation with neighboring SNPs. In our experiments, three window width values were used: 1, 5, and 10, considering neighbors only to the left, neighbors only to the right, and neighbors from both sides. We also analyzed the behavior of the models including all SNPs within the window, and including only those SNPs with evidence of correlation, according to the Exact Fisher's Test.

For each LD mosaic and each of the four LD ranges previously defined, two measures are taken: (1) h, the total relative HTZ frequency, considering the entire dataset, (2) $error$, relative frequency of total errors, considering the entire dataset and the four inference methods, and (3) A quality indicator given by the ratio $error/h$.

The ratio $error/h$ falls in one of three scenarios, with the following interpretation: (1) Values close to zero indicate low error concentration with respect to the HTZ frequency (or "easy to resolve" regions); (2) Values close to 1 indicate a proportion of error frequency with respect to the HTZ frequency in a given region (or "expected error occurrence" regions); and (3) Values that are strictly greater than 1, indicate a high error frequency with respect to the HTZ concentration (or "hard to resolve" regions). These hard/easy to resolve regions are with respect to the haplotype inference methods.

Table 1 presents the results of our analysis considering all four methods, with the Switch Distance metric, where each line represents the average of the ratios $error/h$ obtained with a given LD mosaic estimation, stratified by LD level. An LD mosaic is denoted by $Wx\_d\_test\_FISHER$, where $x \in \{1, 5, 10\}$ (window width); $d \in \{L, R, L\_R\}$ (SNPs considered on the Left, on the Right or both); $test \in \{WITH, WO\}$ (considering or not considering the Exact Fisher's Test). The ratios for all mosaics are not shown due space limitations in this paper. The last line presents the average ratio obtained by all mosaics estimated by LD level. Independently of the criteria (window size and direction) used, methods tend to present similar behavior in regard to their response (ratio) within each particular LD concentrations region. In our experiments, all the LD mosaics presented ratios strictly greater than 1 in regions of low LD (Column Low in Table 1), indicating that all mosaics reported the occurrence of more errors in regions of low LD. For almost all mosaics, in all LD regions, ratios called considering the Exact Fisher's Test presented values higher or equal to the corresponding ratios taking all SNPs within the window, which indicates that this is a good filter for noisy SNPs. We note that mosaic W10_L_WITH_FISHER presented the best behavior; it has the highest ratio in the low LD regions and a ratio close to the smallest observed in high LD regions. Hence, this is the mosaic of choice for the LD extraction.

Table 1. Average ratios quality obtained by each LD mosaic for Switch Distance, considering the four LD ranges (Low, Mid-Low, Mid-High and High).

Mosaic rule - LD range	Low	Mid-Low	Mid-High	High
W10_L_WITH_FISHER	**4.50**	**1.05**	**0.33**	**0.17**
Average[a]	2.54	1.03	0.54	0.31

[a] Average for 18 mosaics considered.

All insights gathered through this LD window analysis suggest that, in order to attain better results, the SNPs Base should have MLR models estimated in a LD-stratified fashion.

4.2 MLR Models with Selected Regressors

The MLR models were estimated with the variables selected using the Step-Wise method, which applies Akaike Information Criterion (AIC) to choose an optimal subset of regressors. To avoid the introduction of bias, the bidirectional elimination was chosen.

The results of Step-Wise and the coefficients of estimated models for the Individuals Base are shown in Table 2. Our results show that for this base, the most important features are CSV and NS0, selected for all estimated models. The second most important feature is NS1, considered for all models except SE-PTG. DHZ and LB2 are selected for one model each, ER-PTG and SE-PTG, respectively.

Table 3 presents the MLR models estimated for the SNPs Base, considering the four different LD levels and the regressors selected by the Step-Wise Method. The combination of features HTZ and LD was selected for most models. For method PTG, HTZ was the only feature selected in three of the four LD levels, providing evidence that the accuracy of PTG depends heavily on the distribution of ambiguous sites in the dataset (expressed by HTZ).

Table 2. MLR models estimated for Individuals Base using the Step-Wise Method.

Method	Model
ER-fastPHASE	$3.214 \times 10^{-3} + 2.078 \times 10^{-6}CSV - 3.499 \times 10^{-6}NS0 - 3.696 \times 10^{-6}NS1$
ER-Haplorec	$2.776 \times 10^{-3} + 2.559 \times 10^{-6}CSV - 3.046 \times 10^{-6}NS0 - 2.712 \times 10^{-6}NS1$
ER-PTG	$4.619 \times 10^{-3} + 1.173 \times 10^{-6}CSV - 6.689 \times 10^{-7}NB2 + 1.112 \times 10^{-7}DHZ - 4.716 \times 10^{-6}NS0 - 5.157 \times 10^{-6}NS1$
ER-Beagle	$3.985 \times 10^{-3} + 2.731 \times 10^{-6}CSV - 4.020 \times 10^{-6}NS0 - 5.564 \times 10^{-6}NS1$
SE-fastPHASE	$1.150 \times 10^{-2} - 3.456 \times 10^{-6}CSV - 1.072 \times 10^{-6}NB2 - 1.731 \times 10^{-6}NS0 - 1.458 \times 10^{-6}NS1$
SE-Haplorec	$1.226 \times 10^{-2} - 3.384 \times 10^{-6}CSV - 2.292 \times 10^{-6}NB2 - 2.626 \times 10^{-6}NS0 - 2.493 \times 10^{-6}NS1$
SE-PTG	$8.798 \times 10^{-3} - 6.357 \times 10^{-6}CSV - 3.129 \times 10^{-6}NB2 - 2.752 \times 10^{-4}LB2 - 2.321 \times 10^{-6}NS0$
SE-Beagle	$1.225 \times 10^{-2} - 4.237 \times 10^{-6}CSV - 2.753 \times 10^{-6}NB2 - 2.379 \times 10^{-6}NS0 - 2.615 \times 10^{-6}NS1$

Table 3. MLR models estimated for the SNPs Base considering different LD levels, using features selected by the Step-Wise Method.

LD Level	Method	Model
Low	SD-fastPHASE	$0.095 + 0.271HTZ - 10.698LD$
	SD-Haplorec	$0.089 + 0.275HTZ - 10.312LD$
	SD-Beagle	$0.066 + 0.305HTZ - 11.134LD$
	SD-PTG	$0.015 + 0.4775HTZ$
Mid-Low	SD-fastPHASE	$3.112 + 0.0499HTZ - 8.881LD$
	SD-Haplorec	$3.687 + 0.059HTZ - 10.489LD$
	SD-Beagle	$3.803 + 0.062HTZ - 10.964LD$
	SD-PTG	$1.539 + 0.478HTZ - 4.660LD$
Mid-High	SD-fastPHASE	$0.811 + 0.012HTZ - 1.291LD$
	SD-Haplorec	$0.827 + 0.014HTZ - 1.320LD$
	SD-Beagle	$0.856 + 0.010HTZ - 1.140LD$
	SD-PTG	$0.045 + 0.444HTZ$
High	SD-fastPHASE	$0.547 + 0.003HTZ - 0.556LD$
	SD-Haplorec	$0.835 - 0.742LD$
	SD-Beagle	$0.718 + 0.0044HTZ - 0.651LD$
	SD-PTG	$0.320 + 0.436HTZ$

The coefficients (contribution) of HTZ were positive in all models, indicating that there is a positive correlation between this feature and inference error. LD presented negative coefficients in all models in all LD Levels, however in a smaller degree. The results show that, as the LD level increases, the absolute value of the LD coefficient decreases.

An important observation is that the noises (indicated by the intercept) are greater for Mid-Low LD than for the other LD levels. That characterizes these regions as a "hard scenario" for predicting errors in Haplotype Inference methods (not for the HI methods themselves), which is consistent with the average ratio 1.05 found for the W10_L_WITH_FISHER mosaic in the Mid-Low LD level (See Table 1).

4.3   MLR Models Accuracy

The accuracy of the HI error prediction models was analyzed for each one of the MLR models estimated. A set of hypothesis tests was designed to assess the closeness between the values predicted (P) by each MLR model and the corresponding actual (A) response values of the HI method. Since there is no t-test for alternative hypothesis $A = P$, we apply the three t-tests below, none of which had the null hypothesis rejected. The three Null (H_0) and Alternative Hypothesis (H_a) were: (1) $H_0 : A \geq P, H_a : A < P$; (2) $H_0 : A \leq P, H_a : A > P$; and (3) $H_0 : A = P, H_a : A \neq P$. These t-tests had 95% of confidence, with the p-values for the hypothesis ranging from 0.4037 to 0.5585 (hence much higher than 0.05). Since the null hypothesis cannot be rejected, the p-values are high, and the number of samples is sufficiently large, there is evidence that the actual values are statistically equal to the corresponding average predicted value.

5   Discussion

We demonstrate that the correlation between LD level and the occurrence of HI errors varies along the genotypes. We also present evidence that considering a window with the 10 SNPs immediately to the left, and eliminating the non-informative SNPs through Fisher's Test is more suitable when seeking a correlation between LD and Inference Errors. It is known that eliminating non-informative SNPs is a promising strategy for haplotype inference, since it reduces noise. But, as far as we know, a quantitative assessment of the impact of that filtering on the correlation between LD and the occurrence of errors in the haplotype inference results has never been reported in the literature. The fact that SNPs in the window to the left of a SNP are apparently more informative than the SNPs on the window to the right suggests that the direction in which the methods proceed during the resolution process impacts the occurrence of errors. Hence, it would be wise to attempted to reduce the errors caused by the choice of direction in the resolution process.

We delineate scenarios, based on LD measures, that reveal a higher or smaller propensity of the HI methods to present inference errors. The absence of correlation between SNPs in a given region (low LD) translates into less information for the HI methods, which tend to present more errors in these regions. In this work we show that, within low LD level regions, the relative error frequency is more than four times bigger than the relative frequency of HTZ. We also show that when a SNP has LD-value between 0.25 and 0.5, the error frequency is proportional to the HTZ level (concentration of sites to be resolved), an evidence that in those regions the LD-value is not informative, hence new alternative approaches for HI in these regions are necessary. For regions where LD level > 0.5, a smaller number of errors occur, characterizing these regions as easy to resolve (the detailed results of the analysis are shown in Table 1).

Due to the correlation between LD levels and the occurrence of HI errors, specific MLR models were estimated for each particular LD interval (low, mid-low, mid-high and high), which considerably improved the prediction performance of models estimated without LD levels. We identify the Low LD regions as the easy to predict scenario, and the Mid-Low LD regions as a hard to predict scenario for MLR models based on SNPs data. It is important to stress that the error that is being discussed here is the Haplotype Inference error by the methods included in our study, not errors in our prediction models.

Considering individuals (genotypes) to predict inference errors showed to be far more promising than considering SNPs. Although the SNPs Base produced good results, the Individuals Base presented more accurate results in most scenarios. It is important to note that, based on the Switch Error metric, all models for the Individuals base were highly, almost perfectly accurate. In this context, information on conservation and locality (HTZ blocks), which pertains only to individuals, seem to have considerable impact on the accuracy observed in the results, suggesting that these features should have a more central role in the development of new techniques.

The MLR models were developed based on seven biologically relevant features, and also for features selected using the Step-Wise method. From these models, NS0 was selected as the most relevant variable, being used in all models for the Individuals base. The second most important feature was NS1, used in all but one of the models. Considering only the Individuals Base, CSV was selected for all models. CSV tries to capture conservation information on the sequences, which is an important evidence in favor of the known biological parsimony principle. The experiments show statistical evidence that almost all the MLR models in Individuals can efficiently infer errors in the HI algorithms studied, although in some specific cases high relative error and deviation measures can be observed.

6   Conclusion

In this work we present the results of extensive statistical analyses of the influence that features of the genotype sequences may have on the errors of haplotype inference methods. We show that SNPs to the left (upstream region) of a given SNP are more informative than SNPs on the right (downstream region) to estimate LD and to predict HI error, which is compatible with well known transcription regulatory (Genetics) models.

We also build precise MLR models to predict Switch error for the four HI methods studied. A particularly exciting result for prediction of SE in the known and popular method fastPHASE; this model has a median of relative error of 0.9, meaning that 50% of the predictions for this model had relative error at most 0.9%.

We believe that the insights provided by our analysis can be used for a more effective choice of algorithms, and can also be exploited in the design of better approaches for the Haplotype Inference Problem.

References

1. Lin, D., Wang, L., Li, Y.: Haplotype-based statistical inference for population-based case-control and cross-sectional studies with complex sample designs. J. Surv. Stat. Methodol 4(2), 188–214 (2016)
2. Laehnemann, D., Borkhardt, A., McHardy, A.C.: Denoising dna deep sequencing data-high-throughput sequencing errors and their correction. Brief. Bioinf. 17(1), 154–179 (2016)
3. O'Connell, J., et al.: A general approach for haplotype phasing across the full spectrum of relatedness. PLoS Genetics 10(4), e1004234 (2014)
4. Howie, B.N., Donnelly, P., Marchini, J.: A flexible and accurate genotype imputation method for the next generation of genome-wide association studies. PLoS Genet. 5(6), e1000529 (2009)
5. Browning, B.L., Browning, S.R.: A fast, powerful method for detecting identity by descent. Am. J. Hum. Genet. 88(2), 173–182 (2011)
6. Eronen, L., Geerts, F., Toivonen, H.: Haplorec: efficient and accurate large-scale reconstruction of haplotypes. BMC Bioinf. 7, 542 (2006)
7. Scheet, P., Stephens, M.: A fast and flexible statistical model for large-scale population genotype data: applications to inferring missing genotypes and haplotypic phase. Am. J. Hum. Genet. 78(4), 629–644 (2006)
8. Li, Z., Zhou, W., Zhang, X.S., Chen, L.: A parsimonious tree-grow method for haplotype inference. Bioinformatics 21, 3475–3481 (2005)
9. Rosa, R.S., Guimarães, K.S.: Insights on haplotype inference on large genotype datasets. In: Ferreira, C.E., Miyano, S., Stadler, P.F. (eds.) BSB 2010. LNCS, vol. 6268, pp. 47–58. Springer, Heidelberg (2010). doi:10.1007/978-3-642-15060-9_5
10. Stephens, J.C., et al.: Haplotype variation and linkage disequilibrium in 313 human genes. Science 293(5529), 489–493 (2001)
11. The International HapMap Consortium: The international hapmap consortium. Nature 426, 789–796 (2003)
12. Niu, T., Qin, Z.S., Xu, X., Liu, J.S.: Bayesian haplotype inference for multiple' linked single-nucleotide polymorphisms. Am. J. Hum. Genet. 70, 157–169 (2002)
13. Lin, S., Cutler, D.J., Zwick, M.E., Chakravarti, A.: Haplotype inference in random population samples. Am. J. Hum. Genet. 71(5), 1129–1137 (2002)
14. Montgomery, D., Runger, G.: Applied statistics and probability for engineers, 4th edn. LTC, São Paulo (2003)

Detecting Change Points in fMRI Data via Bayesian Inference and Genetic Algorithm Model

Xiuchun Xiao[1,3], Bing Liu[2], Jing Zhang[2(✉)], Xueli Xiao[3], and Yi Pan[3,4(✉)]

[1] College of Electronic and Information Engineering,
Guangdong Ocean University, Zhanjiang 524025, China
[2] Department of Mathematics and Statistics,
Georgia State University, Atlanta, GA 30303, USA
jing.maria.zhang@gmail.com
[3] Department of Computer Science,
Georgia State University, Atlanta, GA 30303, USA
yipan@gsu.edu
[4] Department of Biology, Georgia State University, Atlanta, GA 30303, USA

Abstract. Dynamic functional connectivity detection in fMRI has been recently proved to be powerful for exploring brain conditions, and a variety of methods have been proposed. This paper mainly investigates the field of change point detection based on Bayesian inference and genetic algorithm. We define different indicator vectors as different individuals, which represent some possible change point distributions, and use Bayesian posterior probability to evaluate their fitness. Accordingly, we also present an improved genetic algorithm, which is applied to evolve the individuals toward the best one. Then, the most possible change points distribution could be resolved. The method has been applied to several synthesized data and simulation results reveal that the proposed method can detect change points in fMRI datasets with higher precision and lower time consumption.

Keywords: fMRI · Change point detection · Genetic algorithm · Bayesian inference

1 Introduction

Functional magnetic resonance imaging (fMRI) is a functional neuroimaging method that measures human brain activities by quantifying the blood flow using MRI technology [1]. Multiple recent researches on neuronal network-level activities using fMRI dataset have invoked increasing number of attentions [2–4].

Recently, several neuroscience studies reflect that moment-by-moment functional switching is commonly involved in brain dynamic interactions between connections from higher to lower-order cortical areas and intrinsic cortical circuits, and the brain may undergo a series of state-change while performing a task [2, 5]. Therefore, exploring functional dynamics may be very useful for revealing the mechanism of the human brain since a task is performed in cortical areas.

© Springer International Publishing AG 2017
Z. Cai et al. (Eds.): ISBRA 2017, LNBI 10330, pp. 314–324, 2017.
DOI: 10.1007/978-3-319-59575-7_28

Functional dynamics detection in fMRI has been proved to be powerful for exploring brain conditions. It is an ongoing challenge in the area of neuroscience. To solve this problem and some relative tasks, a variety of methods have been proposed [2–9].

In order to detect the temporal boundary, an abrupt change of multivariate functional interactions in the fMRI (also called change point), Lian et al. proposed a Bayesian connectivity change point model (BCCPM) based on Markov Chain and Monte Carlo (MCMC) scheme (called MCMC-Based-BCCPM) [2]. This method has been successfully applied to detect change points in fMRI dataset. Unfortunately, the MCMC scheme is a high consuming method. It's a sequential algorithms and not easy to be speeded up with GPU or multi-processor computers. In [7, 8], Bayesian magnitude change point model (BMCPM), dynamic Bayesian variable partition model (DBVPM), Bayesian connectivity bi-partition change point model (BCBCPM) were proposed, respectively. However, all of these methods use MCMC scheme or a hierarchical two-level Metropolis-Hastings scheme as their optimization or search strategy.

To develop an algorithm that may be parallelized in GPU or multi-processor computers, we investigate the field of most Bayesian inference based change point detection methods [2, 3, 7, 8]. For the fact that genetic algorithm could be a good performance parallel algorithm and is very fit for optimization problems [10, 11], we choose it as our optimization strategy.

We define different indicator vectors to represent different distributions of change points, use these indicator vectors as different individuals in genetic algorithm, and then calculate the Bayesian posterior probability to evaluate their fitness. Finally, an improved specific evolutional procedure is applied to evolve the individuals toward the best one. The best individual then represents the most likely change point distribution in the corresponding fMRI dataset.

The rest of this paper is organized as follows. Section 2 describes the main theory and method, including Bayesian connectivity change point model (BCCPM) and improved genetic algorithms, which are applied to calculate posterior probability and to solve optimization and search problems, respectively. Section 3 shows our experimental results, with some comparisons between MCMC-Based-BCCPM [2] and the proposed method. Section 4 concludes our method and discusses further work in the future.

2 Theory and Method

2.1 Bayesian Connectivity Change Point Model [2, 3, 7, 8]

Given an R × T dataset $X = (x_1, x_2, \cdots, x_T)$, in which T is the number of observations and R is the number of ROIs, we are interested in if there are some differences in the joint probabilities within these ROIs between different time periods.

We define a block indicator vector as,

$$\vec{I} = (I_1, I_2, \cdots, I_T), \tag{1}$$

where $I_k = 1$ if the k-th observation x_k is a change point of the start of a temporal block, $I_k = 0$ otherwise.

Now, suppose a set of vectors $x_1, x_2, \cdots, x_T$ i.i.d. (independent and identically distributed) from R-dimensional multivariate normal distribution, i.e., $x \sim N(\mu, \sum)$ t = 1, 2, ..., T, where T denotes the number of vectors, R denotes the dimension of vectors $x_1, x_2, \cdots, x_T$, μ denotes the R-dimensional mean vector, and $\sum$ denotes the R × R covariance matrix. The conjugate prior distribution of $(\mu, \sum)$ is N-Inv-Wishart $(\mu_0, \Lambda_0/\kappa_0, v_0, \Lambda_0)$ [12], and the posterior distribution of $(\mu, \sum)$ based on the data $X = (x_1, x_2, \cdots, x_T)$ is also N-Inv-Wishart $(\mu_T, \Lambda_T/\kappa_T, v_T, \Lambda_0)$ [12]. Therefore, we can calculate the probability of $x_1, x_2, \cdots, x_T$ as follows,

$$p(x_1, x_2, \cdots, x_T) = \frac{p(x_1, x_2, \cdots, x_S; \mu, \sum)}{p(\mu, \sum; x_1, x_2, \cdots, x_T)}$$

$$= (\frac{1}{2\pi})^{mT/2} (\frac{K_0}{K_T})^{R/2} \frac{\Gamma_R(\frac{V_T}{2})}{\Gamma_R(\frac{V_0}{2})} \frac{(det(\Lambda_0))V_0/2}{(det(\Lambda_T))V_T/2} 2^{RT/2} \qquad (2)$$

where $\Gamma_R(z)$ is the multivariate gamma function:

$$\Gamma_R(z) = \pi^{R(R-1)/4} \prod_{j=1}^{R} \Gamma(z + (1-j)/2) \qquad (3)$$

Consider the block indicator vector in Eq. (1), the likelihood of the data matrix $X = (x_1, x_2, \cdots, x_T)$ is:

$$p(X|\vec{I}) = \prod_{b=1}^{\sum I_k} p(X_b) \qquad (4)$$

where, X_b is the temporal observations that belong to b-th block and $p(X_b)$ can be calculated according to Eq. (2). The temporal blocks are independent from each other; therefore, the posterior distribution of the configuration is:

$$p(\vec{I}|X) \propto p(\vec{I})p(X|\vec{I}), \text{ where } p(\vec{I}) = \prod_{t=1}^{T} p(\vec{I}_t) \text{ and } p(\vec{I}_t) \text{ is Bern } (0.5).$$

It is worth noting that Eq. (4) will be regarded as the fitness function of the proposed genetic algorithm to calculate the fitness of every new individual generated by the evolutional operators.

2.2   Genetic Algorithm

Genetic algorithm (GA) is one kind of evolutionary method inspired by the process of natural selection. It is commonly used to solve optimization and search problems by relying on bio-inspired operators such as selection, crossover, and mutation (Liu 2016). Genetic algorithm is a self adaptive search algorithm and can automatically achieve the optimal solution of the problem.

As aforementioned, in the field of change point detection, a possible distribution of the change points in fMRI dataset can be represented as an indicator vector. We can define different indicator vectors as different individuals of genetic algorithm, and use Bayesian posterior probability to evaluate their fitness. Then, by performing some evolutionary operators such as selection, crossover, and mutation to evolve the individuals toward the best one, which represents the most possible change points distribution.

(1) Algorithm framework

The proposed genetic algorithm is more general but has some extra steps which is very different from it. All steps are described as follows:

Algorithm 1

Step 1 Initialize all parameters which may be used in the algorithm, and randomly produce a given number of indicator vectors, which represent different individuals. These individuals are regarded as the 0-th generation.

Step2 If maximal iterative number is reached, save all the results of our algorithm, else perform step 3.

Step3 Calculate fitness values of the individuals in current generation using Bayesian posterior probability, then sort them from high to low according to the fitness values.

Step 4 Copy a few individuals with highest fitness values directly into the next generation.

Step 5 Randomly select some individuals with relative higher fitness values, and select two arbitrary positions to perform crossover operation. Some new individuals are produced.

Step 6 Randomly select a portion of new individuals and mutate some random indicators from 1 to 0, or from 0 to 1.

Step 7 Go to Step 2.

Figure 1 shows the structure and pipeline of the proposed algorithm, and the flow-chart of the program is illustrated with "black" solid line. Additionally, in order to demonstrate how the data flows in our processing procedure, its directions are illustrated with "green" solid line.

(2) Excellent individuals' survival strategy

Commonly, the general genetic algorithm may select some excellent individuals to produce the next generation by using evolutionary operators such as crossover and mutation. However, in our method, we copy a few best individuals directly to the next generation. This strategy could make sure the optimal solution will not be changed by the coming evolutional operators and also could make them exert more influence on generating the new individuals. In fact, we think this strategy is fair because the excellent individuals may have long life or survive long in the natural world.

(3) Selection, crossover, and mutation strategies

The selection operator maybe very important for genetic algorithm since the next generation seriously rely on the selection strategy. In order to make those good

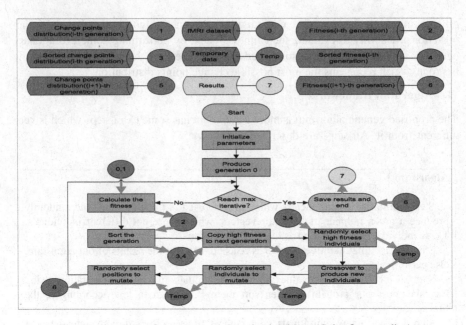

Fig. 1. Flowchart of the proposed Genetic algorithm. (Color figure online)

individuals have more chance to be selected, we propose a selection method based on sorted individuals.

Crossover operator is usually simple. In fact, two points crossover strategy can work well in most situations. Therefore, we will explore this design. The mutation operation is also very simple. However, which individual and which position in this individual should mutate may be the two main aspects involved into our consideration.

Our selection, crossover, and mutation strategies can be summarized as follows:

Algorithm 2
(some details of Algorithm 1)

Selection and crossover:
Step 1 Randomly produce an integer $n \in 1\sim n0$.
Step 2 Randomly produce two different integers a1 and a2 $\in 1\sim n$, and select the a1-th and a2-th individuals in the sorted generation i.
Step 3 Randomly produce two different integers b1 and b2 $\in 1\sim N$.
Step 4 Crossover the selected individuals at the selected positions.
Mutation:
Step 5 Randomly produce a float number $u \in [0,1]$.
Step 6 if $u > u0$, go to step 8.
Step 7 Randomly produce s integers $c1, c2, ..., cs \in [0,N]$
Step 8 Change the c1-th, c2-th,..., and cs-th position from 1 to 0 or from 0 to 1.
Step 9 If all the individuals have been generated, end; otherwise, go to Step1.

As mentioned in Fig. 1, Algorithms 1 and 2, we can easily see that the sorted individuals with fitness values exert very important influence on the whole procedure of our method. Figure 2 illustrated the influence more intuitively. Carefully inspect Fig. 2(b) to (c), we can see that all the evolutional operators used to produce individuals from the i-th generation to (i + 1)-th generation should use the sorted fitness results.

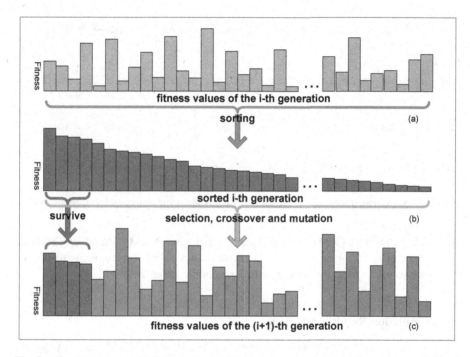

Fig. 2. The fitness values' influence: sorted fitness values exert more influence on evolutional operators in each iteration.

3   Experimental Results

In this section, several simulation datasets are generated to evaluate and validate the GA-Based-BCCPM and MCMC-Based-BCCPM proposed in [2].

3.1   Simulation Datasets

In order to verify that the proposed method can effectively find the change points, six different structure of dynamic networks (Fig. 3(a–f)) are generated [2]. These six different structure of dynamic networks include little amount of change point distribution mode (such as Fig. 3(a)–(c)), and also include the mode with a lot of change point distribution (such as Fig. 3(d)–(f)). We do several experiments to validate the proposed

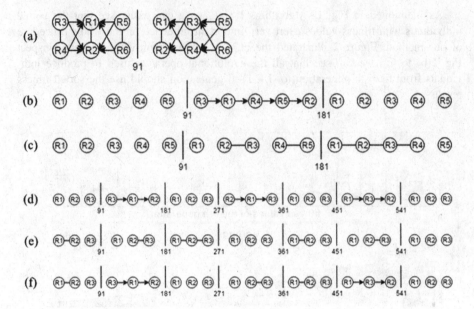

Fig. 3. Six different structure of dynamic networks and their change-point-distributions.

model (GA-Based-BCCPM), and also compare it with MCMC-Based-BCCPM. In Fig. 3, the real positions of all the change points of six different dynamic networks are shown by using a vertical solid line with a position number below it.

3.2 Simulation Results and Comparison

We perform several experiments between MCMC-Based-BCCPM and GA-Based-BCCPM for network (a)–(f) illustrated in Fig. 3. We repeat every simulation experiments for 5 times and save all results to calculate their average performance. For the purpose of fairness, all parameters of BCCPM are set as same values. The iterative number of GA-Based-BCCPM is set as 100 while GA-Based-BCCPM is set as 20000 to synchronously achieve good convergence and detection results for both the two methods.

Figure 4 illustrates the convergence curve of the MCMC-Based and GA-Based-BCCPM. Figure 4(a)–(f) denote the convergence curves of network (a)–(f), respectively; and the results of MCMC-Based and GA-Based-BCCPM are listed on the left and right column, respectively. Carefully observe the results of left and right columns, we can see the convergence curve of MCMC-Based-BCCPM vibrates even when it has reached its highest peak, while GA-Based-BCCPM only climbs for the highest peak. Obviously, the good results of proposed method may benefit from step 4 in Algorithm 1, since this step will always copy individuals with highest fitness values directly into the next generation.

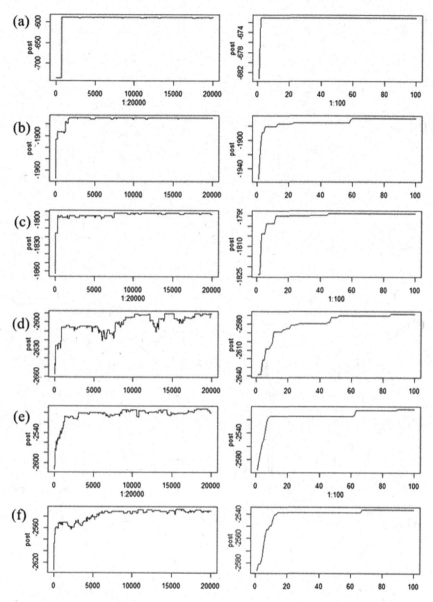

Fig. 4. Convergence curve for network (a)–(f) (Left column is results of MCMC-Based-BCCPM. Right column is results of GA-Based-BCCPM)

Figure 5 illustrates Change point detection results for network (a)–(f) of the MCMC-Based and GA-Based-BCCPM. Figure 5(a)–(f) denote the detected positions of change points of network (a)–(f), respectively; and the results of MCMC-Based and GA-Based-BCCPM are listed on the left and right columns, respectively. Carefully check all the detected change point position of left and right columns in Fig. 5 and real

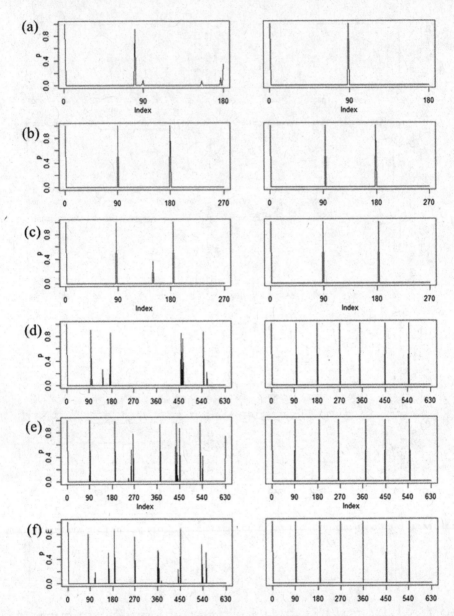

Fig. 5. Change point detection results for network (a)–(f) (Left column is results of MCMC-Based-BCCPM. Right column is results of GA-Based-BCCPM)

positions in Fig. 3, we can conclude that GA-Based-BCCPM outperforms MCMC-Based-MCCPM for all six networks in detection precision. Even more, GA-Based-BCCPM does not miss or mistake change points in any network, while MCMC-Based-MCCPM mistake some positions in network (a), (c), (e), and (f); and miss some change points in network (d).

Finally, we take the time consumption into our consideration, the running time of MCMC-Based and GA-Based-BCCPM are listed in Table 1. The environment of our experiment is as follows: operating system: Windows 10 Pro; system type: 64-bit operating system x64-based processor; CPU: Intel(R) Core(TM) i7-6600U CPU@2.6 GHz 2.81 Hz; memory: 12 GB. The last row of Table 1 is the average running time and the other rows are networks (a)-(f). It is very easy to see that GA-Based-BCCPM is better than MCMC-Based-BCCPM in all networks, and of course the average time consumption.

Table 1. Comparison of time consumption (ms)

	MCMC-Based-BCCPM	GA-Based-BCCPM
Network (a)	5796.8	3631.2
Network (b)	6206.0	3568.6
Network (c)	7825.2	4391.0
Network (d)	6268.8	4231.4
Network (e)	6218.2	4231.2
Network (f)	6472.0	4256.4
Average	6464.5	4051.6

4   Conclusion and Future Work

This paper mainly presents a new change point detection method for fMRI data based on Bayesian inference and genetic algorithm. We define different indicator vectors as different individuals and use Bayesian posterior probability to evaluate their fitness. Then, an improved genetic algorithm is applied to evolve the individuals toward the best one. The simulation experiments show that our method can detect the change points with higher precision and lower time consumption on several simulated datasets.

In the future, we will apply our method on real fMRI dataset, and combine it with other Bayesian inference model, such as BMCPM, DBVPM, and BCBCPM. Furthermore, for the potential parallel realization of the genetic algorithm, the proposed method could be easily realized in a parallel mode and run efficiently within GPU or multi-processors computers, thus, we will also investigate the GPU or multi-processors version of it.

Acknowledgements. The authors are grateful for support from Georgia State University Brains-Behavior Seed grant. This research was supported by the Molecular Basis of Disease(MBD) at Georgia State University.

References

1. https://en.wikipedia.org/wiki/Functional_magnetic_resonance_imaging
2. Lian, Z., Li, X., Xing, J., et al.: Exploring functional brain dynamics via a Bayesian connectivity change point model. In: Proceedings of the IEEE 11th International Symposium on Biomedical Imaging (ISBI 2014), Beijing, China, pp. 21–24. IEEE, April 2014

3. Guo, X., Liu, B., Chen, L., Chen, G., Pan, Y., and Zhang, J.: Bayesian inference for functional dynamics exploring in fmri data. Comput. Math. Methods Med. (2016)
4. Ou, J., Lian, Z., Xie, L., et al.: Atomic dynamic functional interaction patterns for characterization of ADHD. Hum. Brain Mapp. **35**(7), 5262–5278 (2014)
5. Gilbert, C.D., et al.: Brain states: top-down influences in sensory processing. Neuron **54**, 677–696 (2007)
6. Calhoun, V.D., Kiehl, K.A., Pearlson, G.D.: Modulation of temporally coherent brain networks estimated using ICA at rest and during cognitive tasks. Hum. Brain Mapp. **29**(7), 828–838 (2008)
7. Lian, Z., Li, X., Zhang, H., et al.: Detecting cell assembly interaction patterns via Bayesian based change-point detection and graph inference model. In: Proceedings of the IEEE 11th International Symposium on Biomedical Imaging (ISBI 2014), Beijing, China, pp. 21–24. IEEE, April 2014
8. Lian, Z., Lv, J., Xing, J., et al.: Generalized fMRI activation detection via Bayesian magnitude change point model. In: Proceedings of the IEEE 11th International Symposium on Biomedical Imaging (ISBI 2014), Beijing, China, pp. 21–24. IEEE, April 2014
9. Zhang, J., Li, X., Li, C., et al.: Inferring functional interaction and transition patterns via dynamic bayesian variable partition models. Hum. Brain Mapp. **35**(7), 3314–3331 (2014)
10. Liu, Z., Kong, Y., Su, B.: An improved genetic algorithm based on the shortest path problem. In: 2016 IEEE International Conference on Information and Automation (ICIA), Ningbo, China, pp. 328–332 (2016)
11. Pei, S., et al.: Codelet scheduling by genetic algorithm. In: 2016 IEEE Trustcom/BigDataSE/ISPA, Tianjin, China, pp. 1492–1499 (2016)
12. Gelman, A., et al.: Bayesian Data Analysis. Texts in Statistical Science, 2nd edn. Chapman & Hall/CRC, New York/Boca Raton (2003)

Extracting Depression Symptoms from Social Networks and Web Blogs via Text Mining

Long Ma$^{(\boxtimes)}$, Zhibo Wang, and Yanqing Zhang

Department of Computer Science, Georgia State University, Atlanta, USA
{lma5, zwang2}@student.gsu.edu, yzhang@gsu.edu

Abstract. Accurate depression diagnosis is a very complex long-term research problem. The current conversation oriented depression diagnosis between a medical doctor and a person is not accurate due to the limited number of known symptoms. To discover more depression symptoms, our research work focuses on extracting entity related to depression from social media such as social networks and web blogs. There are two major advantages of applying text mining tools to new depression symptoms extraction. Firstly, people share their feelings and knowledge on social medias. Secondly, social media produce big volume of data that can be used for research purpose. In our research, we collect data from social media initially, pre-process and analyze the data, finally extract depression symptoms.

Keywords: Depression symptoms · Social media · Text mining · Word clustering · Word embedding · NLP

1 Introduction

Mental illness has been prevalent in the world, depression is one of the most common psychological problem. Untreated depression increases the chance of dangerous behaviors. The significant challenge of detecting depression is the recognition that depressive symptoms may differ from patients' behavior and personality [1]. For clinic depression, doctors may evaluate the patient via the depression test taken by patients. Apparently, these clinical records are restricted due to many factors, such as age, sex; moreover, they are private and expensive. To overcome such limitations of clinical data, it would be beneficial to use text mining tools to extract and analyze depression symptoms from social media, such as Twitter. Social media generates countless data every day because of millions of active users share and communicate in entire community, it changes human interaction [2, 3]. Other than the traditional data, such as literatures, social media data is richer and more accessible [4]. However, investigating this new fast-growth of data requires advanced development tool to discover useful insights. These advanced technologies include Natural Language Processing (NLP) [5], data mining, machine learning, social media analysis and so on. In our research, the goal is to extract and summarize the uncommon but potentially helpful factors that depressive symptom performed from the social media data. Finally, the extracted depression symptoms will be used as references when manually recognizing the clinical depression by humans.

© Springer International Publishing AG 2017
Z. Cai et al. (Eds.): ISBRA 2017, LNBI 10330, pp. 325–330, 2017.
DOI: 10.1007/978-3-319-59575-7_29

Earlier work for driving the depression symptoms on literatures help people learn the knowledge of depression detection. Wang *et al.* [6] applied the Latent Dirichlet Allocation (LDA) [7] as the topic categorizing tool to many of texts on adolescent substance use. Through separating the collections of articles into distinct themes by LDA, the known depressive facts were captured. To extract the key entities, Ma *et al.* [8] has proposed a hybrid method to group entities that share sematic similarities. Wang and Ma's work show their methods performed well on structured data, our work introduces an unsupervised learning approach that could be used for unstructured data.

The rest of this paper is organized as follows: The second section introduces the method for data collection. The next section illustrates how to preprocess raw data. The forth section discusses the experiment on the result analysis. The last section concludes our work and introduces the future research.

2   Data Collection

2.1   Tweets (TW)

Twitter rapidly has become one of the most popular social media since it launched, it advises 313 million active users who produce 6,000 tweets on Twitter every second as June, 2016[1]. In favor of gathering the depression related data, we keep monitoring each streaming tweet that includes the word "depression" in entire Twitter community. Totally, we roughly have gathered 54-million of tweets that discussed the depression relevant field.

2.2   Professional Twitter Accounts (PTA)

Another extension of Twitter data collection is that we gather each tweet that has been posted by professional mental health account. The purpose of collecting these specific tweets is that PTAs are more knowledgeable and professional on mental health field, especially in the depression fields. Starting to web scraping the initial webpage[2], thousands of professional mental health tweets have been accumulated at the end.

2.3   Depression Blogs (DB)

Other than collecting data from the professional Twitter accounts and active Twitter users, extra data resource comes from the depression related web blogs. Similarly, web scraping begins at the specific webpage[3] to gather data. These blogs and their deep links are almost referring to the depressive symptoms and relative treatments.

---

[1] https://about.twitter.com/company.

[2] http://treat-depression.com/top-mental-health-accounts-to-follow-on-twitter.

[3] http://www.healthline.com/health/depression/best-blogs-of-the-year.

3   Methods

3.1   Data Preprocessing

Because the data we have collected from the tweets and web blogs are biased and noisy, cleaning data is our first task. Generally, the special characters, such as retweet tags "@RT: xxx" and link address "http://www.", contain less information, they are removed at the beginning. In the next step, stop words and punctuations are removed by stop word list that has been aggregated online. Non-words are very common in social media data due to any types of typo or acronyms, for instance, "hrt", "lmao". These words are filtered by the NLTK toolkit [9]. Finally, we had the raw data cleaned. Table 1 shows the number of words have left after each step of data preprocessing procedure.

Table 1. Word counts

Steps	TW	PTA	DB
Raw data	54 M	18 M	46 M
Special character removal	7 M	2.6 M	8 M
Stop words removal	3 M	1.2 M	3.4 M
Non-words removal	0.72 M	0.2 M	0.74 M

3.2   Data Analysis

3.2.1   Word Frequency

The basic approach to analyze data is to calculate the word frequency in the documents. In the traditional text mining research, the frequent words are considered as the important words in the nature language analysis. The collected data includes many common words that are semantically related to the depressive symptoms that we are familiar with, e.g. words "anxiety" and "disorder" are universal in the data set.

3.2.2   Word Embedding

To learn the relationships among words in the documents, words should be transformed into vectors. Currently, there are two common strategies to generate word vectors: one-hot encoding and the word embedding. The essential idea of one-hot encoding [10] is to study to associate each word in the vocabulary with vector representation. Building a vocabulary of the size N from the whole collection of corpus, each word is mapped to a vector with length N, where the Nth digit indicates this word's existence or index, e.g. 00001...00, 01000...00. Therefore, words are represented in a high and sparse dimension and every word corresponds to a point in the vector space, as a result, it is difficult to capture the "relationship" among words. However, Bengio *et al.* [11] developed a neural network language model (NNLM) that learns a probability distribution over words in the corpus and the model is trained to produce the vector representations of word. In contrast to the one-hot encoding, the NNLM generates word vectors with low and dense vector, it is easier to capture the word's property.

Mikolov *et al.* [12] extended the Bengio' work and built a new neural network language leaning model, Word2Ve, using different learning methods: Continues Bag of Words (CBOW) and Skip-gram. The CBOW predicts the current word given the neighboring words in the surrounding window, while, the Skip-gram learns the context words by giving a word in the input layer and predict its surrounding words in the output layer. The reason that we use the Word2Vec due to it consume less time than Benhio's NNLM while training.

3.2.3   Word Clustering

We applied the Word2Vec to generate the good quality of word vectors through training the whole data set. After that, each word has multiple degrees of similarity and it can be computed via a linear calculation. For instance, vector ("Pairs") – vector ("France") + vector ("Italy") produces a word vector that is assumed as similar as "Rome". In our work, we could find the similar depression facts by given a depression symptom or a similar word. Similar words tend to be close to each other, to group semantic similar words, we use the K-means [13] to partition the N objects into K ($K \ll N$) clusters depends on their geometric locations. Thus, the word clustering technique could help us understand the relationship between two words.

4   Experiments

4.1   Word Frequency

We calculate the word frequency on three data sets individually. Table 2 displays the portion of results of the 50 most frequent words in each data set. The most frequent words might not reveal enough depressive symptoms as we expected, but we found many useful facts about depression that are not very common in data. We exhibit the portion of these depression facts in the Table 3.

4.2   Word Similarity

Semantic similar words are extracted via given a word from the vocabulary. For example, given a word "depression", the Table 4 shows the sample of the depression's similar words and corresponding cosine similarities. The larger value of cosine similarity, the closer between two words.

Table 2. Word counts

TW	PTA	DB
Depression	Depression	Depression
Anxiety	Risk	Family
Suicide	Environment	Paranoia
Stress	Treatment	Race
Postpartum	Season	Business

Table 3. Frequent facts

Data sets	Frequent facts
TW	Lupus, bipolar, autism, jealousy, marijuana
PTA	Discomfort, fear, inability, strategist, army
DB	Sensitivity, darkness, rainy, Japan, drunk

Table 4. Similar words of "depression"

TW		PTA		DB	
Similar words	Cosine similarity	Similar words	Cosine similarity	Similar words	Cosine similarity
Anxiety	0.566	Disorder	0.969	Stress	0.965
Asthma	0.285	Family	0.904	Heroin	0.919
Narcissism	0.279	Anxiety	0.897	Abuse	0.891
Hysteria	0.178	Pain	0.858	Vaccine	0.883

4.3 Word Clustering

In our experiment, the value of K is defined as 5. To illustrate the word clustering in our data sets, we show one word group and its contents on three data sets in the Fig. 1. It is easy to find semantic similar words are accumulated to the same cluster. This approach is helpful to discover some unfamiliar but truly depression relevant symptoms. More data we collected, more information could be captured in on social medias.

Fig. 1. Word clusters and contents on TW, PTA, DB respectively

5 Conclusion

Clinic depression has been a serious mental illness since past decades which negatively affects human's health. it is difficult to confirm human's depression symptoms from their behaviors via restricted clinic records. Our proposed methods and experiments illustrate that social network and web blogs provide rich information for depression symptoms extraction from a distinctive perspective. Current advanced natural language processing approach, like Word2Vec, can be helpful for medical uses. In the future, we will collect other types of data, e.g. image and video from other social networks. Additionally, advanced entity selection technique would be used to select more accurate and meaningful depression symptoms.

References

1. Griffin, J.M., Fuhrer, R., Stansfeld, S.A., Marmot, M.: The importance of low control at work and home on depression and anxiety: do these effects vary by gender and social class? Soc. Sci. Med. **54**(5), 783–798 (2002)
2. Cai, Z., He, Z., Guan, X., Li, Y.: Collective data-sanitization for preventing sensitive information inference attacks in social networks. IEEE Trans. Dependable Secure Comput. (2016)
3. He, Z., Cai, Z., Han, Q., Tong, W., Sun, L., Li, Y.: An energy efficient privacy-preserving content sharing scheme in mobile social networks. Pers. Ubiquit. Comput. **20**(5), 833–846 (2016)
4. Mathioudakis, M., Koudas, N: Twittermonitor: trend detection over the twitter stream. In: Proceedings of the 2010 ACM SIGMOD International Conference on Management of Data, pp. 1155–1158. ACM (2010)
5. Chowdhury, G.: G: Natural language processing. Ann. Rev. Inf. Sci. Technol. **37**(1), 51–89 (2003)
6. Wang, S.H., Ding, Y., Zhao, W., Huang, Y.H., Perkins, R., Zou, W., Chen, J.J.: Text mining for identifying topics in the literatures about adolescent substance use and depression. BMC Public Health **16**(1), 279 (2016)
7. Blei, D.M., Ng, A.Y.: Jordan, M.I: Latent dirichlet allocation. J. Mach. Learn. Res. **3**(Jan), 93–1022 (2003)
8. Ma, L., Zhang, Y.: Using Word2Vec to process big text data. In: 2015 IEEE International Conference on Big Data (Big Data), pp. 2895–2897 (2015)
9. Bird, S.: NLTK: the natural language toolkit. In: Proceedings of the COLING/ACL on Interactive Presentation Sessions, pp. 69–72. Association for Computational Linguistics (2016)
10. Turian, J., Ratinov, L., Bengio, Y.: Word representations: a simple and general method for semi-supervised learning. In: Proceedings of the 48th Annual Meeting of the Association for Computational Linguistics, pp. 384–394. Association for Computational Linguistics (2010)
11. Bengio, Y., Ducharme, R., Vincent, P., Jauvin, C.: A neural probabilistic language model. J. Mach. Learning Res. **3**(Feb), 1137–1155 (2003)
12. Mikolov, T., Chen, K., Corrado, G., Dean, J.: Efficient estimation of word representations in vector space. arXiv preprint arXiv:1301.3781 (2013)
13. Witten, I.H., Frank, E., Hall, M.A., Pal, C.J.: Data Mining: Practical Machine Learning Tools and Techniques. Morgan Kaufmann, Burlington (2016)

A Probabilistic Approach to Multiple-Instance Learning

Silu Zhang[(⊠)], Yixin Chen, and Dawn Wilkins

Department of Computer and Information Science,
The University of Mississippi, University, MS 38677, USA
{ychen,dwilkins}@cs.olemiss.edu, szhang6@go.olemiss.edu

Abstract. This paper introduced a probabilistic approach to the multiple-instance learning (MIL) problem with two Bayes classification algorithms. The first algorithm, named Instance-Vote, provides a simple approach for posterior probability estimation. The second algorithm, Embedded Kernel Density Estimation (EKDE), enables data visualization during classification. Both algorithms were evaluated using MUSK benchmark data sets and the results are highly competitive with existing methods.

Keywords: Multiple-instance learning · Non-linear dimensionality reduction · Data visualization

1 Introduction

Machine Learning approaches have been widely applied in drug activity prediction [2,12], i.e., predicting whether an unknown drug will bind to a target (protein) based on existing knowledge of drugs-protein interactions. The multiple-instance learning (MIL) problem arises when a drug has more than one conformation and several of those can bind to the target. However, only the labels of drugs in the training set are given: a drug is positive if at least one of its conformations binds to the target, and negative otherwise. The label of each conformation is unknown. The task is to predict the label of an unseen drug (i.e., bag) given its conformations (i.e., instances). MIL is also applied in gene function prediction at the isoform level [5]. In this context, a gene is considered as a bag, which consists of multiple isoforms, referred as instances.

Of existing MIL algorithms, one class is based upon learning the labels of instances and then labelling the bag using instance label information. The assumption typically used is that a bag is positive if it has at least one positive instance and negative if all of its instances are negative [1,4]. A different assumption is that all instances contribute equally and independently to a bag's label, and the bag label was generated by combining the instance-level probability estimates [10,11,13]. There are also many methods that convert the MIL problem to a supervised learning problem using feature mapping [3]. However,

© Springer International Publishing AG 2017
Z. Cai et al. (Eds.): ISBRA 2017, LNBI 10330, pp. 331–336, 2017.
DOI: 10.1007/978-3-319-59575-7_30

feature mapping usually results in increased dimensionality, and a commonly used approach to overcome this problem is feature selection.

In this paper, we develop two Bayes classifiers for MIL. The first approach, named Instance-Vote, attaches the bag label to its instances for all bags in the training set. For any new bag, we simply use a k-NN classifier to predict the label for each instance in the bag, followed by estimating the probability of the bag being positive via the percentage of positive instances in the bag. The second algorithm, named as EKDE (Embedded Kernel Density Estimation) converts the MIL to a supervised learning problem by mapping each bag into an instance-defined space. Instead of feature selection, a non-linear dimensionality reduction method, t-SNE (t-Distributed Stochastic Neighbor Embedding), is then used to reduce the dimension to 1 or 2. The advantage of this approach is the capability of data visualization. The class conditional probability densities are then estimated in this low dimensional space by kernel density estimation (KDE). The classification is based on the posterior probability according to Bayes' theorem.

2   Methodology

2.1   MIL via Instance-Vote

A New Interpolation of Instance Label. The main challenge of MIL is that the label of instances are unknown. The classical MIL assumption treats a bag as negative if none of its instances is positive. Here we relax the assumption by allowing negative bags to contain positive instances. In order to predict instance labels, we assign bag label to all its instances in the training set. We then use k-NN classifier to predict instance labels. This above process of generating instance-based training data clearly introduces noise into instance labels. However, the noise can be accounted for by the following voting model and the choice of threshold parameter.

Voting for Bag Label. To classify a bag, all its instances cast a vote based on the instance label. We assume that the posterior probability of a bag being positive (or negative) is a monotonically non-decreasing function of the probability of a randomly chosen instance from the bag being positive (or negative), i.e.,

$$\Pr(y = +|B) = f(\Pr(x_i \in +|B))$$

where y is the label of bag B, x_i is a randomly chosen instance from the bag, f is an unknown monotonically non-decreasing function. The maximum likelihood estimate of $\Pr(x_i \in +|B)$ is obtained as $\frac{m^+}{m}$, where m^+ is the number of positive instances in the bag and m is the total number of instances in the bag. We use a simple Bayes decision rule for classification, i.e.,

$$y = \begin{cases} + & \text{if } \Pr(x_i \in +|B) > \theta, \\ - & \text{otherwise,} \end{cases}$$

where θ is the threshold parameter.

2.2   Embedded Kernel Density Estimation

In this approach, we convert the MIL problem to supervised learning via feature mapping. We aim to find the probability distributions of the two classes using KDE and then apply the Bayes decision rule. However, KDE does not perform well for high dimensional data, since data points are too sparse in high dimensional space. The solution is to learn an embedding of the data and apply KDE in the low dimensional latent space ($d = 1$ or 2). Therefore, we name this approach as Embedded Kernel Density Estimation (EKDE).

Feature Mapping. We adopt the same method described in [3] considering its good performance. Each bag is represented by all the instances in the training set via a similarity measurement. The similarity of a bag B_i and an instance x^k is defined as:

$$s(B_i, x^k) = \max_j \exp(-\frac{\left\| x_{ij} - x^k \right\|^2}{\sigma^2}),$$

where x_{ij} is the j'th instance in bag B_i with $j = 1, \ldots, n_i$, n_i is the number of instances in bag B_i, and σ is a predefined scaling factor. Then bag B_i can be represented as: $[s(B_i, x^1), s(B_i, x^2), \ldots (B_i, x^n)]$, where n is the total number of instances in the training set, i.e., $\sum_{i=1}^{l} n_i = n$, where l is the total number of bags in the training set. The dimension after feature mapping is now n, which can be considerably large. Therefore, dimensionality reduction is desired.

Dimensionality Reduction and Visualization. Among all existing dimensionality reduction techniques, t-SNE was chosen due to its prominent performance [8]. Specifically, we chose the parametric t-SNE since it provides a mapping function from the original space to the low dimensional space [7]. The dimension of latent space was set to 1 or 2 such that KDE can be reliably implemented and visualization of the data can also be achieved. Although not required by classification, visualization is beneficial for data analysis.

Probability Density Estimation and Classification. According to Bayes' theorem, given a bag represented as x, the posterior probabilities can be computed as

$$\Pr(y = +|x) = \frac{p(x|y = +)\Pr(y = +)}{p(x)}, \Pr(y = -|x) = \frac{p(x|y = -)\Pr(y = -)}{p(x)},$$

where y is the bag label. Assuming bags being i.i.d., the maximum likelihood estimates of $\Pr(y = +)$ and $\Pr(y = -)$ are $\Pr(y = +) = \frac{l^+}{l}, \Pr(y = -) = \frac{l^-}{l}$, where $l^+(l^-)$ is the number of positive(negative) bags in training set. The class conditional densities $p(x|y = +)$ and $p(x|y = -)$ can be estimated by KDE using training data after dimensionality reduction. $p(x)$ is a constant respect to y. The classifier can make predictions on the bag label y by setting a threshold θ for the odd ratio (OR):

$$y = \begin{cases} + & \text{if OR} > \theta, \\ - & \text{otherwise,} \end{cases}$$

where $\text{OR} = \frac{\Pr(y=+|x)}{\Pr(y=-|x)}$.

3 Experimental Results

3.1 Data Sets

The benchmark datasets MUSK1 and MUSK2 are used in our study. In these two datasets, each molecule (bag) has more than one conformation (instance). The label of the molecule is "musk" (positive) if any of its conformations is a musk or "non-musk" if none of its conformations is a musk.

3.2 Experimental Setup

For the Instance-Vote algorithm, different values of k were tested for k-NN classifier instance classification. For the EKDE algorithm, the setup of feature mapping is same as [3]. We used the implementation of parametric t-SNE that is publicly available at [9]. A Gaussian kernel was used in KDE and the optimal bandwidth was determined by 10-fold cross validation.

3.3 Results

Both algorithms were tested using 10-fold cross-validation at the bag level. We use area under ROC (receiver operating characteristics) curve (AUC) for evaluation since it is a preferred measurement over accuracy [6]. Each experiment was performed 10 times and the average AUC was used for comparison with other algorithms.

In the testing of Instance-Vote algorithm, $k = 3$ gives the optimal cross-validation result for both data sets. The AUC obtained is shown in Table 1. The result is surprisingly good considering the simpleness of this algorithm. This may suggest that noise introduced during labelling instances in the training set is not significant. From the chemistry point of view, it is reasonable that many conformations of a musk molecule can preserve the musk property.

We next present the results of the EKDE algorithm. After feature mapping, the data dimensions are 476 (MUSK1) and 6598 (MUSK2), as determined by the total number of instances in the training sets. The dimension is then reduced to 1 or 2 by applying parametric t-SNE. Due to the limit of space, we only show the visualization results in 2D (Fig. 1). The two classes are separated well for both data sets with minor overlapping in MUSK2, thanks to the superiority of t-SNE on preserving the local structure. The AUC results are shown in Table 1.

For comparison, we also include the results of various existing MIL algorithms that use AUC as the measure for evaluation (Table 1). Among all of the listed method, Instance-Vote is the simplest and has comparable results with the others. The EKDE algorithm outperforms others on MUSK1 and is comparable with those on MUSK2.

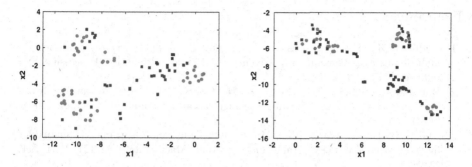

Fig. 1. Visualization of MUSK1 (left) and MUSK2 (right) data sets in 2D. Positive and negative bags are presented as red circles and blue squares, respectively. (Color figure online)

Table 1. AUC obtained by the proposed algorithms and other methods on the MUSK data set (All listed algorithms were evaluated by 10-fold cross-validation.).

Algorithms	MUSK1	MUSK2
Instance-Vote	0.921	0.856
EKDE ($d = 1$)	**0.954**	0.859
EKDE ($d = 2$)	0.941	0.865
MI RVM [11]	0.942	**0.987**
RVM [11]	0.951	0.985
MI Boost [11,13]	0.899	0.964
MI LR [10,11]	0.846	0.795
DD(1) [10]	0.895	0.903
DD(3) [10]	0.883	0.850
DD(5) [10]	0.861	0.838

4   Conclusions

In this paper, we introduced two Bayes algorithms to solve the multiple-instance problem. The Instance-Vote algorithm acquires the label of each instance in the training set from its associated bag and predict an unseen bag label base on the percentage of predicted positive instances in the bag. The EKDE algorithm performs KDE after feature mapping in the embedded low dimensional space with the help of parametric t-SNE. In this approach, both classification and data visualization can be achieved. We have shown that both algorithms are competitive with other MIL algorithms on MUSK benchmark data sets.

Acknowledgements. This work was supported by the Department of Computer and Information Science, University of Mississippi.

References

1. Andrews, S., Tsochantaridis, I., Hofmann, T.: Support vector machines for multiple-instance learning. In: Advances in Neural Information Processing Systems, pp. 577–584 (2003)
2. Burbidge, R., Trotter, M., Buxton, B., Holden, S.: Drug design by machine learning: support vector machines for pharmaceutical data analysis. Comput. Chem. **26**(1), 5–14 (2001)
3. Chen, Y., Bi, J., Wang, J.Z.: Miles: multiple-instance learning via embedded instance selection. IEEE Trans. Pattern Anal. Mach. Intell. **28**(12), 1931–1947 (2006)
4. Dietterich, T.G., Lathrop, R.H., Lozano-Pérez, T.: Solving the multiple instance problem with axis-parallel rectangles. Artif. Intell. **89**(1), 31–71 (1997)
5. Eksi, R., Li, H.D., Menon, R., Wen, Y., Omenn, G.S., Kretzler, M., Guan, Y.: Systematically differentiating functions for alternatively spliced isoforms through integrating RNA-seq data. PLoS Comput. Biol. **9**(11), e1003314 (2013)
6. Huang, J., Ling, C.X.: Using AUC and accuracy in evaluating learning algorithms. IEEE Trans. knowl. Data Eng. **17**(3), 299–310 (2005)
7. van der Maaten, L.: Learning a parametric embedding by preserving local structure. RBM **500**(500), 26 (2009)
8. van der Maaten, L., Hinton, G.: Visualizing data using t-SNE. J. Mach. Learn. Res. **9**(Nov), 2579–2605 (2008)
9. Matten, L.: t-SNE. https://lvdmaaten.github.io/tsne/
10. Ray, S., Craven, M.: Supervised versus multiple instance learning: an empirical comparison. In: Proceedings of the 22nd International Conference on Machine Learning, pp. 697–704. ACM (2005)
11. Raykar, V.C., Krishnapuram, B., Bi, J., Dundar, M., Rao, R.B.: Bayesian multiple instance learning: automatic feature selection and inductive transfer. In: Proceedings of the 25th International Conference on Machine Learning, pp. 808–815. ACM (2008)
12. Warmuth, M.K., Liao, J., Rätsch, G., Mathieson, M., Putta, S., Lemmen, C.: Active learning with support vector machines in the drug discovery process. J. Chem. Inf. Comput. Sci. **43**(2), 667–673 (2003)
13. Xu, X., Frank, E.: Logistic regression and boosting for labeled bags of instances. In: Dai, H., Srikant, R., Zhang, C. (eds.) PAKDD 2004. LNCS, vol. 3056, pp. 272–281. Springer, Heidelberg (2004). doi:10.1007/978-3-540-24775-3_35

Net2Image: A Network Representation Method for Identifying Cancer-Related Genes

Bolin Chen, Yuqiong Jin, and Xuequn Shang[✉]

School of Computer Science, Northwestern Polytechnical University,
Xi'an 710072, Shaanxi, China
shang@nwpu.edu.cn

Abstract. Although many machine learning algorithms have been proposed to identify cancer-related genes, their prediction accuracy is still limited due to the complex relationship between cancers and genes. To improve the prediction accuracy, many deep learning based tools have been developed, and they have shown their efficiency to handle complex relationships. To use those tools, a deliberate data representation method is indispensable, since majority tools only take those image-like data as inputs. In this study, we propose a novel network representation method, called *Net2Image*, to transfer topological networks into image-like datasets. The local topological information of individual vertices from six biomolecular networks and one DNA methylation dataset are encoded as $80 * 6$ matrices. They are then employed as inputs to train the model for identifying cancer-related genes using TensorFlow. The numerical experiments show that the proposed method can achieve very high prediction accuracy, which outperforms many existing methods.

Keywords: Deep learning · Biomolecular network · Cancer-related gene · Multiple data integration

1 Introduction

The identification of cancer-related genes plays essential roles in understanding the mechanism of cancer pathogenesis. To achieve this, many machine learning methods have been proposed from various computational aspects by using the assumption of "guilty-by-association" [1] and the strategy of multiple data integration [2], such as those regression methods [3], the global network based method (PRINCE [4]), the combining gene expression and protein interaction method (CGI [5]), the random walk with restart (RWR [6]), the Markov random field based method (MRF [2]), and data integration rank based method (DIR [7]) *etc.* Although those algorithms have achieved big improvement for understanding the pathogenic mechanism of many genetic diseases, their prediction accuracy is still limited, and novel efficient and powerful algorithms need to be further developed.

B. Chen and Y. Jin—Equally contributing authors.

© Springer International Publishing AG 2017
Z. Cai et al. (Eds.): ISBRA 2017, LNBI 10330, pp. 337–343, 2017.
DOI: 10.1007/978-3-319-59575-7_31

Deep learning based methods have gained big success in many areas. They can model the complex relationship between diverse variables by using the combination of multiple nonlinear functions. However, deep-learning based studies in computational biology areas still mainly follow the basic strategy of those image processing methods. They either focus on medical image process directly, such as the skin cancer image process [8], or use a method to transfer various sequential data into images, such as the DeepBind method [9].

In this study, we propose a novel network representation method to transfer the local topological information of individual vertices into encoded matrices. Six biomolecular networks and one DNA methylation dataset are integrated to generate those matrices. They are then been used as inputs of TensorFlow to train the model of for identifying cancer-related genes in this study.

2   Materials and Methods

2.1   Problem Formulation

The problem of cancer-related gene identification is to prioritize all unknown genes in terms of their associations with cancers. This can be described as a *two-class classification problem* (i.e., taking the value of 1 or 0 to indicate if the gene is related to cancers or not). Suppose human genome consists of a set of N genes $G = \{g_1, g_2, \ldots, g_N\}$, where let $g_1, g_2, \ldots, g_n$ be genes that we do not know their associations with any genetic disorders, $g_{n+1}, g_{n+2}, \ldots, g_{n+m}$ be genes that we know their associations with cancers, and $g_{n+m+1}, g_{n+m+2}, \ldots, g_{n+m+k}$ be genes that we know their associations with other genetic disorders except cancers. Obviously, we have $N = n + m + k$.

A typical machine learning method usually takes genes in $g_{n+1}, g_{n+2}, \ldots, g_{n+m}$ as positive instances, and randomly select roughly equal number of genes in $g_1, g_2, \ldots, g_n$ as the negative instances. However, we argue that it makes more sense if the negatives instances are selected from $g_{n+m+1}, g_{n+m+2}, \ldots, g_{n+m+k}$. Since those genes have been well studied by many researchers, and no association relationship has been found between any of them and cancers. They are more likely to be non-cancer-related genes compared with those unknown genes.

2.2   Network Representation

A typical deep learning framework, such as the TensorFlow, usually takes pixels of individual images as inputs. Hence, the objective of network representation is to find a reasonable set of local topological features for individual vertices.

Given a configuration for all genes, let $N_1(g_i)$ denote the *neighbor set* of gene g_i in a network, and $N_1^+(g_i)$ and $N_1^-(g_i)$ denote the subset of $N_1(g_i)$ whose elements are labelled with 1 and 0, respectively. Several topological indices can be employed to conduct the network representation. Take the degree $d(g_i)$ and the direct neighborhood information $N_1(g_i)$ for example. We use the following elements to represent the part features of gene g_i.

$$x_1(g_i) = d(g_i) = |N_1(g_i)|$$
$$x_2(g_i) = min\{d(g_j)\}, \text{ where } g_j \in N_1(g_i)$$
$$x_3(g_i) = median\{d(g_j)\}, \text{ where } g_j \in N_1(g_i) \qquad (1)$$
$$x_4(g_i) = ave\{d(g_j)\}, \text{ where } g_j \in N_1(g_i)$$
$$x_5(g_i) = max\{d(g_j)\}, \text{ where } g_j \in N_1(g_i)$$

We can similarly get $x_6(g_i)$ to $x_{10}(g_i)$ by using the set of $N_1^+(g_i)$, and $x_{11}(g_i)$ to $x_{15}(g_i)$ by using the set of $N_1^-(g_i)$. Hence, we obtain 15 raw features by using the degree and the direct neighborhood information. Similarly, we can further generalize this idea by using the degree and the second-order of neighbors, the clustering coefficient and the direct neighbors, and the clustering coefficient and the second-order of neighbors to generate the raw features of $x_{16}(g_i)$ to $x_{30}(g_i)$, $x_{31}(g_i)$ to $x_{45}(g_i)$ and $x_{46}(g_i)$ to $x_{60}(g_i)$, respectively.

Moreover, we would like to describe the location of individual vertices within a network using the farthest neighbor set. In [10], Goh et al. have shown that disease-related genes tend to locate at the peripheral position but rather those central positions. For the set of $N_f(g_i)$, we let

$$x_{61}(g_i) = dist(g_i, g_j), \text{ where } g_j \in N_f(g_i)$$
$$x_{62}(g_i) = min\{dist(g_j, g_k)\}, \text{ where } g_j \in N_f(g_i) \text{ and } g_k \in N_f(g_j)$$
$$x_{63}(g_i) = median\{dist(g_j, g_k)\}, \text{ where } g_j \in N_f(g_i) \text{ and } g_k \in N_f(g_j) \qquad (2)$$
$$x_{64}(g_i) = ave\{dist(g_j, g_k)\}, \text{ where } g_j \in N_f(g_i) \text{ and } g_k \in N_f(g_j)$$
$$x_{65}(g_i) = max\{dist(g_j, g_k)\}, \text{ where } g_j \in N_f(g_i) \text{ and } g_k \in N_f(g_j)$$

where $dist(u, v)$ represents the length of shortest path between the vertex u and v. The reason we use $dist(g_j, g_k)$, but rather $dist(g_i, g_j)$ is due to the fact that the value of $dist(g_i, g_j)$ between g_i and any $g_j \in N_f(g_j)$ is a constant. There is no different among the value of $min(\cdot)$, $median(\cdot)$, $ave(\cdot)$ and $max(\cdot)$ under this situation. However, the furthest neighbors of $N_f(g_j)$, where $g_j \in N_f(g_i)$ confess the location information of g_i. It should have obvious difference between x_{61} and x_{62} to x_{65} if it locates in the central position, while the difference would be tiny if it locates in the peripheral area. Similarly, the raw features of $x_{66}(g_i)$ to $x_{70}(g_i)$, and $x_{71}(g_i)$ to $x_{75}(g_i)$ can be obtained by using the $N_f^+(g_i)$ and $N_f^-(g_i)$, respectively.

In addition, DNA methylation level of individual genes also contribute to the features related to cancers [11]. We randomly selected five methylation sites for each gene, and calculated the rest raw features as follows.

$$x_{76}(g_i) = meth_1(g_i)$$
$$x_{77}(g_i) = meth_2(g_i)$$
$$x_{78}(g_i) = meth_3(g_i) \qquad (3)$$
$$x_{79}(g_i) = meth_4(g_i)$$
$$x_{80}(g_i) = meth_5(g_i)$$

2.3   The Evaluation Process Based on TensorFlow

To evaluate the prediction accuracy of M_0, we use the leave-one-out cross validation paradigm as follows. Suppose $g_{t1}, g_{t2}, \ldots, g_{tk}$ are those genes in the training set, including both positive instances and negative instances. The leave-one-out experiment leaves one gene g_{tj} out per time, and uses the rest genes to train the model using TensorFlow. The fine trained model can be denoted as M_j when g_{tj} is left out in such experiment. The receiver operating characteristic (ROC) curve and the area under the ROC curve (AUC) are employed to demonstrate the performance of the proposed method.

The close similarity between M_0 and each $M_1, M_2, \ldots, M_k$ makes it possible to use the pretrained result of M_0 as the initial value to train $M_1, M_2, \ldots, M_k$ in the evaluation stage. This does not mean that a gene's information is used in the experiment when the gene is left out. We only select the initial values as close to the true values as possible, thereby increasing the training speed.

3   Experiments and Results

3.1   Data Sources

In this study, six biomolecular networks and one DNA methylation dataset are employed to construct the raw features of individual genes, which include four PPI networks, one pathway co-occurrence network and one gene co-expression network. The availability of those datasets are described in Table 1.

Table 1. The summarized information of the integrated datasets

Database	Node no.	Version	URL
HPRD	9465	Release 9	http://hprd.org/download
BioGrid	16085	3.4.143	http://thebiogrid.org/download.php
IntAct	14214	4.2.3.2	http://www.ebi.ac.uk/intact/downloads
InWeb_IM	17653	2016_09_12	https://www.intomics.com/inbiomap
Pathways	16007	c2.all.v5.2	http://software.broadinstitute.org/gsea/msigdb
Expressions	15484	E-TABM-305	http://www.ebi.ac.uk/arrayexpress/
Methylation	21227	GSE36064	-

The node entries in different datasets are not the same. In this study, we selected genes appear in at least six datasets and resulted in 9189 identical vertices as the candidate genes. After taking the intersection with individual datasets, we obtained the valid number of entries as shown in Table 2.

Table 2. The number of valid entries of the integrated datasets

	HPRD	BioGrid	IntAct	InWeb_IM	Pathway	Expression	Methylation
Node no.	7933	9156	8967	9184	9015	6108	7765
Edge no.	31794	146831	76019	107242	1465417	103447	-

The known gene-disease associations are obtained from the paper of Goh, et al. [10] and the OMIM database (downloaded on Dec 13, 2016). The former dataset in [10] gives a very good classification for known genetic disorders. To keep updating this classification list, we manually checked each new entry from the OMIM database, and obtained a set of 2550 disease genes related to known genetic disorders. Among them, 263 genes are related to individual cancers.

3.2 The Performance of the Proposed Method and the Comparison with Previous Methods

The performance of the proposed method is evaluated using the ROC curve and the AUC value. Figure 1 gives the comparison between the proposed Net2Image

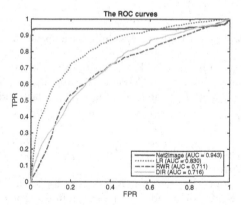

Fig. 1. The cross-validation results of four methods. The magenta line represents the ROC curve of the proposed Net2Image method. The green line represents the ROC curve of the logistic regression (LR) method. The cyan line represents the ROC curve of the random work with result (RWR) method. The red line represents the ROC curve of the data integration rank (DIR) method. AUC values are listed in parentheses. (Color figure online)

method and several existing algorithms, where the ROC curves of those existing methods are adopted from [3]. We can see from the figure that Net2Image significantly outperform those algorithms, such as (1) logistic regression (LR), (2) random walk with restart (RWR), and (3) data integration rank (DIR) in terms of the AUC values. The AUC value of Net2Image is 0.943, which is very promising for identifying cancer-related genes.

4   Conclusions

In this paper, we have proposed a novel network representation method to transfer network topological information into image-like matrices. Our proposed method has the following three technical innovations. (1) It gives a method to handle network inputs for various deep learning based tools. (2) The raw features of individual vertices is related to all genes, but rather only those genes in the training set. It is more reliable than normal machine learning based methods. (3) The initial labels of unknown genes are estimated together with mode in the pretraining stage. Prior information is not necessary for those unknown genes.

Compared with previous algorithms, the proposed Net2Image method also achieve a very high AUC value. It can predict the cancer-related genes with the accuracy at 0.934, which is very promising comparing with exist methods.

Acknowledgement. This work was supported by the National Natural Science Foundation of China under Grant Nos. 61602386 and 61332014 and the Foundation of top university visiting for excellent youth scholars of Northwestern Polytechnical University.

References

1. Altshuler, D., Daly, M., Kruglyak, L.: Guilt by association. Nat. Genet. **26**(2), 135–137 (2000)
2. Chen, B., Wang, J., Li, M., Wu, F.X.: Identifying disease genes by integrating multiple data sources. BMC Med. Genomics **7**(Suppl 2), S2 (2014)
3. Chen, B., Li, M., Wang, J., Shang, X., Wu, F.X.: A fast and high performance multiple data integration algorithm for identifying human disease genes. BMC Med. Genomics **8**(Suppl 3), S2 (2015)
4. Vanunu, O., Magger, O., Ruppin, E., Shlomi, T., Sharan, R.: Associating genes and protein complexes with disease via network propagation. PLoS Comput. Biol. **6**(1), e1000641 (2010)
5. Ma, X., Lee, H., Wang, L., Sun, F.: CGI: a new approach for prioritizing genes by combining gene expression and protein-protein interaction data. Bioinformatics **23**(2), 215–221 (2007)
6. Köhler, S., Bauer, S., Horn, D., Robinson, P.N.: Walking the interactome for prioritization of candidate disease genes. Am. J. Hum. Genet. **82**(4), 949–958 (2008)
7. Chen, Y., Wang, W., Zhou, Y., Shields, R., et al.: In silico gene prioritization by integrating multiple data sources. PLoS One **6**(6), e21137 (2011)
8. Esteva, A., Kuprel, B., Novoa, R.A., Ko, J., et al.: Dermatologist-level classification of skin cancer with deep neural networks. Nature **542**(7639), 115–118 (2017)

9. Alipanahi, B., Delong, A., Weirauch, M.T., Frey, B.J.: Predicting the sequence specificities of DNA-and RNA-binding proteins by deep learning. Nat. Biotechnol. **33**(8), 831–838 (2015)

10. Goh, K.I., Cusick, M.E., Valle, D., Childs, B., et al.: The human disease network. Proc. Natl. Acad. Sci. **104**(21), 8685–8690 (2007)

11. Robertson, K.D.: DNA methylation and human disease. Nat. Rev. Genet. **6**(8), 597–610 (2005)

Computational Prediction of Influenza Neuraminidase Inhibitors Using Machine Learning Algorithms and Recursive Feature Elimination Method

Li Zhang[1,2], Haixin Ai[1,2,3], Qi Zhao[4], Junfeng Zhu[1], Wen Chen[5],
Xuewei Wu[1], Liangchao Huang[5], Zimo Yin[5], Jian Zhao[1],
and Hongsheng Liu[1,2,3(✉)]

[1] School of Life Science, Liaoning University, Shenyang 110036, China
liuhongsheng@lnu.edu.cn
[2] Research Center for Computer Simulating and Information
Processing of Bio-macromolecules of Liaoning Province,
Shenyang 110036, China
[3] Engineering Laboratory for Molecular Simulation and Designing of Drug
Molecules of Liaoning, Shenyang 110036, China
[4] School of Mathematics, Liaoning University, Shenyang 110036, China
[5] School of Information, Liaoning University, Shenyang 110036, China

Abstract. Recent outbreaks of highly pathogenic influenza have highlighted
the need to develop novel anti-influenza therapeutics. Neuraminidase has
become the most important target for the treatment of influenza virus. In this
study, classification models were developed from a large training dataset containing 457 neuraminidase inhibitors and 358 non-inhibitors using random forest
and support vector machine algorithms. Recursive feature elimination (RFE)
method was used to improve the accuracy of the models by selecting the most
relevant molecular descriptors. The performances of the models were evaluated
by five-fold cross-validation and independent validation. The accuracies of all
the models are over 86% in both validation methods. This work suggests
machine learning algorithms combined with RFE method can be used to build
useful models for predicting influenza neuraminidase inhibitors.

Keywords: Machine learning · Neuraminidase inhibitor · Feature selection

1 Introduction

The flu continues to be a serious public health threat that causes severe morbidity and
mortality throughout the world every year. In recent years, the emerging of new avian
influenza virus [1] and drug-resistant strains [2] have exacerbated this threat.

The neuraminidase (NA) of the influenza virus removes sialic acid from the gly-
coprotein on the surface of the host cell, resulting in the release of the virion from the
infected cells, which makes NA a key protein in the life cycle of the virus, and has
become the most important anti-influenza drug target [3]. Many NA inhibitors have

© Springer International Publishing AG 2017
Z. Cai et al. (Eds.): ISBRA 2017, LNBI 10330, pp. 344–349, 2017.
DOI: 10.1007/978-3-319-59575-7_32

been developed, such as oseltamivir and zanamivir. Unfortunately, the strains with oseltamivir resistance have been widely spread around the world, and the use of zanamivir has been limited by its poor oral bioavailability [1].

In recent years, computer-aided drug design (CADD) has been widely used in the development of various new drugs. Many new methods, such as molecular docking [4, 5], pharmacophore modeling [6], quantitative structure-activity relationships (QSAR) [7], and machine learning [8, 9], have been applied to screen or design new potential anti-influenza molecules. To date, several machine learning models have been built to discriminate NA inhibitors and non-inhibitors [8, 9].

Feature selection methods can improve the accuracy, reduce the complexity and increase the generalization ability of the classification model generated with machine learning models. Recursive feature elimination (RFE) has been used in many bioinformatics studies [10]. However, in the recent studies of prediction of NA inhibitors using machine learning methods, feature selection methods dose not used to select the most relevant features from a large poll of molecular features [8, 9].

The aim of this study is to build machine learning models for classification of inhibitor and non-inhibitor of influenza A virus neuraminidase using the most relevant features selected by recursive feature elimination and other feature selection methods.

2   Materials and Methods

2.1   Dataset

The molecules with known inhibitory activity (IC50 values) for influenza A virus neuraminidase were collected from search of literatures and bindingDB [11] database. As the inhibition assays were not exactly identical, there may be some deviations in the IC50 values. Therefore, we categorized the molecules with IC50 < 10 μM as inhibitors, and those with IC50 > 50 μM as non-inhibitors. The molecules in the grey area (10 μM < IC50 < 50 μM) were winkled out, which may reduce the possible influence on the accuracy of the model [12]. The dataset were then split into a training set and an independent validation set by the ratio of 4:1 using stratified sampling. The training set, containing 366 inhibitors and 286 non-inhibitors, was used to develop prediction models. The independent validation set, containing 91 inhibitors and 72 non-inhibitors, was used to evaluate the performance of the final model.

The 3D structures of the molecules were generated by Corina. Then, a total of 1845 molecular descriptors were generated by the PaDEL-Descriptor [13] software (version 2.21). Salt was also removed using this software.

2.2   Methods for Model Building

Three machine learning algorithms (k-nearest neighbors (kNN), random forest (RF) and support vector machine (SVM)) were used to develop classification models for prediction of NA inhibitors. The kNN, RF, and SVM algorithms were all executed in R (version 3.3.1). Important parameters for the algorithms were tuned using grid search method implemented in R package *caret* [14] (version 6.0–71).

2.3 Feature Selection

The descriptors with zero variation in the training set and the highly correlated descriptors (Pearson's correlation coefficients > 0.75) were removed using the R package *caret*. And then the recursive feature elimination (RFE) method was used to further remove the redundant features.

2.4 Evaluation of Prediction Performance

The performance of the models was evaluated by five-fold cross-validation and independent validation methods [15–17]. The performance of the models was assessed by the quantity of accuracy (Q), specificity (SP), sensitivity (SE), and Mathew's correlation coefficient (MCC). The definitions and equations can be found in [18, 19]. In addition, the area under the ROC (receiver operating characteristic) curve (AUC) was also used to assess the performance of the models [20–24].

3 Results and Discussion

3.1 Performance of the Models

Based on the training set, three machine learning models, namely kNN-All, RF-All, and SVM-All, were built with all molecular descriptors generated by PaDEL-Descriptor. The five-fold cross-validation results are shown in Table 1. The accuracy of kNN-All, RF-All, and SVM-All are 80.7%, 86.3% and 87%, respectively. It was observed that the specificity of these models needs to be improved.

Table 1. Performance of the models as evaluated by five-fold cross-validation.

Model	Descriptors	Q (%)	SE (%)	SP (%)	MCC	AUC
kNN-All	1845	80.7	85.5	74.5	0.606	0.866
RF-All	1845	86.3	89.4	82.3	0.722	0.940
SVM-All	1845	87.0	88.7	84.7	0.737	0.929
RF 215	215	86.7	88.3	84.6	0.731	0.937
SVM-215	215	87.2	86.6	88.1	0.745	0.933
RF-RFE	58	87.0	88.1	85.6	0.737	0.942
SVM-RFE	28	86.4	87.2	85.5	0.726	0.920

As the accuracy of the model produced by kNN is low, the models built by RF and SVM were further refined by feature selection methods. After removing the zero variation descriptors and the highly correlated descriptors, 215 molecular descriptors were remained. Two models, namely RF-215 and SVM-215, were built based on these selected descriptors. As shown in Table 1, the overall accuracy of the models built by RF and SVM are 86.7% and 87.2%, which are all improved. The specificity of the two models also improved, although the sensitivity has declined.

After applying the RFE feature selection method, two optimal models (RF-RFE and SVM-RFE) were obtained. As shown in Table 1, the accuracy of the RF model was improved from 86.7% to 87.0%, but the accuracy of the SVM model was reduced from 87.2% to 86.4%. The specificity of the RF model was also improved. It can be seen that the RFE method reduces the number of molecular descriptors in the SVM and RF models to 58 and 28 in cases where the model accuracy is less affected.

The results of external validation are given in Table 2. The overall accuracy of these models ranges from 90.2–92.6%, indicating that these models was not over-fitted, and could be applied to new compounds for screening new NA inhibitors.

Table 2. Evaluation of the models using an independent validation set.

Model	TP	FN	TN	FP	SE (%)	SP (%)	Q (%)	MCC	AUC
RF-215	85	6	66	6	93.4	91.7	92.6	0.851	0.977
SVM-215	83	8	66	6	91.2	91.7	91.4	0.827	0.973
RF-RFE	86	5	65	7	94.5	90.3	92.6	0.851	0.974
SVM-RFE	82	9	65	7	90.1	90.3	90.2	0.802	0.952

As demonstrated in a series of recent publications [18, 19] in developing new prediction methods, user-friendly and publicly accessible web-servers will significantly enhance their impacts, we shall make efforts in our future work to provide a web-server for the prediction method presented in this paper.

3.2 Relevance Molecular Descriptors for Predicting Neuraminidase Inhibitors

In this study, a total of 58 and 28 molecular descriptors are selected for the RF-RFE and SVM-RFE model by various feature selection methods. There are 24 molecular descriptors both included in RF-RFE and SVM-RFE model. The normalized mean value and their standard errors of 24 selected molecular descriptors of NA inhibitors

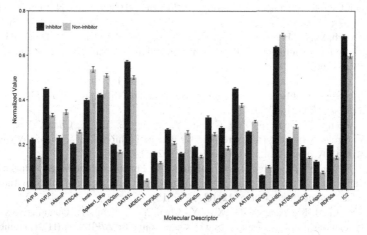

Fig. 1. Comparison of the 24 selected molecular descriptors. The values of the molecular descriptors are normalized to the range of 0–1 using Min-Max normalization technique.

and non-inhibitors are represented in Fig. 1. It can be found that the descriptor has a significant difference between the inhibitor and the non-inhibitor, indicating that these descriptors may be related to the mechanism of action of the inhibitors.

4   Conclusion

In this work, machine learning models capable of discriminate NA inhibitors and non-inhibitors, were developed from a large training set containing 457 NA inhibitors and 358 non-inhibitors using random forest and support vector machine algorithms and various feature selection methods. The models were validated by five-fold cross-validation and independent validation. The results shown that machine learning algorithms can used to predict NA inhibitor with a high accuracy. Feature selection methods are found to be useful in improving the accuracy and reduce the complexity of the models. The molecular descriptors selected by RFE method can provide clues to interpret the mechanism of the known NA inhibitors. The models are potentially useful tools for screening new NA inhibitors from large chemical libraries.

Acknowledgments. This work was supported by the National Natural Science Foundation of China (No: 31570160), Innovation Team Project (No: LT2015011) from Education Department of Liaoning Province, Large-scale Equipment Shared Services Project (No: F15165400) and Applied Basic Research Project (No: F16205151) from Science and Technology Bureau of Shenyang. This project was supported by Engineering Laboratory for Molecular Simulation and Designing of Drug Molecules of Liaoning.

References

1. Hurt, A.C.: The epidemiology and spread of drug resistant human influenza viruses. Curr. Opin. Virol. **8**, 22–29 (2014)
2. Gao, R., Cao, B., Hu, Y., Feng, Z., Wang, D., Hu, W., Chen, J., Jie, Z., Qiu, H., Xu, K.: Human infection with a novel avian-origin influenza A (H7N9) virus. N. Engl. J. Med. **368**, 1888–1897 (2013)
3. Matrosovich, M.N., Matrosovich, T.Y., Gray, T., Roberts, N.A., Klenk, H.-D.: Neuraminidase is important for the initiation of influenza virus infection in human airway epithelium. J. Virol. **78**, 12665–12667 (2004)
4. Ai, H., Zhang, L., Chang, A.K., Wei, H., Che, Y., Liu, H.: Virtual screening of potential inhibitors from TCM for the CPSF30 binding site on the NS1A protein of influenza A virus. J. Mol. Model. **20**, 2142 (2014)
5. Ai, H., Zheng, F., Deng, F., Zhu, C., Gu, Y., Zhang, L., Li, X., Chang, A.K., Zhao, J., Zhu, J.: Structure-based virtual screening for potential inhibitors of influenza A virus RNA polymerase PA subunit. Int. J. Pept. Res. Ther. **21**, 149–156 (2015)
6. Batool, S., Mushtaq, G., Kamal, W., Kamal, M.A.: Pharmacophore-based virtual screening for identification of novel neuraminidase inhibitors and verification of inhibitory activity by molecular docking. Med. Chem. **12**, 63–73 (2016)
7. Cong, Y., Li, B.-K., Yang, X.-G., Xue, Y., Chen, Y.-Z., Zeng, Y.: Quantitative structure–activity relationship study of influenza virus neuraminidase A/PR/8/34 (H1N1) inhibitors by genetic algorithm feature selection and support vector regression. Chemometr. Intell. Lab. **127**, 35–42 (2013)

8. Lian, W., Fang, J., Li, C., Pang, X., Liu, A.-L., Du, G.-H.: Discovery of Influenza A virus neuraminidase inhibitors using support vector machine and Naïve Bayesian models. Mol. Divers. **20**, 439–451 (2016)

9. Li, Y., Kong, Y., Zhang, M., Yan, A., Liu, Z.: Using support vector machine (SVM) for classification of selectivity of H1N1 neuraminidase inhibitors. Mol. Inform. **35**, 116–124 (2016)

10. Tao, P., Liu, T., Li, X., Chen, L.: Prediction of protein structural class using tri-gram probabilities of position-specific scoring matrix and recursive feature elimination. Amino Acids **47**, 461–468 (2015)

11. Gilson, M.K., Liu, T., Baitaluk, M., Nicola, G., Hwang, L., Chong, J.: BindingDB in 2015: a public database for medicinal chemistry, computational chemistry and systems pharmacology. Nucleic Acids Res. **44**, D1045–D1053 (2016)

12. Li, B.-K., Cong, Y., Yang, X.-G., Xue, Y., Chen, Y.-Z.: In silico prediction of spleen tyrosine kinase inhibitors using machine learning approaches and an optimized molecular descriptor subset generated by recursive feature elimination method. Comput. Biol. Med. **43**, 395–404 (2013)

13. Yap, C.W.: PaDEL-descriptor: an open source software to calculate molecular descriptors and fingerprints. J. Comput. Chem. **32**, 1466–1474 (2011)

14. Kuhn, M.: Caret package. J. Stat. Softw. **28**, 1–26 (2008)

15. Chen, X., Yan, C.C., Zhang, X., Zhang, X., Dai, F., Yin, J., Zhang, Y.: Drug–target interaction prediction: databases, web servers and computational models. Brief. Bioinform. **17**, 696–712 (2016)

16. Chen, X., Ren, B., Chen, M., Wang, Q., Zhang, L., Yan, G.: NLLSS: predicting synergistic drug combinations based on semi-supervised learning. PLoS Comput. Biol. **12**, e1004975 (2016)

17. Chen, X., Yan, C.C., Zhang, X., You, Z.-H.: Long non-coding RNAs and complex diseases: from experimental results to computational models. Brief. Bioinform. bbw060 (2016). doi:10.1093/bib/bbw060

18. Chen, W., Feng, P., Yang, H., Ding, H., Lin, H., Chou, K.-C.: iRNA-AI: identifying the adenosine to inosine editing sites in RNA sequences. Oncotarget **8**, 4208–4217 (2017)

19. Chen, W., Tang, H., Ye, J., Lin, H., Chou, K.-C.: iRNA-PseU: identifying RNA pseudouridine sites. Mol. Ther. Nucleic Acids **5**, e332 (2016)

20. Chen, X., Huang, Y.-A., You, Z.-H., Yan, G.-Y., Wang, X.-S.: A novel approach based on KATZ measure to predict associations of human microbiota with non-infectious diseases. Bioinformatics **33**, 733–739 (2017)

21. Huang, Z.-A., Chen, X., Zhu, Z., Liu, H., Yan, G.-Y., You, Z.-H., Wen, Z.: PBHMDA: Path-based human microbe-disease association prediction. Front. Microbiol. **8**, 233 (2017)

22. Chen, X., Huang, Y.-A., Wang, X.-S., You, Z.-H., Chan, K.: FMLNCSIM: fuzzy measure-based lncRNA functional similarity calculation model. Oncotarget **7**, 45948–45958 (2016)

23. Chen, X., You, Z., Yan, G., Gong, D.: IRWRLDA: improved random walk with restart for lncRNA-disease association prediction. Oncotarget **7**, 57919–57931 (2016)

24. Chen, W., Ding, H., Feng, P., Lin, H., Chou, K.-C.: iACP: a sequence-based tool for identifying anticancer peptides. Oncotarget **7**, 16895 (2016)

Differential Privacy Preserving Genomic Data Releasing via Factor Graph

Zaobo He[1(✉)], Yingshu Li[1], and Jinbao Wang[2]

[1] Department of Computer Science, Georgia State University, Atlanta, GA, USA
zhe4@student.gsu.edu
[2] The Academy of Fundamental and Interdisciplinary Sciences,
Harbin Institute of Technology, Harbin, Heilongjiang, China

Abstract. Privacy preserving data releasing is an important problem for reconciling data openness with individual privacy. The state-of-the-art approach for privacy preserving data release is *differential privacy*, which offers powerful privacy guarantee without confining assumptions about the background knowledge about attackers. For genomic data with huge-dimensional attributes, however, current approaches based on differential privacy are not effective to handle. Specifically, amount of noise is required to be injected to genomic data with tens of million of SNPs (Single Nucleotide Polymorphism), which would significantly degrade the utility of released data. To address this problem, this paper proposes a differential privacy guaranteed genomic data releasing method. Through executing belief propagation on factor graph, our method can factorize the distribution of sensitive genomic data into a set of local distributions. After injecting differential-privacy noise to these local distributions, synthetic sensitive data can be obtained by sampling on noise version distribution. Synthetic sensitive data and factor graph can be further used to construct approximate distribution of non-sensitive data. Finally, samples non-sensitive genomic data from the approximate distribution to construct a synthetic genomic dataset.

Keywords: Differential privacy · SNP/trait associations · Belief propagation · Factor graph · Data releasing

1 Introduction

With the developing of DNA-genotyping technology, more and more individuals tend to genotype their DNA, in order for genetic services. For example, 23andMe [1], one of the most popular DNA-sequencing service providers, has provided such services for more than 900,000 individuals. With genotyped DNA, individuals can learn about their predispositions to disease. Meanwhile, massive DNA sequences are significantly beneficial to searchers to develop new genetic diagnostic methods or medicines. Furthermore, more and more research groups release the uncovered associations among genotypes, haplotypes, or phenotypes (such as GWAS catalog [3], DisGeNET [2]), which further enrich genetic services and researches.

© Springer International Publishing AG 2017
Z. Cai et al. (Eds.): ISBRA 2017, LNBI 10330, pp. 350–355, 2017.
DOI: 10.1007/978-3-319-59575-7_33

However, individuals privacy are increasingly threaten with more and more genomic data are available online, although significant benefit are brought by them. In this paper, we propose an effective method to address the problem of releasing differentially private kin-genomic data. Focusing on the genotype and trait privacy, we start from exploring factorize the joint conditional distribution of sensitive genomes into sets of local probability distributions by executing belief propagation on factor graph which captures the dependency relationship among family members, SNPs and traits due to family genetic relationship and SNP/trait associations. To guarantee differential privacy, noise is injected to these local probability distributions to construct approximate distribution for sensitive genomes and then synthetic sensitive genomes can be sampled. Then, synthetic sensitive genomes and factor graph are further used to construct approximate distribution of non-sensitive genomes. Finally, synthetic nonsensitive genomes can be sampled and released. Compared with large body of previous works, which mainly focused on improving the output of differential privacy mechanism (such as optimizing specific query results), we study how to factorize a huge-dimensional distribution into a set of local distributions, so that scale of noise can be reduced by injecting into local distributions.

2 Genomic Data Model

Suppose the number of individuals in the target family is m. The SNP set of an individual is denoted as S with size $|S| = n$. The content of an arbitrary SNP $i(i \in S)$ is denoted as s_i, where s_i takes value from: (i) BB (both alleles inherited from parents are major alleles), (ii) Bb (alleles inherited from parents are major allele and minor allele) or (iii) bb (both alleles inherited from parents are minor alleles). We assume some individuals in a target family intend to release their part of SNPs or traits (such as diseases, hair color, height, etc.) in order for genetic services or research purpose. However, privacy concerns drive them just to release part of SNPs or non-sensitive traits whereas sensitive part is kept private. We denote the set of non-sensitive variables (including SNPs and traits) as X_K, while sensitive variables as X_U.

For SNP/trait associations reported by GWAS catalog, the trait set considered are denoted as T, with size $|T| = r$. t_j is defined to be the trait j ($j \in T$) of an individual.

3 Solution

3.1 Solution Overview

This section sketches an overview of our method for releasing genomic data with ϵ-differential privacy guarantee. The proposed method runs in five phases:

Phase 1: Construct a factor graph $\mathcal{G}$ incorporating all variables X, $X = X_U \cup X_K$ for target family, the Mendelian inheritance probabilities $\mathcal{F}(s_i^F, s_i^M, s_i^C)$, and

SNP/trait associations $\mathcal{A}$. Calculate the joint conditional distribution of sensitive variables $p(X_U|X_K, \mathcal{F}, \mathcal{A})$ by executing belief propagation on factor graph $\mathcal{G}$.

Phase 2: Construct ϵ-differential privacy algorithm to generate the noise version of $p(X_U|X_K, \mathcal{F}, \mathcal{A})$ constructed in Phase 1. We denote the noise joint conditional distribution of sensitive variables as $p^*(X_U|X_K, \mathcal{F}, \mathcal{A})$.

Phase 3: Sample sensitive variables from the noise joint conditional distribution $p^*(X_U|X_K, \mathcal{F}, \mathcal{A})$ generated in Phase 2 to generate synthetic sensitive variables $X_U{}^*$.

Phase 4: Calculate the joint conditional distribution of non-sensitive variables $p(X_K|X_U{}^*, \mathcal{F}, \mathcal{A})$ by executing belief propagation on factor graph $\mathcal{G}$ that has incorporated synthetic sensitive variables $X_U{}^*$ generated in Phase 3.

Phase 5: Sample non-sensitive variables from the joint conditional distribution $p(X_K|X_U{}^*, \mathcal{F}, \mathcal{A})$ calculated in Phase 4 to generate synthetic sensitive variables $X_K{}^*$.

Finally, the target family releases the synthetic genomic data X^*, $X^* = X_K{}^* \cup X_U{}^*$. In short, our method is to use synthetic SNPs and traits X^* of individuals to approximate the real SNPs and traits X. Relatively, the joint conditional distribution computation (Phase 1 and Phase 4) and sampling (Phase 3 and Phase 5) are straightforward. However, constructing ϵ-differential privacy algorithm in Phase 2 is non-trivial. In the following subsections, we detail these phases and prove our method satisfies ϵ-differential privacy.

3.2   Generation of Joint Conditional Distribution

According to belief propagation, the joint conditional distribution $p(X_U|X_K, \mathcal{F}, \mathcal{A})$ in Phase 1 and $p(X_K|X_U{}^*, \mathcal{F}, \mathcal{A})$ in Phase 4 in prior section can be transformed into the products of several local functions and each function supports a subset of variables. For example, $p(X_U|X_K, \mathcal{F}, \mathcal{A})$ can be factorized into

$$p(X_U|X_K, \mathcal{F}, \mathcal{A}) = \frac{1}{Z} \prod_{i \in S} \prod_{j \in T} f_i(s_i^C, s_i^F, s_i^M, \mathcal{F}) g_{ij}(s_i, t_j, \mathcal{A}) \qquad (1)$$

where Z ia a constant normalization factor. Note that $f_i(s_i^C, s_i^F, s_i^M, \mathcal{F}) \propto p(s_i^C|s_i^F, s_i^M, \mathcal{F})$, which can be obtained from public statistics, and $g_{ij}(s_i, t_j, \mathcal{A}) \propto p(s_i, t_j)$, which can be obtained from SNP/trait association reported by GWAS catalog. $p(X_K|X_U{}^*, \mathcal{F}, \mathcal{A})$ can be factorized similarly.

3.3   Generation of Noise Joint Conditional Distribution

Given the joint conditional distribution $p(X_U|X_K, \mathcal{F}, \mathcal{A})$, to construct approximate distribution $p^*(X_U|X_K, \mathcal{F}, \mathcal{A})$, we need to inject ϵ-differential privacy noise to such $n \times r$ items, as shown in Eq. (1), where n is the number of SNPs of an individual and r is the number of traits. The calculation of $p(X_K|X_U{}^*, \mathcal{F}, \mathcal{A})$

based on $X_U{}^*$ in Phase 4 in Sect. 3.1, however, does not need any additional information from the original data; namely, the joint conditional distribution of X_K can be derived from $X_U{}^*$ directly.

To drive $p(X_U|X_K, \mathcal{F}, \mathcal{A})$ in an ϵ-differential privacy manner, for any $i \in S$, Laplace noise can be injected into $f_i(s_i^C, s_i^F, s_i^M, \mathcal{F})$ directly, with scale of $6n/m\epsilon$, which guarantees that the noise version of $f_i(s_i^C, s_i^F, s_i^M, \mathcal{F})$, represented as $f_i^*(s_i^C, s_i^F, s_i^M, \mathcal{F})$, satisfies $(\epsilon/2n)$-differential privacy since $f_i(s_i^C, s_i^F, s_i^M, \mathcal{F})$ has sensitivity $3/m$, where n is the number of SNPs of an individual and m is the number of individuals in the target family.

For any $j \in T$, Laplace noise is injected to $g_{ij}(s_i, t_j, \mathcal{A})$, with scale of $4r/m\epsilon$, which guarantees that the noise version of $g_{ij}(s_i, t_j, \mathcal{A})$, represented as $g_{ij}^*(s_i, t_j, \mathcal{A})$, satisfies $(\epsilon/2r)$-differential privacy since $f_i(s_i^C, s_i^F, s_i^M, \mathcal{F})$ has sensitivity $2/m$, where r is the number of traits.

3.4 Privacy Guarantee

According to composition property [6], our method satisfies ϵ-differential privacy. Specifically, noise injected to all family factor nodes and all trait factor nodes are $\epsilon/2$, respectively.

4 Related Works

Privacy preserving genomic data release has received much attention in recent years. [17] proposes methods for differential privacy preserving release of GWAS catalog statistics, including χ^2-statistics, minor allele frequency and p-values. [18] proposes a method sharing data with differential privacy manner by splitting original genomes in a top-down way, and then add noise to each block. [16] states that current methods for identifying high scoring SNPs with differential privacy guarantee have low accuracy and high computational complexity, so that the authors proposed a new neighbor distance definition for performing private GWAS. [13] proposes privacy preserving algorithms for supporting exploratory analysis, including the location of SNPs with strong association with specific disease, correlations among SNPs.

In addition to differential privacy, [15] states how to combat against the statistical analysis attack (Homer's attack [11]), by restricting data release scale. Existing works have shown that personal information is threaten by attackers that usually launch attacks by exploiting data correlations and effective privacy preserving methods have also been proposed, such as location, social attributes [5,7,8,10], or mobile wireless networks [9,19,20]. [4] releases certain number of most crucial SNPs. However, their method makes several unfeasible assumptions, such as taking only χ^2 into consideration, fixed individual size, and the attacker knows the number of SNPs to release as background knowledge. For example, [14] proposes privacy preserving algorithms in order for calculating the statistical information about SNPs involving number and location which significantly imply the association between SNPs and diseases. [12] proposes two privacy metrics, adversary incorrectness and uncertainty, to quantify the privacy loss due to inference attacks on released genomic data.

5   Conclusions

We have proposed a differential-privacy preserving kin-genomic data releasing method. Based on factor graph, which has been proven an effective model to incorporate high-dimensional data and multiple correlations among them, our method can factorize the joint conditional distribution of sensitive genomes into sets of local probability distributions by executing belief propagation on factor graph which captures the dependency relationship among family members, SNPs and traits due to family genetic relationship and SNP/trait associations. A key part of our method is that, to ensure differential privacy, noise can be directly injected into low-dimensional local distributions rather than huge-dimensional genomic data, which significantly improve data utility.

Acknowledgments. This work is partly supported by the National Science Foundation (NSF) of China under grant 61632010, 61602129.

References

1. https://www.23andme.com/
2. Disgenet - a database of gene-disease associations. http://www.disgenet.org/web/DisGeNET/menu
3. The NHGRI-EBI catalog of published genome-wide association studies. https://www.ebi.ac.uk/gwas/docs/about
4. Bhaskar, R., Laxman, S., Smith, A., Thakurta, A.: Discovering frequent patterns in sensitive data. In: Proceedings of the 16th ACM SIGKDD International Conference on Knowledge Discovery and Data Mining, KDD 2010, New York, NY, USA, pp. 503–512 (2010)
5. Cai, Z., He, Z., Guan, X., Li, Y.: Collective data-sanitization for preventing sensitive information inference attacks in social networks. IEEE Trans. Depend. Secur. Comput. **PP**(99), 1 (2016)
6. Dwork, C.: Differential privacy. In: Bugliesi, M., Preneel, B., Sassone, V., Wegener, I. (eds.) ICALP 2006. LNCS, vol. 4052, pp. 1–12. Springer, Heidelberg (2006). doi:10.1007/11787006_1
7. Han, M., Yan, M., Cai, Z., Li, Y.: An exploration of broader influence maximization in timeliness networks with opportunistic selection. J. Netw. Comput. Appl. **63**, 39–49 (2016)
8. Han, M., Yan, M., Cai, Z., Li, Y., Cai, X., Yu, J.: Influence maximization by probing partial communities in dynamic online social networks. In: Transactions on Emerging Telecommunications Technologies (2016)
9. He, Z., Cai, Z., Han, Q., Tong, W., Sun, L., Li, Y.: An energy efficient privacy-preserving content sharing scheme in mobile social networks. Pers. Ubiquit. Comput. **20**(5), 833–846 (2016)

10. He, Z., Cai, Z., Sun, Y., Li, Y., Cheng, X.: Customized privacy preserving for inherent data and latent data. Pers. Ubiquit. Comput. **21**(1), 43–54 (2017)
11. Homer, N., Szelinger, S., Redman, M., Duggan, D., Tembe, W., Muehling, J., Pearson, J.V., Stephan, D.A., Nelson, S.F., Craig, D.W.: Resolving individuals contributing traace amounts of DNA to highly complex mixtures using high-density SNP genotyping microarrays. PLoS Genet. **4**, e1000167 (2008)
12. Humbert, M., Ayday, E., Hubaux, J.P., Telenti, A.: Addressing the concerns of the lacks family: quantification of kin genomic privacy. In: Proceedings of the 2013 ACM SIGSAC Conference on Computer & Communications Security, pp. 1141–1152. ACM (2013)
13. Johnson, A., Shmatikov, V.: Privacy-preserving data exploration in genome-wide association studies. In: Proceedings of the 19th ACM SIGKDD International Conference on Knowledge Discovery and Data Mining, pp. 1079–1087. ACM (2013)
14. Johnson, A., Shmatikov, V.: Privacy-preserving data exploration in genome-wide association studies. In: Proceedings of the 19th ACM SIGKDD International Conference on Knowledge Discovery and Data Mining, KDD 2013, pp, 1079–1087. ACM, New York (2013)
15. Sankararaman, S., Obozinski, G., Jordan, M.I., Halperin, E.: Genomic privacy and limits of individual detection in a pool. Nat. Genet. **41**(9), 965–967 (2009)
16. Simmons, S., Berger, B.: Realizing privacy preserving genome-wide association studies. Bioinformatics **32**(9), 1293–1300 (2016)
17. Uhlerop, C., Slavković, A., Fienberg, S.E.: Privacy-preserving data sharing for genome-wide association studies. J. priv. Confid. **5**(1), 137 (2013)
18. Wang, S., Mohammed, N., Chen, R.: Differentially private genome data dissemination through top-down specialization. BMC Med. Inf. Decis. Mak. **14**(1), S2 (2014)
19. Zhang, L., Cai, Z., Wang, X.: Fakemask: a novel privacy preserving approach for smartphones. IEEE Trans. Netw. Serv. Manag. **13**(2), 335–348 (2016)
20. Zheng, X., Cai, Z., Li, J., Gao, H.: Location-privacy-aware review publication mechanism for local business service systems. In: The 36th Annual IEEE International Conference on Computer Communications (INFOCOM) (2017)

In Silico Simulation of Signal Cascades in Biomedical Networks Based on the Production Rule System

Sangwoo Kim[1] and Hojung Nam[2(✉)]

[1] Severance Biomedical Science Institute,
Yonsei University College of Medicine, 50 Yonsei-ro,
Seodaemun-gu, Seoul 03722, South Korea
swkim@yuhs.ac
[2] School of Electrical Engineering and Computer Science,
Gwangju Institute of Science and Technology,
123 Cheomdangwagi-ro, Buk-gu, Gwangju 61005, South Korea
hjnam@gist.ac.kr

Abstract. Inferring novel findings from known biological knowledge is one of the ultimate goals in systems biology. However, the observation of system-level responses to a given perturbation has not been thoroughly explored due to the lack of proper large-scale inference models. We developed a novel expert system that can be applied to conventional biological networks based on the production rule system which works by transforming networks into a knowledgebase. Testing on large-scale multi-level biomedical networks confirmed the applicability of our system and revealed that hundreds of molecules are affected by the cascades of given signals, thereby activating or repressing key pathways in a cell.

Keywords: Network simulation · Expert system · Production system

1 Introduction

One key application of biological system analysis is to predict the molecular- and system-level responses to a given perturbation; a typical form the question is "What happens if a protein A is activated?" This analysis is a good substitute for *in vivo* or *in vitro* knock-in/knock-out experiments. Based on the accumulated knowledge, every rule that is relevant to protein A is regarded to calculate the direct effects of the perturbation; for instance, A inhibits gene B and phosphorylates protein C. Subsequently, signal cascades from B and C are inspected to find further effects. The signals are propagated through the network edges, possibly triggering the same molecules repeatedly and finally reach a steady state. Once performed accurately, this simulation can be a much faster, less costly and more efficient way to test multiple hypotheses. More importantly, there are also many perturbations that cannot be tested in a living organism due to certain technical or ethical issues. Therefore, developing a reliable knowledgebase and predicting system-level changes are highly practical goals in current biomedical sciences, demanding faster and more efficient models for application.

© Springer International Publishing AG 2017
Z. Cai et al. (Eds.): ISBRA 2017, LNBI 10330, pp. 356–361, 2017.
DOI: 10.1007/978-3-319-59575-7_34

An expert system is an artificial intelligence system initially developed to enhance decision-making abilities when starting with a pile of knowledge. Expert systems consist of two major sub-systems: a knowledgebase and an inference engine. The expert system's knowledgebase stores facts and rules that form the basis of inference. The inference engine checks the current facts to determine if there are rules to be met when deducing novel facts. With a slight transformation, expert systems can be efficiently used to answer biomedical questions, providing several benefits. First, they can deal with a very large-scale knowledgebase and are relatively free from biological complexity. Unlike quantitative analysis, a genome-scale investigation can be easily undertaken, which is essential to track system-level behavior. Second, the final answer reflects a final steady state from the initial perturbation. Because biological systems are densely inter-connected, a change in a single gene or protein may propagate to the entire system. Thus, observing a few directly affected entities is insufficient and may lead to an inaccurate conclusion. Third, multiple and complex perturbations can be tested when setting initial facts differently and repeatedly. We can also set the direction of inference as either forward or backward to uncover the effects of a perturbation or to find factors that consequently cause a perturbation, respectively. Taken together, rule-based expert systems can be a good alternative to other network analysis models, especially for the monitoring of system-level effects.

In this study, we developed a production-based expert system that works on general biological networks. We initially constructed a knowledgebase in the form of an integrated network. Based on the knowledgebase, we designed an actionable production rule system to represent biological networks and implemented it with the Jess rule engine. We applied our system to the artificial switching on and off of the two major oncogenes of the Kirsten rat sarcoma viral oncogene (KRAS) and the epidermal growth factor receptor (EGFR). A more complex example of the administration of a targeted cancer drug, Lapatinib, on an in silico EGFR-mutated model is shown. Finally, we show the overall effects on multiple genes are mixtures of positive and negative signals whose intensity should be regarded by a qualitative manner.

2 Brief Methods

2.1 Production System for a Biomedical Network

A production system is a set of rules that confer the core behavior of an expert system. Given a set of facts, it is the rule that determines how to deal with the fact, how to react to incoming triggers, and how to modify the knowledge as the initial perturbation propagates.

Basic Definitions. A *fact* refers to a unit of knowledge in an expert system. Facts describe a single instance of a template, which is an abstract cast that defines the properties of objects. A *rule* is an "IF ... THEN ..." style of definition of a behavior in a given condition. The 'IF' part is located at the left-hand side (LHS) of a rule. Likewise, the 'THEN' part is called the right-hand side (RHS) and the contents are a set of actions. Based on the knowledgebase, each rule is tested as to whether its LHS is satisfied. Once there is a match, the corresponding RHS part is triggered to execute predefined actions. For such a case, we consider a rule as having been fired. Given a set of facts and rules, an inference engine continuously checks the current status of the system. When there are no more rules to be applied or when no fact has been changed, the engine stops and the system overall becomes stable again.

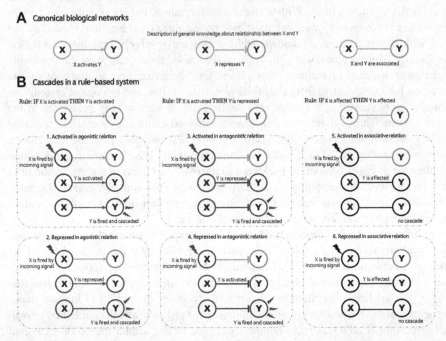

Fig. 1. Conceptual design of the production rule system for biological networks. A. Conventional node and edge view of networks. The relationship between two entities (here, 'X' and 'Y') describes the general state of the system. B. Rule-based view of the same relationship. Unlike the conventional view, the rule-based system is only triggered when a perturbation occurs. There are six possible signal cascades between two entities depending on the type of the incoming signal (triggered up or down) and the type of relationship

Design of an Expert System for Biomedical Inference. Designing a production rule system for biomedical inference requires a proper transformation of biological networks into rule-based knowledge. Notably, there is an essential discrepancy between the two forms (Fig. 1). In biological networks, an edge connecting two nodes describes a general status: a relationship between two molecules in a normal, unperturbed environment (Fig. 1A). Conversely, a rule in a rule-based system describes how a perturbed signal is transmitted. For example, "X activates Y" in a conventional biological network implies (normally) that "X activates Y," whereas it is interpreted as "IF X is perturbed to be activated (or repressed), THEN Y is also activated (or repressed)" in a rule-based system (Fig. 1B). To implement such a rule and its conditional propagation, two types of information are needed for each fact: one to determine whether the rule satisfies all of the conditions and is ready to be triggered and the other for describing the types of incoming signals (agonistic or antagonistic).

2.2   Implementing the Expert System

Based on this design, we implement an expert system that works on a large-scale biomedical network. The actual implementation, however, requires the management of the status of all facts and rules as well as the monitoring of rules which are ready to be fired. Because all rules must be re-evaluated for firing every time any changes arise with regard to facts, the overall calculation is considered infeasible, especially for a real-time simulation. In this study, we use the Jess expert system [1], which implements the Rete algorithm [2]. This system provides an efficient means of matching facts against rules in a pattern-matching production system to overcome the heavy computational burden.

3   Results and Discussions

We applied the developed expert system to a system-level simulation. First, two known oncogenes (KRAS and EGFR) were artificially activated and repressed to monitor the effects on the entire genome. Secondly, multiple perturbations were simulated by mimicking the administration of an anti-cancer drug to a pre-perturbed (EGFR activated) model. Finally, semi-quantitative traits were analyzed by measuring the number of triggering events for the affected gene set.

Oncogene Activation and Repression. The system-level response to the artificial switching on and off of KRAS is shown in Fig. 2A. The simulation shows that the up-regulation of KRAS affected 640 genes (189 activated, 422 inhibited and 29 neutral) due to the single initial perturbation; genes that are visualized as larger in size and with a denser color are strongly affected by the initial perturbation. Likewise, the down-regulation of KRAS affected 572 genes (Fig. 2A right, 245 activated, 283 inhibited and 44 neutral). We found that the JAK/STAT pathway is strongly up-regulated by KRAS activation (STAT1 is activated 13 times in the simulation), while only a partial increase of STAT1 activity is observed despite the continuous JAK2 activation in the KRAS repression model. STAT1 is activated only once, but no further stimulation is observed. The complex crosstalk between JAK/STAT and the receptor tyrosine kinase (RTK)/ Ras/MAPK signaling pathways depicts the stochastic nature of the system-level response to the single perturbation. Notably, a well-known mechanism of the RTK pathway is the JAK-independent tyrosine phosphorylation of STATs [3].

   Similarly, the epidermal growth factor receptor (EGFR) gene was also tested in the same manner (Fig. 2B). EGFR is a well-known receptor tyrosine kinase and an oncogene. The aberrant activation of EGFR is frequently shown in multiple types of cancer including lung cancer [4] and breast cancer [5], especially in the form of gene amplification and DNA mutation. Many cancer drugs, such as gefitinib, erlotinib and lapatinib, have been developed to target the mutant form of EGFR. In our simulation of EGFR activation, 544 genes are affected (382 activated, 134 inhibited and 28 neutral). We found that many other oncogenes were activated as well, including MAPK1, MAPK3 (both activated 32 times), SRC (26 times) and PIK3CA (22 times). When down-regulated, however, most of the previously activated genes were dramatically repressed, including

A

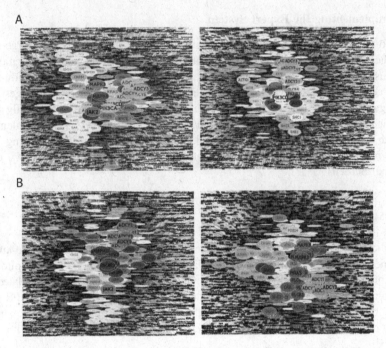

B

Fig. 2. Responses to the switching on and off of KRAS and EGFR oncogenes. A. KRAS was artificially either up-regulated (left) or down-regulated (right). When up-regulated, STAT1 is positively affected (green color), implying activation of the JAK/STAT pathway. When down-regulated, JAK-STAT was still activated but SRC and CRK were strongly down-regulated (red color). B. EGFR activation (left) leads to the up-regulation of the ADCY and MAPK gene families. The MAPK pathway is closely related to cancer and is frequently targeted as a potential therapeutic marker. Switching off EGFR (right) leads to the repression of many oncogenes, including MAPK genes, HRAS and SHC1 (Src homology 2 domain containing a protein). (Color figure online)

PIK3CA (inhibited 48 times), MAPK1/3 (42 times), HRAS (36 times), KRAS (35 times) and NRAS (35 times). The simulation results hold a clue as to how the targeted therapies on a single EGFR gene could be successful compared to the down-regulation effects of KRAS, which is widely known as an undruggable oncogene [6].

4 Conclusion

We developed an expert system that can be used for simulating system-level cascades of a given signal. With the efficient algorithm and the applicability to complex biomedical networks, we expect our system can be a good substitute to other biomedical systems analysis methods when in vivo/in vitro experiments are not available to test multiple hypotheses.

Acknowledgement. This work was supported by the Bio-Synergy Research Project (NRF-2014M3A9C4066449) of the Ministry of Science, ICT and Future Planning through the National Research Foundation.

References

1. Friedman-Hill, E.J.: Jess, the Expert System Shell for the Java Platform (2002)
2. Forgy, C.L.: Rete: a fast algorithm for the many pattern/many object pattern match problem. Artif. Intell. **19**, 17–37 (1982)
3. Rawlings, J.S., Rosler, K.M., Harrison, D.A.: The JAK/STAT signaling pathway. J. Cell Sci. **117**, 1281–1283 (2004)
4. Bethune, G., Bethune, D., Ridgway, N., Xu, Z.: Epidermal growth factor receptor (EGFR) in lung cancer: an overview and update. J. Thorac. Dis. **2**, 48–51 (2010)
5. Bhargava, R., Gerald, W.L., Li, A.R., Pan, Q., Lal, P., Ladanyi, M., Chen, B.: EGFR gene amplification in breast cancer: correlation with epidermal growth factor receptor mRNA and protein expression and HER-2 status and absence of EGFR-activating mutations. Mod. Pathol. **18**, 1027–1033 (2005)
6. Gysin, S., Salt, M., Young, A., McCormick, F.: Therapeutic Strategies for Targeting Ras Proteins. Genes Cancer **2**, 359–372 (2011)

Using the Precision Medicine Analytical Method to Investigate the Impact of the Aerobic Exercise on the Hypertension for the Middle-Aged Women

Wei Zhou[1], Guangdi Liu[2], Jun Luo[1], Tingran Zhang[1(✉)], and Le Zhang[3(✉)]

[1] Key Laboratory of Physical Evaluation and Sports Performance Monitoring of the State Administration of Sports, College of Physical Education, Southwest University, Chongqing 400715, People's Republic of China
sharemix@swu.edu.cn
[2] Library of Chengdu University, Chengdu University, Chengdu, China
liuguangdi1103@126.com
[3] College of Computer and Information Science, Southwest University, Chongqing 400715, People's Republic of China
zhanglcq@swu.edu.cn

Abstract. The purpose of the study is to investigate why acute aerobic exercise will cause the difference of ambulatory blood pressure for middle-aged women at different time. (1) Methods: There are fifteen middle-aged women volunteered for the study. Each participant receives three experimental interventions: (1) a non-exercise control trail; (2) at 06:30 am and (3) 16:30 p.m. 30 min of aerobic exercise with the mean exercise intensity at 60% of heart rate reserve. The experimental order is random and each participant will wear an automated ABP device to monitor the ABP and heart rate changes for 24 h; (2) Results: The systolic blood pressure, diabolic blood pressure, mean arterial pressure after aerobic exercise in the afternoon as well as the daytime DBP and daytime MAP is significantly lower than the systolic blood pressure, DBP and MAP after aerobic exercise in the morning and non-exercise control trail. Especially, systolic blood pressure can continually reduce for 2 h after acute aerobic exercise in the afternoon; and (3) Conclusion: Aerobic exercise in the afternoon can inhibit the rise of 24 h-ABP and morning blood pressure significantly, which can decrease the incidence of cerebrovascular and cardiovascular events.

Keywords: Hypertension · Precision medicine · Systolic blood pressure · Ambulatory blood pressure

---

W. Zhou and G. Liu—Contributed equally to this work.

Electronic supplementary material The online version of this chapter (doi:10.1007/978-3-319-59575-7_35) contains supplementary material, which is available to authorized users.

© Springer International Publishing AG 2017
Z. Cai et al. (Eds.): ISBRA 2017, LNBI 10330, pp. 362–367, 2017.
DOI: 10.1007/978-3-319-59575-7_35

1 Introduction

It is well known that hypertension is a very common and serious disease [1]. According to the statistics of WHO [2], hypertension impacts about 1 billion people worldwide. For example, the death of about 8 million people results from hypertension induced chronic disease for each year. Thus, hypertension is considered to be an important and urgent public health problem of this world. Previously, Zanchetti and Weber consider [3] that only about 31% of patients being able to rely on antihypertensive drugs to control blood pressure. Suggested by the American College of Sports Medicine (ACSM) [4], regular exercise can prevent and treat high blood pressure. However, there are still 25%–35% of patients whose blood pressures have no decline after exercise training [5], although the basic exercise prescription for hypertension prevention and treatment has been established. For this reason, the purpose of this study is to understand the relationship between time of day for exercise and the degree of daily blood pressure, and then we can'improve the formulation of exercise prescriptions.

Currently, several hypertension scientists are doing pilot research work in this area. For example, Fairbrother et al. [6] reported that aerobic exercise in the morning is the most effective time to improve nighttime blood pressure and sleep quality, compared with noon and evening hours. However, it does not strictly specify the exercised time of the participants. Park et al. [7] reveal that if the people whose blood pressure will not fall down in the night have 30 min of moderate intensity exercise in the afternoon, their SBP decrease rate is significantly greater than the people whose blood pressure will fall down in the night. However, the experimental design lacks the simulated non-exercise control conditions. Jones et al. reveal [8] that if the people carry out the excise in the afternoon, their SBP, DBP and MAP decreasing rate are significantly greater than carrying out the excise in the morning. The study design employs casual blood pressure to monitor short-term changes in blood pressure after exercise. However, there are some problems with casual blood pressure measurement itself, such as terminal digit preference, the observation bias and white-coat hypertension, etc. To overcome the shortcomings of the previous research, this study propose the following three innovations: First, it strictly specify the length of the exercise intervention and the duration of the exercise intervention to reduce the impact of the participants' exercise intervention differences on the results; Second, it employs self-control model and the interval between two interventions are more than two days which greatly reduce the impact of the previous excise on the next excise interventions; Finally, 24-h ambulatory blood pressure (ABP) monitoring is used to increase the accuracy of the results.

Fifteen middle-aged women volunteered for the study. Participants completed three randomly assigned conditions: a non-exercise control trail, and 30 min of aerobic exercise at different time of day, with the mean exercise intensity at 60% of heart rate reserve. After all three sessions, participants underwent 24 h ABP and heart rate monitoring with an automated ABP device. A repeated measure analysis of variance (ANOVA) [9] was used to assess the effect of acute aerobic exercise varies with time of day on 24-h ABP and heart rate. The study demonstrates if the participants carry out the oxygen excise in the afternoon, their average 24 h SBP, DBP and MAP are significantly less than the participants carrying out the aerobic exercise and non-exercise in the morning. Moreover, if the participants do aerobic exercise in the afternoon, SBP can keep decreasing for 2 h.

2 Materials and Methods

This section is detailed in the supplementary file.

3 Results

3.1 Diurnal Rhythm Changes of 24 h-SBP, 24 h-DBP and 24 h-MAP After Aerobic Exercise

All the data in this study follow the normal distribution, demonstrated by Shapiro-wilk Test, Fig. 1 and Table 2 of the supplementary file show three experimental interventions: (a) a non-exercise control trail; (b) aerobic exercise in the afternoon; (c) aerobic exercise in the morning. For 06:00 SBP, 07:00 SBP, 08:00 SBP, 17:00 SBP, 18:00 SBP, case (c) is statistically less than case (a). For 17:00 SBP, 18:00 SBP, case (b) is statistically less than case (a). Figure 2 and Table 3 of the supplementary file show three experimental interventions: (a) a non-exercise control trail; (b) aerobic exercise in the afternoon; (c) aerobic exercise in the morning. For 06:00 DBP, 08:00 DBP, 16:00 DBP, case (c) is statistically less than case (a). For 14:00 DBP, case (b) is statistically less than case (a). Figure 3 and Table 4 of the supplementary file show three experimental interventions: (a) a non-exercise control trail; (b) aerobic exercise in the afternoon; (c) aerobic exercise in the morning. For 06:00 MAP, 07:00 MAP, 8:00 MAP, 9:00 MAP, 14:00 MAP, 17:00 MAP case (c) is statistically less than case (a).

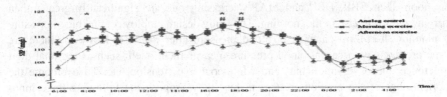

Fig. 1. 24 h systolic blood pressure change of different duration aerobic exercise. (a) * indicate simulation group (SG) and afternoon's exercise group (AEG) are significantly different (p < 0.05); (b) # indicate morning exercise group (MEG) and afternoon's exercise group (AEG) are significantly different (p < 0.05).

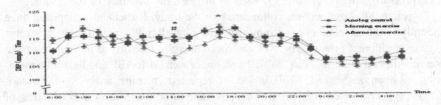

Fig. 2. 24 h diastolic blood pressure change of different duration aerobic exercise. (a) * indicate SG and AEG are significantly different (p < 0.05); (b) # indicate MEG and AEG are significantly different (p < 0.05).

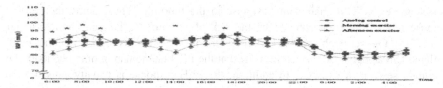

Fig. 3. Mean 24 h arterial blood pressure change of different duration aerobic exercise. (a) * indicate SG and AEG are significantly different (p < 0.05); (b) # indicate MEG and AEG are significantly different (p < 0.05).

4 Discussion and Conclusion

The study is to explore the mechanism why acute aerobic exercise will cause the difference of ambulatory blood pressure (ABP) for middle-aged women at different time. And then, we can improve the formulation of exercise prescriptions based on this explored mechanism.

Table 1 and 4 of the supplementary file and Figs. 1, 2 and 3 show three experimental interventions: (a) a non-exercise control trail; (b) aerobic exercise in the afternoon; (c) aerobic exercise in the morning. For average value of 24 h-SBP 24 h-DBP, 24 h-MAP,06:00 SBP, 07:00 SBP, 08:00 SBP, 17:00 SBP, 18:00 SBP, 06:00 DBP, 08:00 DBP, 16:00 DBP, 06:00 MAP, 07:00 MAP, 8:00 MAP, 9:00 MAP, 14:00 MAP, 17:00 MAP case (c) is statistically less than case (a) and (b). These results turns out those aerobic exercises can effectively decrease the blood pressure on day-time. We consider since the overall sympathetic keep high activity in the day time, it will greatly reduce the sympathetic nerve system activity after exercise. And then, it reduces peripheral vascular resistance and BP [13]. Thus, this result has potential clinical advantages in preventing CVD and hypertension. It is also consistent with the previous reports of the Pinto et al. [14] and Whelton et al. [15], which indicate that even blood pressure has slightly decreased, it can dramatically improve the health of participants as well as when the SBP drops 3–5 mmHg, it will reduce the incidence of coronary artery disease by 5–9% percent and the risk of myocardial infarction by 8–14% and all-cause mortality by 4–7%. Also, Fung et al. [16] demonstrate that exercise is a non-pharmacological treatment to decrease the blood pressure and reduce the risk factors of CVD. Furthermore, the proper exercise time can improve antihypertensive effect [17]. Tables 2 and 4 of the supplementary file show three experimental interventions: (a) a non-exercise control trail; (b) aerobic exercise in the afternoon; (c) aerobic exercise in the morning. For an hour after exercise SBP, two hours after exercise SBP, three hours after exercise SBP, 17:00 MAP, case (c) is statistically less than case (a). For eleven hour after exercise SBP, twelve hours after exercise SBP, three hours after exercise SBP, case (c) is statistically less than case (a) and (b). Those results turn out that the fall of SBP can last more than two hours after aerobic exercise in the afternoon, whereas it has no post-exercise hypotension (PEH) phenomenon after aerobic exercise in the morning. The results imply that the total peripheral resistance (TPR) is the main reason for the difference in blood pressure response caused by various time periods of exercise and doing exercise in the afternoon can decrease the TPR than doing exercise in the morning. Thus, doing the aerobic exercise in the afternoon can reduce the risk of cerebrovascular and cardiovascular events [18]. Our research results are consistent with the previous reports of Jones et al. [19] and Ciolac et al. [20] which indicate that the PEH phenomenon only occurs in the afternoon session of exercise. In addition, the PEH response is usually caused by a persistent decrease in vascular resistance. Hamer [21] demonstrate that the afternoon session of exercise can improve antihypertensive effect. Though this research turns out that aerobic exercises can effectively decrease the blood pressure on daytime and the fall of SBP can last more than two hours after aerobic exercise in the afternoon, but it still has several shortcomings. For example, it does not screen out the prehypertension patient, when we choose the candidates of the participant. Also, we do not develop a predictive model to prevent the occurrence of the hypertension. Therefore, our further study will develop such a prediction model that can investigate the relation between time of day for exercise and the degree of daily blood pressure. And then, we can improve the formulation of exercise prescriptions by using the model. Furthermore, it is important for us to locate the response of patients with different blood pressure levels (hypertension, prehypertension or normal blood pressure) after exercise accurately, as well as use the data as a reference for blood pressure prevention and treatment in the future.

References

1. Kjeldsen, S.E.: 2003 European society of hypertension-European society of cardiology guidelines for the management of arterial hypertension. J. Hypertens. **31**(28), 2159–2219 (2013)
2. Whitworth, J.A.: 2003 World Health Organization (WHO)/International Society of Hypertension (ISH) statement on management of hypertension. J. Hypertens. **21**(11), 1983–1992 (2003)
3. Zanchetti, A., Waeber, B.: Hypertension: which aspects of hypertension should we impact on and how? J. Hypertens. **24**, S2–S5 (2006)
4. Pescatello, L.S., Franklin, B.A., Fagard, R., Farquhar, W.B., Kelley, G.A., Ray, C.A.: American college of sports medicine position stand. Exercise and hypertension. Med. Sci. Sports Exerc. **36**(3), 533–553 (2004)
5. Alley, J.R., Mazzochi, J.W., Smith, C.J., Morris, D.M., Collier, S.R.: Effects of resistance exercise timing on sleep architecture and nocturnal blood pressure. J. Strength Cond. Res. **29**(5), 1378–1385 (2015)
6. Fairbrother, K., Cartner, B., Alley, J.R., Curry, C.D., Dickinson, D.L., Morris, D.M., Collier, S.R.: Effects of exercise timing on sleep architecture and nocturnal blood pressure in prehypertensives. Vascul. Health Risk Manag. **10**, 691–698 (2014)
7. Park, S., Jastremski, C., Wallace, J.: Time of day for exercise on blood pressure reduction in dipping and nondipping hypertension. J. Hum. Hypertens. **19**(8), 597–605 (2005)
8. Jones, H., Pritchard, C., George, K., Edwards, B., Atkinson, G.: The acute post-exercise response of blood pressure varies with time of day. Eur. J. Appl. Physiol. **104**(3), 481–489 (2008)

9. Cardoso Jr., C.G., Gomides, R.S., Queiroz, A.C.C., Pinto, L.G., Lobo, F.D.S., Tinucci, T., Mion Jr., D., Forjaz, C.L.D.M.: Acute and chronic effects of aerobic and resistance exercise on ambulatory blood pressure. Clinics **65**(3), 317–325 (2010)

10. Quinn, T.J.: Twenty-four hour, ambulatory blood pressure responses following acute exercise: impact of exercise intensity. J. Hum. Hypertens. **14**(9), 547–553 (2000)

11. Pescatello, L.S., Guidry, M.A., Blanchard, B.E., Kerr, A., Taylor, A.L., Johnson, A.N., Maresh, C.M., Rodriguez, N., Thompson, P.D.: Exercise intensity alters postexercise hypotension. J. Hypertens. **22**(10), 1881–1888 (2004)

12. Wang, Y., Wang, Q.J.: The prevalence of prehypertension and hypertension among US adults according to the new joint national committee guidelines: new challenges of the old problem. Arch. Intern. Med. **164**(19), 2126–2134 (2004)

13. Lehmkuhl, L.A.A., Park, S., Zakutansky, D., Jastremski, C.A., Wallace, J.P.: Reproducibility of postexercise ambulatory blood pressure in Stage I hypertension. J. Hum. Hypertens. **19**(8), 589–595 (2005)

14. Pinto, A., Di Raimondo, D., Tuttolomondo, A., Fernandez, P., Arna, V., Licata, G.: Twenty-four hour ambulatory blood pressure monitoring to evaluate effects on blood pressure of physical activity in hypertensive patients. Clin. J. Sport Med. **16**(3), 238–243 (2006)

15. Whelton, S.P., Chin, A., Xin, X., He, J.: Effect of aerobic exercise on blood pressure: a meta-analysis of randomized, controlled trials. Ann. Intern. Med. **136**(7), 493–503 (2002)

16. Fung, M.M., Peters, K., Redline, S., Ziegler, M.G., Ancoli-Israel, S., Barrett-Connor, E., Stone, K.L., Osteoporotic Fractures in Men Research Group: Decreased slow wave sleep increases risk of developing hypertension in elderly men. Hypertension **58**(4), 596–603 (2011)

17. Pescatello, L.S., Kulikowich, J.M.: The aftereffects of dynamic exercise on ambulatory blood pressure. Med. Sci. Sports Exerc. **33**(11), 1855–1861 (2001)

18. Giles, T.D.: Circadian rhythm of blood pressure and the relation to cardiovascular events. J. Hypertens. **24**, S11–S16 (2006)

19. Jones, H., George, K., Edwards, B., Atkinson, G.: Effects of time of day on post-exercise blood pressure: circadian or sleep-related influences? Chronobiol. Int. **25**(6), 987–998 (2008)

20. Ciolac, E.G., Guimarães, G.V., Bortolotto, L.A., Doria, E.L., Bocchi, E.A.: Acute aerobic exercise reduces 24-h ambulatory blood pressure levels in long-term-treated hypertensive patients. Clinics **63**(6), 753–758 (2008)

21. Hamer, M.: The anti-hypertensive effects of exercise. Sports Med. **36**(2), 109–116 (2006)

Understanding Protein-Protein Interface Formation Mechanism in a New Probability Way at Amino Acid Level

Yongxiao Yang and Xinqi Gong[✉]

Institute for Mathematical Sciences, Renmin University of China,
Beijing 100872, China
{yongxiaoyang, xinqigong}@ruc.edu.cn

Abstract. Although many studies about near native protein-protein interface recognition have been done in the past thirty years, the formation mechanism of protein-protein interface is still ambiguous. Here, we propose a new probability way to understand protein-protein interface formation mechanism at amino acid level. The probability of two surface residues from different monomers as a true interface residue pair in the complex is estimated by their geometric and physicochemical properties in the structures of protein monomers. The residue pairs with different probabilities combine together to form a protein-protein interface. The probabilities of residue pairs on candidate interfaces are integrated for near native interface recognition. Five simple probability based discriminants are constructed based on the distances and contact areas between residues. The performances are comparable to the ones of the sophisticated methods developed previously. The idea proposed in this work will make positive influence on the future study of protein-protein interactions.

Keywords: Protein-protein interface formation mechanism · Probability way · Amino acid level · Geometric features · Neural network

1 Introduction

Protein-protein interactions play an important role in many cellular processes. Interacting protein monomers can be associated in protein-protein complexes by the interfaces. The structures of protein-protein complexes are the key to understand the mechanism behind the cellular processes. The principle of interface formation is the core to obtain the structures of protein-protein complexes from their known components [1].

The ways to understand the interface formation mechanism vary with the perspectives. From the geometric view, Fischer and Koshland proposed the key-lock theory and the induced fit theory for protein-ligand interaction in 1890 and 1958 respectively [2–4]. From the physical perspective, the native interface is the one with the lowest binding free energy. Additionally, different knowledge-based potentials were developed and employed to calculate the interaction potential on the interface of protein-protein complex [5]. From the view of data mining, the principle of interface

© Springer International Publishing AG 2017
Z. Cai et al. (Eds.): ISBRA 2017, LNBI 10330, pp. 368–372, 2017.
DOI: 10.1007/978-3-319-59575-7_36

formation can be extracted from the known structures of protein-protein complexes determined by biologists.

Despite of the remarkable endeavor to elucidate the formation mechanism of protein-protein interface, there is still no perfect or universal theories which can be used to predict all the protein-protein complex correctly. Protein-protein interfaces are composed of many different interface residue pairs. These residue pairs make different contributions to the interface stability. The residue pairs are usually transformed from the surface residues of different protein monomers. The surface residues may make interface residues pairs with different probabilities. If the probabilities can be estimated based on the properties of the surface residues, the interface stability could be also assessed by integrating these probabilities.

In this work, nine features are adopted to characterize the surface residues. The performances of different feature combinations on interface residue pair prediction are investigated by neural networks to look for the best ones. The predicted values are used to evaluate the possibilities of the surface residues as true interface residue pairs. Five mathematical expressions are constructed by integrating the predicted values and used to assess the candidate interfaces generated by docking algorithm.

2   Materials and Methods

2.1   Datasets and Features

In protein-protein docking benchmark version 5.0 [6], there are 67 dimers satisfy our request at unbound state. These dimers are divided into three subsets (training, validation and test sets) according to the version of benchmark. Five geometric and four physicochemical descriptors are used to characterize the surface residues of protein monomer. The five geometric features are absEA (absolute Exterior solvent accessible Area), relEA (relative Exterior solvent accessible Area), EC (Exterior Contact area with other residues), EV (Exterior Void area, which don't contact the other residues and water molecules), and IC (Interior Contact area between the atoms of surface residue). The four physicochemical features are H1 (Hydropathy index 1), H2 (Hydropathy index 1), pK_a1 (standard) and pK_a2 (computation) [7–9]. Additionally, we use *ftdock* [10] to generate 10000 candidate interfaces using the monomer structures at unbound state for every dimer respectively. There are 30, 20, and 12 dimers for which near native interface are generated by docking in the training, validation, test set respectively.

2.2   Neural Network Models and Discriminants

In order to explore the performances of different combinations of the nine features, we trained 5621000 pattern recognition neural networks [11]. 45667 models were selected to test the new discriminants for near native protein-protein interface recognition according to the performances for interface residue pair prediction.

The five discriminants are constructed as follows:

$$D1 = \sum_{A_{ij} \neq 0} p_{ij} \tag{1}$$

$$D2 = \sum (A_{ij} \cdot p_{ij}) \tag{2}$$

$$D3 = \sum p_{ij} \tag{3}$$

$$D4 = \sum \frac{p_{ij}}{r_{ij}} \tag{4}$$

$$D5 = \sum \frac{p_{ij}}{r_{ij}^2} \tag{5}$$

where A_{ij} is the contact area between the ith and jth interface residues from the receptor and ligand respectively; p_{ij} is the predicted value of them as a true interface residue pair which is calculated using the trained neural network models and the features of the two residues; r_{ij} is the distance between this interface residue pair, it is represented by the minimum distance between any two atoms from the ith and jth interface residues respectively.

3 Results

3.1 The Overall Results of the Five Discriminants

When the mean percentages of positive dimers (those have at least one near native interfaces among the retained candidate interfaces) in the three sets are equal or greater than 5, 10, 15, 20, 25, 30, 35, 40, 45, and 50 respectively, and the standard deviations of the percentages are equal or less than 5, the minimum numbers of retained interfaces are recorded. This minimum number (n) of retained candidate interfaces (MNRCI) is recorded when the mean percentage is equal or greater than p.

Table 1. The overall results of the five discriminants

Discriminant	MNRCI (p)[a]									
	5	10	15	20	25	30	35	40	45	50
D1	1	2	5	7	10	22	28	46	55	–
D2	1	1	4	7	9	15	20	65	65	–
D3	2	7	11	16	26	40	44	63	78	–
D4	1	3	7	9	15	25	29	37	46	–
D5	1	3	6	9	11	25	25	46	52	99

[a]MNRCI (p): Minimum Number of Retained Candidate Interfaces when the mean percentage of positive dimers in the three datasets is equal or greater than p.

As shown in Table 1, when the mean percentage is less than 40, D2 is the best discriminants; when the mean percentage is equal and greater than 40, D4 is the best one. The reason may be that D2 considers the contact areas which focus on "short range interaction" and D4 considers the distances which incorporate "short and long range interaction".

3.2   The Best Discriminants for Near Native Interface Recognition

The discriminants are constructed and not trained. The near native interface recognition capability can be evaluated by the number of positive dimers when the number of retained candidate interfaces is N. The performances of the best discriminants and ZDOCK3.0.2 [12] are compared in the whole set.

As shown in Table 2, the result of D2 is a little worse than the one of ZDOCK3.0.2 when the top 10 candidate interfaces are retained. In consideration of the discriminants at amino acid level and the scoring items adopted by ZDOCK3.0.2 at atom level [12], it could be accepted. D2 represents the effective interface area of the candidate interface. The features adopted in the model are EC, EV, IC and H1. The three different areas could reflect the flexibility of an amino acid; H1 reflects the hydrophobic ability of amino acid, what is more, it is the only one feature that can discriminate different amino acid types. The probability of a surface residue pair as a true interface one is calculated based on the flexibility and hydrophobic ability of the two surface residues. This is a new way for understanding the mechanism of residue-residue interaction. The effective interface area may be taken as a criterion for judging the stability of an interface. D1 is the sum of probabilities of strictly contacting interface residue pairs in the decoys as true interface residue pairs in the experimental structure. The features used here are relEA, EC, IC, H1, pK_a1 and pK_a2, which indicates that the electrostatics is not ignorable when the high percentage of positive complexes is obtained. The result of D1 is comparable to the one of ZDOCK3.0.2.

Table 2. The comparison of the results of the best discriminants, ZDOCK3.0.2

Method	Features	NRCI[a]	NPD[b]
ZDOCK3.0.2	Shape complementarity, electrostatics, knowledge-based pair potentials	10	25
		100	39
D2	EC, EV, IC, H1	10	20
D1	relEA, EC, IC, H1, pKa1, pKa2	100	38

[a]NRCI: Number of Retained Candidate Interfaces; [b]NPC: Number of Positive Dimers.

4   Conclusions

In this work, we propose a new way to understand the protein-protein interface formation mechanism at amino acid level. The surface residues are described by nine simple features. The features reflect the flexibility, hydrophobic ability, amino acid type and electrostatics of the surface residues. They are used to estimate the possibilities of surface residue pairs as true interface ones. The estimated values are integrated to evaluate the interfaces and discriminate the near native interfaces from non-near native ones. The effective interface area and the possibility of the candidate interface as a near native one are estimated by D2 and D1 respectively. The results of the two best models can be bracketed to the ones reported before. More effective descriptors, more accurate estimated values and more powerful integrative ways will give better results. The results will give some new perspective for near native protein-protein interface recognition. The way to understand the interface formation mechanism of protein-protein interaction will make positive influences on the future research.

Acknowledgments. This research was supported by National Natural Science Fundation of China (31670725), and State Key Laboratory of Membrane Biology to Xinqi Gong. Experiments run on Renda Xing Cloud that currently has 64 physical nodes.

References

1. Xue, L.C., Dobbs, D., Bonvin, A.M., Honavar, V.: Computational prediction of protein interfaces: a review of data driven methods. FEBS Lett. **589**(23), 3516–3526 (2015). doi:10.1016/j.febslet.2015.10.003
2. Fischer, E.: Ueber die optischen Isomeren des Traubenzuckers, der Gluconsäure und der Zuckersäure. Berichte der deutschen chemischen Gesellschaft **23**(2), 2611–2624 (1890). doi:10.1002/cber.189002302157
3. Fischer, E.: Einfluss der configuration auf die Wirkung der Enzyme. Berichte der deutschen chemischen Gesellschaft **27**(3), 2985–2993 (1894). doi:10.1002/cber.18940270364
4. Koshland, D.E.: Application of a theory of enzyme specificity to protein synthesis. Proc. Natl. Acad. Sci. U.S.A. **44**(2), 98–104 (1958). doi:dx.doi.org/10.1073/pnas.44.2.98
5. Moont, G., Gabb, H.A., Sternberg, M.J.: Use of pair potentials across protein interfaces in screening predicted docked complexes. Proteins **35**(3), 364–373 (1999). doi:10.1002/(SICI)1097-0134(19990515)35:3<364::AID-PROT11>3.0.CO;2-4
6. Vreven, T., Moal, I.H., Vangone, A., Pierce, B.G., Kastritis, P.L., Torchala, M., Chaleil, R., Jimenez-Garcia, B., Bates, P.A., Fernandez-Recio, J., Bonvin, A.M., Weng, Z.: Updates to the integrated protein-protein interaction benchmarks: docking benchmark version 5 and affinity benchmark version 2. J. Mol. Biol. **427**(19), 3031–3041 (2015). doi:10.1016/j.jmb.2015.07.016
7. Kyte, J., Doolittle, R.F.: A simple method for displaying the hydropathic character of a protein. J. Mol. Biol. **157**(1), 105–132 (1982). doi:10.1016/0022-2836(82)90515-0
8. Eisenberg, D.: Three-dimensional structure of membrane and surface proteins. Annu. Rev. Biochem. **53**, 595–623 (1984). doi:10.1146/annurev.bi.53.070184.003115
9. Olsson, M.H., Sondergaard, C.R., Rostkowski, M., Jensen, J.H.: PROPKA3: consistent treatment of internal and surface residues in empirical pKa predictions. J. Chem. Theory Comput. **7**(2), 525–537 (2011). doi:10.1021/ct100578z
10. Gabb, H.A., Jackson, R.M., Sternberg, M.J.: Modelling protein docking using shape complementarity, electrostatics and biochemical information. J. Mol. Biol. **272**(1), 106–120 (1997). doi:10.1006/jmbi.1997.1203
11. Kishore, R., Kaur, M.T.: Backpropagation algorithm: an artificial neural network approach for pattern recognition. Int. J. Sci. Eng. Res. **3**(6), 1–4 (2012)
12. Pierce, B.G., Hourai, Y., Weng, Z.: Accelerating protein docking in ZDOCK using an advanced 3D convolution library. PLOS ONE **6**(9), e24657 (2011). doi:10.1371/journal.pone.0024657

Detecting Potential Adverse Drug Reactions Using Association Rules and Embedding Models

Kai Guo, Hongfei Lin[✉], Bo Xu, Zhihao Yang, Jian Wang,
Yuanyuan Sun, and Kan Xu

School of Computer Science and Technology, Dalian University of Technology,
Dalian, China
hflin@dlut.edu.cn

Abstract. Adverse drug reactions (ADRs) may occur following a single dose or prolonged administration of a drug or result from the combination of two or more drugs. Given the restrictions of the traditional methods like clinical trials, it's difficult to detect the ADRs in a timely manner. Many countries have built spontaneous adverse drug event reporting systems, which provide a large amount of adverse drug event reports for research purpose. In this paper, we utilize the association rule mining to reconstruct the data from adverse drug event reports, and apply modified embedding models to calculate the relevance of the drug and adverse reactions to detect potential ADRs. We examine the effectiveness of methods by conducting experiments on two drugs: Gadoversetamide and Rofecoxib, finding 6 potential drug reactions, which can be further verified by biomedical data.

Keywords: Adverse drug reactions · Embedding model · Association rules

1 Introduction

Adverse drug reactions (ADRs) have been one of the most important reasons which harm the public health. Given the restrictions of the clinical trials before drugs sold to the public, it cannot find all the ADRs for drugs. Adverse Drug Event Systems have been established to collect adverse drug event reports and supply important data for the researchers. This way bears substantial significance to break the limit of time and money cost, and accelerates the progress of adverse drug reactions mining.

Most of the researches about adverse effect events are based on electronic medical records and adverse effect event reports. For example, Yang et al. [1] applied the association rules to mining the adverse drug reactions from spontaneous reporting system. Harpaz et al. [2] mined the adverse effect between drugs on the FDA data by association rules. Kuo et al. [3] adopted Apriori association analysis algorithm for the detection of adverse drug reactions in health care data.

In this work, we propose a novel method to detect ADRs using association rules and embedding models, which have been prove effective in many other text mining tasks. In our method, association rules are used to measure the similarity between drug mentions

Z. Cai et al. (Eds.): ISBRA 2017, LNBI 10330, pp. 373–378, 2017.
DOI: 10.1007/978-3-319-59575-7_37

and potential adverse reactions for reconstructing the data, and modified embedding models are used to representation terms of the mentions and reactions, which could capture much semantic and syntactic information for producing a good performance. Experimental results show that our method is effective to detect ADRs for drugs.

2 Method

2.1 Pre-processing

OpenFDA [4] provides the adverse drug event reports, which contain the drug and other medical information. We clean the data by dropping the non-alphabetic characters like *+&/ and removing the words in bracket to improve the accuracy of the experiment.

MetaMap can map biomedical text to the UMLS Metathesaurus and discover Metathesaurus concepts referred to in text. We apply this tool to get the standard format of the drug names. MetaMap can recognize 133 kinds of semantic types, but only 22 kinds of them are drug-related.

Due to the linguistic features or other errors, the same entity may have a variety of expression, for example "rash skin", "spots" and "exanthemas" are match to "rash". We utilize the database of SIDER (Side Effect Resource) [5] as the dictionary to recognize the side reactions, and expand ADRs terms rely on CHV (The Consumer Health Vocabulary) [6], Data Reconstruction based on Association Rules.

2.2 Data Reconstruction Based on Association Rules

An adverse reaction report may contain multiple drugs with a variety of ADRs, so one-to-one relationships between the drug and adverse reaction tend to be uncertain. Therefore, we propose to refine the adverse reactions using association rules.

Let $X = \{x_1, x_2, x_3...x_m\}$ be a set of items, association rules can be defined as $A \Rightarrow B$, and $A \subset X, B \subset X, A \cap B =$, where both A and B are the subset of X, and there're two basic measures used in associations mining: support and confidence. The support and confidence of Drug-A $\Rightarrow$ Adverse-B can be defined as the frequency of the items appears in the dataset:

$$support(A \Rightarrow B) = P(A \cup B) = \frac{count(AUB)}{totalcount} \qquad (1)$$

$$confidence(A \Rightarrow B) = \frac{support(AUB)}{support(A)} = \frac{count(AUB)}{count(A)} \qquad (2)$$

count (AUB) is the number of threads that contain Drug-A and Adverse-B, the total count is the total number of threads in the dataset, and support (A) is number of Drug-A in the whole dataset, so support measures the probability of the threads. We filter the drug and ADRs entity with the value of support lower than 100, side reaction with the value of support lower than 50. According to the support and confidence, we can measure the importance of the result of association rules, and process the data in accordance with the threshold value.

According to the results of the process above, each report has been divided into small adverse reaction reports, and each of them is made up of one drug A and adverse reaction B. For example, 3 kinds of drugs and 4 kinds of adverse reactions can be divided to 12 kinds of relations. We can calculate the value of support and confidence for every relationship, the part of results shown in Table 1.

We reconstruct the dataset for the next step of experiment according to the confidence value of drug and adverse reaction, and filters the tuple which has the value of confidence lower than 0.1, if the threshold is too low, the noise data can't be dropped completely, and the higher threshold will filter some valuable data.

And the number of tuples is proportional to their value of confidence, this method of data building greatly expands range of adverse reactions mining, breaks the limit of the ordinary way which only goals to the certain kinds of drugs or adverse reactions.

Table 1. The confidence value of drug and adverse reaction

Drug, adverse reaction	Confidence
Gadoversetamide, nephrogenic systemic fibrosis	0.8377
Trasylol, anxiety	0.7192
Propoxyphene napsylate, arrhythmia	0.6223
Ivermectin, asthenia	0.5261
Iletin, blood glucose increased	0.5074

2.3   Detecting Potential ADRs Using Modified Embedding Model

It's difficult to reveal the potential relations between drugs and adverse using the traditional methods. As the open source tools released by Google, word2vec [7] has been used widely in many fields as an efficient method for learning high quality distributed vector representations that capture a large number of precise syntactic and semantic word relationships. In this paper, word2vec is applied in the field of potential adverse drug reaction mining to identify potential unknown relationship to expand the scope of recognition.

We regard the tuple of one drug and one adverse reaction as an item set, and training the distributed vector on it without the information of sequence. Before the training, every entity is represented as a vector generated randomly, and the model tunes the vector to change the inner product of the two vectors depending on the co-occurrence relationship, and then generates the distributed vector at last. It was found that if an entity occurs only in a certain kind or several kinds of drugs, there are few opportunities for tuning this vector, and leads to a bad performance. So we filter this kind of adverse reaction to approve the result. The objective function used in this paper is:

$$max \sum_{r_m \in R} \sum_{d_i \in V_m} \sum_{d_i \neq d_j} \log p(d_i | d_j) \tag{3}$$

the conditional probability and the relevance of drug d_i and d_j is defined as

$$p(d_i|d_j) = \frac{\exp\left(v_{d_i}^T \cdot v_{d_j}\right)}{\sum_{d \in V_m} \exp(v_d^T \cdot v_{d_i})} \tag{4}$$

$$\text{relevance } (d_i, d_j) = \cos(d_i, d_j) = \frac{v_{d_i}^T \cdot v_{d_j}}{||v_{d_i}|| \cdot ||v_{d_j}||} \tag{5}$$

3   Result

To examine the effectiveness of the method proposed in this paper, we conduct experiments on two kinds of drugs: Gadoversetamide and Rofecoxib, and confirmed the detected ADRs by biomedical literatures. Since our method is general, it can also be applied to detect ADRs for other kinds of drugs.

3.1   Result Based on Gadoversetamide

Gadoversetamide is a gadolinium-based MRI contrast agent, particularly for imaging of the brain, spine and liver. We detect top-10 poteintial adverse reactions of Gadoversetamide based on the proposed method, and report the results in Table 2, where the column "SIDER"indicates whether the corresponding adverse reacions have been recorded in SIDER database.

Table 2. Gadoversetamide's potential adverse reactions

Adverse reaction	Relevance	SIDER	Adverse reaction	Relevance	SIDER
Nausea	0.9717	Yes	Nephrogenic system fibrosis	0.8921	Yes
Rash	0.9696	Yes	Diarrhea	0.8911	No
Stress	0.9472	No	Atrial fibrillation	0.8593	No
Vomiting	0.9326	Yes	Abdominal pain	0.8378	Yes
Pneumonia	0.9286	Yes	Postural dizziness	0.8114	No

Stress refers to a sense of tension and mental stress in patients which is a common adverse reaction. According to the result calculated by this paper, Gadoversetamide may have a strong correlation with the Stress, and this result also been confirmed by a report published in the eHealthMe, this report studied 2,590 people who have side effects while taking Gadoversetamide, and 1,182 of them have Stress (Tension), especially for people who are female, 50–59 old.

Diarrhea as another potential adverse reaction mined by this paper, also confirmed in related biomedical report. Broome [8] points out that Nephrogenic Systemic Fibrosis is a common adverse reactions associated with Diarrhea, and it has been verified that Gadoversetamide has a strong correlation with Nephrogenic System Fibrosis. We can confirmed that Diarrhea is the potential adverse reaction of Gadoversetamide.

As the result, the method proposed in this paper mined 6 kinds of ADRs which have been recorded in the SIDER, which can prove the effectiveness of this method for adverse reaction mining. More importantly, 2 kinds of ADRs which are not recorded in SIDER have been proved as potential adverse reactions.

3.2 Result Based on Rofecoxib

Rofecoxib is a nonsteroidal antiinflammatory drug (NSAID) was marketed by Merck & Co. to treat osteoarthritis and acute pain conditions,. We detect top-10 poteintial adverse reactions of Rofecoxib based on our method, and report the results in Table 3.

Acute Renal Failure also called acute kidney injury, generally it occurs because of damage to the kidney tissue. Previous study in [9] indicates that Celecoxib and Rofecoxib have a strong correlation with Renal Failure through a detail experimental study.

Table 3. Rofecoxib'spotential adverse reactions

Adverse reactions	Relevance	SIDER	Adverse reactions	Relevance	SIDER
Acute renal failure	0.7239	No	Mental depression	0.7019	No
Nausea	0.7188	Yes	Pneumonia	0.6952	Yes
Rash	0.7064	Yes	Atrial fibrillation	0.6822	Yes
Vomiting	0.7063	Yes	Atrial dilatation	0.6632	No
Peripheral edema	0.7054	Yes	Oral pain	0.6587	No

Atrial Dilatation refers to enlargement of the heart. According to the research, Atrial Dilatation is a common complication of Atrial Fibrillation, and Atrial Fibrillation is an adverse reaction of Refecoxib, so we can indicate that Atrial Dilatation is a potential sider reaction of Refecoxib, this conclusion also been confirmed by the biomedical literatures. Campbell et al. [10] in the study of patient with Acute Congestive Heart Failure caused by Rofecoxib, indicated that this patient also has the symptoms of Atrial Dilatation. Therefore, we can determine Atrial Dilatation is one of the potential adverse reaction of Refecoxib.

Oral Pain is a common symptom, and we can know that Oral Lesion is an adverse reaction of Refecoxib. Because the Oral Pain is a common complication of Oral Lesion, we can confirm that Oral Pain is a potential adverse reaction.

In conclusion, among the 10 kinds of potential adverse reactions mined by the paper, 6 of them have been recorded by the SIDER, and the other 3 of them can be confirmed by the biomedical literatures, while remaining one of them have not been confirmed temporary, still need be analyzed in the future work. As a result, we can indicate that the combing of Apriori and distributed vector can effectively identify potential unknown relationship to expand the scope of recognition, and have good performance in the field potential adverse reaction mining, but still has room for improvement.

4   Conclusion and Future Work

In the paper, we propose to detect potential adverse drug reactions using assioation rules and embeding models. In the method, we utilize biomedical tools to extract and clean the data from adverse events reports, reconstruct the data based on the association rules, and build the trainning data with the weights of each sample. Then we train the a modified word2vec model based on biomedical data and calculate the relevance between the vectors of drugs and reactions to effectivly mine the potential adverse reactions. Experimental results show our method can effectively detect potential adverse reactions.

Acknowledgements. This work is partially supported by grant from the Natural Science Foundation of China (Nos. 61572102, 61402075, 61602078, 61562080), the Fundamental Research Funds for the Central Universities the National Key Research Development Program of China (No. 2016YFB1001103).

References

1. Yang, W., Xie, Y.M., Xiang, Y.Y.: Apply association rules to analysis adverse drug reactions of shuxuening injection based on spontaneous reporting system data. Zhongguo Zhong yao za zhi = Zhongguo zhongyao zazhi = China J. Chin. Materia Med. **39**(18), 3616–3620 (2014)
2. Harpaz, R., Haerian, K., Chase, H.S., et al.: Statistical mining of potential drug interaction adverse effects in FDA's spontaneous reporting system. AMIA. Ann. Symp. Proc. **2010**(7), 281–285 (2010)
3. Kuo, M.H., Kushniruk, A.W., Borycki, E.M., et al.: Application of the Apriori algorithm for adverse drug reaction detection. Stud. Health Technol. Inf. **148**(148), 95–101 (2009)
4. Kass-Hout, T.: OpenFDA: innovative initiative opens door to wealth of FDA's publicly available data. Food Drug Adm. (2014). https://blogs.fda.gov/fdavoice/index.php/2014/06/openfda-innovative-initiative-opens-door-to-wealth-of-fdas-publicly-available-data/?source=govdelivery&utm_medium=email&utm_source=govdelivery#sthash.WcQ6vf0U.dpuf
5. Kuhn, M., Campillos, M., Letunic, I., et al.: A side effect resource to capture phenotypic effects of drugs. Mol. Syst. Biol. **6**(1), 343 (2010)
6. Zeng, Q.T., Tse, T.: Exploring and developing consumer health vocabularies. J. Am. Med. Inform. Assoc. **13**(1), 24–29 (2006)
7. Mikolov, T., Sutskever, I., Chen, K., et al.: Distributed representations of words and phrases and their compositionality. Adv. Neural. Inf. Process. Syst. **26**, 3111–3119 (2013)
8. Broome, D.R.: Nephrogenic systemic fibrosis associated with gadolinium based contrast agents: A summary of the medical literature reporting. Eur. J. Radiol. **66**(2), 230–234 (2008)
9. Ahmad, S.R., Kortepeter, C., Brinker, A., et al.: Renal failure associated with the use of celecoxib and rofecoxib. Drug Saf. **25**(7), 537–544 (2002)
10. Campbell, R.J., Sneed, K.B.: Acute congestive heart failure induced by rofecoxib. J. Am. Board Fam. Pract. **17**(2), 131–135 (2004)

Genome-Wide Analysis of Response Regulator Genes in *Solanum lycopersicum*

Jun Cui[1], Ning Jiang[1], Jun Meng[2(✉)], and Yushi Luan[1(✉)]

[1] School of Life Science and Biotechnology, Dalian University of Technology,
Dalian, China
luanyush@dlut.edu.cn
[2] School of Computer Science and Technology,
Dalian University of Technology, Dalian, China
mengjun@dlut.edu.cn

Abstract. Using whole genome - wide analysis, we identified 40 response regulator (RR) genes in *Solanum lycopersicum*. They can be divided into 7 subgroups according the structure characteristics and the sequence similarity and topology. The analyses of gene structure, protein motif, chromosome distribution, gene duplication and comparative phylogenetic analysis are performed in detail. The transcription levels of SlRRs in biotic stresses are further analyzed to obtain the functions information of these genes. Furthermore, qRT - PCR analysis shows 11 SlRRs which may be involved in tomato - *Phytophthora infestans* interaction, played different roles between resistant and sensitive tomato. Our systematic analyses provide insights into the characterization of SlRRs in tomato and basis for further functional studies of these genes.

Keywords: Genome-wide analysis · Tomato · Response regulator · Gene expression · *Phytophthora infestans*

1 Introduction

The two-component system (TCS), also known as the histidyl-aspartyl (His-Asp) phosphorelay systems is involved in cytokinin signal transduction and plays important roles in various biological process [1]. A simple TCS involves a histidine (His) sensor kinase and a response regulator (RR) [2]. Studies have demonstrated that plants RRs are involved in various stresses. For example, RRs are also involved in drought or water stress as demonstrated in *Arabidopsis* and soybean [3, 4]. *Arabidopsis* RR2 can be interacted with the salicylic acid response factor TGA3, and the expression of *PR1* was induced, which resisted to *Pseudomonas syringae* pv. tomato DC3000 [5].

Solanum lycopersicum is either a major crop plant or, a model system for plant - pathogen interaction, whose genome project was initiated by the Tomato Genome Consortium in 2012 [6]. Tomato genome also is important and useful resource to study

This work is supported in part by the National Science Foundation of China under Grant Nos. 31471880 and 61472061.

Z. Cai et al. (Eds.): ISBRA 2017, LNBI 10330, pp. 379–384, 2017.
DOI: 10.1007/978-3-319-59575-7_38

the evolution and tomato- pathogen interaction. Although quite a few studies have been reported on RRs in the model and crop plants [4, 7, 8], the reports on tomato RR (SlRR) are extremely rare and their functions remain unclear. To gain the characterization of the SlRR family and their functions on response to *Phytophhora infestans* stress, whole genome - wide analysis was conducted to identify the SlRRs. The detailed analyses of the gene structure, protein motif, genomic distribution and comparative phylogenetic analysis were performed. 11 SlRRs might be involved in tomato-*P. infestans* interaction and played different roles in between resistant and sensitive tomato.

2   Materials and Methods

2.1   Identification of SlRRs in Tomato

32 *Arabidopsis thaliana* RR (ARR) protein sequences were downloaded from TAIR (http://www.arabidopsis.org/) and used as queries to search against the Tomato Genome SL 2.40 by BLAST. The candidate SlRRs were identified with an e-value cutoff of 1e-5. The reciprocal BLASTP searches were conducted by using the candidate SlRRs as queries to verify the veracity of candidate proteins. The putative SlRRs were examined for the receiver domain using PROSITE (http://prosite.expasy.org/). The theoretical isoelectric point and molecular weight of SlRRs were identified by Prot-Param tool (http://web.expasy.org/protparam/).

2.2   Chromosomal Location, Sequence Analysis and Phylogenetic Analysis

The *SlRRs* were located on tomato chromosome according to their positions given in the database with the Mapchart software [9]. The gene structures were generated based on the information of cDNA sequences, genomic sequences and intron/exon distri-bution patterns. Protein motifs of SlRRs were identified using MEME with an e-value $\leq$ 1e-10 [10]. To compare the evolutionary relationship and identify the subgroup, the maximum likehood (ML) tree was constructed using MEGA 6.0 [11].

2.3   Expression Profiles of SlRR Genes in Tomato

The expression profiles were determined by analyzing the RNA-Seq data. The RNA-Seq data came from our previous works including resistant tomato (*S. pimpinellifolium L3708*), sensitive tomato (*S. lycopersicum Zaofen No. 2*) and tomato challenged with *P. infestans*. Others were from tomato challenged with *P. syringae* [12] and tomato yellow leaf curl virus [13]. The calculation of fold change was per-formed according to the method of Cui et al. [14].

2.4   Plant Materials, Treatment and qRT - PCR Analysis

Inoculated tomatoes (*S. pimpinellifolium L3708* and *S. lycopersicum Zaofen No. 2*) with *Phytophthora infestans* were performed according to the method of Cui et al. [14]. The extraction of total RNA, synthetization and qRT - PCR were performed using RNAiso Plus (TaKaRa, China) TransStart®Top Green qPCR SuperMix kit (Trans, China), respectively.

3   Results

3.1   The SlRR Gene Family in Tomato

A total of 40 SlRR sequences were obtained from *S. lycopersicum* genome, which were identified to contain receiver domain. The 40 *SlRR* genes were subsequently renamed from *SlRR0* to *SlRR39* according to their order on the chromosomes, respectively. The ORF lengths of *SlRR* ranged from 426 bp to 2193 bp, encoding peptides ranges from 141 to 730 amino acids. Their molecular weights ranged from 15.8 kDa to 79.8 kDa and the isoelectric points ranged from 4.83 to 9.06.

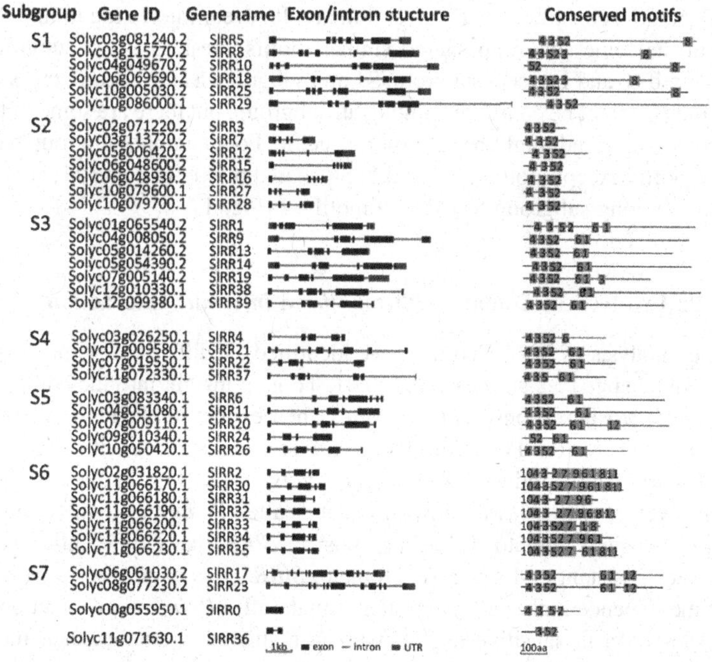

Fig. 1. The subgroup, intron pattern and conserved protein motifs of SlRRs.

3.2   Chromosomal Distribution and Phylogenetic Analysis of SlRR Gene Family

Genome chromosomal location analyses revealed that *SlRRs* were unevenly distributed across all 12 chromosomes, except SlRR0. Chromosome 11 had the largest number (8) of *SlRRs*. Based on structure characteristics and sequence similarity and topology, we subdivided the 40 SlRRs into 7 subgroups (S1–S7) except for 2 members (SlRR0 and SlRR36) using ML method (Fig. 1). Analysis of the evolutionary relationship between tomato and *Arabidopsis* RRs revealed that there was not equal representation. S6, SlRR0 and SlRR36 were not clustered together with ARRs. While, subgroup S2 and S4 fit into the same clade with A - type ARRs, subgroup S3 with B - type ARRs, and subgroup S1, S5 and S7 with pseudo - type ARRs.

3.3   Gene Structure and Motif Composition of SlRRs

The exon-intron organizations of 40 *SlRRs* were identified based on their number and distribution. A detail illusion of exon-intron structures is shown in Fig. 1. The coding sequences were disrupted by introns, and the intron numbers ranged from 0 to 11. *SlRR17* and *SlRR23* each contained a maximum of 11 introns and *SlRR0* had no intron.

12 conserved motifs were identified in SlRRs using the MEME tool (Fig. 1). The motifs 4, 3, 5 and 2 constitute the receiver domain. It is very interesting to find that most members in the same subgroup shared similar motifs. Besides, most subgroups also shared the motifs 6 and 1 except subgroup S1 and subgroup S2. In a word, this suggested that these motifs might be conserved among all subgroup, but there are some of the other motifs were variable and might be subgroup-specific. For example, subgroup S1 did not shared any motifs except motif 4, 3, 5and 2; S6 shared motif 10, 7 and 11; subgroup S7 and SlRR20 (among subgroup S5) shared motif 12 (Fig. 1).

3.4   SlRRs Involved in Tomato - *Phytophthora infestans* Interaction

With further analysis of RNA-Seq data, we identified 11 differential expression *SlRRs* based on | foldchange | >2 and p value < 0.01. To gain insight into the comprehensive roles of these *SlRRs* in response to *P. infestans*, their expression levels were detected at the indicated times by qRT-PCR from *P infestan*- resistant and - sensitive tomatoes. All *SlRRs* was significantly induced after *P. infestans* stress (Fig. 2). For example, the expression level of *SlRR31* was down-regulated after *P. infestans* infection in both resistant and sensitive tomato. However, most of *SlRRs* have differential expression trends between resistant and sensitive - tomato. *SlRR5* was expressed down-regulated with the time-dependent change in resistant tomato after *P. infestans* infection. While, its expression level in sensitive tomato was up-regulated gradually, then moderately down-regulated.

4 Discussion

4.1 RRs in Tomato Genome

In this study, the members of subgroup S2 was divided into the same clade with A - type RRs. - type *RRs*, *ARR4* and *ARR5* were also found to be induced by drought, salt, and low temperature [3], suggesting A - type RRs should be as a molecular link between stress and cytokinin signaling. We found that subgroup S3 belonged to B - type ARRs. In previous studies, B-type ARRs showed their nuclear localization in *Arabidopsis*, which was an indication that they are transcriptional factors [7]. Subgroup S1 and S5 belonged pseudo-type RRs by protein structure and comparative phylogenetic analysis. The previous research showed that APRRs might be related to *Arabidopsis* circadian clock because expressions of several *APRRs* were controlled by circadian rhythm [7].

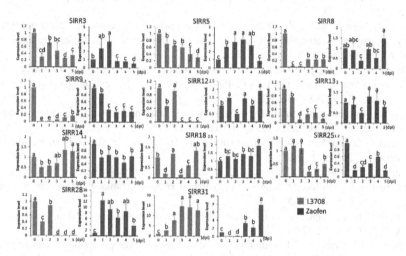

Fig. 2. Expression patterns of *SlRRs* involved in response to *P. infestans* attack in tomato. X - Axis represents different time point; Y - axis represents corresponding relative expression (n = 3 per each time point). Tomato *actin* expression was used as a control.

4.2 Expression Profiling of RR Genes Under *P. infestans* Infection

In *Arabidopsis*, drought significantly induced the expression of A-type *ARRs*, *ARR5*, *ARR7*, and *ARR15*, whereas almost all A - type *RRs* in rice genes were suppressed by drought stress [15]. Compared with intensive studies on the functions of RR in abiotic stresses, studies on biotic stresses were very limited. In plants, the cytokinin could be changed after some pathogen infection, which induced the differential expression of RR genes because cytokinin signal transduction was performed though two-component system involved RRs [5]. Meanwhile, in *Arabidopsis-P. syringae* interaction, the quantity of cytokinin-activated transcription factor ARR2 was up-regulated and the salicylic acid response factor TGA3 specifically interacted with ARR2 and recruited it to the PR1 promoter, which induced expression of *PR1*. These resisted to

P. syringae [5]. In this study, 11 *SlRRs* were identified as differential expression *SlRRs* after further analysis of these RNA-Seq data. The qRT-PCR results indicated that the expression levels of these SlRRs were changed after *P. infestans* infection and these expression showed a time-dependent response. Notably, most of them had difference overall expression trends between resistant and sensitive tomato (Fig. 2).

References

1. Hutchison, C.E., Kieber, J.J.: Cytokinin signaling in *Arabidopsis*. Plant Cell **14**, S47–S59 (2002)
2. West, A.H., Stock, A.M.: Histidine kinases and response regulator proteins in two-component signaling systems. Trends Biochem. Sci. **26**, 369–376 (2001)
3. Urao, T., Yakubov, B., Yamaguchi-Shinozaki, K., et al.: Stress-responsive expression of genes for two component response regulator-like proteins in *Arabidopsis thaliana*. FEBS Lett. **427**, 175–178 (1998)
4. Le, D.T., Nishiyama, R., Watanabe, Y., et al.: Genome-wide expression profiling of soybean two-component system genes in soybean root and shoot tissues under dehydration stress. DNA Res. **18**, 17–29 (2011)
5. Choi, J., Huh, S.U., Kojima, M., et al.: The cytokinin-activated transcription factor ARR2 promotes plant immunity via TGA3/NPR1-dependent salicylic acid signaling in *Arabidopsis*. Dev. Cell **19**, 284–295 (2010)
6. Tomato Genome Consortium: The tomato genome sequence provides insights into fleshy fruit evolution. Nature **485**, 635–641 (2012)
7. Hwang, I., Sheen, J.: Two-component circuitry in *Arabidopsis* cytokinin signal transduction. Nature **413**, 383–389 (2001)
8. Gahlaut, V., Mathur, S., Dhariwal, R., et al.: A multi-step phosphorelay two-component system impacts on tolerance against dehydration stress in common wheat. Funct. Integr. Genomics **14**, 707–716 (2014)
9. Voorrips, R.E.: MapChart: software for the graphical presentation of linkage maps and QTLs. J. Hered. **93**, 77–78 (2002)
10. Bailey, T.L., Gribskov, M.: Combining evidence using p-values: application to sequence homology searches. Bioinformatics **4**, 48–54 (1998)
11. Tamura, K., Stecher, G., Peterson, D., et al.: MEGA6: molecular evolutionary genetics analysis version 6.0. Mol. Biol. Evol. **30**, 2725–2729 (2013)
12. Yang, Y.X., Wang, M.M., Yin, Y.L., et al.: RNA-seq analysis reveals the role of red light in resistance against *Pseudomonas syringae* pv. tomato DC3000 in tomato plants. BMC Genom. **16**, 120 (2015)
13. Chen, T., Lv, Y., Zhao, T., et al.: Comparative transcriptome profiling of a resistant vs. susceptible tomato (*Solanum lycopersicum*) cultivar in response to infection by tomato yellow leaf curl virus. PLoS ONE **8**, e80816 (2013)
14. Cui, J., Luan, Y., Jiang, N., et al.: Comparative transcriptome analysis between resistant and susceptible tomato allows the identification of lncRNA16397 conferring resistance to Phytophthora infestans by co-expressing glutaredoxin. Plant J. **89**, 577–589 (2016)
15. Kang, N.Y., Cho, C., Kim, N.Y., et al.: Cytokinin receptor-dependent and receptor-independent pathways in the dehydration response of *Arabidopsis thaliana*. J. Plant Physiol. **169**, 1382–1391 (2012)

A Fully Automatic Geometric Parameters Determining Method for Electron Tomography

Yu Chen[1,2], Zihao Wang[1,2], Lun Li[1,3], Xiaohua Wan[1], Fei Sun[2,4,5], and Fa Zhang[1(✉)]

[1] Key Lab of Intelligent Information Processing and Advanced Computing Research Lab, Institute of Computing Technology, Chinese Academy of Sciences, Beijing, China
zhangfa@ict.ac.cn
[2] University of Chinese Academy of Sciences, Beijing, China
[3] School of Mathematical Sciences, University of Chinese Academy of Sciences, Beijing, China
[4] National Key Laboratory of Biomacromolecules, CAS Center for Excellence in Biomacromolecules, Institute of Biophysics, Chinese Academy of Sciences, Beijing 100101, China
[5] Center for Biological Imaging, Institute of Biophysics, Chinese Academy of Sciences, Beijing 100101, China

Abstract. Electron tomography (ET) is a promising technique for investigating in situ three-dimensional (3D) structure of proteins and protein complexes. To obtain a high-resolution 3D ET reconstruction, alignment and geometric parameters determination of ET tilt series are necessary. However, the common geometric parameters determining methods depend on human intervention, which are not only fairly subjective and easily introduce errors but also labor intensive for high-throughput tomographic reconstructions. To overcome these problems, in this paper, we presented a fully automatic geometric parameters determining method. Taking advantage of the high-contrast reprojections of ICON and a series of image processing and edge recognition techniques, our method achieves a high-precision full automation for geometric parameters determining. Experimental results on the resin embedded dataset show that our method has a high accuracy comparable to the common 'manual positioning' method.

Keywords: Electron tomography · Geometric parameters determination · Human intervention · Full automation · Comparable accuracy

1 Introduction

Electron tomography (ET) is a promising technique for investigating in situ three-dimensional (3D) structure of proteins and protein complexes [1,2]. In ET, a series of two-dimensional (2D) projection micrographs (tilt series) are taken in different orientations and then used to reconstruct the 3D density of the

© Springer International Publishing AG 2017
Z. Cai et al. (Eds.): ISBRA 2017, LNBI 10330, pp. 385–389, 2017.
DOI: 10.1007/978-3-319-59575-7_39

ultrastructure based on the projection-slice theorem [3]. Usually, the mechanical instability, and inevitable transformation and deformations of the sample occur during data collection will affect the projection environment, leading to the mismatch of tilt series. To obtain a high-resolution 3D reconstruction, the projection parameters of the tilt series should be calibrated accurately first.

Recent years, the topic of alignment in ET has been widely discussed and many high-precision alignment algorithms have been proposed. A problem related to the alignment is the determination of geometric parameters, which describe the geometry of the 3D reconstruction with respect to a fixed coordinate system, including the direction of the tilt axis (azimuthal angle), the tilt angle offset, the thickness of sample and the z-shift of reconstruction [4]. The most common way to determine the geometric parameters depends on human intervention, refer to as 'manual positioning' in this paper, such as in IMOD (a successful and widely used ET tool) [4]. In manual positioning, three 2D reconstructed slices (from top, middle and bottom of the 3D reconstruction) will be selected first and then a boundary model, containing manually selecting position features of the 3D reconstruction, will be created accordingly and used to calculate the geometric parameters. Although manual positioning has a high accuracy, two key issues remain to be solved. Firstly, as high resolution sub-volume averaging demands on high-throughput tomographic reconstructions [2], the need of human intervention in manual positioning will become a bottleneck of high efficient automatic ET alignment and reconstruction; secondly, manual selecting position features is fairly subjective and easily introduces errors especially for Cryo-ET reconstruction, in which the extremely low SNR and the ray artifacts caused by missing wedge will make the position features hard to identify.

To overcome these problems, we propose a fully automatic ET geometric parameters determining method. Based on high contrast reprojections of ICON [5], our method has two advantages. Firstly, the reconstruction of ICON suffers less from ray artifacts and has a higher SNR. Taking advantage of the reprojections of ICON, position features selected by our method are much clearer, which is essential for high-precision fully automatic geometric parameters determination. Secondly, ICON can partially restore the unsampled information in ET reconstruction. Thus, our method can generate a clear 90-degree reprojection of the reconstructed volume (such 90-degree reprojection is normally too blurred to identified for traditional ET reconstruction algorithms), and then the azimuthal angle can be directly determined.

2   Method

2.1   Geometry

The geometry of ET reconstruction is defined as Fig. 1. The projection coordinate system (X, Y, Z) is fixed with respect to the microscope; Y-axis is the tilt axis and Z-axis is the optical axis. Volume V is the 3D density of the specimen with thickness T, reconstructed from the aligned tilt series. The alignment procedure will align the tilt axis of tilt series perpendicular to X-axis [6]. Thus, the reconstructed specimen coordinate system (X', Y', Z') can be defined relative to (X, Y, Z) by a rotation about X-axis by an angle θ_{az} (azimuthal angle) (Y'-axis is the actual tilt axis), a rotation about Y'-axis by an angle θ_{to} (tilt angle offset) and a shift d_z along Z-axis (z-shift). Usually, the azimuthal angle and the tilt angle offset are small for an aligned tilt series, so we approximately regard the tilt angle offset as a rotation about Y-axis. Here, we define a 90-degree reprojection as the reprojection obtained by rotating the reconstruction around Y-axis by 90° and define a X_{90}-reprojection as the reprojection obtained by rotating the reconstruction around X-axis by 90°. All 90-degree reprojections presented in this paper are rotated in-plane by −90° and only the central areas are displayed.

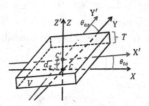

Fig. 1. The geometry of ET reconstruction.

2.2 Geometric Parameters Determination

There are four geometric parameters to be determinated, including z-shift, thickness, tilt angle offset and azimuthal angle.

We model the determination of z-shift and thickness as an optimization problem (see Eqs. (1) and (2)).

$$f(x, z, t) = \begin{cases} 1, & z - \lfloor \frac{t}{2} \rfloor <= x <= z + \lceil \frac{t}{2} \rceil \\ 0, & otherwise \end{cases} \tag{1}$$

$$\mathbf{max}_{z,t} corrcoef(reprojection_{1D}, f(z,t)) \quad \textbf{s.t.} \quad z, t < s \tag{2}$$

where $reprojection_{1D}$ is the 1D reprojection of a X_{90}-reprojection along X-axis; $corrcoef(reprojection_{1D}, f(z,t))$ calculates the normalized correlation coefficient (NCC) between $reprojection_{1D}$ and $f(z,t)$; s is the size of $reprojection_{1D}$; z and s are used to calculate z-shift and t is thickness.

To determine the tilt angle offset, we first generate a template according to the thickness and rotate it with a certain angular step to generate a series of templates, representing different tilt angles. And then we calculate the NCCs between the X_{90}-reprojection and templates. Instead of using the tilt angle with the maximum NCC as the tilt angle offset, we calculate the tilt angle offset using Eq. (3).

To determine the azimuthal angle, we use the 90-degree reprojection instead of the X_{90}-reprojection and the same procedure is used.

$$\theta_{to} = \sum_{t \in \Psi} \theta_t \cdot \frac{NCC_t}{\sum_{k \in \Psi} NCC_k}, \quad \Psi = \{z | NCC_z > mean(NCC)\} \tag{3}$$

where NCC_t is the NCC of tilt angle θ_t; θ_{to} is the tilt angle offset; $mean(NCC)$ is the mean value of all $NCCs$.

3   Results and Discussion

We tested our method using the resin embedded ET dataset downloaded from IMODs website [4]. The geometric parameters determination module TOMO-PITCH in IMOD is used to illustrate the performance of our method.

The boundary model used in TOMOPITCH are shown in Fig. 2(A–C) and the reprojections used in our method are shown in Fig. 2(D, E). Although the sample is slightly bent, the correction of reprojections by our method is good (Fig. 2(F, G)), which demonstrates the robustness of our method to the deformation of sample. And the absolute differences between TOMOPITCH and our method are small (Table 1) which demonstrates the comparable accuracy of our method to TOMOPITCH.

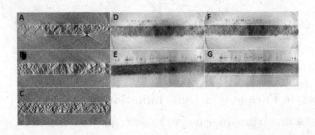

Fig. 2. Test our method using resin embedded ET dataset. (A–C) the boundary model used in TOMOPITCH; (D) X_{90}-reprojection used in our method; (E) 90-reprojection used in our method; (F) corrected X_{90}-reprojection by our method; (G) corrected 90-reprojection by our method.

Table 1. The geometric parameters of resin embedded ET dataset determined using TOMOPITCH and our method.

	θ_{az}	θ_{to}	Thickness	z-shift
TOMOPITCH	2.39	0.17	64	0.8
Our method	2.66	0.5	66	2
Abs (difference)	0.27	0.33	2	1.2

For visual validation, we reconstructed the tilt series using WBP and then corrected the reconstruction using the geometric parameters from TOMOPITCH and our method, respectively. Figure 3(A) shows the $281th$ XY-slice ($256th$ XY-slice is the central slice) of uncorrected tomogram, because of the azimuthal angle, the bottom part of slice is out of the sample. Figure 3(B–C) show the same XY-slices of corrected tomograms, both corrections generate flat reconstructions and our method is visually identical with TOMOPITCH.

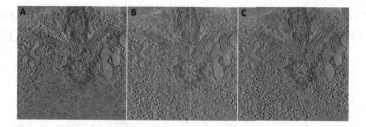

Fig. 3. The corrected reconstruction of TOMOPITCH and our method. (A) the 281*th* XY-slice of uncorrected reconstruction; (B) the 281*th* XY-slice of corrected reconstruction by TOMOPITCH; (C) the 281*th* XY-slice of corrected reconstruction by our method.

4 Conclusion

We proposed a fully automatic ET geometric parameters determining method to solve two key issues of the common manual positioning method by introducing ICON reprojections and a series of image process techniques. The experimental results demonstrate that our method has a high accuracy, which is comparable to TOMOPITCH.

Acknowledgments. This research is supported by the NSFC projects Grant Nos. U1611263, U1611261, 61232001, 61472397, 61502455, 61672493 and Special Program for Applied Research on Super Computation of the NSFC-Guangdong Joint Fund (the second phase), the Strategic Priority Research Program of Chinese Academy of Sciences (Grant No. XDB08030202), the National Basic Research Program (973 Program) of Ministry of Science and Technology of China (2014CB910700).

References

1. Asano, S., Engel, B.D., Baumeister, W.: In situ cryo-electron tomography: a post-reductionist approach to structural biology. J. Mol. Biol. **428**(2), 332–343 (2015)
2. Briggs, J.A.: Structural biology in situ - the potential of subtomogram averaging. Curr. Opin. Struct. Biol. **23**(2), 261–267 (2013)
3. Mersereau, R.M., Oppenheim, A.V.: Digital reconstruction of multidimensional signals from their projections. Proc. IEEE **62**(10), 1319–1338 (1972)
4. Kremer, J.R., Mastronarde, D.N., Mcintosh, J.R.: Computer visualization of three-dimensional image data using IMOD. J. Struct. Biol. **116**(1), 71–76 (1996)
5. Deng, Y., Chen, Y., Zhang, Y., Wang, S., Zhang, F., Sun, F.: ICON: 3D reconstruction with 'missing-information' restoration in biological electron tomography. J. Struct. Biol. **195**(1), 100 (2016)
6. Renmin, H., Liansan, W., Zhiyong, L., Fei, S., Fa, Z.: A novel fully automatic scheme for fiducial marker-based alignment in electron tomography. J. Struct. Biol. **192**(3), 403–417 (2015)

Evaluating the Impact of Encoding Schemes on Deep Auto-Encoders for DNA Annotation

Ning Yu[1]([✉]), Zeng Yu[2], Feng Gu[3], and Yi Pan[4]

[1] Department of Informatics, University of South Carolina Upstate,
800 University Way, Spartanburg, SC 29303, USA
nyu@uscupstate.edu
[2] School of Information Science and Technology, Southwest Jiaotong University,
Chengdu 611756, Sichuan, China
yuzeng2005@163.com
[3] Department of Computer Science, College of Staten Island, 2800 Victory Blvd.,
Staten Island, NY 10314, USA
Feng.Gu@csi.cuny.edu
[4] Department of Computer Science, Georgia State University, 25 Park Place,
Atlanta, GA 30303, USA
yipan@cs.gsu.edu
http://www.uscupstate.edu

Abstract. Deep Neural Networks show their promise over traditional neural network on DNA genomic analysis. However, due to the uncertainty of DNA sequence data, it performs differently in various encoding schemes. In this article we focus on the comparison of different schemes on various auto-encoder algorithms in DNA annotation and analyze their impacts on deep learning. We also aim to find the best encoding schemes used on deep auto-encoder algorithms for DNA annotation.

Keywords: Deep neural network · Auto-encoders · Encoding schemes · DNA genomic analysis

1 Introduction

Data representation is an important component in current deep learning research, such as natural language processing, language translation and genome analysis [1]. Letter/character based representation of genome sequence is readable and understandable for human being [2,3] but a problem for a machine, especially for numeric machine learning. We can easily quantify man-made quantitative factors such as voltage, current, pixel and coordinate but we indeed have some problems in quantifying human biological particles such as DNA nucleotides. Although many encoding schemes have been developed for bioinformatics, encoding those DNA nucleotides on deep learning has not been discussed. We cannot avoid the numeric representations for those biological units. Inappropriate encoding schemes can directly lead to numeric bias and signal loss. Data

© Springer International Publishing AG 2017
Z. Cai et al. (Eds.): ISBRA 2017, LNBI 10330, pp. 390–395, 2017.
DOI: 10.1007/978-3-319-59575-7_40

representation problems of DNA sequences have emerged as an important issue when the deep learning technology is highlighted.

In this article, we study the impacts of encoding schemes on several variant algorithms in auto-encoders by applying deep neural networks to gene annotation. Four variant auto-encoder algorithms include orthodox auto-encoder, denoising auto-encoder, hidden-layer denoising auto-encoder and double denoising auto-encoder. Nine typical encoding schemes are studied, typically representing the three categories. These encoding schemes are DAX [2], Arbitrary [2], EIIP [4], Neural [5], Complementary [5], Enthalpy [6], Entropy [7], Statistic [7], and Galois [8]. Basically, DAX, Arbitrary, Neural and Galois are binary linear code; EIIP, Enthalpy, Entropy and Statistic are bio-chemical mapping; Complementary is Cartesian coordinate coding. For DNA genome analysis in deep neural networks, direct mapping schemes such as DAX, EIIP and Complementary have the better performance than those pre-processed schemes such as Enthalpy, Entropy and Galois. It is perhaps because direct mapping does not wrap any information of DNA sequences while pre-processed schemes have hidden some information by encoding them together. Experiments show that Complementary can beat other schemes in more than half of cases and it is regarded as one of the best encoding schemes in genomic data representation.

The rest of the paper is organized as follows. Section 2 presents the details of auto-encoder and its variants. Section 3 provides the comparison and evaluation results on different encoding schemes and auto-encoder algorithms, and Sect. 4 gives the conclusion.

2 Auto-Encoder and Variants

Auto-encoder is an artificial neural network that can be used to constitute a multiple-layer percetron architectures for deep learning machine shown in Fig. 1(a). The hidden layer h and the iterative estimation x^* can be calculated through weights. The iteration becomes stable when it has the minimum distance between x and x^*. The preliminary ideas of shallow/deep neural network had been discussed for long time since 90s, however, mature concepts of deep learning including deep neural network were proposed in mid-2000s [9–11]. Since then, it has been applied to life sciences and shown tremendous promise [12–15].

The simplest auto-encoder is based on a feedforward, non-recurrent neural network similar to the multiple-layer perceptron (MLP). The difference is that the output layer of auto-encoder has the same number of nodes as the input layer and an auto-encoder is trained to reconstruct their own inputs instead of being trained to predict the output value. Thus, training the neighboring set of two layers minimizes the errors between layers and eliminates the problem of error propagation that often occurs in conventional neural network.

A denoising auto-encoder partially corrupts input data and uses the corrupted data for training in order to recover the original undistorted input. This technique can robustly obtain a corrupted input that will be useful for recovering the corresponding clean input. To train an auto-encoder for denoising data, it is necessary to perform preliminary stochastic mapping in order to corrupt the data and use as input for a normal autoencoder, with the only exception being that the loss should be still computed for the initial input instead of the corrupted one. Different from input-layer denoising, hidden-layer denoising auto-encoder model (HDAE) corrupts the units in hidden layer instead of input-layer and reconstructs the hidden layer.

Figure 1(b) illustrates the architecture of double denoising auto-encoder. An example x is stochastically corrupted to $\tilde{x}$. The auto-encoder then maps it to hidden representation h via encoding and attempts to reconstruct x via decoding, producing reconstruction x^*. Reconstruction error is measured by loss $L(x, x^*)$. Meanwhile, the hidden representation h is also stochastically corrupted to $\tilde{h}$ and then $\tilde{h}$ is mapped to an intermediate reconstructed input $\bar{x}$ via decoding and attempts to reconstruct h via encoding, producing reconstruction h^*. Reconstruction error is also measured by loss $L(h, h^*)$.

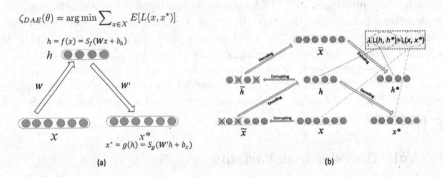

Fig. 1. Auto-encoders. (a) Auto-encoder. (b) Double denoising auto-encoder.

3   Comparison and Evaluation

The data sets are the standard benchmark from fruitfly.org for predicting gene splicing sites on human genome sequences [16]. The data set I is the Acceptor locations containing 6,877 sequences with 90 features. The data set II is the Donor locations including 6,246 sequences with 15 features. The Acceptor data sets have 70bp in the intron (ending with AG) and 20bp of the following exon. The Donor data sets have 7bp of the exon and 8bp of the following intron (starting with GT). The standard data sets contain real and fake splice sites and a window of upstream/downstream 40bp around the actual splice sites D (Donor) A (Acceptor). The data set of cleaned 269 genes is divided into a test and a training data set [16].

Figure 2(a), (b) shows the performance of an auto-encoder. Complementary scheme shows the superiority over other schemes in Fig. 2(a) where the data set has more features than those in Fig. 2(b). DAX scheme shows the best performance in Fig. 2(b).

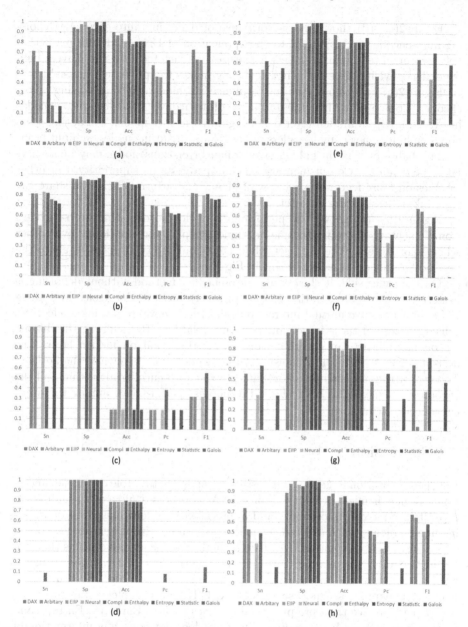

Fig. 2. Comparative results. (a) and (b) Auto-encoder on acceptor and donor data set; (c) and (d) denoising auto-encoder on acceptor and donor data set; (e) and (f) hidden-layer auto-encoder on acceptor and donor data set; (g) and (h) double denoising auto-encoder on acceptor and donor data set.

Figure 2(c), (d) shows the performance of denoising auto-encoder. Denoising auto-encoder seems not fit to the application of DNA structure prediction because corrupted input data (DNA features) at each location may have a high dependency with others such that denoising makes the prediction messed.

Figure 2(e), (f) shows the performance of hidden-layer denoising auto-encoder. Compared with the performance of input-layer denoising auto-encoder in Fig. 2(c), (d), in hidden-layer denoising auto-encoder model, corrupting some nodes on hidden layers makes a less impact than corrupting nodes on input layer. It is probably because in hidden layer some correlations/nodes may be so trivial to be denoised. Complementary scheme manifests its superiority over other schemes on more-feature data set while DAX and arbitrary schemes share the top rank on less-feature data set.

Figure 2(g), (h) shows the performance of double denoising auto-encoder. Complementary encoding scheme continues keeping its superiority over other schemes in large-feature data set while DAX and arbitrary schemes share the best performances on measurement in Fig. 2(h).

On the other side, it shows that auto-encoder method without denoising is better than other three variants. It is probably because the nucleotides along DNA sequence have mutual interactions and the removal of noising nucleotides causes the loss of these importantly mutual relations. Particularly, the denoising on input layer can generate more harms than that on hidden layer. The over-fitting issues occur in Denoising Auto-encoder more frequently than others. The over-fitting occurrence is correlated to the auto-encoder algorithm and the encoding schemes. The DAE shows the poorest performance among the auto-encoder algorithms.

4   Conclusion

Data representation in genome analysis plays an important role due to the uncertainty of bio-chemical properties along DNA sequences. Unlike other man-made quantitative factors, uncertainty and imprecision have long existed in quantifying human biology such as DNA sequences. We summarize the existing encoding schemes of DNA sequences and discuss several auto-encoder algorithms that can be applied in DNA genomic analysis. By experiments on DNA gene annotation, we compare and analyze those typical encoding schemes. Eventually, we find direct mapping schemes such as DAX, EIIP and Complementary have the better performance than pre-processed schemes such as Enthalpy, Entropy and Galois. It is perhaps because direct mapping does not wrap any information of DNA sequences while pre-processed schemes have hidden some information by encoding them together. Experiments also show that Complementary can beat other schemes in more than half of cases and it is regarded as one of the best encoding schemes in genomic data representation. The evaluation and assessment provide the important evidence to choose the proper data representation for using deep learning methods on DNA genome data analysis.

References

1. Wu, X., Cai, Z., Wan, X.-F., Hoang, T., Goebel, R., Lin, G.: Nucleotide composition string selection in HIV-1 subtyping using whole genomes. Bioinformatics 23(14), 1744–1752 (2007)
2. Yu, N., Guo, X., Gu, F., Pan, Y.: DNA AS X: an information-coding-based model to improve the sensitivity in comparative gene analysis. In: Harrison, R., Li, Y., Măndoiu, I. (eds.) ISBRA 2015. LNCS, vol. 9096, pp. 366–377. Springer, Cham (2015). doi:10.1007/978-3-319-19048-8_31
3. Wu, J., Wan, X.F., Xu, L., Lin, G., Cai, Z., Goebel, R.: Identifying a few foot-and-mouth disease virus signature nucleotide strings for computational genotyping. BMC Bioinform. 9, 279 (2008)
4. Nair, A., Sreenadhan, S.: A coding measure scheme employing electron-ion interaction pseudopotential (EIIP). Bioinformation 1(6), 197–202 (2006)
5. Arniker, S.B., Kwan, H.K., Law, N.F., Lun, D.P.K.: DNA numerical representation and neural network based human promoter prediction system. In: 2011 Annual IEEE India Conference, pp. 1–4, December 2011
6. Kauer, G., Blöcker, H.: Applying signal theory to the analysis of biomolecules. Bioinformatics 19(16), 2016–2021 (2003)
7. Jabbari, K., Bernardi, G.: Cytosine methylation and CpG, TpG (CpA) and TpA frequencies. Gene 26(333), 143–149 (2004)
8. Rosen, G.L.: Signal processing for bibiological-inspired gradient source localization and DNA sequence analysis. Ph.D. dissertation, Georgia Institute of Technology, School of Electrical and Computer Engineering, August 2006
9. Hinton, G., Dayan, P., Frey, B., Neal, R.: The "wake-sleep" algorithm for unsupervised neural networks. Science 268(5214), 1158–1161 (1995)
10. Hintonemail, G.E.: Learning multiple layers of representation. Trends Cogn. Sci. 11(10), 428–434 (2007)
11. Deng, L., Hinton, G., Kingsbury, B.: New types of deep neural network learning for speech recognition and related applications: an overview. In: 2013 IEEE International Conference on Acoustics, Speech and Signal Processing (ICASSP), pp. 8599–8603, May 2013
12. Bengio, Y., Courville, A., Vincent, P.: Representation learning: a review and new perspectives. IEEE Trans. Pattern Anal. Mach. Intell. 35(8), 1798–1828 (2013)
13. Di Lena, P., Nagata, K., Baldi, P.: Deep architectures for protein contact map prediction. Bioinformatics 28(19), 2449–2457 (2012)
14. Eickholt, J., Cheng, J.: Predicting protein residueresidue contacts using deep networks and boosting. Bioinformatics 28(23), 3066–3072 (2012)
15. Leung, M.K.K., Xiong, H.Y., Lee, L.J., Frey, B.J.: Deep learning of the tissue-regulated splicing code. Bioinformatics 30(12), i121–i129 (2014)
16. Reese, M., Eeckman, F., Kulp, D., Haussler, D.: Improved splice site detection in genie. J. Comput. Biol. 4(3), 311–323 (1997)

Metabolic Analysis of Metatranscriptomic Data from Planktonic Communities

Igor Mandric[1], Sergey Knyazev[1], Cory Padilla[2], Frank Stewart[2], Ion I. Măndoiu[3], and Alex Zelikovsky[1(✉)]

[1] Department of Computer Science, Georgia State University, Atlanta, GA, USA
{imandric1,skniazev1}@student.gsu.edu, alexz@cs.gsu.edu
[2] School of Biological Sciences, Georgia Institute of Technology, Atlanta, GA, USA
cpadilla7@gatech.edu, frank.stewart@biology.gatech.edu
[3] Computer Science and Engineering Department,
University of Connecticut, Storrs, CT, USA
ion@engr.uconn.edu

Abstract. This paper describes an enhanced method for analyzing microbial metatranscriptomic (community RNA-seq) data using Expectation - Maximization (EM)-based differentiation and quantification of predicted gene, enzyme, and metabolic pathway activity. Here, we demonstrate the method by analyzing the metatranscriptome of planktonic communities in surface waters from the Northern Louisiana Shelf (Gulf of Mexico) during contrasting light and dark conditions. The analysis reveals that the level of transcripts encoding proteins of oxidative phosphorylation varys little between day and night. In contrast, transcripts of pyrimidine metabolism are significantly more abundant at night, whereas those of carbon fixation by photosynthetic organisms increase 2-fold in abundance from night to day.

1 Introduction

RNA-seq is a standard method for comparative analysis of gene transcription across different conditions. It supplanted a widely used microarray approach, enabling analysis of a much larger number of genes, including those represented in pools of transcripts from complex multi-species communities (metatranscriptomes). RNA-seq allows researchers to determine and compare gene transcription levels, as well as the transcriptional activity of distinct metabolic pathways. Diverse bioinformatic tools have been developed to facilitate comparisons of RNA-seq data [1–10]. Such tools include web-based services with automated pipelines that allow assessment of the metabolic properties represented in RNA-seq datasets. For example, the MAP platform [11] predicts genes expressed in samples, while also provides information about gene classification into orthology groups (see Fig. 1). Unfortunately, such pipelines fail to quantify transcripts in concert with the annotation step. We therefore propose an enhanced pipeline that combines the biochemical annotation with quantification analysis. For this purposes, we propose to use an expectation-maximization (EM) technique similar to one from IsoEM2 [12]. We tested our algorithm using metatranscriptome

© Springer International Publishing AG 2017
Z. Cai et al. (Eds.): ISBRA 2017, LNBI 10330, pp. 396–402, 2017.
DOI: 10.1007/978-3-319-59575-7_41

data from marine bacterioplankton sampled during both the day and nighttime, and therefore likely exhibiting predictable variation in community transcription patterns.

2   Methods

In this section we describe the procedure of inferring metabolic pathway activity levels from RNA-Seq data for naturally occurring microbial communities. We also apply differential pathway activity level analysis similar to the non-parametric statistical approach described in [13], which was successfully applied for gene differential expression.

A general meta-omic pipeline is described on Fig. 1. Several metatranscriptomic samples are sequenced on an Illumina Hi-Seq (2×150 bp) and the resulting reads are assembled into a set of contigs. Genes detected on the contigs are mapped against protein databases and enzymatic functions are inferred. Finally, the representation of metabolic pathways is inferred based on the presence/absence of enzymes within each pathway. The above generic pipeline has been described in [11]. This paper proposes to enhance the above pipeline with the inference of metabolic pathway activity levels using repeated maximum likelihood inference and resolution by the Expectation - Maximization (EM) algorithm. The proposed inferences are depicted in red on Fig. 1.

Inference of Pathway Activity Levels. The first step is to estimate the abundances of the assembled contigs. The abundances can be inferred by any RNA-seq quantification tool. Here, we suggest using IsoEM2 [12], as this method is sufficiently fast to handle Illumina Hiseq data and more accurate than kallisto [14]. The next proposed step is to estimate the abundance of enzymes based on contig abundances. For this step we propose so-called *1-st EM*. The *2-nd EM* is used to infer metabolic pathway activity levels based on inferred enzyme abundances and databases of metabolic pathways. The 1-st and the 2-nd EM's can be also integrated into a single *direct EM* that directly infers pathway activity

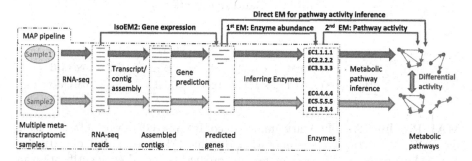

Fig. 1. The pipeline MAP and the enhanced pipeline for quantification and differential analysis of the metabolic pathway activity. The quantification enhancements are drawn in red. (Color figure online)

levels from contig abundances. All componentsm (1-st EM, 2-nd EM and direct EM) are built with similarities to IsoEM2 methodology.

Differential Analysis of Pathway Activity. Using the estimates of pathway activity levels in the differential pathway activity analysis requires estimating uncertainty. The extension of our bootstrapping approach introduced in [15] is useful for the direct maximum likelihood model since the pathway activity levels are inferred directly from RNA-seq reads that can be resampled. The current version of IsoEM2 allows the user to generate bootstrapped samples from the RNA-Seq reads and to infer abundance estimates, based on Fragments Per Kilobase of transcript per Million mapped reads (FPKM). We estimate pathway activity level for each of the bootstrapped samples and then run a differential expression (DE) analysis similar to the one described in [13].

3   Results

In this section we apply our analysis pipeline to two conditions (day. night) of a planktonic marine microbial community. We describe a subset of the most abundant pathways and conduct a differential pathway activity level analysis that highlights statistically significant functional features from the repertoire of metabolic processes occurring in the community.

Datasets. The samples were collected from surface waters (2 m depth) at 12:30 and 23:55 (local time) at a station on the Northern Louisiana Shelf (Gulf of Mexico) in July 2015. Seawater ($\sim$1 L) was pumped directly onto a 0.22 um Sterivex filter, preserved in 1.8 ml of RNA-later and flash frozen. Samples were stored a $-80°$ C until extraction. RNA was isolated from the samples by a phenol-chloroform method following the Mirvana RNA kit protocol. Samples were treated with DNase to remove residual DNA signal from the metatranscriptome. The RNA-Seq data were generated via Illumina HiSeq 2500 sequencing at the Department of Energy - Joint Genome Institute (DOE-JGI). Detailed information about the two samples is provided in the Table 1.

Table 1. Dataset description

Sample				Reads			Contigs	
Name	Depth	Code	Time	Length	Count	Insert size	Total	Total length
Day	2 m	177_2 m	12:30 PM	2 × 151 bp	89.4 M	195 ± 49	94.7 k	58.3 MB
Night	2 m	240_2 m	11:55 PM	2 × 151 bp	91.4 M	187 ± 49	108 k	68.1 MB

MAP Pipeline. A preliminary annotation of RNA-seq data was obtained using the DOE-JGI Metagenome Annotation Pipeline (MAP v.4) (JGI portal) [11]. The MAP processing consists of feature prediction including identification of protein-coding genes. In this pipeline, the MEGAHIT metagenome assembler is used to first assemble RNA-Seq reads into scaffolds. Further, several software suites (GeneMark.hmm, MetaGeneAnnotator, Prodigal, FragGeneScan) are used to predict genes on assembled scaffolds. The MAP pipeline also annotates genes according to EC numbers, which are a necessary input in our maximum likelihood model. The annotations are obtained via homology searches (using USEARCH) against a non-redundant proteins sequence database (max-hits = 50, e-value = 0.1) where each protein is assigned to a KEGG Orthology group (KO). The top 5 hits for each KO, with the condition that the identity score is at least 30% and 70% of the protein length is matched, are used. The KO IDs are translated into EC numbers using KEGG KO to EC mapping.

The Enhanced Quantification Pipeline. Our enhanced pipeline is depicted in red on Fig. 1. We start our analysis from the RNA-Seq metatranscriptomic reads. First, we find the abundance estimates (frequencies) for each metatranscriptomic gene/transcript by applying Maximum Likelihood abundance estimation. For this purpose we use IsoEM2. The custom GTF annotation file needed for supplying each run of IsoEM2 was prepared by using the fastaToGTF script from the same software suite. Next, we use FPKM estimates as the weights of each transcript for inferring abundances of each EC number. We use transcripts to EC notation alignments as provided by the MAP pipeline.

Highly Active Pathways. Table 2 shows the 10 most active pathways in the Day sample sorted in descending order of their activity level, i.e., the number of reads attributed by the proposed maximum likelihood model. The 11th pathway listed (ko0061) is among the 10 most active at night but is not among the 10 most active in the day. Similarly, the pathway ko00195 is among the most 10 active at night but is not among the 10 most active in the day. All other 9 pathways are among the most active during both night and day.

Differential Pathway Analysis. In Table 3 there is a list of all metabolic pathways which are up-regulated at noon with at least 1.7 fold change, 95% confidence and at least 1000 reads assigned by EM. The values of abundances are given at 95% confidence interval upper boundary (therefore, they are slightly greater than in the Table 2). In Table 4 there is a list of all metabolic pathways

Table 2. 10 most abundant pathways in the day and night samples.

Pathway		Abundance reads $\times 10^3$	
Code	Description	Day	Night
ko00190	Oxidative phosphorylation (Energy metabolism)	2260	2700
ko00710	Carbon fixation in photosynthetic organisms (Energy metabolism)	837	422
ko00240	Pyrimidine metabolism (Nucleotide metabolism)	644	1110
ko00270	Cysteine and methionine metabolism (Amino acid metabolism)	568	176
ko00020	Citrate cycle - TCA cycle (Carbohydrate metabolism)	525	411
ko00900	Terpenoid backbone biosynthesis (Metabolism of terpenoids and polyketides)	508	261
ko01230	Biosynthesis of amino acids	333	471
ko00195	Photosynthesis (Energy metabolism)	327	63
ko00230	Purine metabolism (Nucleotide metabolism)	318	618
ko00630	Glyoxylate and dicarboxylate metabolism (Carbohydrate metabolism)	299	530
ko00061	Fatty acid biosynthesis (Lipid metabolism)	37	179

which are up-regulated at noon with at least 1.7 fold change, 95% confidence and at least 1000 reads assigned by EM.

Discussion. The results in Tables 2, 3 and 4 are reflective of planktonic microbial communities driven by a diurnal cycle. During the daytime, pathways mediating photosynthesis, carbon fixation, and the building blocks for amino acid biosynthesis are the most abundant. At night there is an increase in nucleotide and lipid generation, probably for new cell production. In general, the community appears to be gaining energy and substrates during the day and expending them at night by generating crucial cellular components. This is supported by the differential expression between the day and night transcript pools, with energy (photosynthesis) and small organic molecule synthesis (e.g., fructose, glutamine-glutamate, glycosaminoglycan, etc.) being up-regulated during the day and the synthesis of larger biomolecules at night (e.g. lipid metabolism, amino acids, and carotenoids). There is a clear shift in energy sources between day and night. While oxidative phosphorylation is highly transcribed at both time points, it is clear that photosynthesis elevates some of this energy requirement. This is evidenced by a slight decrease of oxidative phosphorylation and increase of TCA-related transcripts during the day, potentially replenishing the NADH/NADPH reserves for the use of the electron transport chain at night. As predcited, these results indicate a community undergoing diel cycling, thereby providing validation of our proposed EM-based pipeline and suggesting this method as an valuable tool for coupled annotation and quantification of metabolic pathways in community RNA-seq data.

Table 3. Up-regulated pathways in the day sample

Pathway		Reads in 10^3	
Code	Description	Day	Night
ko00051	Fructose and mannose metabolism (Carbohydrate metabolism)	326	34.1
ko00195	Photosynthesis (Energy metabolism)	488	93.1
ko00261	Monobactam biosynthesis (Biosynthesis of other secondary metabolites)	237	44.5
ko00410	beta-Alanine metabolism (Metabolism of other amino acids)	10.0	0.01
ko00471	D-Glutamine and D-glutamate metabolism	6.79	0
ko00532	Glycosaminoglycan biosynthesis - chondroitin sulfate / dermatan sulfate	28.8	3.65
ko00533	Glycosaminoglycan biosynthesis - keratan sulfate	22.9	0.609
ko00604	Glycosphingolipid biosynthesis - ganglio series	4.17	0
ko00660	C5-Branched dibasic acid metabolism (Carbohydrate metabolism)	4.39	0.01
ko00930	Caprolactam degradation (Xenobiotics biodegradation and metabolism)	3.80	0.883
ko00332	Carbapenem biosynthesis (Biosynthesis of other secondary metabolites)	10.3	1.54
ko00565	Ether lipid metabolism (Lipid metabolism)	10.4	0.682
ko00590	Arachidonic acid metabolism (Lipid metabolism)	51.8	19.4
ko00270	Cysteine and methionine metabolism (Amino acid metabolism)	787	246
ko00514	Other types of O-glycan biosynthesis (Glycan biosynthesis and metabolism)	7.75	2.96
ko00450	Selenocompound metabolism (Metabolism of other amino acids)	201	80.2
ko00710	Carbon fixation in photosynthetic organisms(Energy metabolism)	1000	487
ko00983	Drug metabolism - other enzymes (Xenobiotics biodegradation & metabolism)	58.3	16.5
ko00520	Amino sugar and nucleotide sugar metabolism (Carbohydrate metabolism)	265	123

Table 4. Up-regulated pathways in the night sample

Pathway		Reads in 10^3	
Code	Description	Day	Night
ko00053	Ascorbate and aldarate metabolism (Carbohydrate metabolism)	0	1.88
ko00061	Fatty acid biosynthesis (Lipid metabolism)	55.9	270
ko00120	Primary bile acid biosynthesis (Lipid metabolism)	2.75	116
ko00140	Steroid hormone biosynthesis (Lipid metabolism)	0	4.11
ko00232	Caffeine metabolism (Biosynthesis of other secondary metabolites)	0	1.05
ko00260	Glycine, serine and threonine metabolism (Amino acid metabolism)	49.3	227
ko00311	Penicillin and cephalosporin biosynthesis	0	2.74
ko00365	Furfural degradation (Xenobiotics biodegradation and metabolism)	0	2.12
ko00430	Taurine and hypotaurine metabolism (Metabolism of other amino acids)	3.19	62.3
ko00472	D-Arginine and D-ornithine metabolism (Metabolism of other amino acids)	0	1.25
ko00780	Biotin metabolism (Metabolism of cofactors and vitamins)	7.05	48.6
ko00906	Carotenoid biosynthesis (Metabolism of terpenoids and polyketides)	0	26.2
ko00984	Steroid degradation (Xenobiotics biodegradation and metabolism)	0	2.07
ko00362	Benzoate degradation (Xenobiotics biodegradation and metabolism)	3.58	16.7
ko00592	alpha-Linolenic acid metabolism (Lipid metabolism)	0.19	2.89
ko00072	Synthesis and degradation of ketone bodies (Lipid metabolism)	2.67	11.6
ko00364	Fluorobenzoate degradation (Xenobiotics biodegradation and metabolism)	0.180	2.96
ko01051	Biosynthesis of ansamycins (Metabolism of terpenoids and polyketides)	0	3.38
ko00760	Nicotinate and nicotinamide metabolism (Mcofactors and vitamins)	30.2	103
ko00281	Geraniol degradation (Metabolism of terpenoids and polyketides)	1.57	170
ko00627	Aminobenzoate degradation (Xenobiotics biodegradation and metabolism)	0.949	4.06
ko00730	Thiamine metabolism (Metabolism of cofactors and vitamins)	10.4	35.4
ko00643	Styrene degradation (Xenobiotics biodegradation and metabolism)	0.958	22.6
ko01200	Carbon metabolism	13.7	86.9
ko00220	Arginine biosynthesis (Amino acid metabolism)	3.53	11.0
ko00440	Phosphonate and phosphinate metabolism	1.30	5.33
ko00905	Brassinosteroid biosynthesis (Metabolism of terpenoids and polyketides)	2.00	35.6
ko00941	Flavonoid biosynthesis (Biosynthesis of other secondary metabolites)	2.84	6.03
ko00720	Carbon fixation pathways in prokaryotes (Energy metabolism)	1.36	15.9
ko00290	Valine, leucine and isoleucine biosynthesis (Amino acid metabolism)	68.0	193
ko00403	Indole diterpene alkaloid biosynthesis	0	2.68
ko01053	Biosynthesis of siderophore group nonribosomal peptides	0	1.16
ko00920	Sulfur metabolism (Energy metabolism)	47.7	135
ko00625	Chloroalkane and chloroalkene degradation	24.3	51.8

Acknowledgements. IM, SK and AZ were partially supported from NSF Grants 1564899 and 16119110, IM and SK were partially supported by GSU Molecular Basis of Disease Fellowship, IIM was partially supported from NSF Grants 1564936 and 1618347, CP and FS were partially supported by NSF Grants 1151698, 1558916, and 1564559, and Simons Foundation award 346253.

References

1. Donato, M., Xu, Z., Tomoiaga, A., Granneman, J.G., MacKenzie, R.G., Bao, R., Than, N.G., Westfall, P.H., Romero, R., Draghici, S.: Analysis and correction of crosstalk effects in pathway analysis. Genome Res. **23**(11), 1885–1893 (2013)

2. Efron, B., Tibshirani, R.: On testing the significance of sets of genes. Ann. Appl. Stat. **1**, 107–129 (2007)
3. Huson, D.H., Mitra, S., Ruscheweyh, H.-J., Weber, N., Schuster, S.C.: Integrative analysis of environmental sequences using MEGAN4. Genome Res. **21**(9), 1552–1560 (2011)
4. Konwar, K.M., Hanson, N.W., Pagé, A.P., Hallam, S.J.: MetaPathways: a modular pipeline for constructing pathway/genome databases from environmental sequence information. BMC Bioinform. **14**(1), 202 (2013)
5. Mitrea, C., Taghavi, Z., Bokanizad, B., Hanoudi, S., Tagett, R., Donato, M., Voichita, C., Dr, S.: Methods and approaches in the topology-based analysis of biological pathways. Front. Physiol. **4**, 278 (2013)
6. Sharon, I., Bercovici, S., Pinter, R.Y., Shlomi, T.: Pathway-based functional analysis of metagenomes. J. Comput. Biol. **18**(3), 495–505 (2011)
7. Subramanian, A., Tamayo, P., Mootha, V.K., Mukherjee, S., Ebert, B.L., Gillette, M.A., Paulovich, A., Pomeroy, S.L., Golub, T.R., Lander, E.S., et al.: Gene set enrichment analysis: a knowledge-based approach for interpreting genome-wide expression profiles. Proc. Nat. Acad. Sci. U.S.A. **102**(43), 15545–15550 (2005)
8. Tarca, A.L., Draghici, S., Bhatti, G., Romero, R.: Down-weighting overlapping genes improves gene set analysis. BMC Bioinform. **13**(1), 136 (2012)
9. Temate-Tiagueu, Y., Seesi, S.A., Mathew, M., Mandric, I., Rodriguez, A., Bean, K., Cheng, Q., Glebova, O., Măndoiu, I., Lopanik, N.B., Zelikovsky, A.: Inferring metabolic pathway activity levels from rna-seq data. BMC Genom. **17**(5), 542 (2016). doi:10.1186/s12864-016-2823-y
10. Ye, Y., Doak, T.G.: A parsimony approach to biological pathway reconstruction/inference for genomes and metagenomes. PLoS Comput. Biol. **5**(8), 1000465 (2009)
11. Huntemann, M., Ivanova, N.N., Mavromatis, K., Tripp, H.J., Paez-Espino, D., Tennessen, K., Palaniappan, K., Szeto, E., Pillay, M., Chen, I.-M.A., et al.: The standard operating procedure of the DOE-JGI metagenome annotation pipeline (MAP v. 4). Stan. Genomic Sci. **11**(1), 17 (2016)
12. Mandric, I., Temate-Tiagueu, Y., Shcheglova, T., Seesi, S.A., Zelikovsky, A., Mandoiu, I.: Fast bootstrapping-based estimation of confidence intervals of expression levels and differential expression from RNA-SEQ data. Bioinformatics (to appear)
13. Al Seesi, S., Tiagueu, Y.T., Zelikovsky, A., Măndoiu, I.I.: Bootstrap-based differential gene expression analysis for RNA-SEQ data with and without replicates. BMC Genom. **15**(8), 2 (2014)
14. Bray, N.L., Pimentel, H., Melsted, P., Pachter, L.: Near-optimal probabilistic RNA-SEQ quantification. Nat. Biotechnol. **34**(5), 525–527 (2016)
15. Al Seesi, S., Mangul, S., Caciula, A., Zelikovsky, A., Măndoiu, I.: Transcriptome reconstruction and quantification from RNA sequencing data. Genome Anal.: Curr. Proced. Appl. 39 (2014)

NemoLib: A Java Library for Efficient Network Motif Detection

Andrew Andersen$^{(\boxtimes)}$ and Wooyoung Kim$^{(\boxtimes)}$

Division of Computing and Software Systems, School of Science, Technology,
Engineering, and Mathematics, University of Washington Bothell, 18115 Campus
Way NE, Bothell, WA 98011-8246, USA
{drewda,kimw6}@uw.edu

Abstract. A network motif is defined as an overabundant subgraph
pattern in a network and has been applied in various biological and
medical problems. Various network motif detection algorithms and tools
are currently available. However, most existing software programs are
outdated, incompatible with modern operating systems, or do not provide sufficient operation instructions. Furthermore, most tools provide
limited information regarding network motifs, which necessitates post-
processing program to apply to real problems. Consequently, the lack of
usability brings a certain amount of skepticism about the relevance of
network motifs in investigating real biological problems. Therefore, this
paper introduces NemoLib (network motif library) as a general purpose
tool for detection and analysis of network motifs. NemoLib is highly
programmable Java library which provides for extensibility.

Keywords: NemoLib · ESU · Biological network · Network motif

1 Introduction

Rapid technological development in molecular biology has led to an explosion of
omics research and a subsequent need to model the omic data in a useful way.
Biological networks have been proven to be useful to model biological systems,
with graph nodes representing molecules in the system and edges representing
the interactions between those molecules [7].

Various graph theory topics are applied to resolve real biological problems
[2], but this paper focuses on network motifs as one of the graph theory methods: unique and frequent subgraph patterns that appear in a particular network.
Analysis of network motifs have led to practical uses to predict protein-protein
interactions [1], to determine protein functions [4], to detect breast-cancer susceptibility genes [22], to investigate evolutionary conservation [19], and to discover essential proteins [11]. Furthermore, a broad spectrum of applications has
been explored: 'motif clustering' [5], 'motif themes' [21], 'relative graphlet frequency distances'[15], 'motif modes' [12], and 'MotifScores' [20].

© Springer International Publishing AG 2017
Z. Cai et al. (Eds.): ISBRA 2017, LNBI 10330, pp. 403–407, 2017.
DOI: 10.1007/978-3-319-59575-7_42

However, discovering and classifying network motifs are computationally expensive, requiring nonpolynomial time to perform. The process involves enumerating or searching for millions of subgraphs in the input graph and classifying them through canonical labeling or isomorphic testing. Then, a network motif's uniqueness is established through rigorous statistical testing in a large pool of randomly-generated networks of the same order and size. Consequently, various heuristic methods and parallel algorithms have been proposed that alleviate the performance concerns of exhaustive search methods [10].

Many motif search programs are also available [16]: MFinder [9], FANMOD [18], Kavosh [8], Mavisto [17], NeMoFinder [3], Grochow's [6], and MODA [14]. However, most of them are not functional. Some of these programs are unable to execute on modern operating systems, or there are no explanations of how to use them. Furthermore, most tools fail to provide the instances of network motifs that are often necessary to discover biological significance, thus requiring the more expensive post-processing. In sum, none of the existing systems provide efficient frameworks for network motif detection. Therefore, we present the Network Motif Library (NemoLib), a modern network motif detection library designed for extensibility and sustainability.

2   NemoLib: Network Motif Library

2.1   NemoLib Overview

The NemoLib library intends to abstract away much of the complexity of building a network motif detection tool while allowing for customizability. The design of the library follows the object-oriented principle of "open for extension, closed for modification" by exposing an API of common tools used for network motif detection in the form of Java classes. Additionally, NemoLib is available through a public GitHub repository where it can be accessed and updated by the open-source community, and it uses Apache Maven as a build tool and dependency management system to ensure all NemoLib dependencies are up to date. Presently the NemoLib library has been tested to work in Linux and Windows environments.

2.2   NemoLib Design and Components

The library includes some standard classes one might expect in a network motif detection library, such as Graph to represent a target network and Subgraph to represent instances of subgraphs appearing in a network. However, it also includes a host of additional classes to reduce the redundancy of rewriting common network motif detection tasks.

The TargetNetworkAnalyzer class executes a subgraph enumeration algorithm on the target network and produces a mapping of each size-k subgraph label to its relative frequency in the target network. Additionally, the TargetNetworkAnalyzer class can produce a SubgraphEnumerationResult object, described below.

The RandomGraphAnalyzer class generates a pool of random graphs of the same size and order as the target graph. Each of those graphs is then enumerated, and the relative frequency of each size-k subgraph label is appended to a list which is mapped to the corresponding label name. The resulting map of relative frequency lists is returned to the client.

Finally, the RelativeFrequencyAnalyzer class takes as parameters the maps produced by the TargetNetworkAnalyzer and RandomGraphAnalyzer classes and determines the p-value and z-score for each label in the target network. The p-value is used to determine whether any label should be identified as a network motif based on a threshold set by the user.

The library also includes two Java interface classes: SubgraphEnumerator and SubgraphEnumerationResult. The SubgraphEnumerator interface can be applied to any enumeration algorithm. Presently, both the ESU and RandESU algorithms have been implemented in the library using the SubgraphEnumerator interface. An object implementing the SubgraphEnumerator is passed as a parameter to each of the TargetGraphAnalyzer and RandomGraphAnalyzer classes and is used to enumerate the subgraphs. The SubgraphEnumerationResult stores the results of enumeration in a manner specified by the client. Some examples of classes implementing the SubgraphEnumerationResult included in the library are SubgraphCount, which maps subgraphs to absolute subgraph frequencies; SubgraphProfile, which maps labels first to target network vertices then to absolute subgraph frequencies; and SubgraphCollection, which maps Subgraph objects to label frequencies.

3   Conclusion and Future Study

The NemoLib library fills a void in the network motif detection software community by applying modern software engineering techniques to existing algorithms. Its combination of programmability and flexibility should make it an ideal tool for testing novel network motif detection algorithms and for learning about network motif detection.

One limitation of the library at present is the reliance on the nauty program developed by McKay [13]. While nauty executes perfectly well on most environments, it is not written in Java and must be called externally from the library, creating a dependency that cannot be managed by Maven. Development of a Java version of the nauty algorithm would eliminate this concern.

A primary area of future study we plan to focus on is parallel and multithreaded client programs implementing the library. We would also like to add a GUI or create a web application that utilizes the library as a service, allowing researchers to upload a file containing a description of a network, select configuration parameters, and receive the results. Finally, we would like to add the option for execution results to be produced in a modern, portable data format like XML or JSON.

The NemoLib library can be accessed at https://github.com/drewandersen/nemolib.

References

1. Albert, I., Albert, R.: Conserved network motifs allow protein-protein interaction prediction. Bioinformatics **20**(18), 3346–3352 (2004)
2. Callebaut, W.: Scientific perspectivism: a philosopher of science's response to the challenge of big data biology. Stud. Hist. Philos. Biol. Biomed. Sci. **43**(1), 69–80 (2012)
3. Chen, J., Hsu, W., Lee, M., Ng, S.: NeMoFinder: dissecting genome-wide protein-protein interactions with meso-scale network motifs. In: Proceedings of the 12th ACM SIGKDD International Conference on Knowledge Discovery and Data Mining, New York, NY, pp. 106–115 (2006)
4. Chen, J., Hsu, W., Lee, M.L., Ng, S.K.: Labeling network motifs in protein interactomes for protein function prediction. In: International Conference on Data Engineering, pp. 546–555 (2007). Biological networks like PPI (protein-protein interaction) contain small networks with higher occurrences than those expected by chance. Small network design can help uncover the design of the complex network
5. Dobrin, R., Beg, Q.K., Barabasi, A.L., Oltvai, Z.N.: Aggregation of topological motifs in the Escherichia coli transcriptional regulatory network. BMC Bioinform. **5**, 10 (2004). The authors presents that the two motif types of feed-forward and bi-fan are aggregate into homologous motif clusters in the transcriptional regulatory network of the bacterium, Escherichia coli
6. Grochow, J.A., Kellis, M.: Network motif discovery using subgraph enumeration and symmetry-breaking. In: Speed, T., Huang, H. (eds.) RECOMB 2007. LNCS, vol. 4453, pp. 92–106. Springer, Heidelberg (2007). doi:10.1007/978-3-540-71681-5_7
7. Junker, B.H., Schreiber, F.: Analysis of Biological Networks. Wiley, Hoboken (2008)
8. Kashani, Z., Ahrabian, H., Elahi, E., Nowzari-Dalini, A., Ansari, E., Asadi, S., Mohammadi, S., Schreiber, F., Masoudi-Nejad, A.: Kavosh: a new algorithm for finding network motifs. BMC Bioinform. **10**(1), 318 (2009). 19799800
9. Kashtan, N., Itzkovitz, S., Milo, R., Alon, U.: Efficient sampling algorithm for estimating sub-graph concentrations and detecting network motifs. Bioinformatics **20**, 1746–1758 (2004). 15001476
10. Kim, W., Diko, M., Rawson, K.: Network motif detection: algorithms, parallel and cloud computing, and related tools. Tsinghua Sci. Technol. **18**(5), 469–489 (2013)
11. Kim, W., Li, M., Wang, J., Pan, Y.: Essential protein discovery based on network motif and gene ontology. In: Proceedings of IEEE Bioinformatics and Biomedicine, pp. 470–475 (2011). This paper is an application of biological network motifs to detect essential proteins in a PPI network
12. Lee, W.P., Jeng, B.C., Pai, T.W., Tsai, C.P., Yu, C.Y., Tzou, W.S.: Differential evolutionary conservation of motif modes in the yeast protein interaction network. BMC Genomics **7**(1), 89 (2006). 16638125
13. McKay, B.: Practical graph isomorphism. Congr. Numer. **30**, 45–87 (1981)
14. Omidi, S., Schreiber, F., Masoudi-Nejad, A.: Moda: an efficient algorithm for network motif discovery in biological networks. Genes Genet. Syst. **84**(5), 385–395 (2009). This paper, while not explicitly stating it, seems to discuss how to find network motifs in parrallel. This incorporates their algorithm, MODA, which utilizes recognition of pattern growth to expand beyond their 8 node target limit. Well, it is not a parallel algorithm. (added by wkim)

15. Przulj, N., Corneil, D.G., Jurisica, I.: Modeling interactome: scale-free or geometric? Bioinformatics **20**(18), 3508–3515 (2004)
16. Ribeiro, P., Silva, F., Kaiser, M.: Strategies for network motifs discovery. In: Fifth IEEE International Conference on e-Science, e-Science 2009, pp. 80–87 (2009). iD: 1
17. Schreiber, F., Schwobbermeyer, H.: MAVisto: a tool for the exploration of network motifs. Bioinformatics **21**, 3572–3574 (2005). 16020473 Serial Algorithm (added 1/23/2013)
18. Wernicke, S., Rasche, F.: FANMOD: a tool for fast network motif detection. Bioinformatics **22**, 1152–1153 (2006). 16455747
19. Wuchty, S., Oltvai, Z.N., Barabasi, A.L.: Evolutionary conservation of motif constituents in the yeast protein interaction network. Nat. Genet. **35**(2), 176–179 (2003). http://dx.doi.org/10.1038/ng1242
20. Xie, Z.R., Hwang, M.J.: An interaction-motif-based scoring function for protein-ligand docking. BMC Bioinform. **11**(1), 298 (2010)
21. Zhang, L., King, O., Wong, S., Goldberg, D., Tong, A., Lesage, G., Andrews, B., Bussey, H., Boone, C., Roth, F.: Motifs, themes and thematic maps of an integrated Saccharomyces cerevisiae interaction network. J. Biol. **4**(2), 6 (2005). 15982408
22. Zhang, Y., Xuan, J., de los Reyes, B.G., Clarke, R., Ressom, H.W.: Network motif-based identification of breast cancer susceptibility genes. In: 2008 30th Annual International Conference of the IEEE Engineering in Medicine and Biology Society, pp. 5696–5699. IEEE, August 2008

A Genetic Algorithm for Finding Discriminative Functional Motifs in Long Non-coding RNAs

Brian L. Gudenas and Liangjiang Wang[✉]

Department of Genetics and Biochemistry,
Clemson University, Clemson, SC, USA
{bgudena, liangjw}@clemson.edu

Abstract. Long non-coding RNAs (lncRNAs), each with >200 nucleotides in length, constitute a large portion of the human transcriptome. Although recent studies indicate that lncRNAs play key roles in gene regulation, development and disease, the RNA functional motifs are still poorly understood. Most of the existing algorithms for motif finding are severely limited in scalability with regards to sequence and motif size. In this study, we propose a novel genetic algorithm for discriminative motif identification capable of handling large input sequences and motif sizes by utilizing genetic operators to learn and evolve in response to the input sequences. We utilize our method on long non-coding RNA (lncRNA) transcripts as a test case to identify functional motifs associated with subcellular localization. Our methodology shows high accuracy and the ability to identify functional motifs associated with subcellular localization in lncRNAs, which recapitulates a previous experimental study.

Keywords: Genetic algorithm · Pattern discovery · Functional motif · LncRNA

1 Introduction

The identification and subsequent functional annotation of short reoccurring motifs within molecular sequences has been integral for the field of genetics. Generally, the first step regarding functional annotation of a novel protein or RNA sequence is the identification of known functional motifs within the primary sequence. Once identified, these motifs allow the functional inference of the previously uncharacterized sequence. Initially, functional motifs were identified assuming a random uniform background nucleotide model, however the nucleotides within a gene have structure; therefore, the utilization of real sequences as a background set is advantageous. The use of a positive and negative sequence set is known as discriminative motif identification. Popular algorithms for discriminative motif discovery include discriminative regular expression motif elicitation (DREME) [1]. However, the DREME algorithm suggests input sequences are less than 500 nucleotides and that the motif width is less than or equal to 8 nucleotides.

Recently, tens of thousands of long non-coding RNAs (lncRNAs) have been discovered in primates, the vast majority of which are functionally uncharacterized.

© Springer International Publishing AG 2017
Z. Cai et al. (Eds.): ISBRA 2017, LNBI 10330, pp. 408–413, 2017.
DOI: 10.1007/978-3-319-59575-7_43

LncRNAs are poorly conserved across species and can perform a myriad of diverse functions, adding to the complexity of their functional annotation. Unlike mRNAs, lncRNAs can localize in many different places within the cell, which can provide insights into their functionality. Localization motifs have been identified in lncRNA transcripts which regulate subcellular localization. Thus it may be possible to identify motifs within lncRNA transcripts, genome-wide, associated with subcellular localization, providing a valuable first step in the functional annotation of human lncRNAs. One method to do this would be to identify two distinct sets of lncRNAs, one enriched in the cytoplasm and the other enriched in the nucleus, followed by finding motifs overrepresented in one set of transcripts but not the other. However, there are tens of thousands of human lncRNAs, which have a median transcript length of 592 bp and could possibly contain long functional motifs [2]. Due to these issues and limitations, previous approaches are unsuitable for our purposes of the identification of functional motifs in full length lncRNAs genome-wide.

Genetic algorithms, which mimic biological evolution to stochastically evolve a population of solutions over time, for motif finding have been utilized previously, such as MDGA [3]. MDGA represents a solution as a vector of indices which indicate the starting position of the motif in each sequence and therefore fail to utilize all the information present in a sequence, such as multiple motif occurrences. Furthermore, all the methods mentioned previously were developed for the identification of transcription factors binding sites, which are small motifs and are contained in very short sequences, such as ChIP-Seq peaks (<100 nucleotides). We propose to create a novel genetic algorithm (GA) for discriminative motif discovery to identify long functional motifs in full-length lncRNA transcripts, a use-case previous methods are incapable of.

2   Proposed Approach

2.1   Representation and Population Initialization

We represent a solution as a position weight matrix (PWM) of length w, which is therefore not dependent on the input sequence size. A fixed number of individuals is initialized to create a population by creating randomized PWMs utilizing the conjugate Dirichlet distribution.

2.2   Fitness

The fitness function must identify similar sequence motifs in the positive set which are underrepresented in the negative set. A useful metric for the identification of informative similar sequences is the information content (IC), the total information content of a PWM M is as follows [4].

$$IC(M) = \sum_{i=1}^{w} \sum_{\beta \in \{A,C,G,T\}} M_{\beta i} \log_2 \frac{M_{\beta i}}{p_\beta} \tag{1}$$

Where W is the motif width, $M_{\beta i}$ is the frequency of nucleotide β of column i and p_β is the background probability of nucleotide β. We chose to augment this popular metric to incorporate our goal of minimizing the matches of the PWM in the negative sequence set. We define the fitness score to be the information content of the PWM divided by the total matches of the PWM in a small random subset S_β^- of negative set S^-.

$$Fitness = \frac{IC(M)}{\left(\frac{\sum_{S \in S_\beta^-} \sigma_M(S)}{|S_\beta^-|}\right)} \tag{2}$$

Where $\sigma_M(S)$ equals the match score of motif M in sequence set S and $\left|S_\beta^-\right|$ is the total sequence length scanned, used as a normalization factor. Using a sliding-window across each sequence we calculate a score as the log-likelihood PWM score divided by the maximal PWM score.

2.3   Selection

During each iteration, selection occurs to determine which solutions survive into the next generation. A solution's probability of survival is approximately proportional to their fitness score. Linear-rank selection then occurs with replacement to generate a new population of solutions.

2.4   Crossover and Mutation

Crossover occurs by randomly selecting two parent individuals which are then recombined to create two novel children solutions. Therefore, crossover allows the recombination of solutions to further explore the solution space. To avoid the positional bias of the traditionally used 1-point crossover, we chose to utilize random uniform crossover. In random uniform crossover, we randomly choose a crossover number c between (1: $w - 1$), then we randomly select c columns of the parental solution PWMs to be switched. The resulting children solutions then replace the initial parental solutions used for crossover.

Mutation occurs by randomly selecting solutions which are then altered stochastically to allow further exploration of the search space, thereby avoiding local maxima. For each solution M selected for mutation, we randomly select a fixed proportion β of the positive sequence set S^+, then we score the solution across all possible windows of the sequences in S_β^+. Based on the maximal scoring position of M in each sequence of S_β^+ we then update the PWM to form M' using the alignment of the selected positions.

2.5   Implementation

The program was implemented in the R statistical language utilizing the framework from the R package GA [5, 6]. The pseudo-code is as follows:

1. Set Input Parameters:
   S = Full sequence set ordered by positive/negative
   W = Motif Width; P = Population size; $i = 0$
   R = Number of generations to terminate if best solution not improved
2. INITIALIZE Population: Randomly create P PWMs
3. Fitness EVALUATION: **BEST** = max(Fitness(P))
4. Genetic Operators: While ($i < R$)
   ELITISM: Save top 5% of fittest solutions for next generation
   SELECTION: Linear-rank selection to create new P
   Random uniform CROSSOVER of random subset of P
   MUTATE random subset of P
   Evaluate fitness of new P
   TERMINATION: If (max(fitness(P)) == BEST) {
   $i = i + 1$ } else {
   BEST = max(Fitness(P))
   $i = 0$ }

3 Results

To evaluate our genetic algorithm (GA) we first begin with synthetic data in order to obtain complete control over the sequence attributes. We created a set S^+ of 100 independent and identically distributed sequences over a range of lengths with a single randomly implanted motif of length **15,** each with **3** random mutations per sequence. Our negative sequence set S^- is simply a dinucleotide shuffle of S^+. To assess the sensitivity of the GA we evaluate its performance over a large range of different sequence lengths N, because as N increases the noise to signal ratio increases (Fig. 1).

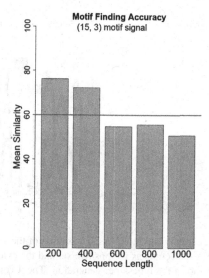

Fig. 1. Motif identification accuracy.

Motif similarity is represented by the log-likelihood PWM score divided by the maximal log-likelihood PWM score, averaged over 11 trials for each of the different sequence lengths. The red line is the threshold for a significant match as defined by Hansen [7].

3.1 Differentially Localized LncRNAs

Lastly, we wanted to identify possible functional motifs in lncRNAs which are differentially localized within the cell. For this test case we used published results from which lncRNA transcript abundances were quantified from fractionated cellular compartments, either from the nucleus or the cytoplasm [2]. We extracted the sequences for all lncRNAs quantified between the nuclear and cytoplasmic fractions, using sequences enriched in the nucleus as our positive sequence set and the cytoplasmic lncRNAs as the negative set. This data resulted in a set of 1749 lncRNA transcripts with a total sequence length of 2.42 megabases, of which 981 sequences are enriched in the nucleus while 768 are enriched in the cytoplasm. Using the nuclear transcripts as our positive set and the cytoplasmic transcripts as our negative set we ran the GA on this large dataset to identify 14-mers associated with subcellular localization (Fig. 2).

Notably, the best motif identified contains a core pentamer RNA motif previously identified to dictate nuclear localization of lncRNAs [8]. The pentamer RNA sequence motif was found to be AGCCC with the restriction sites of (G or C) at −3 and (T or A) at −8, which the motif we identified contains all of, except the −8 restriction site [8]. Next, we calculated the total matches of the identified motif in each sequence set, finding 3,133 instances of the motif in the nuclear set and only 1,871 in the negative set. Furthermore, the counts of this motif in each lncRNA shows a small yet significant correlation with the nuclear/cytoplasm FPKM ratio (Pearson's Correlation Coefficient = 0.14, p-value = 4.31×10^{-9}).

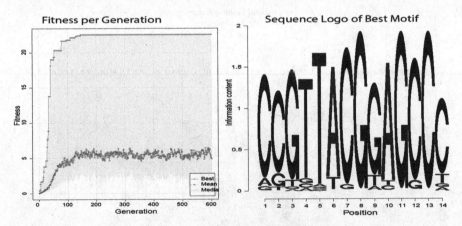

Fig. 2. Motif found in nuclear enriched lncRNAs. The output of the genetic algorithm ran with the nuclear enriched lncRNA transcripts as the positive set and the cytoplasmic lncRNAs as the negative set. The first plot shows the fitness per generation. The second figure is the sequence logo of the best motif found.

4 Conclusion

Based on our preliminary observations we have shown that our genetic algorithm is capable of identifying discriminative motifs in large sequence sets, such as sequence sets containing thousands of lncRNA transcripts. We demonstrated that our algorithm can achieve high accuracy in synthetic tests despite a high noise to signal ratio. In addition, utilizing entire lncRNA transcripts derived from the transcript quantification of fractionated cells we have identified a motif enriched in nuclear transcripts. Remarkably, the motif identified recapitulates an experimentally identified lncRNA localization motif identified in an independent study [8]. The counts of this motif in the lncRNA transcripts also shows a significant positive correlation with subcellular localization. This preliminary work shows that it is possible to computationally identify functional motifs in previously uncharacterized lncRNAs.

Our method is capable of handling large input sequence sets as well as identifying arbitrarily large motifs. In addition, to speed up the motif identification process over a range of motif widths, solutions from the final population can be used as a seed population for identifying motifs of size $w + 1$. This procedure can be done iteratively and will speed up the motif identification procedure because the initial population will already contain informative seed motifs to be expanded upon. Furthermore, the genetic algorithm framework could likely be improved upon through augmentation of the genetic operators.

References

1. Bailey, T.L.: DREME: motif discovery in transcription factor ChIP-seq data. Bioinformatics 27, 1653–1659 (2011). doi:10.1093/bioinformatics/btr261
2. Derrien, T., Johnson, R., Bussotti, G., et al.: The GENCODE v7 catalog of human long noncoding RNAs: analysis of their gene structure, evolution, and expression. Genome Res. 22, 1775–1789 (2012). doi:10.1101/gr.132159.111
3. Che, D., Song, Y., Rasheed, K.: MDGA: Motif discovery using a genetic algorithm. In: Proceedings 2005 Conference on Genetic and Evolutionary Computation, pp. 447–452 (2005)
4. Shannon, C.E.: A mathematical theory of communication. Bell Syst. Tech. J. 27, 379–423 (1948). doi:10.1145/584091.584093
5. R Core Team. R: a language and environment for statistical computing. R Foundation for Statistical Computing (2015). http://www.r-project.org/
6. Scrucca, L.: GA: a package for genetic algorithms in R. J. Stat. Softw. 53(4), 1–37 (2013)
7. Hansen, L., Mariño-ramírez, L., Landsman, D.: Differences in local genomic context of bound and unbound motifs. Gene 506, 125–134 (2012). doi:10.1016/j.gene.2012.06.005
8. Zhang, B., Gunawardane, L., Niazi, F., et al.: A novel RNA motif mediates the strict nuclear localization of a long noncoding RNA. Mol. Cell. Biol. 34, 2318–2329 (2014). doi:10.1128/MCB.01673-13

Heterogeneous Cancer Cell Line Data Fusion for Identifying Novel Response Determinants in Precision Medicine

Wojciech Czaja and Jeremiah Emidih[(⊠)]

Department of Mathematics, University of Maryland, College Park, MD 20740, USA
jemidih@math.umd.edu

Abstract. The most commonly used anticancer treatments include DNA-targeted, genome specific drugs such as topoisomerase inhibitors. Unfortunately, many patients do not respond to these cytotoxic drugs since it is not yet possible to identify genomic parameters for response or non-response. Understanding genomic networks which determine treatment response is thus a cornerstone to effective use of DNA-targeted drugs. Large cancer cell line databases with extensive genomic and drug response information provide opportunities to examine these networks and recover genomic determinants for response to DNA-targeted drugs. We utilize a novel data fusion framework based on Laplacian eigenmap embeddings to identify potential response determinants to the topoisomerase inhibitor Topotecan.

1 Background

Genomic analysis is required for precise medicine to find the right drug for the right patient. In order to do this accurately, one must integrate terabytes of data over dozens of databases, filled with data of varying quality and content.

Cancer cell line databases have collected data on over 1,300 cell lines representing the broad spectrum of human cancers. Specifically, there is extensive analysis of genomic and drug response data on the cell lines in the National Cancer Institute (NCI-60) database and the Broad Institute Cancer Therapeutics Response Portal (CTRP) database [1,13].

Topoisomerase inhibitors are among the most effective and widely used anticancer drugs, and in spite of known molecular pathways for DNA repair, it is not currently possible to predict how cells respond to these DNA-targeted drugs [10]. Using standard statistical tools, the Pommier laboratory (LMP) of the NCI Developmental Therapeutics Branch (NCI-DTB) discovered new factors to determine topoisomerase inhibitor response from databases based on the NCI-60 [1,14,16]. These include previously unsuspected response determinants like the putative helicase SLFN11 [15,16].

The NCI-DTB uses CellMiner web-based suites to build prediction functions from these response determinants based on the elastic net regression algorithm, which was applied previously for drug response prediction for other databases

© Springer International Publishing AG 2017
Z. Cai et al. (Eds.): ISBRA 2017, LNBI 10330, pp. 414–419, 2017.
DOI: 10.1007/978-3-319-59575-7_44

[8,9,12]. Elastic net is a linear regression model which adds a quadratic penalty to the loss function from LASSO regression. An issue with this method though is that highly correlated response determinants, such as genes that are co-expressed, can be swapped to obtain different predictive models which are very similar in predictive performance.

Thus, the goal of this work is to find novel response determinants to topoisomerase inhibitors in a method independent of features selected by elastic net. In particular, we shall use a nonlinear feature selection model to avoid the aforementioned problem with a variety of equivalent predictive functions. The features to be selected are among the cell line database information which gives gene expression values indicating log2 probe intensity using the Affymetrix HG U133 Plus 2.0 Microarray platform [13]. The main databases considered, NCI-60 and CTRP, have drug response and gene expression information for the cell lines, so the feature selection method must appropriately integrate the two types of data. Our approach will attempt this by examining the expression of genes on cell lines that are particularly sensitive or resistant to the topoisomerase inhibitor Topotecan.

2 Method

The goal is to utilize a nonlinear feature selection and data integration, or *data fusion*, to find predictive response determinants. To do this, we reformulate the problem from a mathematical perspective: given a data set which is expressed in two different modalities, find a similarity metric and elements of the set whose expressions differ beyond an appropriate threshold.

We start with a data set X of size $|X| = m$ that is expressed in two modalities a and b, which we represent with data matrices $A \in \mathbb{R}^{m \times n}$ and $B \in \mathbb{R}^{m \times p}$, respectively. For a data point $x \in X$, let $A(x)$ denote its expression in modality a and similarly for $B(x)$. In general, n is not equal to p, which makes naive comparison of n-dimensional $A(x)$ and p-dimensional $B(x)$ difficult.

Our method uses three main steps: embed the data matrices onto feature manifolds, create a joint representational feature space, and use a stable preimage mapping to facilitate comparison.

2.1 Manifold Learning Data Representation for Biological Process-Based Feature Selection

Laplacian eigenmaps (LE) is a nonlinear topology-preserving transformation technique that is designed to embed data onto a manifold based on local neighborhoods of data points [3]. We use LE to transform the data set X into two feature spaces Γ_a and Γ_b which capture intrinsic information of X based on the modalities a and b. The mappings are given by $\phi_a : X \to \Gamma_a$ and $\phi_b : X \to \Gamma_b$, where $m_a = dim(\Gamma_a)$ and $m_b = dim(\Gamma_b)$ are chosen to be greater than n and p, respectively, to avoid unnecessary loss of information due to dimension reduction.

2.2   Data Fusion Through Joint Feature Space Representation

The embedded points in Γ_a are now of the form $\phi_a(x) = \left(\phi_a^i(x)\right)_{i=1}^{m_a}$, with a similar form for points in Γ_b. At this step, we can embed the points in Γ_a into Γ_b using the Coifman-Hirn rotation operator $\mathcal{O}_{ab}$. For a data point $x \in X$, $\mathcal{O}_{ab}(x)$ is given by

$$\mathcal{O}_{ab}(x) = (\sum_{i=1}^{m_a} x_i \langle \phi_a^i, \phi_b^j \rangle)_{j=1}^{m_b},$$

where $x_i = \phi_a^i(x)$. Now, each data point in X is represented by two points in Γ_b, one for each modality. This technique, matching features using a spectral rotation in the feature space, has been used previously for heterogeneous image data [2,4,6,7].

2.3   Inverse Mapping for Finding Potential Response Determinants

LE is a nonlinear, continuous embedding of points in Euclidean space into a smooth manifold, so finding pre-images of points is highly non-trivial, especially when the points are not in the convex hull of the training points used to form the manifold. Alexander Cloninger, Wojciech Czaja, and Timothy Doster solve this issue by developing a fast and accurate pre-imaging algorithm for LE which exploits the sparsity of certain matrices constructed in the LE process [5]. By

Let $\widetilde{\phi}_b : \Gamma_b \to \mathbb{R}^p$ denote this approximate pre-image, and let $y = \mathcal{O}_{ab}(x)$ for a data point $x \in X$. Then $B(x)$ and $\widetilde{\phi}_b(y)$ are both data vector representations in $\mathbb{R}^p$ of the same point x which can be easily compared using a preferred norm. We denote the data matrix of these pre-image points as $\widetilde{A} \in \mathbb{R}^{m \times p}$. An advantage of comparing points in the original space $\mathbb{R}^p$ as opposed to comparing in the representational feature space Γ_b is that qualitative meanings of points in the original space may change in the feature space.

This entire process is symmetric in modalities a and b, providing a multitude of options for comparing points.

3   Preliminary Results and Future Work

The CTRP dataset we used contains 1283 genes expressed over the 792 cell lines that have response data for Topotecan. The genes chosen are known to have broad relevance to cancer and pharmacology. They were also filtered such that the upper quartile gene expression across cell lines was over 5 and the range of gene expressions across cell lines was over 2. This curation is for three reasons: lowly expressed genes may not affect the cell in substantive ways, genes with low variability are not reliable response determinants, and the pre-image mapping uses an optimization technique which is sensitive to very small values. Let G be the set of these genes.

The cell lines were subsetted and classified by having Topotecan responses 2 standard deviations above and below the mean response, creating a set of

13 resistant cell lines and a set of 18 sensitive cell lines, respectively. We represent gene expression along resistant and sensitive cell lines as data matrices $R \in \mathbb{R}^{1283 \times 13}$ and $S \in \mathbb{R}^{1283 \times 18}$, respectively. Within the context of the feature selection algorithm, sensitivity and resistance to Topotecan are the modalities over which G is described.

We then obtain LE embeddings $\phi_r(G)$ and $\phi_s(G)$, where $m_r = m_s = 40$ was chosen for computational ease. Following the rest of the algorithm twice, embedding Γ_r into Γ_s and vice versa, we obtain two additional data matrices:

$$\widetilde{R} = \widetilde{\phi_r} \circ \mathcal{O}_{sr} \circ \phi_s(G) \in \mathbb{R}^{1283 \times 18}$$

and

$$\widetilde{S} = \widetilde{\phi_s} \circ \mathcal{O}_{rs} \circ \phi_r(G) \in \mathbb{R}^{1283 \times 13}.$$

We then computed the match distances for each gene $g \in G$, given by $d_r(g) = \left\|\widetilde{S}(g) - R(g)\right\|_2$ and $d_s(g) = \left\|\widetilde{R}(g) - S(g)\right\|_2$. Let δ_r, δ_s be the mean plus 1 standard deviations of the sets $d_r(G)$ and $d_s(G)$, respectively. Potential response determinants were thus elements of the set $G' = \{g \in G | d_r(g) > \delta_r \text{ and } d_s(g) > \delta_s\}$.

G' contains 38 genes and includes SLFN11, which was previously discovered to be a response determinant using statistical methods and then further validated. This provides a promising start toward finding response determinants with low ambiguity that give more accurate predictive models (Fig. 1).

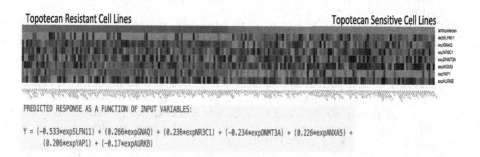

Fig. 1. A heatmap view of drug activity and gene expression data for 200 CTRP cell lines with the most extreme responses to Topotecan. Higher and lower values are represented by red and green, respectively. The LASSO-derived regression model augments the predictive capacity of SLFN11 expression, with a (10-fold cross-validation) predicted vs. observed drug response value Pearson's correlation of 0.67, versus $r = 0.45$ for SLFN11 alone. (Color figure online)

We plan on repeating this method on DNA-targeted drugs such as MK-1775 and the PARP inhibitor Olaparib. Eventually, we aim to make the entire data integration paradigm more robust by examining other operators on the feature spaces and other nonlinear embeddings based on data-dependent graphs.

Acknowledgments. The work presented in this paper was supported in part by DTRA through grant HDTRA 1-13-1-0015 and by ARO through grant W911NF1610008. We are assisted by genomic research conducted by Yves Pommier, Vinodh Rajapakse, and the LMP at the NCI-Developmental Therapeutics Branch. Alexander Cloninger and Timothy Doster are also specially acknowledged for making the data fusion model viable through their creation of a stable pre-image algorithm for Laplacian eigenmap embeddings.

References

1. Abaan, O.D., Polley, E.C., Davis, S.R., Zhu, Y.J., Bilke, S., Walker, R.L., Pineda, M., Gindin, Y., Jiang, Y., Reinhold, W.C., Holbeck, S.L.: The exomes of the NCI-60 panel: a genomic resource for cancer biology and systems pharmacology. Cancer Res. **73**(14), 4372–4382 (2013)
2. Benedetto, J., Cloninger, A., Czaja, W., Doster, T., Kochersberger, K., Manning, B., McCullough, T., McLane, M.: Operator based integration of information in multimodal radiological search mission with applications to anomaly detection. In: SPIE Defense+ Security, p. 90731A. International Society for Optics and Photonics, May 2014
3. Belkin, M., Niyogi, P.: Laplacian eigenmaps for dimensionality reduction and data representation. Neural Comput. **15**(6), 1373–1396 (2003)
4. Cloninger, A., Czaja, W., Doster, T. Operator analysis and diffusion based embeddings for heterogeneous data fusion. In: 2014 IEEE International Geoscience and Remote Sensing Symposium (IGARSS), pp. 1249–1252. IEEE, July 2014
5. Cloninger, A., Czaja, W., Doster, T.: The pre-image problem for Laplacian eigenmaps utilizing l_1 regularization with applications to data fusion. Inverse Probl. (2016)
6. Czaja, W., Hafftka, A., Manning, B., Weinberg, D.: Randomized approximations of operators and their spectral decomposition for diffusion based embeddings of heterogeneous data. In: 2015 3rd International Workshop on Compressed Sensing Theory and its Applications to Radar, Sonar and Remote Sensing (CoSeRa), pp. 75–79. IEEE, June 2015
7. Czaja, W., Manning, B., McLean, L., Murphy, J.M.: Fusion of aerial gamma-ray survey and remote sensing data for a deeper understanding of radionuclide fate after radiological incidents: examples from the Fukushima Dai-Ichi response. J. Radioanal. Nucl. Chem. **307**(3), 2397–2401 (2016)
8. Garnett, M.J., Edelman, E.J., Heidorn, S.J., Greenman, C.D., Dastur, A., Lau, K.W., Greninger, P., Thompson, I.R., Luo, X., Soares, J., Liu, Q.: Systematic identification of genomic markers of drug sensitivity in cancer cells. Nature **483**(7391), 570–575 (2012)
9. Luna, A., Rajapakse, V.N., Sousa, F.G., Gao, J., Schultz, N., Varma, S., Reinhold, W., Sander, C., Pommier, Y.: rcellminer: exploring molecular profiles and drug response of the NCI-60 cell lines in R. Bioinformatics, btv701 (2015)
10. Pommier, Y.: Drugging topoisomerases: lessons and challenges. ACS Chem. Biol. **8**(1), 82 (2013)
11. Rajapakse, V.N., Czaja, W., Pommier, Y.G., Reinhold, W.C., Varma, S.: Predicting expression-related features of chromosomal domain organization with network-structured analysis of gene expression and chromosomal location. In: Proceedings of the ACM Conference on Bioinformatics, Computational Biology and Biomedicine, pp. 226–233. ACM, October 2012

12. Reinhold, W.C., Sunshine, M., Varma, S., Doroshow, J.H., Pommier, Y.: Using cell miner 1.6 for systems pharmacology and genomic analysis of the NCI-60. Clin. Cancer Res. clincanres-0335 (2015)

13. Rees, M.G., Seashore-Ludlow, B., Cheah, J.H., Adams, D.J., Price, E.V., Gill, S., Javaid, S., Coletti, M.E., Jones, V.L., Bodycombe, N.E., Soule, C.K.: Correlating chemical sensitivity and basal gene expression reveals mechanism of action. Nat. Chem. Biol. **12**(2), 109–116 (2016)

14. Sousa, F.G., Matuo, R., Tang, S.W., Rajapakse, V.N., Luna, A., Sander, C., Varma, S., Simon, P.H., Doroshow, J.H., Reinhold, W.C., Pommier, Y.: Alterations of DNA repair genes in the NCI-60 cell lines and their predictive value for anticancer drug activity. DNA Repair **28**, 107–115 (2015)

15. Tang, S.W., Bilke, S., Cao, L., Murai, J., Sousa, F.G., Yamade, M., Rajapakse, V., Varma, S., Helman, L.J., Khan, J., Meltzer, P.S.: SLFN11 is a transcriptional target of EWS-FLI1 and a determinant of drug response in Ewing Sarcoma. Clin. Cancer Res. **21**(18), 4184–4193 (2015)

16. Zoppoli, G., Regairaz, M., Leo, E., Reinhold, W.C., Varma, S., Ballestrero, A., Doroshow, J.H., Pommier, Y.: Putative DNA/RNA helicase Schlafen-11 (SLFN11) sensitizes cancer cells to DNA-damaging agents. Proc. Natl. Acad. Sci. **109**(37), 15030–15035 (2012)

Agent-Based in Silico Evolution of HCV Quasispecies

Alexander Artyomenko[1](✉), Pelin B. Icer[1], Pavel Skums[1,2],
Sumathi Ramachandran[2], Yury Khudyakov[2], and Alex Zelikovsky[1]

[1] Department of Computer Science, Georgia State University, Atlanta, GA, USA
{aartyomenko,alexz}@cs.gsu.edu, pskums@gsu.edu
[2] Centers for Disease Control and Prevention, Atlanta, GA, USA

Abstract. Intra-host genetic diversity of hepatitis C virus (HCV) plays crucial role in disease progression and treatment outcome. Development of new treatment strategies, generation and validation of new biomedical hypothesis, development of algorithms and models for analysis of viral data and understanding of viral evolution require studying of thousands of intra-host viral populations. Since such amounts of experimental data are not readily available, simulated data are required. However, to the best of our knowledge, currently, there is no a general framework for generation of realistic intra-host HCV populations, which takes into account complex interactions between virus and host, impact of dynamic selection pressures and statistical effects, such as bottleneck and genetic drift.

In this paper, we propose a general framework for agent-based simulation of intra-host evolution of HCV quasispecies, which takes into account aforementioned factors. We performed a series of simulations and compared properties of simulated populations with corresponding properties of populations reconstructed from sequencing reads. Main population properties used here are positional k-mer entropy and hamming distances between variants within the population. Our simulations showed that presence of immune response lead to decrease of relative k-entropy and increase of standard deviation of hamming distances moving it closed to values observed in chronically infected patients.

1 Introduction

Intra-host viral evolution is a subject of great interest in epidemiology. Understanding of molecular mechanisms of immune escape and drug resistance is very important for devising successful public health interventions to control infections. Next-generation sequencinng revealed existence of numerous closely related genetic variants (quasispecies) of viruses in each infected host. The underlaying mechanisms defining the structure of intra-host viral populations are not clear and, to our knowledge, there are no models for realistic simulation of such structures.

In this paper, we propose the first agent-based simulation of intra-host viral evolution, which accounts for random mutations, immune response and

© Springer International Publishing AG 2017
Z. Cai et al. (Eds.): ISBRA 2017, LNBI 10330, pp. 420–424, 2017.
DOI: 10.1007/978-3-319-59575-7_45

immune cross-reactivity. In order to validate the proposed model and simulations, the simulated viral populations are compared with experimentally obtained sequences of intra-host viral populations as well as with randomized sequences. Our comparisons consider divergence, diversity and titer dynamics as well as entropy and hamming distance distributions.

The next section proposes a model of intra-host viral evolution and immune response. Experimental sequence data sets and validation results are discussed in the last section.

2   Model

The proposed simplified model (Fig. 1) consists of HCV virions, T-cells, and antibodies in the host blood, and liver cells (hepatocytes), which can be infected by virions or killed by T-cells. The model has two main components: viral infection and immune system.

Viral Infection

HCV infection is modeled using 2 components: set of virions in the blood (i.e. set of viral variants with counts), and set of infected liver cells. Virion is an instance of a viral variant in the host's blood. It consists of RNA (DNA) molecule, capsid, and lipids. Viral variant v is a class of virions sharing the same RNA (DNA) sequence. Variant count $q(v)$ is a number of virions for a given variant v. B is a set of viral variants (with variant counts) in the host's blood. Virions in the blood can infect healthy liver cells and infected liver cells produce new virions to the blood.

Immune System

In the proposed model immune system consists of B-cells, T-cells, and antibodies. Any virion in the blood can activate B-cell to produce antibody. The

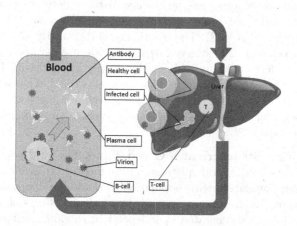

Fig. 1. Model components and their interaction. Virions and plasma cells are in blood. Host's liver cells (hepatocytes) can be infected or uninfected. Antibodies produced by B-cells neutralize virus. T-cells can attack and kill infected cells.

corresponding viral variant after delay will be added to the set of viral variants targeted by antibodies AB. T-cells can attack infected liver cells and kill them.

Simulation Procedure

Simulation requires an initial founder sequence, simulation time T, and other optional parameters. For initialization, program creates a virion using a founder sequence. After that it repeats following three steps T times. On the step one each live virion in the blood is either dies with probability dv_b, or activates B-cell with probability p_{bc}, or with probability p_c randomly infects a host's cell if it is uninfected, or stays in blood otherwise. If a virion interacts with B-cell, the process of antibody production is activated after d iterations. At the second step, each host's infected cell with the probability dc_i is either eliminated by immune system and replaced with a healthy cell, or produces a new sequence with mutation rate μ and releases new virion to the blood. At the third step, each active antibody eliminates virions with the probability p_{ab}, which depends on the number of sequences targeted by other active antibodies within a single mutation from the specific sequence targeted by that antibody. Thus the model contains the simple implementation of cross-immunoreactivity.

3   Results and Discussion

Viral quasispecies populations simulated by the proposed tool were compared with the experimental HCV data obtained from acutely and chronically infected individuals [1,2]. For all experimental samples, HCV HVR1 and NS5A regions were sequenced using 454 GS Junior System (454 Life Sciences, Branford, CT) or End-point limiting-dilution real-time PCR [3]. Below we analyze evolution and structure of simulated and experimental viral populations.

Viral Population Dynamics. We estimated the dynamics of simulated intrahost viral population divergence (the average distance between variants existing at a given time and the founder variant), nucleotide diversity [4] and number of virions in the blood (Fig. 2). The divergence and diversity steadily grow over time, which agrees with the observed dynamics of these parameters during first several years of infection for individuals chronically infected with HCV [2]. The dynamic of viral concentration in the blood suggests rapid expansion during the early stage of infection followed by the rapid decline after seroconversion and consecutive oscillatory slow growth or steady average size. Similar behavior is observed for real data [2].

Viral Population Structure at a Given Time. The viral population structure is shaped by selection and epistasis [5]. We measure their effect by comparing the observed population with population of randomized variants preserving allele structure but having erased epistasis and reduced effects of selection. The randomized variants are generated by independent random shuffling of states in each position. We measure diversity of viral variants by k-entropy which takes into account frequency and number of different viral variants as well as the standard deviation of the distribution of pairwise distances between viral variants.

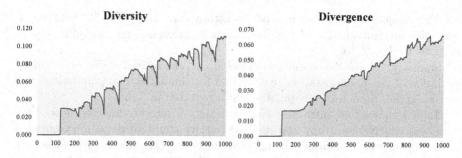

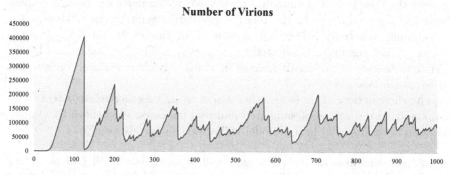

Fig. 2. Diversity, divergence, and number of virions over time in simulation.

We define k-entropy as follows. For each k sequential positions (k-mer) starting with position i we count the k-mer entropy

$$e_i = \sum_{h \in H} -f_i(h)log_2(f_i(h))$$

where h is a k-mer starting from the i-th position, H is set of all distinct k-mers and $f_i(h)$ is relative frequency of k-mer h. The k-entropy is the average of k-mer entropies over all positions.

Table 1. Median values of ratio R_d for $k = 2, 10$ and ratio R_d. Datasets include 3 types of simulations: without immune response, with immune response but without cross-immunoreactivity, and with immune response and cross-immunoreactivity, and 2 types of real datasets: recently infected hosts and chronically infected hosts.

Datasets		Median R_e (k = 2)	Median R_e (k = 10)	Median R_d
Simulated	No immune response	1.004	1.008	0.565
	No cross-immunoreactivity	0.987	0.883	0.955
	Cross-immunoreactivity	0.986	0.875	0.939
Real	Recently infected hosts	0.987	0.961	0.927
	Chronically infected hosts	0.907	0.642	2.185

The Table 1 reports the relative k-entropy, i.e., the ratio R_e between k-entropies for simulated/real populations and corresponding randomized populations for $k = 2, 10$.

Similarly, the relative standard deviation, i.e., the ratio R_d between standard deviation of the pairs-distances distribution for simulated/real populations and corresponding randomized populations also reported in Table 1.

Table 1 shows that intra-host HCV variants in recently infected hosts have increased relative k-entropy and reduced relative standard deviation compared with quasispecies in chronically infected hosts. In the simulated datasets we observe that the lack of immune response leads to similarly increase of relative k-entropy and reduction of the relative standard deviation. Even though the cross-immunoreactivity is known to greatly affect viral evolution [6,7] our simulations are not significantly affected by its presence. This may be caused by the oversimplification of the existing model in which only closely related variants are cross-immunoreactive.

The current simulation model also cannot bring together relative $k$entropies and standard deviations of simulated and real datasets. We believe that incorporation of epistasis [5] in our model will erase this inconsistency.

Acknowledgements. AA and AZ were partially supported from NSF Grants 1564899 and 16119110, AA was partially supported by GSU Molecular Basis of Disease Fellowship.

References

1. Astrakhantseva, I.V., Campo, D.S., Araujo, A., Teo, C.G., Khudyakov, Y., Kamili, S.: Differences in variability of hypervariable region 1 of hepatitis C virus (HCV) between acute and chronic stages of HCV infection. Silico Biol. **11**(5, 6), 163–173 (2011)
2. Ramachandran, S., Campo, D.S., Dimitrova, Z.E., Xia, G.-L., Purdy, M.A., Khudyakov, Y.E.: Temporal variations in the hepatitis C virus intrahost population during chronic infection. J. Virol. **85**(13), 6369–6380 (2011)
3. Ramachandran, S., Xia, G.L., Ganova-Raeva, L.M., Nainan, O.V., Khudyakov, Y.: End-point limiting-dilution real-time PCR assay for evaluation of hepatitis C virus quasispecies in serum: performance under optimal and suboptimal conditions. J. Virol. Methods **151**(2), 217–224 (2008)
4. Nei, M., Li, W.H.: Mathematical model for studying genetic variation in terms of restriction endonucleases. Proc. Nat. Acad. Sci. **76**(10), 5269–5273 (1979)
5. Sanjuán, R., Cuevas, J.M., Moya, A., Elena, S.F.: Epistasis and the adaptability of an RNA virus. Genetics **170**(3), 1001–1008 (2005)
6. Iwasa, Y., Michor, F., Nowak, M.: Some basic properties of immune selection. J. Theor. Biol. **229**(2), 179–188 (2004)
7. Skums, P., Bunimovich, L., Khudyakov, Y.: Antigenic cooperation among intrahost HCV variants organized into a complex network of cross-immunoreactivity. Proc. Nat. Acad. Sci. **112**(21), 6653–6658 (2015)

Modeling the Spread of HIV and HCV Infections Based on Identification and Characterization of High-Risk Communities Using Social Media

Deeptanshu Jha[1], Pavel Skums[2,3(✉)], Alex Zelikovsky[2],
Yury Khudyakov[3], and Rahul Singh[1(✉)]

[1] Department of Computer Science,
San Francisco State University, San Francisco, CA 94132, USA
rahul@sfsu.edu
[2] Department of Computer Science,
Georgia State University, Atlanta, GA 30302, USA
pskums@gsu.edu
[3] Centers for Disease Control and Prevention, Atlanta, GA 30333, USA

Abstract. Epidemiological dynamics of diseases, which may be transmitted due to sexual behavior or injecting drug use, can vary across demographic, socio-behavioral, and geographic population groups. Typically, studies modeling infection dissemination in such settings use simulated data and employ simplified contact networks. Here, we demonstrate feasibility of simulating HIV/HCV epidemics over a real-world contact network inferred using social media mining. Such networks can lead to more realistic modeling of disease transmission patterns in high-risk population than what is possible at the current state-of-the-art. In particular, we studied how topological characteristics of transmission networks are reflected by viral phylogenies.

1 Introduction

Diseases caused by RNA viruses such as HIV and hepatitis C virus (HCV), constitute major causes of debilitation and mortality in the world. Amongst these, HCV alone infects nearly 3% of the world population. In the USA, HCV infection is the most common chronic blood-borne disease and one of the leading causes of liver failure. Since 2007 in the USA, HCV infection has exceeded the mortality from HIV infection [1].

Effective epidemiological modeling focused on measuring a disease and its dissemination among human populations is essential for designing successful public health interventions. The design and implementation of such models is stymied by multiple challenges for diseases whose etiological agents are RNA viruses. These include:

- *Genetic variability and nature of the etiological agent*: High genetic heterogeneity stemming from error-prone replication [2], is a distinguishing feature of RNA viruses. Infected hosts hold reservoirs of genetically related variants (called

Z. Cai et al. (Eds.): ISBRA 2017, LNBI 10330, pp. 425–430, 2017.
DOI: 10.1007/978-3-319-59575-7_46

quasispecies) whose diversity complicates understanding disease progression and epidemic spread. HCV infection can be initially asymptomatic and lead to late-stage clinical complications.

- *Impact of human behavior*: Injecting drug use and sexual behavior are established risk factors for transmission of HCV and HIV [3, 4]. The transmission is facilitated by sharing needles and other drug preparation equipment or by sexual behavior. Unfortunately, addiction (in particular opioid addiction) has become a national health crisis. It is estimated that each injection drug user is likely to infect about 20 other users within a short period of time [5].

- *Limitations of molecular and epidemiological data gathering and analysis*: In spite of the advent of techniques such as Next-Generation Sequencing (NGS), which can generate large quantities of viral genetic data, development of robust epidemiological models based on biological sampling of viral variants is limited, owing to complexity of molecular data. Moreover, traditional epidemiological data collection techniques rely on explicit reporting of indicator-based data from patient records or surveys. Such a methodology is limited in capturing all contextual details of infected hosts, which is crucial for modeling transmission, and somewhat time-lags in detection of infection spread during data collection/processing.

In this paper, we outline an approach for identifying at-risk communities and inter-relationships that may exist therein based on information extracted from the social media platform *Twitter* and simulate epidemics over these networks. It is well known that human contact is one of the significant factors in the transmission of infectious diseases [6]. In particular, for diseases caused by HIV and HCV, geographical co-location, social ties, and high-risk behavior are especially crucial in disease transmission. The proposed approach estimates these factors through the analysis of Twitter microblogs. It should be noted *a priori* that social media signals may not directly reflect factors that cause disease spread as personal health impediments. Especially, diseases caused by HIV or HCV are rarely discussed openly in social media. However, the (latent) causative factors are often reflected in online behavior and relationships.

Epidemiological dynamics of HIV and HCV infections can vary across demographic, socio-behavioral, and geographic population groups leading to different infection patterns defined by structures of corresponding contact networks. Typically, studies modeling infection dissemination employ simplified contact network models [4]. We simulate epidemics over the contact network inferred by social media mining and study how properties of that network translated into topologies of transmission networks are reflected by viral phylogenies.

2 Method

Our approach consists of the following steps: (1) data sampling, identification of HCV, HIV, and addiction-related informational handles and distinguishing these from handles of users who follow these, (2) grouping users by their geo-locations, and (3) relating users to each other so as to form a host network, which would underlie any inferred transmission network. We briefly describe these steps in the following.

User accounts on Twitter can be classified into a number of categories including information handles, user accounts, organizational handles, celebrity handles. As there may be any number of informational handles dedicated to HCV, HIV and drug addiction, we use a content-driven classification tree to distinguish personal user accounts from informational handles and then successively identify addiction/recovery handles followed by HIV and finally HCV informational handles. That is, in this classification tree a binary classification problem is solved at each level to identify the corresponding account. Once we have identified personal user accounts, informational HCV handle, informational HIV handles, and informational addiction/recovery handles we expand the set of user related to these handles. Let U_{HCV} denote the set of users who follow HCV-related handles, U_{HIV} be the set of users who follow HIV-related handles, and U_{AR} be the set of users who follow addiction/recovery-handles. Further, for a given user u, let $L(u)$ denote the set of users who are followed by u and $F(u)$ be the set of users who follow u. The aforementioned user sets are then expanded following Eq. (1). In this equation the extension of the functions $L(.)$ and $F(.)$ to sets is straightforward.

$$U_i = U_i \cup L(U_i) \cup F(U_i), \ i \in \{HCV, HIV, AR\} \tag{1}$$

$$w(u, v) = \frac{1}{m} \sum_{i=1}^{m} \frac{u_i \cap v_i}{u_i \cup v_i} \tag{2}$$

Next, users are grouped by their geographic location. For this purpose, we use the primary location specified in the user profile. Currently, we exclude users who do not specify location information explicitly. Finally, we identify users who mutually follow each other and term them friends. These users are represented using a weighted graph, henceforth called the host network, where each node represents a user and each edge connects friends. The weights on the edges indicate the extent of "closeness" between the corresponding vertices with the assumption the closely related vertices are more likely to facilitate infection transmission than those that are distantly related. The weighting is accomplished using the Jaccard similarity of tweet content, descriptions and mutual friends and followers of the corresponding two users based on Eq. (2). In this equation, the index i enumerates the features used for determining the closeness of vertices (users) u and v. We use a number of features that capture the similarity of the tweet contents and the similarity of the social network of each user. Finally, the connected components of the network are identified and network communities are identified using the Louvain modularity detection algorithm [7].

The virus spread over the host network is simulated using SI model. The virus is transmitted along each edge at a rate proportional to its similarity weight estimated using Eq. (2). The simulation stops, when all vertices are infected. The epidemic history is represented by a transmission tree, which is a rooted binary labeled tree with leafs representing infected hosts and interior nodes representing transmission events: an interior node with a label x and its children with labels x and y represent an infection of a host y by a host x. Once a transmission tree is constructed, it is transformed into a transmission network, which indicates who infected whom: the vertices of transmission network are infected hosts, and the arcs connect hosts linked by transmission. We employ a strict molecular clock assumption, under which genetic distances between viral strains sampled from infected hosts should be proportional to the time after transmission event. Therefore, the distance between each pair of hosts is defined as the doubled simulated time to their most recent common ancestor in the transmission tree. We also assume that for each pair of hosts, the possible direction of transmission is known (based on the diversity of intra-host viral populations [10]) and agrees with the difference between infection times of these hosts. For each simulation, the weighted adjacency matrix was used to construct phylogenetic trees using UPGMA, Complete, and WPGMA linkage methods, as implemented in Matlab (MathWorks Inc., Natick). For each phylogenetic tree, the labels of internal nodes were inferred using possible directions of transmissions and the corresponding transmission network was built. The transmission networks inferred from phylogenies were compared with the actual transmission networks.

3 Results

We collected Twitter data collected over approximately two weeks focusing on HIV, HCV, and addiction. We identified 1974 users in New York city (NYC) who followed HIV/HCV/Addiction handles (Fig. 1, top row, left). We identified a subnetwork of 746 mutual followers, which contained one large connected component involving 689 users, with a diameter of 11 and average path length of 4.425. We identified 39 clusters (henceforth termed communities) in this component (Fig. 1, top row, middle). The communities which had at least 10 users were manually examined and were all found to have distinct themes such as "HIV/Homosexuality", "Drug Addiction", "Epidemiology", "LGBT Advocacy", and "Public Health". Especially important in our problem context was the identification of the HIV/Homosexuality community, which represents an at-risk group (Fig. 1, top row, right). This community consisted of thirty male users of which nineteen were found to have self-identified as homosexuals and four were found to have self-identified as HIV positive.

   Phylogenies do not accurately reflect the actual transmission history; our results indicate that on average, transmission networks corresponding to phylogenetic trees contain $\sim 26.6\%$ of real transmission links and $\sim 25.1\%$ of real transmission ancestries (pairs of ancestor-descendant). Our results underline the observation that for transmission history inference more specific methods are required, which rely on additional data in the form of epidemiological information or network theory-based analysis [8, 9]. However, phylogenies are much better in preserving more general properties of transmission networks (Fig. 1, middle and bottom rows). The middle row in Fig. 1 shows the results of comparison of degree (left) and distance (right) distributions of real and phylogeny-based transmission networks. As a measure of comparison, we used p-value of Kolmogorov-Smirnov test. For degree distributions, average p-values for 3 linkage methods were 0.7184, 0.8368 and 0.6227 respectively, indicating the level of confidence that degrees of both networks follow the same distribution. The distance distribution is reflected by phylogenies less well: the average p-values were 0.4717, 0.1280 and 0.5126 respectively, with the hypothesis that distances of both networks follow the same distribution being rejected at 5% level in 25.3% of cases. It is also

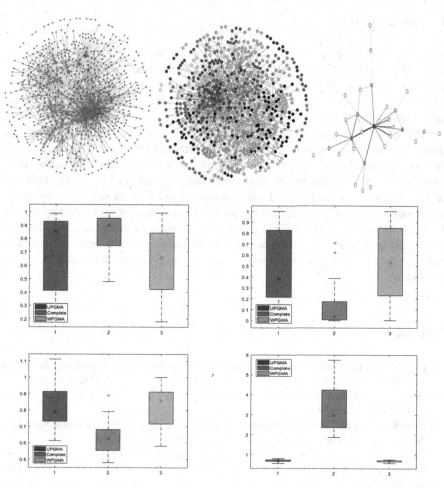

Fig. 1. Top row: (left) user network from NYC. (Middle) communities identified in the network are shown in distinct colors. (Right) the HIV/Homosexuality community identified from the social media data. Middle row: degree and distance distribution of transmission networks reflected by phylogenies. Bottom row: diameter and s-metric reflected by phylogenies. (Color figure online)

worth noting, that complete linkage reflected the degree distributions better than the other two methods, while being worse in reflecting distance distributions. The average ratios of diameters of phylogeny-based and real transmission networks for 3 linkage methods are 0.8095, 0.6406 and 0.8138, respectively. All 3 methods tend to underestimate the diameter, although the estimates from UPGMA and WPGMA stay relatively close to 1. UPGMA and WPGMA also underestimate the value of s-metric [9], while complete linkage significantly overestimates it (the average ratios were 0.7005, 3.3671 and 0.6890, respectively).

4 Conclusions

In this paper, we have described techniques that allow simulating HIV/HCV epidemics over a real-world contact network inferred using social media mining. The proposed approach opens an opportunity for modeling context-sensitive aspects of human behavior that can be crucial for understanding the epidemiological dynamics of diseases caused by HIV or HCV. We have presented preliminary results that show how information derived from social media can be used to create contact networks and simulate the spread of diseases. Methods for gathering data from social media for use in rigorous epidemiological modeling are currently at infancy; many outstanding challenges have yet to be studied in depth, including the design of sampling strategies, dealing with the non-uniform and sparse nature of the data, modeling the epistemic uncertainty in the data, and combining such evidence in existing models of disease dynamics. However, social media also allows capture of data and relationships that can be invaluable in modeling of disease spread. Preliminary results from our research, including those presented in this paper, underline the potential of this new source of information.

References

1. Ly, K.N., Xing, J., Klevens, R.M., et al.: The increasing burden of mortality from viral hepatitis in the United States between 1999–2007. Ann. Intern. Med. **156**, 271–278 (2012)
2. Drake, J.W., Holland, J.J.: Mutation rates among RNA viruses. Proc. Natl. Acad. Sci. **96**(24), 13910–13913 (1999)
3. Nelson, P.K., et al.: Global epidemiology of hepatitis B and hepatitis C in people who inject drugs: results of systematic reviews. Lancet **378**(9791), 571–583 (2011)
4. Villandre, L., et al.: Assessment of overlap of phylogenetic transmission clusters and communities in simple sexual contact networks: applications to HIV-1. PLoS ONE **11**(2), e0148459 (2016)
5. Magiorkinis, G., et al.: Integrating phylodynamics and epidemiology to estimate transmission diversity in viral epidemics. PLoS Comput. Biol. **9**(1), e1002876 (2013)
6. Clayton, D., Hills, M.: Statistical Models in Epidemiology. Oxford University Press, Oxford (2013)
7. Blondel, V.D., Guillaume, J.-L., Lambiotte, R., Lefebvre, E.: Fast unfolding of communities in large networks. J. Stat. Mech.: Theor. Exp. **10** (2008). P10008
8. Jombart, T., et al.: Bayesian reconstruction of disease outbreaks by combining epidemiologic and genomic data. PLoS Comput. Biol. **10**(1), e1003457 (2014)
9. Lun, L., et al.: Towards a theory of scale-free graphs: definition, properties, and implications. Internet Math. **2**(4), 431–523 (2005)
10. Campo, D.S., et al.: Accurate genetic detection of hepatitis C virus transmissions in outbreak settings. J. Infect. Dis. **213**(6), 957–965 (2016)

Author Index

Printed in the United States
By Bookmasters